An Introduction to the Representation Theory of Groups

An Introduction to the Representation Theory of Groups

Emmanuel Kowalski

Graduate Studies in Mathematics

Volume 155

American Mathematical Society
Providence, Rhode Island

2010 *Mathematics Subject Classification.* Primary 20-01, 20Cxx, 22A25.

For additional information and updates on this book, visit
www.ams.org/bookpages/gsm-155

Library of Congress Cataloging-in-Publication Data

Kowalski, Emmanuel, 1969–
An introduction to the representation theory of groups / Emmanuel Kowalski.
pages cm. — (Graduate studies in mathematics ; volume 155)
Includes bibliographical references and index.
ISBN 978-1-4704-0966-1 (alk. paper)
1. Lie groups. 2. Representations of groups. 3. Group algebras. I. Title.

QA387.K69 2014
515′.7223—dc23

2014012974

∞ The paper used in this book is acid-free and falls within the guidelines
established to ensure permanence and durability.
Visit the AMS home page at http://www.ams.org/

10 9 8 7 6 5 4 3 2 1 19 18 17 16 15 14

Contents

Chapter 1. Introduction and motivation 1
§1.1. Presentation 3
§1.2. Four motivating statements 4
§1.3. Prerequisites and notation 8

Chapter 2. The language of representation theory 13
§2.1. Basic language 13
§2.2. Formalism: changing the space 21
§2.3. Formalism: changing the group 42
§2.4. Formalism: changing the field 65
§2.5. Matrix representations 68
§2.6. Examples 70
§2.7. Some general results 80
§2.8. Some Clifford theory 121
§2.9. Conclusion 124

Chapter 3. Variants 127
§3.1. Representations of algebras 127
§3.2. Representations of Lie algebras 132
§3.3. Topological groups 139
§3.4. Unitary representations 145

Chapter 4. Linear representations of finite groups 159
§4.1. Maschke's Theorem 159

§4.2. Applications of Maschke's Theorem 163
§4.3. Decomposition of representations 169
§4.4. Harmonic analysis on finite groups 190
§4.5. Finite abelian groups 200
§4.6. The character table 208
§4.7. Applications 240
§4.8. Further topics 262

Chapter 5. Abstract representation theory of compact groups 269
§5.1. An example: the circle group 269
§5.2. The Haar measure and the regular representation of a locally compact group 272
§5.3. The analogue of the group algebra 288
§5.4. The Peter–Weyl Theorem 294
§5.5. Characters and matrix coefficients for compact groups 304
§5.6. Some first examples 310

Chapter 6. Applications of representations of compact groups 319
§6.1. Compact Lie groups are matrix groups 319
§6.2. The Frobenius–Schur indicator 324
§6.3. The Larsen alternative 332
§6.4. The hydrogen atom 344

Chapter 7. Other groups: a few examples 355
§7.1. Algebraic groups 355
§7.2. Locally compact groups: general remarks 369
§7.3. Locally compact abelian groups 371
§7.4. A non-abelian example: $\mathrm{SL}_2(\mathbf{R})$ 376

Appendix A. Some useful facts 409
§A.1. Algebraic integers 409
§A.2. The spectral theorem 414
§A.3. The Stone–Weierstrass Theorem 420

Bibliography 421

Index 425

Chapter 1

Introduction and motivation

This book is intended to provide a basic introduction to some of the fundamental ideas and results of *representation theory*. In this preliminary chapter, we start with some motivating remarks and provide a general overview of the rest of the text; we also include some notes on the prerequisites—which are not uniform for all parts of the notes—and discuss the basic notation that we use.

In writing this text, the objective has never been to give the shortest or slickest proof. To the extent that the author's knowledge makes this possible, the goal is rather to explain the ideas and the mechanism of thought that can lead to an understanding of "why" something is true, and not simply to the quickest line-by-line check that it holds.

The point of view is that representation theory is a fundamental theory, both for its own sake and as a tool in many other fields of mathematics; the more one knows, understands, and breathes representation theory, the better. This style (or its most ideal form) is perhaps best summarized by P. Sarnak's advice in the *Princeton Companion to Mathematics* [**24**, p. 1008]:

> One of the troubles with recent accounts of certain topics is that they can become too slick. As each new author finds cleverer proofs or treatments of a theory, the treatment evolves toward the one that contains the "shortest proofs." Unfortunately, these are often in a form that causes the new student to ponder, "How did anyone think of this?" By going back to the original sources one can

> usually see the subject evolving naturally and understand how it has reached its modern form. (There will remain those unexpected and brilliant steps at which one can only marvel at the genius of the inventor, but there are far fewer of these than you might think.) As an example, I usually recommend reading Weyl's original papers on the representation theory of compact Lie groups and the derivation of his character formula, alongside one of the many modern treatments.

So the text sometimes gives two proofs of the same result, even in cases where the arguments are fairly closely related; one may be easy to motivate ("how would one try to prove such a thing?"), while the other may recover the result by a slicker exploitation of the formalism of representation theory. To give an example, we first consider Burnside's irreducibility criterion, and its developments, using an argument roughly similar to the original one, before showing how Frobenius reciprocity leads to a quicker line of reasoning (see Sections 2.7.3 and 2.7.4).

Finally, although I have tried to illustrate many aspects of representation theory, there remains many topics that are barely mentioned or omitted altogether. Maybe the most important are:

- The representation theory of anything else than groups; in particular, Lie algebras and their representations only make passing appearances, and correspondingly those aspects of representation theory that really depend on these techniques are not developed in any detail. Here, the book [**20**] by Fulton and Harris is an outstanding resource, and the book [**18**] by Etingof, Golberg, Hensel, Liu, Schwendner, Vaintrob, and Yudovina illustrates different aspects, such as the representations of quivers.
- In a related direction, since it really depends on Lie algebraic methods, the precise classification of representations of compact Lie groups, through the theory of highest weight representations, is not considered beyond the case of $\mathrm{SU}_2(\mathbf{C})$; this is however covered in great detail in many other texts, such as [**20**] again, the book [**37**] of Knapp (especially Chapter V), or the book [**35**] of Kirillov.

Acknowledgments. The notes were prepared in parallel with the course "Representation Theory" that I taught at ETH Zürich during the Spring Semester 2011. Thanks are obviously due to all the students who attended the course for their remarks and interest, in particular M. Lüthy, M Rüst, I. Schwabacher, M. Scheuss, and M. Tornier, and to the assistants in charge of the exercise sessions, in particular J. Ditchen who coordinated those. Thanks also to "Anonymous Rex" for a comment on a blog post, to U. Schapira for his comments and questions during the class, and to A. Venkatesh for showing me his own notes for a (more advanced) representation theory class, from which I derived much insight.

Thanks to the reviewers for the original book proposal for suggestions and comments—in particular for some well-deserved critical comments concerning certain of the choices of notation in the first version of the text, and for pointing out that Proposition 2.3.23 is false over non-algebraically closed fields.

Finally, many thanks to E. Dunne for reading the whole manuscript carefully and making many suggestions and corrections!

1.1. Presentation

A (linear) representation of a group G is, to begin with, simply a *homomorphism*

$$\varrho : G \longrightarrow \mathrm{GL}(E),$$

where E is a vector space over some field k and $\mathrm{GL}(E)$ is the group of invertible k-linear maps on E. Thus one can guess that this should be a useful notion by noting how it involves the simplest and most ubiquitous algebraic structure, that of a group, with the powerful and flexible tools of linear algebra. Or, in other words, such a map attempts to "represent" the elements of G as symmetries of the vector space E (note that ϱ might fail to be injective, so that G is not mapped to an isomorphic group).

But even a first guess would probably not lead one to imagine how widespread and influential the concepts of representation theory turn out to be in current mathematics. Few fields of mathematics, or of mathematical physics (or chemistry), do not make use of these ideas, and many depend on representations in an essential way. We will try to illustrate this wide influence with examples, taken in particular from number theory and from basic quantum mechanics; already in Section 1.2 below we state four results, where representation theory does not appear in the statements although it is a fundamental tool in the proofs. Moreover, it should be said that representation theory is now a field of mathematics in its own right, which can be pursued without having immediate applications in mind; it does not require external influences to expand with new questions, results and concepts—but we will barely scratch such aspects.

The next chapter starts by presenting the fundamental vocabulary that is the foundation of representation theory and by illustrating it with examples. In Chapter 3, we then present a number of short sections concerning variants of the definition of representations: restrictions can be imposed on the group

G, on the type of fields or vector spaces E allowed, or additional regularity assumptions may be imposed on ϱ when this makes sense. One can also replace groups by other objects: we will mention associative algebras and Lie algebras. These variants are all important topics in their own right, but some will only reappear briefly in the rest of the book.

Continuing, Chapter 4 is an introduction to the simplest case of representation theory: the linear representations of finite groups in finite-dimensional complex vector spaces. This is also historically the first case that was studied in depth by Dirichlet (for finite abelian groups), then Frobenius, Schur, Burnside, and many others. It is a beautiful theory and has many important applications. It can also serve as a "blueprint" to many generalizations: various facts, which are extremely elementary for finite groups, remain valid, when properly framed, for important classes of infinite groups.

Among these, the compact topological groups are undoubtedly those closest to finite groups, and we consider them in Chapter 5. Then Chapter 6 presents some concrete examples of applications involving compact Lie groups (compact matrix groups, such as unitary groups $U_n(\mathbf{C})$)—the most important being perhaps the way representation theory explains a lot about the way the most basic atom, hydrogen, behaves in the real world....

The final Chapter 7 has again a survey flavor, and it is intended to serve as an introduction to two other important classes of groups: algebraic groups, on the one hand, and non-compact locally compact groups, on the other hand. This last case is illustrated through the fundamental example of the group $SL_2(\mathbf{R})$ of two-by-two real matrices with determinant 1. We use it primarily to illustrate some of the striking new phenomena that arise when compactness is missing.

In Appendix A, we have gathered statements and sketches of proofs for certain facts, especially the Spectral Theorem for compact self-adjoint linear operators, which are needed for rigorous treatments of unitary representations of topological groups.

Throughout, we also present some examples by means of exercises. These are usually not particularly difficult, but we hope they will help the reader to get acquainted with the way of thinking that representation theory often suggests for certain problems.

1.2. Four motivating statements

Below are four results, taken in very different fields, which we will discuss again later (or sometimes only sketch when very different ideas are also needed). The statements do not mention representation theory, in fact two of them do not even mention groups explicitly. Yet they are proved using these tools, and they serve as striking illustrations of what can be done using representation theory.

Example 1.2.1 (Primes in arithmetic progressions). Historically, the first triumph of representation theory is the proof by Dirichlet of the existence of infinitely many prime numbers in an arithmetic progression, whenever this is not clearly impossible:

Theorem 1.2.2 (Dirichlet). *Let $q \geqslant 1$ be an integer, and let $a \geqslant 1$ be an integer coprime with q. Then there exist infinitely many prime numbers p such that*

$$p \equiv a \,(\mathrm{mod}\, q),$$

i.e., such that p is of the form $p = nq + a$ for some $n \geqslant 1$.

For instance, taking $q = 10^k$ to be a power of 10, we can say that, for whichever ending pattern of digits $\boldsymbol{d} = d_{k-1}d_{k-2}\cdots d_0$ we might choose, with $d_i \in \{0, 1, 2, 3, 4, 5, 6, 7, 8, 9\}$, *provided* the last digit d_0 is not one of $\{0, 2, 4, 5, 6, 8\}$, there exist infinitely many prime numbers p with a decimal expansion where $\boldsymbol{d}$ are the final digits. To illustrate this, taking $q = 1000$, $\boldsymbol{d} = 237$, we find

$$\begin{gathered} 1237,\ 2237,\ 5237,\ 7237,\ 8237,\ 19237,\ 25237,\ 26237,\ 31237,\ 32237,\\ 38237,\ 40237,\ 43237,\ 46237,\ 47237,\ 52237,\ 56237,\ 58237,\ 64237,\\ 70237,\ 71237,\ 73237,\ 77237,\ 82237,\ 85237,\ 88237,\ 89237,\ 91237,\ 92237 \end{gathered}$$

to be those prime numbers ending with 237 which are $\leqslant 100000$.

We will present the idea of the proof of this theorem in Chapter 4. As we will see, a crucial ingredient (but not the only one) is the simplest type of representation theory: that of groups that are both finite and commutative. In some sense, there is no better example to guess the power of representation theory than to see how even the simplest instance leads to such remarkable results.

Example 1.2.3 (The hydrogen atom). According to current knowledge, about 75% of the observable weight of the universe is accounted for by hydrogen atoms. In quantum mechanics, the possible states of an (isolated) hydrogen atom are described in terms of combinations of "pure" states, and the latter are determined by *discrete* data, traditionally called "quantum numbers"—so that the possible energy values of the system, for instance, form a discrete set of numbers, rather than a continuous interval.

Precisely, in non-relativistic theory, there are four quantum numbers for a given pure state of hydrogen, denoted (n, ℓ, m, s)—*principal*, *angular momentum*, *magnetic*, and *spin*" are their usual names—which are all integers, except for s, with the restrictions

$$n \geqslant 1, \quad 0 \leqslant \ell \leqslant n-1, \quad -\ell \leqslant m \leqslant \ell, \quad s \in \{-1/2, 1/2\}.$$

It is rather striking that much of this quantum mechanical model of the hydrogen atom can be "explained" qualitatively by an analysis of the representation theory of the underlying symmetry group (see [**64**] or [**58**]) leading in particular to a natural explanation of the intricate structure of these four quantum numbers! We will attempt to explain the easiest part of this story, which only involves the magnetic and angular momentum quantum numbers, in Section 6.4.

Example 1.2.4 ("Word" problems)**.** For a prime number p, consider the finite group $\mathrm{SL}_2(\mathbf{F}_p)$ of square matrices of size 2 with determinant 1, and with coefficients in the finite field $\mathbf{F}_p = \mathbf{Z}/p\mathbf{Z}$. This group is generated by the two elements

$$s_1 = \begin{pmatrix} 1 & 1 \\ 0 & 1 \end{pmatrix}, \qquad s_2 = \begin{pmatrix} 1 & 0 \\ 1 & 1 \end{pmatrix} \tag{1.1}$$

(this is a fairly easy fact from elementary group theory, see, e.g., [**51**, Th. 8.8] for $K = \mathbf{F}_p$ or Exercise 4.6.20). Certainly the group is also generated by the elements of the set $S = \{s_1, s_1^{-1}, s_2, s_2^{-1}\}$, and in particular, for any $g \in \mathrm{SL}_2(\mathbf{F}_p)$, there exist an integer $k \geqslant 1$ and elements $g_1, \ldots, g_k$, each of which belongs to S, such that

$$g = g_1 \cdots g_k.$$

Given g, let $\ell(g)$ be the *smallest* k for which such a representation exists. One may ask, how large can $\ell(g)$ be when g varies over $\mathrm{SL}_2(\mathbf{F}_p)$? The following result gives an answer:

Theorem 1.2.5 (Selberg, Brooks, Burger)**.** *There exists a constant $C \geqslant 0$, independent of p, such that, with notation as above, we have*

$$\ell(g) \leqslant C \log p$$

for all $g \in \mathrm{SL}_2(\mathbf{F}_p)$.

All proofs of this result depend crucially on ideas of representation theory, among other tools. And while it may seem to be rather simple and not particularly worth notice, the following *open* question should suggest that there is something very subtle here.

Problem. *Find an* efficient algorithm *that, given p and $g \in \mathrm{SL}_2(\mathbf{F}_p)$, explicitly gives $k \leqslant C \log p$ and a sequence $(g_1, \ldots, g_k)$ in S such that*

$$g = g_1 \cdots g_k.$$

For instance, what would you do with

$$g = \begin{pmatrix} 1 & (p-1)/2 \\ 0 & 1 \end{pmatrix}$$

(for $p \geqslant 3$)? Of course, one can take $k = (p-1)/2$ and $g_i = s_1$ for all i, but when p is large, this is much larger than what the theorem claims is possible!

We will not prove Theorem 1.2.5, nor really say much more about the known proofs. However, in Section 4.7.1, we present more elementary results of Gowers [**23**] (and Nikolov and Pyber [**47**]) which are much in the same spirit, and we use the same crucial ingredient concerning representations of $\mathrm{SL}_2(\mathbf{F}_p)$. The book [**13**] of Davidoff, Sarnak, and Valette gives a complete elementary proof and is fully accessible to readers of this book.

In these three first examples, it turns out that representation theory appears in a similar manner: it is used to analyze functions on a group, in a way which is close to the theory of Fourier series or Fourier integrals; indeed, both of these can also be understood in terms of representation theory for the groups $\mathbf{R}/\mathbf{Z}$ and $\mathbf{R}$, respectively (see Section 7.3). The next motivating example is purely algebraic.

Example 1.2.6 (Burnside's $p^a q^b$ theorem)**.** Recall that a group G is called *solvable* if there is an increasing sequence of subgroups

$$1 \lhd G_k \lhd G_{k-1} \lhd \cdots \lhd G_1 \lhd G = G_0,$$

each normal in the next (but not necessarily in G), such that each successive quotient G_k/G_{k+1} is an abelian group.

Theorem 1.2.7 (Burnside)**.** *Let G be a finite group. If the order of G is divisible by at most two distinct prime numbers, then G is solvable.*

This beautiful result is sharp in some sense: it is well known that the symmetric group $\mathfrak{S}_5$ of order $5! = 120$ is *not* solvable, and since 120 is divisible only by the primes 2, 3 and 5, we see that the analogue statement with 2 prime factors replaced with 3 is not true. (Also it is clear that the converse is not true either: any abelian group is solvable, and there are such groups of any order.)

This theorem of Burnside will be proved using representation theory of finite groups in Section 4.7.2 of Chapter 4, in much the same way as Burnside proceeded in the early 20th century. It was only in the late 1960s that a proof not using representation theory was found, first by Goldschmidt when the primes p and q are odd, and then independently by Bender and Matsuyama for the general case. There is a full account of this in [**29**, §7D], and although it is not altogether overwhelming in length, the reader who compares them will probably agree that the proof based on representation theory is significantly easier to digest.

Remark 1.2.8. There are even more striking results which are much more difficult. For instance, the famous "Odd-order Theorem" of Feit and Thompson states that if G has *odd* order, then G is necessarily solvable.

1.3. Prerequisites and notation

In Chapters 2 and 4, we depend only on the content of a basic graduate course in algebra: basic group theory, abstract linear algebra over fields, polynomial rings, finite fields, modules over rings, bilinear forms, and the tensor product and its variants. In later chapters, other structures are involved: groups are considered with a topology, measure spaces and integration theory is involved, as well as basic Hilbert space theory and functional analysis. All these are used at the level of introductory graduate courses.

We will use the following notation:

(1) For a set X, $|X| \in [0, +\infty]$ denotes its cardinality, with $|X| = \infty$ if X is infinite. There is no distinction in this text between the various infinite cardinals.

(2) We denote by $\mathbf{R}^{+,\times}$ the interval $]0, +\infty[$ seen as a subgroup of the multiplicative group $\mathbf{R}^{\times}$.

(3) If k is a field and $d \geqslant 1$ an integer, an element of $\mathrm{GL}_d(k)$ (or of $\mathrm{GL}(E)$ where E is a finite-dimensional k-vector space) is called *unipotent* if there exists $n \geqslant 1$ such that $(u - \mathrm{Id}_k)^n = 0$.

(4) Given a ring A, with a unit $1 \in A$, and A-modules M and N, we denote by $\mathrm{Hom}(M, N)$ or $\mathrm{Hom}_A(M, N)$ the space of A-linear maps from M to N.

(5) If E is a vector space over a field k, E' denotes the dual space $\mathrm{Hom}_k(E, k)$. We often use the duality bracket notation for evaluating linear maps on vectors, i.e., for $v \in E$ and $\lambda \in E'$, we write

$$\langle \lambda, v \rangle = \lambda(v).$$

(6) For $f : M \to N$, a map of A-modules, $\mathrm{Ker}(f)$ and $\mathrm{Im}(f)$ denote the kernel and the image of f, respectively.

(7) A *projection* $f : M \longrightarrow M$ is a linear map such that $f \circ f = f$. If f is such a projection, we have $M = \mathrm{Im}(f) \oplus \mathrm{Ker}(f)$; we also say that f is the projection on $\mathrm{Im}(f)$ with kernel $\mathrm{Ker}(f)$.

(8) Given A and M, N as above, $M \otimes N$ or $M \otimes_A N$ denotes the tensor product of M and N. Recall that $M \otimes N$ can be characterized up to

isomorphism by the existence of canonical isomorphisms

$$\operatorname{Hom}_A(M \otimes N, N_1) \simeq \operatorname{Bil}(M \times N, N_1)$$

for any A-module N_1, where the right-hand side is the A-module of all A-bilinear maps

$$\beta : M \times N \to N_1.$$

In particular, there is a bilinear map

$$\beta_0 : M \times N \longrightarrow M \otimes N$$

that corresponds to $N_1 = M \otimes N$ and to the identity map in $\operatorname{Hom}_A(M \otimes N, N_1)$. One writes $v \otimes w$ instead of $\beta_0(v, w)$.

The elements of the type $v \otimes w$ in $M \otimes N$ are called *pure tensors*. Note that usually not all elements in the tensor product are pure tensors and that one can have $v \otimes w = v' \otimes w'$ even if $(v, w) \neq (v', w')$.

If $A = k$ is a field and (e_i), (f_j) are bases of the k-vector spaces M and N, respectively, then $(e_i \otimes f_j)$ is a basis of $M \otimes N$. Moreover, any $v \in M \otimes N$ has a *unique* expression

$$v = \sum_j v_j \otimes f_j$$

with $v_j \in M$ for all j.

(9) Given a ring A and A-modules given with linear maps

$$M' \xrightarrow{f} M \xrightarrow{g} M',$$

the sequence is said to be *exact* if $\operatorname{Im}(f) = \operatorname{Ker}(g)$ in M. In particular, a sequence

$$0 \longrightarrow M' \xrightarrow{f} M$$

is exact if and only if $\operatorname{Ker}(f) = 0$, which means that f is injective, and a sequence

$$M \xrightarrow{g} M'' \longrightarrow 0$$

is exact if and only if $\operatorname{Im}(g) = \operatorname{Ker}(0) = M''$, i.e., if and only if g is surjective.

A sequence

$$0 \longrightarrow M' \xrightarrow{f} M \xrightarrow{g} M'' \longrightarrow 0,$$

where all three intermediate 3-term sequences are exact, is called a *short exact sequence*. This means that f is injective, g is surjective and the image of f coincides with the kernel of g. It is also usual to say that M is an *extension* of M'' by M'. Note that there is no typo here: this is indeed the standard terminology, instead of speaking of extensions of M'.

(10) Given a vector space E over a field k and a family $(F_i)_{i\in I}$ of linear subspaces of E, we say that the subspaces F_i *are in direct sum* if the subspace they span is a direct sum of the F_i, or in other words, if

$$F_i \cap \Big(\sum_{\substack{j\in I\\ j\neq i}} F_j\Big) = 0$$

for all $i \in I$ (equivalently, any family $(f_i)_{i\in I}$ of vectors in F_i, which are zero for all but finitely many indices i, is linearly independent).

(11) Given a group G, we denote by $[G, G]$ the *commutator group* (or *derived subgroup*) of G, which is generated by all commutators $[g, h] = ghg^{-1}h^{-1}$. Note that not all elements of $[G, G]$ are themselves commutators; see Remark 4.4.5 for examples. The subgroup $[G, G]$ is normal in G, and the quotient group $G/[G, G]$ is abelian; it is called the *abelianization* of G.

(12) We denote by $\mathbf{F}_p$ the finite field $\mathbf{Z}/p\mathbf{Z}$, for p prime and, more generally, by $\mathbf{F}_q$ a finite field with q elements, where $q = p^n$, $n \geqslant 1$, is a power of p. In Chapter 4, we need some simple facts about these, in particular the fact that for each $n \geqslant 1$, there is—up to isomorphism—a unique extension $k/\mathbf{F}_p$ of degree n, i.e., a finite field k of order $q = p^n$. An element $x \in k$ is in $\mathbf{F}_p$ if and only if $x^p = x$ (e.g., because the equation $X^p - X = 0$ has at most p roots, and all $x \in \mathbf{F}_p$ are roots). The group homomorphism

$$N = N_{k/\mathbf{F}_p} : \begin{cases} k^\times & \longrightarrow & \mathbf{F}_p^\times \\ x & \mapsto & \prod_{j=0}^{n-1} x^{p^j} \end{cases}$$

(called the *norm* from k to $\mathbf{F}_p$) is well defined and surjective. Indeed, it is well defined because one checks that $N(x)^p = N(x)$, and surjective, e.g., because the kernel is defined by a non-zero polynomial equation of degree at most $1+p+p+\cdots+p^{n-1} = (p^n-1)/(p-1)$, and hence contains at most that many elements, so the image has at least $p-1$ elements. Moreover, the kernel of the norm is the set of all x which can be written as y/y^p for some $y \in k^\times$.

Similarly, the homomorphism of abelian groups

$$\mathrm{Tr} = \mathrm{Tr}_{k/\mathbf{F}_p} : \begin{cases} \mathbf{F}_q & \longrightarrow & \mathbf{F}_p \\ x & \mapsto & x + x^p + \cdots + x^{p^{n-1}} \end{cases}$$

is well defined and is surjective; it is called the *trace* from k to $\mathbf{F}_p$.

(13) When considering a normed vector space E, we usually denote the norm by $\|v\|$, and sometimes write $\|v\|_E$, when more than one space (or norm) are considered simultaneously.

(14) When considering a Hilbert space H, we speak synonymously of an inner product or of a positive-definite hermitian form, which we denote $\langle\cdot,\cdot\rangle$ or $\langle\cdot,\cdot\rangle_{\mathrm{H}}$ if more than one space might be understood. We use the convention

that a hermitian form is linear in the first variable and conjugate-linear in the other, i.e., we have

$$\langle \alpha v, w\rangle = \alpha\langle v, w\rangle, \qquad \langle v, \alpha w\rangle = \bar{\alpha}\langle v, w\rangle,$$

for two vectors v, w and a scalar $\alpha \in \mathbf{C}$. We recall that a Hilbert space is *separable* if it has a finite or countable orthonormal basis. If $T : \mathrm{H}_1 \longrightarrow \mathrm{H}_2$ is a continuous (synonymously, bounded) linear operator between Hilbert spaces, the *adjoint* of T is the unique linear operator $T^* : \mathrm{H}_2 \longrightarrow \mathrm{H}_1$ such that

$$\langle T(v_1), v_2\rangle_{\mathrm{H}_2} = \langle v_1, T^*(v_2)\rangle_{\mathrm{H}_1}$$

for all $v_1 \in \mathrm{H}_1$ and $v_2 \in \mathrm{H}_2$. The operator T is called *self-adjoint* if and only if $T^* = T$ and *unitary* if and only if $TT^* = T^*T = \mathrm{Id}$.

(15) We will always consider Hausdorff topological spaces, except if explicitly mentioned otherwise (this will only happen in Section 7.1).

(16) A *Borel measure* on a topological space X is a measure defined on the σ-algebra of Borel sets. A *Radon measure* is a Borel measure which is finite on compact subsets of X, and which satisfies the regularity conditions

$$\mu(A) = \inf\{\mu(U) \mid U \supset A,\ U \text{ open}\} \quad \text{for all Borel sets } A,$$
$$\mu(U) = \sup\{\mu(K) \mid K \subset U,\ K \text{ compact}\} \quad \text{for all open sets } U$$

(see, e.g., [**19**, §7.1]); if X is σ-compact (for instance, if X is a separable metric space), then in fact these regularity conditions are automatically satisfied (see, e.g., [**19**, Th. 7.8]).

(17) The support of a Borel measure μ is the set in X defined as the complement of the union of all open sets U in X such that $\mu(U) = 0$. This definition is useful if either X has a countable basis of open sets, for instance $X = \mathbf{R}$, or if μ is a Radon measure, since in those cases the support of μ is closed; see, e.g., [**19**, Exercise 2, p. 208].

(18) The integral of a non-negative measurable function f, or of an integrable function f, with respect to μ, is denoted by either of the following:

$$\int_X f(x)d\mu(x) = \int_X f d\mu.$$

(19) If $\varphi : X \longrightarrow Y$ is a measurable map between two measure spaces and μ is a measure on X, then the *image measure* $\nu = \varphi_*\mu$ on Y is defined by

$$\nu(B) = \mu(\varphi^{-1}(B))$$

for $B \subset Y$ measurable, or equivalently by the integration formula

$$\int_Y f(y)d\nu(y) = \int_X f(\varphi(x))d\mu(x)$$

for any $f : Y \longrightarrow \mathbf{C}$ which is integrable (or measurable and $\geqslant 0$).

(20) Finally, a probability measure μ on an arbitrary measure space X is a measure such that $\mu(X) = 1$; the measure $\mu(A)$ of a measurable subset $A \subset X$ is then also called the *probability* of A.

Chapter 2

The language of representation theory

2.1. Basic language

We begin by restating formally the following definition:

Definition 2.1.1 (Linear representation). Let G be a group, and let k be a field. A *linear representation of* G, defined over k, is a group homomorphism

$$\varrho : G \longrightarrow \mathrm{GL}(E),$$

where E is a k-vector space. The dimension of E is called the *dimension* of ϱ, or sometimes its *degree* or *rank*. We will denote it $\dim(\varrho)$.

Remark 2.1.2. It is also customary to just say that ϱ is a k-representation of G, and to omit mentioning the field k if it is clear from context. Similarly, when the homomorphism ϱ is clear from context, one may say only that "E is a representation of G". Another common alternative notation is "let (ϱ, E) be a k-representation of G".

Given a representation $\varrho : G \longrightarrow \mathrm{GL}(E)$ and an element $g \in G$, we usually write

$$\varrho(g)v$$

for the image of $v \in E$ under the linear transformation $\varrho(g)$. Such vectors are also sometimes called *G-translates of* v (or simply *translates of* v, when the context is clear). Similarly, when ϱ is clearly understood, one may simply write

$$gv = \varrho(g)v \quad \text{or} \quad g \cdot v = \varrho(g)v,$$

and these notations are all frequently used.

The basic rules that ϱ satisfies are then the relations

$$\varrho(1)v = v \quad (gh)v = \varrho(gh)v = \varrho(g)(\varrho(h)v) = g(hv), \tag{2.1}$$

$$g^{-1}(gv) = \varrho(g^{-1})(\varrho(g)v) = v \tag{2.2}$$

for all g, $h \in G$ and $v \in E$, in addition to the linearity of $\varrho(g)$ for a given g.

This notation emphasizes the fact that ϱ is also the same as a left action of the group G on the vector space E, the action being through linear maps (instead of arbitrary bijections of E). In this viewpoint, one thinks of ϱ as the equivalent data of the map

$$\begin{cases} G \times E & \longrightarrow & E \\ (g, v) & \mapsto & g \cdot v. \end{cases}$$

It should be clear already that representations exist in plenty—they are not among those mathematical objects that are characterized by their rarity. For instance, obviously, any subgroup G of $\mathrm{GL}(E)$ can be thought of as being given with a natural ("tautological" is the adjective commonly used) representation

$$G \hookrightarrow \mathrm{GL}(E).$$

In a different style, for any group G and field k, we can form a vector space, denoted $k(G)$, with a basis $(e_g)_{g\in G}$ indexed by the elements of G (i.e., the k-vector space freely generated by the set G; if G is infinite, note that $k(G)$ is infinite dimensional). Then we may let G act linearly on $k(G)$ by describing a transformation $\pi_G(g)$ through its action on the basis vectors: we define

$$\pi_G(g)e_h = e_{gh} \tag{2.3}$$

for all $g \in G$ and all basis vectors e_h. Then to check that π_G is a linear representation of G on E, it is enough to check (2.1). This is a simple exercise; we give details merely for completeness, but readers should attempt to perform this check, at least in a first reading. First, it is clear that $\pi_G(1)$ acts as identity on the basis vectors, and hence is the identity transformation. Now, given g_1, $g_2 \in G$ and a basis vector e_h, its image under $\pi_G(g_1g_2)$ is $e_{g_1g_2h}$ by definition. And since $\pi_G(g_2)e_h$ is the basis vector e_{g_2h}, we also have

$$\pi_G(g_1)(\pi_G(g_2)e_h) = e_{g_1g_2h} = \pi_G(g_1g_2)e_h,$$

which, with h being arbitrary, means that $\pi_G(g_1g_2) = \pi_G(g_1)\pi_G(g_2)$. By taking $g_2 = g_1^{-1}$, this confirms that π_G is a homomorphism into $\mathrm{GL}(k(G))$.

Another easily defined representation is the *right-regular representation*, or simply *regular* representation ϱ_G of G over k: let $C_k(G)$ be the space[1] of

[1] The notation is not completely standard.

all functions

$$f : G \to k$$

(with pointwise addition and scalar multiplication of functions; we will often write $C(G)$ for $C_k(G)$ when the field is clear in context). One defines $\varrho_G(g)$ acting on $C_k(G)$ by the rule

$$\varrho_G(g)f(x) = f(xg)$$

for all $f \in C_k(G)$, $g \in G$, where $x \in G$ is the point at which the new function $\varrho_G(g)f \in C_k(G)$ is evaluated. It is again a simple matter—one that the reader should attempt, if only because the order of evaluation might seem to be wrong!—to check that ϱ_G is a representation: for $f \in E$, g, $h \in G$, we get that $\varrho_G(gh)f$ maps x to

$$\varrho_G(gh)f(x) = f(xgh),$$

while, $\varrho_G(h)f$ being the function $f_1 : y \mapsto f(yh)$, we see that $\varrho_G(g)\varrho_G(h)f = \varrho_G(g)f_1$ maps x to

$$f_1(xg) = f((xg)h) = f(xgh),$$

which completes the check that $\varrho_G(gh) = \varrho_G(g)\varrho_G(h)$.

Exercise 2.1.3. (1) Show that the formula $\lambda_G(g)f(x) = f(g^{-1}x)$ also defines a representation of G on $C_k(G)$. It is called the *left-regular representation* λ_G *of* G (over k).

(2) Show that the formula

$$\varrho(g, h)f(x) = f(g^{-1}xh)$$

defines a representation ϱ of $G \times G$ on $C_k(G)$.

In the previous examples, the representation map ϱ is injective (it is clear in the second case and easily checked in the third). This is certainly not always the case: indeed, for any group G and field k, there is also a *trivial* representation of G of degree 1 defined over k, which simply maps every $g \in G$ to $1 \in k^\times = \mathrm{GL}(k)$. This is not injective unless $G = 1$. Note that one should not dismiss this trivial representation as obviously uninteresting: as we will see quite soon, it does have an important role to play.

Still we record the names of these two types of representations:

Definition 2.1.4 (Faithful and trivial representations)**.** Let G be a group, and let k be a field.

(1) A representation ϱ of G defined over k is *faithful* if ϱ is injective, i.e., if $\mathrm{Ker}(\varrho) = \{1\}$ in G.

(2) A representation ϱ of G on a k-vector space E is *trivial* if $\varrho(g) = 1$ is the identity map of E for all $g \in G$, i.e., if $\mathrm{Ker}(\varrho) = G$.

Remark 2.1.5. Sometimes only the representation of degree 1 (with $E = k$) mapping g to $1 \in k^\times$ is called "the" trivial representation. We will denote by $\mathbf{1}$ this one-dimensional representation (when G and k are clear in context, or $\mathbf{1}_G$ if only k is).

These examples are extremely general. Before continuing, here are others which are extremely specific, but still very important. We take $k = \mathbf{C}$. Then we have the exponential $z \mapsto e^z$, which is a group homomorphism from $(\mathbf{C}, +)$ to $(\mathbf{C}^\times, \cdot)$, or in other words, to $\mathrm{GL}_1(\mathbf{C}) = \mathrm{GL}(\mathbf{C})$. This means the exponential is a one-dimensional representation (over $\mathbf{C}$) of the additive group of the complex numbers. One can find variants:

- If $G = \mathbf{R}$ or $\mathbf{C}$, then for any $s \in \mathbf{C}$, the map

(2.4)
$$\chi_s : x \mapsto e^{sx}$$

  is a one-dimensional representation.
- If $G = \mathbf{R}/\mathbf{Z}$, then for any $m \in \mathbf{Z}$, the map

(2.5)
$$e_m : x \mapsto e^{2i\pi mx}$$

  is a one-dimensional representation of G (one must check that this is well defined on $\mathbf{R}/\mathbf{Z}$, but this is the case since $e^{2i\pi mn} = 1$ for any $n \in \mathbf{Z}$; indeed, no other representation χ_s of $\mathbf{R}$, for $s \in \mathbf{C}$, has this property since $\chi_s(1) = 1$ means $e^s = 1$).
- If $q \geqslant 1$ is an integer and $G = \mathbf{Z}/q\mathbf{Z}$ if the additive group of integers modulo q, then for any $m \in \mathbf{Z}/q\mathbf{Z}$, the map

(2.6)
$$x \mapsto e^{2i\pi mx/q}$$

  is well defined and it is a one-dimensional representation of G. Indeed, note that $e^{2i\pi\tilde{m}x/q}$ is independent of the choice of a representative $\tilde{m} \in \mathbf{Z}$ of $m \in \mathbf{Z}/q\mathbf{Z}$, since replacing $\tilde{m}$ by $\tilde{m} + kq$ just multiplies the value by $e^{2i\pi xk} = 1$.

More examples, many of which are defined without the intermediate results and language, can be found in Section 2.6, and some readers may want to read that section first (or at least partly) to have some more concrete examples in mind.

Although one can thus see that there are "many" representations in a certain sense, as soon as we try to "compare" them, the impression emerges that this abundance is, for given G and field k, of the same type as the abundance of vector spaces (in contrast with, for instance, the similarly striking abundance of k-algebras). Although they may arise in every corner, many of them are actually the same. In other words, quite often, the representations of G over k can be classified in a useful way. To go into this, we must explain how to relate possibly different representations.

Definition 2.1.6 (Morphism of representations). Let G be a group, and let k be a field. A *morphism*, or homomorphism, between representations ϱ_1 and ϱ_2 of G, both defined over k and acting on the vector spaces E_1 and E_2, respectively, is a k-linear map

$$\Phi : E_1 \longrightarrow E_2$$

such that

$$\Phi(\varrho_1(g)v) = \varrho_2(g)(\Phi(v)) \in E_2,$$

for all $g \in G$ and $v \in E_1$. One also says that Φ *intertwines* ϱ_1 and ϱ_2, or is an *an intertwining operator*, or *intertwiner*, between them, and one may denote this by $\varrho_1 \xrightarrow{\Phi} \varrho_2$.

This definition is also better visualized as saying that, for all $g \in G$, the square diagram

$$\begin{array}{ccc} E_1 & \xrightarrow{\Phi} & E_2 \\ \varrho_1(g) \downarrow & & \downarrow \varrho_2(g) \\ E_1 & \xrightarrow{\Phi} & E_2 \end{array}$$

of linear maps commutes or, even more concisely, by omitting the mention of the representations and writing

$$\Phi(g \cdot v) = g \cdot \Phi(v)$$

for $g \in G$, $v \in E_1$.

It is also easy to see that the set of homomorphisms from ϱ_1 to ϱ_2, as representations of G, is a k-vector subspace of $\mathrm{Hom}(E_1, E_2)$, which we denote $\mathrm{Hom}_G(\varrho_1, \varrho_2)$. (This vector space may of course reduce to 0.)

The following are simple facts, but they are also of crucial importance:

Proposition 2.1.7 (Functoriality). *Let G be a group, and let k be a field.*

(1) *For any representation ϱ of G and a vector space E, the identity map on E is a homomorphism $\varrho \longrightarrow \varrho$.*

(2) *Given representations ϱ_1, ϱ_2, and ϱ_3 on E_1, E_2, and E_3, respectively, and morphisms*

$$E_1 \xrightarrow{\Phi_1} E_2 \xrightarrow{\Phi_2} E_3,$$

the composite $E_1 \xrightarrow{\Phi_2 \circ \Phi_1} E_3$ is a morphism between ϱ_1 and ϱ_3.

Remark 2.1.8 (The category of representations). In the language of category theory (which we will only use incidentally in remarks in this book), this proposition states that the representations of a given group G over a given field k are the objects of a *category* with morphisms given by the intertwining linear maps.

If a morphism Φ is a bijective linear map, its inverse Φ^{-1} is also a morphism (between ϱ_2 and ϱ_1), and it is therefore justified to call Φ an *isomorphism* between ϱ_1 and ϱ_2. Indeed, using the diagram above, we find that the relation

$$\varrho_2(g) \circ \Phi = \Phi \circ \varrho_1(g)$$

is equivalent in that case to

$$\Phi^{-1} \circ \varrho_2(g) = \varrho_1(g) \circ \Phi^{-1},$$

which is the desired fact that Φ^{-1} be an intertwining operator between ϱ_2 and ϱ_1. It is also equivalent to

$$\varrho_1 = \Phi^{-1} \circ \varrho_2 \circ \Phi.$$

As another general example, if a vector subspace $F \subset E$ is stable under all operators $\varrho(g)$ (i.e., $\varrho(g)(F) \subset F$ for all $g \in G$), then the restriction of $\varrho(g)$ to F defines a homomorphism

$$\tilde{\varrho} : G \longrightarrow \mathrm{GL}(F),$$

which is therefore a k-representation of G, and the inclusion linear map

$$i : F \hookrightarrow E$$

is a morphism of representations. One speaks, naturally, of a *subrepresentation* of ϱ or, if the action is clear in context, of E itself.

Example 2.1.9 (Trivial subrepresentations). Consider the case where F is the space of all vectors $v \in E$ which are pointwise invariant under G: $v \in F$ if and only if

$$g \cdot v = v \text{ for all } g \in G.$$

Because G acts by linear maps on E, this subspace F, also denoted $F = E^G$, is a linear subspace of E and a subrepresentation of ϱ. Note that the representation of G on E^G is trivial, in the sense of Definition 2.1.4. This means that if n is the dimension[2] of E^G, and if $\mathbf{1}^n = k^n$ denotes the k-vector space of dimension n with a trivial action of G, we have an isomorphism

$$\mathbf{1}^n \xrightarrow{\sim} E^G$$

(by fixing any basis of E^G). Of course, it is possible—and is frequently the case—that $E^G = 0$.

This space of invariants is the largest subrepresentation of E (for inclusion) which is trivial. More individually, any non-zero vector $v \in E$ which is invariant under G defines a trivial subrepresentation of dimension 1, i.e., an injective morphism

$$\begin{cases} \mathbf{1} & \hookrightarrow & E \\ t & \mapsto & tv \end{cases}$$

[2] Which may be finite or infinite.

of representations. This gives a k-linear isomorphism

$$E^G \simeq \mathrm{Hom}_G(\mathbf{1}, E) \tag{2.7}$$

(the reciprocal map sending $\Phi : \mathbf{1} \to E$ to $\Phi(1)$).

Because fixed points or invariant vectors of various kinds are often of great importance, we see here how useful the trivial representation can be. To give a simple, but very useful, example, the invariant subspace of the regular representation is the one-dimensional subspace of constant (k-valued) functions on G: if $\varphi \in C_k(G)^G$, we have

$$\varphi(x) = \varrho_G(g)\varphi(x) = \varphi(xg)$$

for all x and g, and taking $x = 1$ shows that φ is constant.

On the other hand, note that $k(G)^G$ is *zero* if G is infinite, and one dimensional, generated by

$$\sum_{g \in G} e_g \in k(G)$$

if G is finite.

Example 2.1.10 (One-dimensional representations). A one-dimensional k-representation χ of a group G is simply a homomorphism $\chi : G \longrightarrow k^\times$ (this is because, for any one-dimensional k-vector space, there is a canonical isomorphism $k^\times \longrightarrow \mathrm{GL}(V)$, obtained by mapping $\lambda \in k^\times$ to $\lambda\mathrm{Id}$). Generalizing Example 2.1.9, which corresponds to $\chi = 1$, for an arbitrary k-representation $\varrho : G \longrightarrow \mathrm{GL}(V)$, a non-zero intertwiner $\chi \xrightarrow{\Phi} \varrho$ corresponds to the data of a non-zero vector $v \in V$ such that

$$\varrho(g)v = \chi(g)v$$

for all $g \in G$ (the reader should check this elementary fact). This means exactly that v is a *common eigenvector* for all operators $\varrho(g)$. For instance, χ itself, if seen as a k-valued function on G, is an element of $C_k(G)$ which corresponds to an intertwiner $\chi \longrightarrow C_k(G)$, as the reader should check.

Example 2.1.11 (Invariants under normal subgroups). Consider again a k-representation ϱ of G, acting on E. The space E^G of invariants is a subrepresentation, obviously trivial, as in Example 2.1.9. A very useful fact is that if we take the vectors invariant under a *normal subgroup* of G, we still obtain a subrepresentation of E, though usually not a trivial one.

Lemma 2.1.12 (Invariants under normal subgroups). *Let k be a field, let G be a group, and let $H \lhd G$ be a normal subgroup. Then for any k-representation ϱ of G acting on E, the subspace*

$$E^H = \{v \in E \mid \varrho(h)v = v \text{ for all } h \in H\}$$

is a subrepresentation of ϱ.

Proof. Let $v \in E^H$ and $g \in G$. We want to check that $w = \varrho(g)v \in E^H$. For this we pick $h \in H$, and we write simply

$$\varrho(h)w = \varrho(hg)v = \varrho(g)\varrho(g^{-1}hg)v,$$

and since $h' = g^{-1}hg$ is in H (because H is normal by assumption) and $v \in E^H$, we get $\varrho(h)w = \varrho(g)v = w$ as desired. □

The reader should look for examples where H is not normal and E^H is not stable under the action of G, as well as for examples where E^H is not a trivial representation of G.

Example 2.1.13 (Regular representation)**.** Consider the two examples of representations π_G and ϱ_G associated to a group G and field k that were discussed just after the Definition 2.1.1. We claim that π_G (acting on $k(G)$) is isomorphic to a subrepresentation of ϱ_G (acting on $C(G)$). To see this, we define $\Phi : k(G) \to C(G)$ by mapping a basis vector e_g, $g \in G$, to the characteristic function of the single point g^{-1}; in other words

$$\Phi(e_g)(x) = \begin{cases} 1 & \text{if } x = g^{-1}, \\ 0 & \text{otherwise.} \end{cases}$$

The linear map defined in this way is injective; indeed, $\Phi(v)$ is the function mapping $g \in G$ to the coefficient of the basis element $e_{g^{-1}}$ in the expression of v, and can only be identically zero if v is itself 0 in $k(G)$. We check now that Φ is a morphism of representations. In $k(G)$, we have $g \cdot e_h = e_{gh}$, and in $C(G)$, we find that $g \cdot \Phi(e_h) = \varrho_G(g)\Phi(e_h)$ maps x to

$$\Phi(e_h)(xg) = \begin{cases} 1 & \text{if } xg = h^{-1}, \text{ i.e., if } x = h^{-1}g^{-1} = (gh)^{-1}, \\ 0 & \text{otherwise,} \end{cases}$$

which precisely says that

$$\Phi(g \cdot e_h) = g \cdot \Phi(e_h).$$

The map Φ is an isomorphism if G is finite, but not otherwise; indeed, the image $\mathrm{Im}(\Phi)$ is always equal to the subspace of functions which are zero except at finitely many points.

Remark 2.1.14. The last example makes it fairly clear that our basic definitions will require some adaptations when infinite groups are considered. Typically, if G has a topological structure—compatible with the group operation—the regular representation will be restricted to functions with a certain amount of smoothness or regularity. We will come back to this in Chapter 3 (and later).

We will now discuss the basic formalism of representation theory—roughly speaking, how to manipulate some given representation or representations to obtain new ones. This involves different aspects, as one may try to operate at the level of the vector space E, or of the group G, or even of the field k. The last is of less importance in this book, but we will mention it briefly nevertheless, and it is very important in certain areas of number theory. The other two are, however, of fundamental importance.

2.2. Formalism: changing the space

This part of formalism is the most straightforward. The basic philosophy is simply that essentially any operation of linear or multilinear algebra can be performed on a space E on which a group G acts in such a way that G has a natural action on the resulting space. This is particularly transparent when interpreting representations of G as modules over the group algebra, as explained in Chapter 3, but we will present the basic examples from scratch. However, before reading further, we suggest that the reader try to come up with the definition of the following objects (where the field k and the group G are always fixed):

- quotients of representations, sum and intersection of subrepresentations;
- the kernel and image of a morphism of representations;
- exact sequences and, in particular, short exact sequences, of representations;
- the direct sum of representations;
- the tensor product of two representations;
- the symmetric powers or alternating powers of a representation;
- given a representation ϱ acting on E, the dual (also called contragredient) of ϱ acting on the linear dual space $E' = \mathrm{Hom}_k(E, k)$, and the associated representation of G acting on the space of k-linear maps $\mathrm{End}_k(E) = \mathrm{Hom}_k(E, E)$.

As will be seen, only the last one may not be entirely obvious, and this is because there are in fact two possible answers (though, as we will explain, one of them is much more interesting and important).

Here is an abstract presentation of the mechanism at work; although we will give full details in each case, it is also useful to see that a single process is at work.

Proposition 2.2.1 (Functorial representations). *Let k be a field, and let G be a group. Let T be any covariant functor on the category of k-vector* *spaces, i.e., any rule assigning a vector space $T(E)$ to any k-vector space E, and a map*

$$T(f) : T(E_1) \to T(E_2)$$

to any linear map $f : E_1 \to E_2$, with the properties that

$$\begin{cases} T(f \circ g) = T(f) \circ T(g), \\ T(1_E) = 1_{T(E)}. \end{cases} \tag{2.8}$$

Then given a k-representation

$$\varrho : G \longrightarrow \mathrm{GL}(E),$$

the vector space $T(E)$ has a linear action

$$\pi = T(\varrho) : G \longrightarrow \mathrm{GL}(T(E))$$

given by

$$\pi(g) = T(\varrho(g)).$$

Moreover, for any homomorphism $\varrho_1 \xrightarrow{\Phi} \varrho_2$ of representations of G, the k-linear map $T(\Phi)$ is a homomorphism $T(\varrho_1) \longrightarrow T(\varrho_2)$, and this construction is compatible with composition and identity. In particular, $T(\varrho)$ depends, up to isomorphism of representations, only on the isomorphism class of ϱ itself.

This is a direct translation of the functoriality property of morphisms of representations noted in Proposition 2.1.7.

2.2.1. Quotients, kernels, images, ... We have defined subrepresentations already. The operation of sum and intersection of subspaces, when applied to subrepresentations, lead to other subrepresentations.

Quotients are equally natural objects to consider. Given a representation ϱ of G on E and a subspace $F \subset E$, which is a *subrepresentation* of E or in other words such that $\varrho(g)$ always leaves F invariant, the quotient vector space E/F also has a natural linear action of G, simply induced by ϱ: given $v \in E/F$ and $g \in G$, the action $g \cdot v$ is the image in E/F of $\varrho(g)\tilde{v}$ for any $\tilde{v} \in E$ mapping to v under the canonical surjective map $E \to E/F$. This is well defined because if $\tilde{v}_1$ is another such vector, we have $\tilde{v}_1 = \tilde{v} + w$ with $w \in F$; hence,

$$\varrho(g)\tilde{v}_1 - \varrho(g)\tilde{v} = \varrho(g)w$$

also lies in F, and has image 0 in E/F.

Another global description of this action is that it is such that the surjective map

$$E \longrightarrow E/F$$

is then a morphism of representations, just like the inclusion map $F \longrightarrow E$ is one.

In the same vein, given a morphism

$$\Phi : E_1 \longrightarrow E_2$$

of k-representations of G, we can see that the standard vector spaces associated to Φ are all themselves representations of G:

- the kernel $\operatorname{Ker}(\Phi) \subset E_1$ is a subrepresentation of E_1;
- the image $\operatorname{Im}(\Phi) \subset E_2$ is a subrepresentation of E_2;
- the natural linear isomorphism

$$E_1/\operatorname{Ker}(\Phi) \simeq \operatorname{Im}(\Phi)$$

(induced by Φ) is an isomorphism of representations;

- The cokernel $\operatorname{Coker}(\Phi) = E_2/\operatorname{Im}(\Phi)$ is a representation of G, as quotient of two representations.

These facts are consequences of the definitions and, specifically, of the linearity of the actions of G.

2.2.2. Coinvariants. If we go back to Example 2.1.9, and in particular the identification (2.7) of the homomorphisms from $\mathbf{1}$ to a representation ϱ, one may ask if there is a similar description of the space

$$\operatorname{Hom}_G(E, \mathbf{1})$$

of homomorphisms *from* ϱ *to* the trivial one.

By definition, an element in this space is a k-linear form $E \xrightarrow{\lambda} k$ such that for all $v \in E$ and $g \in G$, we have

$$\lambda(g \cdot v) = \lambda(v).$$

This condition is equivalent to $\ker(\lambda) \supset E_1$, where E_1 is the subspace of E spanned by all vectors of the form

$$g \cdot v - v, \qquad g \in G,\ v \in E,$$

or equivalently it corresponds to a linear form

$$E/E_1 \longrightarrow k$$

extended to E by composition $E \longrightarrow E/E_1 \longrightarrow k$.

Note that E_1 is also a subrepresentation of E, since

$$h \cdot (g \cdot v - v) = hg \cdot v - h \cdot v = (hgh^{-1})v_1 - v_1 \tag{2.9}$$

with $v_1 = h \cdot v$. Hence E/E_1 has an induced structure of representation of G. In fact, this action on E/E_1 is trivial, since $g \cdot v \equiv v$ modulo E_1 for all g and v.

The space E/E_1 is called the space of *coinvariants* of ϱ, and is denoted E_G. It is the *largest* quotient of ϱ that is a trivial representation of G (like the invariant space, it may well be zero) and by the above, we can write

$$\mathrm{Hom}_G(\varrho, \mathbf{1}) \simeq \mathrm{Hom}_k(E_G, k),$$

which identifies the space of homomorphisms to the trivial representation with the linear dual vector space of the coinvariant space.

Exercise 2.2.2. Show that if $H \lhd G$, the H-coinvariant space E_H has an induced structure of representation of G. (This is the analogue, for the coinvariants, of Lemma 2.1.12.)

2.2.3. Direct sums, exact sequences, irreducibility, and semisimplicity. The simplest operation that can be performed on representations is the direct sum. Given G and k, as usual, and k-representations ϱ_1, ϱ_2 of G on E_1 and E_2, respectively, the direct sum $\varrho_1 \oplus \varrho_2$ is the representation

$$G \longrightarrow \mathrm{GL}(E_1 \oplus E_2)$$

such that

$$g \cdot (v_1 + v_2) = \varrho_1(g)v_1 + \varrho_2(g)v_2,$$

for all $v = v_1 + v_2 \in E_1 \oplus E_2$, or more suggestively

$$g \cdot (v_1 + v_2) = g \cdot v_1 + g \cdot v_2.$$

By definition, we see that the subspaces E_1, E_2 or $E = E_1 \oplus E_2$ are subrepresentations of $\varrho_1 \oplus \varrho_2$, and that

$$(\varrho_1 \oplus \varrho_2)/\varrho_1 \simeq \varrho_2, \tag{2.10}$$

the corresponding isomorphism being induced by $v_1 + v_2 \mapsto v_2$.

One can consider more than two factors: for an arbitrary family $(\varrho_i)_{i \in I}$ of k-representations, with ϱ_i acting on E_i, one can define a representation of G on the direct sum

$$E = \bigoplus_{i \in I} E_i$$

by linearity again from the actions of G on each subspace E_i of E.

Note the general relations

$$\dim(\varrho_1 \oplus \varrho_2) = \dim(\varrho_1) + \dim(\varrho_2), \quad \dim(\bigoplus_{i \in I} \varrho_i) = \sum_{i \in I} \dim(\varrho_i),$$

where, since we do not distinguish between infinite cardinals, the convention that the dimension is infinite if either there is an infinite-dimensional ϱ_i, or if I is infinite and infinitely many summands ϱ_i are non-zero. Equally useful are the natural isomorphisms

$$\mathrm{Hom}_G(\varrho, \varrho_1 \oplus \varrho_2) \simeq \mathrm{Hom}_G(\varrho, \varrho_1) \oplus \mathrm{Hom}_G(\varrho, \varrho_2),$$
$$\mathrm{Hom}_G(\varrho_1 \oplus \varrho_2, \varrho) \simeq \mathrm{Hom}_G(\varrho_1, \varrho) \oplus \mathrm{Hom}_G(\varrho_2, \varrho),$$

and similarly for an arbitrary (finite) number of summands.[3]

Exercise 2.2.3. Let G be a group, let k be a field, and let ϱ_1, ϱ_2 be two k-representations of G, acting on E_1 and E_2, respectively. Show that a k-linear map $\Phi : E_1 \longrightarrow E_2$ is a G-homomorphism if and only if the graph

$$\Gamma = \{(v, \Phi(v)) \mid v \in E_1\} \subset E_1 \oplus E_2$$

of Φ is a subrepresentation of $\varrho_1 \oplus \varrho_2$.

Another generalization of the direct sum, based on (2.10), considers any representation ϱ of G acting on E, with an injection

$$\Phi : \varrho_1 \hookrightarrow \varrho$$

such that

$$\varrho/\varrho_1 = \varrho/\operatorname{Im}(\Phi) \simeq \varrho_2 \tag{2.11}$$

as k-representations. However, although there exists of course always a subspace $E_2 \subset E$ such that

$$E = \operatorname{Im}(\Phi) \oplus E_2 \simeq E_1 \oplus E_2$$

as k-vector spaces, *it is not always the case that E_2 can be found as a subrepresentation of ϱ*. When the complement E_2 can be chosen to be a subrepresentation of E, this subrepresentation (say π_2) is necessarily isomorphic to ϱ_2 (since $\pi_2 \simeq (\varrho_1 \oplus \pi_2)/\varrho_1 \simeq \varrho/\varrho_1 \simeq \varrho_2$, as representations of G).

A useful equivalent criterion for the existence of such a complementary subrepresentation is the following:

Lemma 2.2.4. *Let G be a group, let k be a vector space, and let $\varrho : G \longrightarrow \mathrm{GL}(E)$ be a representation. Let $E_1 \subset E$ be a subrepresentation of E.*

(1) *Suppose $E_2 \subset E$ is a subspace of E complementary to E_1, so that $E = E_1 \oplus E_2$. Then E_2 is a subrepresentation of E if and only if the linear projection map*

$$\Phi \begin{cases} E \longrightarrow E \\ v_1 + v_2 \mapsto v_1 \end{cases}, \qquad v_1 \in E_1,\ v_2 \in E_2,$$

with image E_1 and kernel E_2 is an intertwiner, *i.e., if $\Phi \in \mathrm{Hom}_G(E, E)$.*

[3] Recall that $\mathrm{Hom}_k(\bigoplus E_i, E)$ is not isomorphic to the direct sums of the $\mathrm{Hom}_k(E_i, E)$ if the index set is infinite; e.g. for $E = k$, the dual of a direct sum is the product of the duals, which is different for infinitely many factors.

(2) *There exists a linear complement* E_2 *which is a subrepresentation if and only if there exists an intertwiner* $\Phi \in \mathrm{Hom}_G(E, E)$ *which is a projection, such that* $\mathrm{Im}(\Phi) = E_1$. *In this case,* $E_2 = \ker \Phi$ *is such a complement.*

Proof. Assertion (1) is elementary, and (2) is of course a consequence of (1). This follows by noting first that if Φ is an intertwiner, the kernel $\mathrm{Ker}\,\Phi = E_2$ is a subrepresentation, while conversely, if E_2 is a subrepresentation, we get from $v = v_1 + v_2$ with $v_i \in E_i$ the decompositions $\varrho(g)v = \varrho(g)v_1 + \varrho(g)v_2$ with $\varrho(g)v_i \in E_i$ again, and hence $\Phi(\varrho(g)v) = \varrho(g)v_2 = \varrho(g)\Phi(v)$. □

In certain circumstances, a subrepresentation complementary to ϱ_1 always exists (for instance, for finite groups when k has characteristic 0, as we will discuss in Chapter 4). Here is a standard example where it fails. Consider the additive group $G = \mathbf{R}$, the field $k = \mathbf{R}$, and the representation

$$\varrho \begin{cases} G & \longrightarrow & \mathrm{GL}_2(\mathbf{R}) \\ x & \mapsto & \begin{pmatrix} 1 & x \\ 0 & 1 \end{pmatrix}. \end{cases} \tag{2.12}$$

(We leave it as an exercise to check, if needed, that this is a homomorphism.) In terms of the canonical basis (e_1, e_2) of $\mathbf{R}^2$, this means that

$$x \cdot (\alpha e_1 + \beta e_2) = (\alpha + x\beta)e_1 + \beta e_2.$$

The subspace $E_1 = \mathbf{R}e_1$ is a subrepresentation of $\mathbf{R}^2$, indeed, it is isomorphic to the trivial representation $\mathbf{1}_G$ since e_1 is invariant under the action of G (which is obvious when looking at the matrix representation). We claim that there is *no* subrepresentation E_2 which is complementary to E_1. This can be checked either by a direct computation or more abstractly. For the former approach, note that E_2 would be of dimension 1. Now let $f = \alpha e_1 + \beta e_2$ be any vector in $\mathbf{R}^2$, and assume that $\varrho(x)f \in \mathbf{R}f$ for all $x \in \mathbf{R}$. We take $x = 1$ and deduce that there exists $\lambda \in \mathbf{R}$ such that

$$\varrho(1)f = (\alpha + \beta)e_1 + \beta e_2 = \lambda f = \lambda\alpha e_1 + \lambda\beta e_2.$$

This can only happen if $\beta = 0$, in which case $f \in \mathbf{R}e_1$, so that the line spanned by f is in fact the same as E_1.

The more abstract argument runs as follows: the quotient representation $\mathbf{R}^2/E_1$ is itself the trivial representation (this should be checked from the definition; in terms of the matrix representation, it amounts to the fact that the bottom-right coefficients of $\varrho(x)$ are all equal to 1). Thus, if E_2 were to exist, we would have, by the above, an isomorphism

$$\varrho \simeq \mathbf{1}_G \oplus \mathbf{1}_G,$$

which is a trivial representation of dimension 2. Since ϱ is certainly not trivial, this would be a contradiction.

We come back to the general case where (2.11) holds. Given intertwiners

$$\varrho_1 \longrightarrow \varrho_2 \longrightarrow \varrho_3$$

of k-representations of a group G, one says that the sequence is *an exact sequence (of representations)* if and only if it is exact as a sequence of k-vector spaces. Similarly one can speak of *short exact sequences* of representations, and (by linear algebra) the situation (2.11) can be summarized by a *short exact sequence*

$$0 \to \varrho_1 \longrightarrow \varrho \xrightarrow{\Phi} \varrho_2 \to 0 \ ;$$

again using the terminology for modules, one also says that ϱ is an *extension* of ϱ_2 by ϱ_1.

When there is a morphism $\varrho_2 \xrightarrow{\Psi} \varrho$ such that $\Phi \circ \Psi = \mathrm{Id}$, one says that the exact sequence *splits*. This happens precisely when the space of ϱ contains a subrepresentation complementary to ϱ_1 (necessarily isomorphic to ϱ_2), so that $\varrho \simeq \varrho_1 \oplus \varrho_2$. More generally, a sequence of homomorphisms of k-representations of G is exact if and only if it is exact as a sequence of maps of k-vector spaces.

Any time a natural representation can be written (up to isomorphism) as a direct sum, or even an extension, of smaller representations, this gives very useful information on the representation. Typically, one wishes to perform such decompositions as long as it is possible. The obvious limitation is that a representation ϱ might not have any non-trivial subrepresentation to try to "peel off". This leads to the following very important definitions:

Definition 2.2.5 (Irreducible, semisimple, isotypic representations). Let G be a group, and let k be a field.

(1) A k-representation ϱ of G acting on E is *irreducible* if and only if $E \neq 0$ and there is no subspace of E stable under ϱ, except 0 and E itself (in other words, if there is no subrepresentation of ϱ except 0 and ϱ itself).

(2) A k-representation ϱ of G is *semisimple* if it can be written as a direct sum of subrepresentations, each of which is irreducible:

$$\varrho \simeq \bigoplus_{i \in I} \varrho_i$$

for some index set I and some irreducible representations ϱ_i (some of the ϱ_i may be isomorphic.)

(3) A semisimple k-representation ϱ of G is *isotypic* if it is a direct sum of irreducible subrepresentations that are all isomorphic; if these subrepresentations are all isomorphic to a representation π, then one says that ϱ is π-isotypic.

We will see later that, up to permutation, the irreducible summands of a semisimple representation are uniquely determined by ϱ (up to isomorphism of representations, of course). This is part of the Jordan–Hölder–Noether Theorem 2.7.1.

Not all representations of a group are semisimple, but irreducible representations are still fundamental "building blocks" for representations in general. An essential feature of irreducible representations, which is formalized in Schur's Lemma 2.2.6, is that these "building blocks" are "incommensurable", in some sense: two non-isomorphic irreducible representations can have "no interaction".

Lemma 2.2.6 (Schur's Lemma, I). *Let G be a group, and let k be a field.*

(1) *Given an* irreducible *k-representation π of G and an arbitrary k-representation ϱ of G, any G-homomorphism $\pi \longrightarrow \varrho$ is either 0 or injective, and any G-homomorphism $\varrho \longrightarrow \pi$ is either 0 or surjective.*

(2) *Given irreducible k-representations π and ϱ of G, a homomorphism $\pi \longrightarrow \varrho$ is either 0 or is an isomorphism; in particular, if π and ϱ are not isomorphic, we have*

$$\mathrm{Hom}_G(\pi, \varrho) = 0.$$

Proof. (1) Given a morphism Φ from π to ϱ, we know that its kernel is a subrepresentation of π; but if π is irreducible, the only possibilities are that the kernel be 0 (then Φ is injective) or that it is π itself (then Φ is 0). Similarly, for a morphism from ϱ to π, the image is either 0 or π itself.

(2) From (1), if Φ is non-zero and has irreducible source and target, it must be an isomorphism. (Recalling that, by definition, an irreducible representation is non-zero, we see that these are exclusive alternatives.) □

Although an arbitrary representation of a group may fail to contain irreducible subrepresentations, we can always find one in a finite-dimensional non-zero representation by simply selecting a non-zero subrepresentation of minimal dimension. Hence:

Lemma 2.2.7 (Existence of irreducible subrepresentations). *Let G be a group, let k be a field, and let ϱ be a non-zero k-representation of G. If ϱ is finite dimensional, there exists at least one irreducible subrepresentation of G contained in ϱ.*

Remark 2.2.8 (Cyclic vector). It is tempting to suggest a more general argument by saying that, given a non-zero representation $G \longrightarrow \mathrm{GL}(E)$ and given $v \neq 0$, the linear span of the vectors $\varrho(g)v$, $g \in G$ should be irreducible—it is after all the smallest subrepresentation of G containing v for inclusion (indeed, any $F \subset E$ which is stable under the action of G

and contains v must contain all such vectors, hence also their linear span). However, in general, this space is *not* irreducible.

For instance, consider the group $G = \mathbf{Z}/p\mathbf{Z}$ with p a prime number and the representation on $\mathbf{C}^2$ given by

$$x \cdot (z_1, z_2) = (z_1, e^{2i\pi x/p} z_2).$$

Since the two axes are invariant under this action, it is of course not irreducible. However, taking $v = (1,1) \in \mathbf{C}^2$, we see that the span of all $x \cdot v$ contains $(1,1)$ and

$$1 \cdot (1,1) = (1, e^{2i\pi/p}),$$

and since

$$\begin{vmatrix} 1 & 1 \\ 1 & e^{2i\pi/p} \end{vmatrix} = e^{2i\pi/p} - 1 \neq 0,$$

this vector does *generate* the whole space.

For a given representation $\varrho : G \longrightarrow \mathrm{GL}(E)$ of a group G, if there exists a non-zero vector $v \in E$ such that its translates span E, it is customary to say that ϱ is a *cyclic representation* and that v is then a *cyclic vector* (which is far from unique usually). For a given vector v, the space generated by the vectors $\varrho(g)v$, which is a cyclic subrepresentation of ϱ, is called the representation *generated* by v.

The example above generalizes to any group G and any representation of the type

$$\varrho = \bigoplus_{1 \leqslant i \leqslant k} \varrho_i,$$

where the ϱ_i are pairwise non-isomorphic irreducible representations of G: taking $v = (v_i)$ in the space of ϱ, where each v_i is non-zero, it follows from the linear independence of matrix coefficients (Theorem 2.7.28 below) that v is a cyclic vector for ϱ.

The simplest examples of irreducible k-representations of G are the one-dimensional representations

$$\chi : G \longrightarrow \mathrm{GL}(E),$$

where E is a one-dimensional vector space over k. As we noted already, since $\mathrm{GL}(E)$ is *canonically* isomorphic to $k^\times$ by the homomorphism mapping $\lambda \in k^\times$ to $k\,\mathrm{Id}$, this homomorphism χ is just a homomorphism $\chi : G \longrightarrow k^\times$.

Such homomorphisms are sometimes called *characters* of G, although this clashes with the more general notion of character that we will see below in Definition 2.7.36. We use this terminology in the next easy proposition:

Proposition 2.2.9. *Let G be a group, and let k be a field.*

(1) *A character $G \longrightarrow k^\times$ is irreducible.*

(2) *Two characters* $\chi_1, \chi_2 : G \longrightarrow k^\times$ *are isomorphic if and only if they are equal as functions on* G.

We give the argument but urge the reader to try to check this if the result seems unclear.

Proof. (1) A one-dimensional vector space contains no non-zero proper subspace at all, and must therefore be irreducible.

(2) Since $\mathrm{GL}(E) \simeq \mathrm{GL}(k) = k^\times$ for any one-dimensional space E, we can assume that χ_1 and χ_2 both act by scalar multiplication on k. Then an intertwiner $\Phi : k \longrightarrow k$ is given by $\Phi = \lambda\,\mathrm{Id}$ for some fixed $\lambda \in k$, and it is an isomorphism if and only if $\lambda \neq 0$. The intertwining condition becomes

$$\Phi(\chi_1(g)x) = \chi_2(g)\Phi(x),$$

for all $x \in k$, which implies $\lambda\chi_1(g) = \lambda\chi_2(g)$ for all $g \in G$. Clearly, this is possible with $\lambda \neq 0$ if and only if $\chi_1 = \chi_2$. □

In particular, the trivial representation $\mathbf{1}_G$ is irreducible (and it may well be the only one-dimensional representation of G). Thus any trivial representation on a vector space E is also semisimple, since it can be written as a direct sum of trivial one-dimensional trivial subrepresentations

$$E \simeq \bigoplus_{i \in I} k e_i$$

after choosing a basis $(e_i)_{i \in I}$ of E. This shows, in passing, that the decomposition of a semisimple representation as a sum of irreducible ones is usually not unique, just as the choice of a basis of a vector space is not unique.

On the other hand, the two-dimensional representation in (2.12) is not semisimple. Indeed, since we saw that it is not irreducible, it can only be semisimple if the space $\mathbf{R}^2$ decomposes as a direct sum $\mathbf{R}^2 = \mathbf{R}f_1 \oplus \mathbf{R}f_2$, where each summand is a subrepresentation, but we checked above that $\mathbf{R}e_1$ is the only one-dimensional subrepresentation of ϱ.[4]

The following lemma is also very useful as it shows that semisimple representations are stable under the operations we have already seen.

Lemma 2.2.10 (Stability of semisimplicity). *Let* G *be a group, and let* k *be a field.*

(1) *If* ϱ *is a semisimple* k*-representation of* G*, then any subrepresentation of* ϱ *is also semisimple and any quotient representation of* ϱ *is also semisimple.*

---

[4] Note that we will soon be able to use a more abstract argument: from the Jordan–Hölder–Noether Theorem below, one sees that if ϱ were semisimple, it would be trivial, which it is not.

(2) *If* $\varrho : G \longrightarrow \mathrm{GL}(E)$ *is an arbitrary representation of* G *and if* E_1, E_2 *are semisimple subrepresentations of* E, *then their sum* $E_1 + E_2 \subset E$, *whether it is a direct sum or not, is semisimple.*

One should be careful that, if ϱ acts on E and we have stable subspaces E_i such that G acts on E_i via ϱ_i and

$$E = \bigoplus_{i \in I} E_i,$$

it does *not* follow that any subrepresentation is of the type

$$\bigoplus_{i \in J} E_i$$

for some $J \subset I$. This is false even for G trivial, where the only irreducible representation is the trivial one, and writing a decomposition of E amounts to choosing a basis. Then there are usually many subspaces of E which are *not* literally direct sums of a subset of the basis directions (e.g., $E = k \oplus k$ and $F = \{(x, x) \mid x \in k\}$).

We will deduce the lemma from the following more abstract criterion for semisimplicity, which is interesting in its own right: it gives a useful property of semisimple representations, and it is sometimes easier to check because it does not mention irreducible representations.

Lemma 2.2.11 (Semisimplicity criterion). *Let* G *be a group, and let* k *be a field. A* k*-representation*

$$\varrho : G \longrightarrow \mathrm{GL}(E)$$

of G *is semisimple if and only if, for any subrepresentation* $F_1 \subset E$ *of* ϱ, *there exists a complementary subrepresentation, i.e., a* G*-stable subspace* $F_2 \subset E$ *such that*

$$E = F_1 \oplus F_2.$$

It is useful to give a name to the second property: one says that a representation ϱ is *completely reducible* if, for any subrepresentation ϱ_1 of ϱ, one can find a complementary one ϱ_2 with

$$\varrho = \varrho_1 \oplus \varrho_2.$$

Proof of Lemma 2.2.10. (1) Let ϱ act on E, and let $F \subset E$ be a subrepresentation. We are going to check that the condition of Lemma 2.2.11 applies to F.[5] Thus, let $F_1 \subset F$ be a subrepresentation of F; it is also one of E, hence there exists a subrepresentation $F_2 \subset E$ such that

$$E = F_1 \oplus F_2.$$

[5] I.e., a subrepresentation of a completely reducible one is itself completely reducible.

Now we claim that $F = F_1 \oplus (F \cap F_2)$, which shows that F_1 has also a stable complement $F \cap F_2$ in F, and finishes the proof that F is semisimple. Indeed, F_1 and $(F \cap F_2)$ are certainly in direct sum, and if $v \in F$ and we write $v = v_1 + v_2$ with $v_1 \in F_1$, $v_2 \in F_2$, we also obtain

$$v_2 = v - v_1 \in F \cap F_2$$

because v_1 is also in F. The case of a quotient representation is quite similar, and it is left to the reader to puzzle....

(2) The sum $E_1 + E_2$ of two subrepresentations of E is isomorphic to a quotient of the direct sum $E_1 \oplus E_2$ by the surjective linear map

$$\begin{cases} E_1 \oplus E_2 & \longrightarrow & E_1 + E_2 \\ (v_1, v_2) & \mapsto & v_1 + v_2. \end{cases}$$

Since this map is an intertwiner, and the direct sum of semisimple representations is semisimple (which is easy to see from the definition), it follows from (1) that $E_1 + E_2$ is also semisimple. □

Proof of Lemma 2.2.11. Neither direction of the equivalence is quite obvious. We start with a semisimple representation ϱ, acting on E, written as a direct sum

$$E = \bigoplus_{i \in I} E_i$$

of stable subspaces E_i, on which G acts irreducibly, and we consider a stable subspace F. Now we use a standard trick in set theory: we consider a maximal (for inclusion) subrepresentation $\tilde{F}$ of E such that $F \cap \tilde{F} = 0$, or in other words, such that F and $\tilde{F}$ are in direct sum. Observe that, if the conclusion of the lemma is correct, $\tilde{F}$ must be a full complement of F in E, and we will proceed to check this. However, we first check that $\tilde{F}$ exists. This is easy if E is finite dimensional, and in general it follows from a quite standard application of Zorn's Lemma (see, e.g., [**40**, p. 693]). We give the details, which the reader may wish to supply independently.

Consider the set $\mathcal{D}$ of subrepresentations $\tilde{F}$ of E such that $F \cap \tilde{F} = 0$, ordered by inclusion of subspaces. We wish to find a maximal element of this ordered set. The subspace 0 is an element of $\mathcal{D}$, which is therefore not empty. By Zorn's Lemma, it suffices to show that any totally ordered subset of $\mathcal{D}$ has an upper bound in $\mathcal{D}$. Let $\mathcal{S} \subset \mathcal{D}$ be such a subset. Consider the subset

$$\tilde{F} = \bigcup_{V \in \mathcal{S}} V \subset E.$$

From the fact that $\mathcal{S}$ is totally ordered, it follows that $\tilde{F}$ is a linear subspace of E, and that it is a subrepresentation (e.g., for the last property, for any $g \in G$ and $v \in \tilde{F}$, there exists $V \in \mathcal{S}$ such that $v \in V$, and then $g \cdot v \in V \subset \tilde{F}$). Obviously, we have $V \subset \tilde{F}$ for all $V \in \mathcal{S}$, and hence $\tilde{F}$ is an upper bound of

$\mathcal{S}$ for inclusion, and it is enough to show that $\tilde{F} \in \mathcal{D}$ to conclude that it is also an upper bound for $\mathcal{S}$ in $\mathcal{D}$. But if $v \in \tilde{F} \cap F$, there exists $V \in \mathcal{S}$ such that $v \in V \cap F = 0$, by definition of $\mathcal{D}$, and hence we get the desired result.

Now, given a subrepresentation $\tilde{F}$ obtained by this construction, consider for every i the intersection

$$(F \oplus \tilde{F}) \cap E_i \subset E_i.$$

Since E_i is an irreducible representation of G, this intersection is either 0 or equal to E_i. In fact, it cannot be zero, because this would mean that $\tilde{F} + E_i \supsetneq \tilde{F}$ is a larger subrepresentation in direct sum with F, contradicting the definition of $\tilde{F}$. Hence we see that $E_i \subset F \oplus \tilde{F}$ for *all* i, and this means that $F \oplus \tilde{F} = E$.

Now comes the converse. We assume that ϱ, acting on E, is non-zero and is completely reducible. We first claim that E contains at least one irreducible subrepresentation: if E has finite dimension, this is Lemma 2.2.7, and otherwise it requires some care but can be done, as explained in Exercise 2.2.13 below.

We consider the sum E_1 (not necessarily direct) of all irreducible subrepresentations of E. It is non-zero, as we just observed. In fact, we must have $E_1 = E$ because our assumption implies that $E = E_1 \oplus \tilde{E}_1$ for some other subrepresentation $\tilde{E}_1$, and if $\tilde{E}_1$ were non-zero, it would also contain an irreducible subrepresentation, which contradicts the definition of E_1. Thus E is a *sum* of irreducible subrepresentations, say of E_i, $i \in I$; we proceed to conclude by showing it is a *direct* sum of $(E_i)_{i \in J}$ for some subset $J \subset I$.

First, we again use Zorn's Lemma to show that there exists a maximal subset $J \subset I$ (with respect to inclusion) such that the sum of the E_i, $i \in J$, is a direct sum. Indeed, we say that a subset $J \subset I$ is *direct* if the E_i, $i \in J$, are in direct sum. We order the set $\mathcal{D}$ of direct subsets of I by inclusion. Any singleton $\{i\}$ is direct, and if $\mathcal{J}$ is any totally ordered subset of $\mathcal{D}$, the union $K = \bigcup_{J \in \mathcal{J}} J$ is an upper bound for $\mathcal{D}$ for inclusion. We leave to the reader the exercise of checking that K is also direct, and hence is an upper bound for $\mathcal{J}$ in $\mathcal{D}$, which allows to apply Zorn's Lemma to $\mathcal{D}$.

Now fix a maximal direct set J, and let F be the direct sum of those E_i, $i \in J$. For any $i \notin J$, the intersection $E_i \cap F$ cannot be zero, as this would allow us to replace J by $J \cup \{i\}$, which is larger than J; hence $E_i \subset F$ for *all* $i \in I$, and hence $E = F$, which is a direct sum of irreducible subrepresentations. □

Remark 2.2.12. In the finite-dimensional case, the last argument can be replaced by an easy induction on $\dim(E)$. If E is not irreducible, we use the assumption to write

$$E = F \oplus F'$$

for some irreducible subspace F and complementary representation F'. The proof of Lemma 2.2.10 really shows that F' is also completely reducible, and since $\dim(F') < \dim(E)$, by induction, we get that F' is also semisimple, and we are done.

Exercise 2.2.13 (Existence of irreducible subrepresentation). Let G be a group, let k be a field, and let $\varrho : G \longrightarrow \mathrm{GL}(E)$ be a completely reducible k-representation of G, with $E \neq 0$. We want to show that such a representation E contains an irreducible subrepresentation.

(1) Fix a $v \neq 0$. Using Zorn's Lemma, show that there exists a maximal subrepresentation $E_1 \subset E$ (for inclusion) such that $v \notin E_1$.

(2) Write $E = E_1 \oplus E_2$ for some subrepresentation E_2, using the complete reducibility of E. Show that E_2 is irreducible. (*Hint*: If not, show that $E_2 = E_3 \oplus E_4$ for some non-zero subrepresentations of E_2, and that $v \notin E_1 \oplus E_3$ or $v \notin E_1 \oplus E_4$.)

2.2.4. Tensor product. An equally important construction is the tensor product. Given G and k, and representations ϱ_1 and ϱ_2 of G on k-vector spaces E_1 and E_2, we obtain a representation

$$G \to \mathrm{GL}(E_1 \otimes_k E_2)$$

by sending g to $\varrho_1(g) \otimes \varrho_2(g)$. Thus, by definition, for a pure tensor $v \otimes w \in E_1 \otimes E_2$, we have

$$g \cdot (v \otimes w) = \varrho_1(g)v \otimes \varrho_2(g)w,$$

another pure tensor (but we recall that $E_1 \otimes E_2$ is not simply the set of such pure tensors, although they generate the tensor product as k-vector space).

The algebraic (*functorial*) properties of the tensor operation ensure that this is a group homomorphism. We will denote this representation by $\varrho_1 \otimes \varrho_2$, or sometimes simply by $E_1 \otimes E_2$ when the actions on the vector spaces are clear from context. For the same type of general reasons, all the standard isomorphisms between tensor products, such as

$$E_1 \otimes E_2 \simeq E_2 \otimes E_1, \quad E_1 \otimes (E_2 \otimes E_3) \simeq (E_1 \otimes E_2) \otimes E_3, \quad E \otimes k \simeq E,$$

are in fact isomorphisms of representations of G, where k in the last equation represents the trivial (one-dimensional) representation of G. In particular, one can define, up to isomorphism, a tensor product involving finitely many factors, which is independent of the order of the product.[6]

Similarly, we have

$$\varrho \otimes \Big(\bigoplus_i \varrho_i\Big) \simeq \bigoplus_i (\varrho \otimes \varrho_i).$$

[6] In the theory of automorphic representations, an important role is played by certain special infinite tensor products; see [**11**] or [**3**].

If $\varrho \subset \varrho_1$ is a subrepresentation, tensoring with another representation ϱ_2 gives a subrepresentation

$$\varrho \otimes \varrho_2 \hookrightarrow \varrho_1 \otimes \varrho_2,$$

but one should be careful that, in general, not all subrepresentations of a tensor product are of this form (e.g., because of dimension reasons).

Note, finally, the relation $\dim(\varrho_1 \otimes \varrho_2) = (\dim \varrho_1)(\dim \varrho_2)$. In particular, if χ_1 and χ_2 both have dimension 1, so does the tensor product, and in fact since χ_1 and χ_2 take values in $k^\times$, the tensor product $\chi_1 \otimes \chi_2$ is just the product of functions $g \mapsto \chi_1(g)\chi_2(g) \in k^\times$. It is customary to omit the tensor product in the notation in that case, writing just $\chi_1\chi_2$.

Exercise 2.2.14 (Tensor product by a one-dimensional representation)**.** Let $\varrho : G \longrightarrow \mathrm{GL}(E)$ be a k-representation of a group G, and let $\chi : G \longrightarrow \mathrm{GL}(k) = k^\times$ be a one-dimensional representation.

(1) Show that $\varrho \otimes \chi$ is isomorphic to the representation

$$g \mapsto \chi(g)\varrho(g)$$

of G on the same space E. One sometimes says that $\varrho \otimes \chi$ is obtained by *twisting* ϱ by χ.

(2) Show that $\varrho \otimes \chi$ is irreducible (resp. semisimple) if and only if ϱ is irreducible (resp. semisimple).

(3) Show that $\varrho \otimes (\chi_1\chi_2) \simeq (\varrho \otimes \chi_1) \otimes \chi_2$ for any two one-dimensional representations of G.

2.2.5. Multilinear operations. Besides tensor products, all other multilinear constructions have the functoriality property (Proposition 2.2.1) needed to operate at the level of representations of a group. Thus, if $\varrho : G \longrightarrow \mathrm{GL}(E)$ is a k-representation of G, we can construct the following:

- the symmetric powers $\mathrm{Sym}^m(E)$ of E, for $m \geqslant 0$;
- the alternating powers $\bigwedge^m E$, for $m \geqslant 0$.

In each case, the corresponding operation for endomorphisms of E leads to representations

$$G \longrightarrow \mathrm{GL}(\mathrm{Sym}^m(E)), \qquad G \longrightarrow \mathrm{GL}(\textstyle\bigwedge^m E),$$

which are called the *m-th symmetric power* and *m-th alternating power* of ϱ, respectively. Taking direct sums leads to representations of G on the symmetric and alternating algebras

$$\mathrm{Sym}(E) = \bigoplus_{m \geqslant 0} \mathrm{Sym}^m(E), \qquad \bigwedge E = \bigoplus_{m \geqslant 0} \bigwedge^m E.$$

From elementary multilinear algebra, we recall that if E has finite dimension, the full symmetric algebra $\mathrm{Sym}(E)$ is infinite dimensional, but the alternating algebra is not; indeed, $\bigwedge^m E = 0$ if $m > \dim E$. More generally, the dimensions of the symmetric and alternating powers are given by

$$\dim \mathrm{Sym}^m(E) = \binom{\dim(E) + m - 1}{m}, \qquad \dim \bigwedge^m E = \binom{\dim E}{m}.$$

For instance, if $n = \dim(E)$, we have

$$\dim \mathrm{Sym}^2(E) = \frac{n(n+1)}{2}, \qquad \dim \bigwedge^2 E = \frac{n(n-1)}{2}.$$

Remark 2.2.15 (Symmetric powers as coinvariants)**.** Let E be a k-vector space. For any $m \geqslant 1$, there is a natural representation of the symmetric group $\mathfrak{S}_m$ on the tensor power

$$E^{\otimes m} = E \otimes \cdots \otimes E$$

(with m factors), which is induced by the permutation of the factors, i.e., we have

$$\sigma \cdot (v_1 \otimes \cdots \otimes v_m) = v_{\sigma(1)} \otimes \cdots \otimes v_{\sigma(m)}.$$

The classical definition of the m-th symmetric power is

$$\mathrm{Sym}^m(E) = E^{\otimes m}/F$$

where F is the subspace generated by all vectors of the type

$$(v_1 \otimes \cdots \otimes v_m) - (v_{\sigma(1)} \otimes \cdots \otimes v_{\sigma(m)}) = (v_1 \otimes \cdots \otimes v_m) - \sigma \cdot (v_1 \otimes \cdots \otimes v_m),$$

where $v_i \in E$ and $\sigma \in \mathfrak{S}_m$. In other words, in the terminology and notation of Section 2.2.2, we have

$$\mathrm{Sym}^m(E) = E^{\otimes m}_{\mathfrak{S}_m},$$

the space of *coinvariants* of $E^{\otimes m}$ under this action of the symmetric group.

Exercise 2.2.16. Let G be a group, let k be a field, and let $\varrho : G \longrightarrow \mathrm{GL}(E)$ be a finite-dimensional k-representation of G of dimension $d = \dim(E)$.

(1) Let $\delta = \bigwedge^d \varrho$ be the d-th alternating power of ϱ. Show that δ is a one-dimensional k-representation of G isomorphic to the representation $g \mapsto \det(\varrho(g))$.

(2) Let $F \subset E$ be a linear subspace. Show that F is a subrepresentation of E if and only if the one-dimensional subspace $\bigwedge^{\dim(F)} F$ is a subrepresentation of $\bigwedge^{\dim(F)} E$. (*Hint*: Recall that vectors $(v_1, \ldots, v_k)$ in E are linearly independent if and only if $v_1 \wedge \cdots \wedge v_k$ is non-zero in $\bigwedge^k E$.)

2.2.6. Contragredient, endomorphism spaces. Let ϱ be a k-representation of G, acting on the vector space E. Using the transpose operation, we can then define a representation on the dual space $E' = \mathrm{Hom}_k(E, k)$, which is called the *contragredient* $\check{\varrho}$ of ϱ. More precisely, since the transpose reverses the order of composition,[7] the contragredient is defined by the rule

$$\langle g \cdot \lambda, v \rangle = \langle \lambda, g^{-1} \cdot v \rangle,$$

for $g \in G$, $\lambda \in E'$, and $v \in E$, using duality-bracket notation, or in other words the linear form $\check{\varrho}(g)\lambda$ is the linear form

$$v \mapsto \lambda(\varrho(g^{-1})v).$$

Remark 2.2.17. One way to remember this is to write the definition in the form of the equivalent invariance formula

$$\langle g \cdot \lambda, g \cdot v \rangle = \langle \lambda, v \rangle \tag{2.13}$$

for all $\lambda \in E'$ and $v \in E$.

In fact, this definition turns out to be mostly useful when E is finite dimensional, because it is associated to the *algebraic* dual space of E. When considering topological groups and representations satisfying continuity assumptions, another dual representation can be defined using the topological dual, and it has better properties (see Section 3.3, in particular Lemma 3.3.7).

We check explicitly that the contragredient is a representation, to see that the inverse (which also reverses products) compensates the effect of the transpose:

$$\langle gh \cdot \lambda, v \rangle = \langle \lambda, (gh)^{-1} \cdot v \rangle = \langle \lambda, h^{-1} g^{-1} \cdot v \rangle = \langle h \cdot \lambda, g^{-1} \cdot v \rangle = \langle g \cdot (h \cdot \lambda), v \rangle$$

for all g, $h \in G$ and $v \in E$.

The following proposition shows how the contragredient interacts with some of the other operations previously discussed.

Proposition 2.2.18. *Let k be a field, G a group.*

(1) *The contragredient is functorial: given k-representations ϱ_1 and ϱ_2 of G acting on E_1 and E_2, respectively, and an intertwiner $\Phi : \varrho_1 \longrightarrow \varrho_2$, the* transpose ${}^t\Phi$ *is an intertwiner $\check{\varrho}_1 \longrightarrow \check{\varrho}_2$.*

(2) *For any* finite *family of k-representations (ϱ_i) of G, we have canonical isomorphisms*

$$\Big(\bigoplus_i \varrho_i\Big)\check{} \simeq \bigoplus_i \check{\varrho}_i.$$

[7] Equivalently, in the language of Proposition 2.2.1, the assignment $T(E) = E'$ is a contravariant functor which *reverses* arrows in contrast with (2.8), i.e., $T(f \circ g) = T(g) \circ T(f)$, with $T(f)$ the transpose.

(3) *For any k-representations* ϱ_1 *and* ϱ_2 *of* G*, we have canonical isomorphisms*

$$(\varrho_1 \otimes \varrho_2)\check{} \simeq \check{\varrho}_1 \otimes \check{\varrho}_2.$$

(4) *If a k-representation* ϱ *of* G *is such that its contragredient is irreducible, then so is* ϱ*. Moreover, if* ϱ *is* finite dimensional, *then the converse is true, and in fact more generally, if*

$$\varrho : G \longrightarrow \mathrm{GL}(E)$$

is finite dimensional, there is an inclusion-reversing bijection between subrepresentations F *of* ϱ *and* $\check{F}$ *of its contragredient, given by*

$$F \mapsto F^{\perp} = \{\lambda \in E' \mid \lambda(F) = 0\},$$
$$\check{F} \mapsto {}^{\perp}\check{F} = \{x \in E \mid \lambda(x) = 0 \text{ for all } \lambda \in \check{F}\}.$$

(5) *For any k-representation* ϱ *of* G *of finite dimension, we have a canonical isomorphism* $\check{\check{\varrho}} \simeq \varrho$*.*

(6) *If* ϱ *is one dimensional, then the contragredient of* ϱ *is the one-dimensional representation given by* $\check{\varrho}(g) = \varrho(g^{-1}) = \varrho(g)^{-1}$*.*

Recall that the transpose in (1) is defined by

$$\langle {}^t\Phi(\lambda), v\rangle = \langle \lambda, \Phi(v)\rangle$$

for $\lambda \in E_2'$, $v \in E_1$.

Proof. Almost all these properties are elementary consequences of linear algebra. We only give some details concerning (4) and (6). For the former, we observe that both constructions indicated send, in any case, a subrepresentation of ϱ to one of $\check{\varrho}$ or conversely. This follows by the defining formula

$$\langle \check{\varrho}(g)\lambda, v\rangle = \langle \lambda, \varrho(g^{-1})v\rangle;$$

e.g., if $F \subset E$ is a subrepresentation of ϱ, this implies that $F^{\perp}$ is stable under the contragredient. If ϱ (hence also $\check{\varrho}$) is finite dimensional, standard duality of vector spaces shows that the two operations are inverse to each other. In particular, ϱ is then irreducible if and only if $\check{\varrho}$ is.

Without the finite-dimension assumption, we can still argue in this manner to show that if $\check{\varrho}$ is irreducible, the original representation is also. For any subrepresentation F of ϱ, the subrepresentation $F^{\perp}$ of $\check{\varrho}$ must be either 0 or all of $\check{\varrho}$. In the first case, no linear form vanishes on all of F, and that means that F is the whole space. In the second, all linear forms vanish, and this means $F = 0$. Hence ϱ is irreducible.

Finally, for (6), if ϱ is one dimensional, then so is the contragredient $\check{\varrho}$. If we view ϱ as acting by multiplication on k, then the linear form defined by $\lambda(x) = x$ (for $x \in k$) is a basis of k', and we find that

$$\langle \check{\varrho}(g)\lambda, x\rangle = \lambda(\varrho(g^{-1})x) = \varrho(g^{-1})x,$$

i.e., that $\check{\varrho}(g)\lambda = \varrho(g^{-1})\lambda$, which gives the result.[8] □

Remark 2.2.19. The absence of symmetry in some parts of this lemma is not surprising because dual spaces of infinite-dimensional vector spaces do not behave very well in the absence of topological restrictions (see, e.g., [**9**, §7, no. 5, th. 6].)

Exercise 2.2.20 (Contragredient and invariants). Let G be a group, let k be a field, and let $\varrho : G \longrightarrow \mathrm{GL}(E)$ be a k-representation of G.

(1) Show that there exists a natural isomorphism

$$(E')^G \simeq (E_G)'$$

of k-vector spaces, where the left-hand side is the space of invariants of the contragredient of ϱ acting on E' and the right-hand side is the dual of the coinvariant space of E (see Section 2.2.2).

(2) If $\dim(E) < +\infty$, show that there exists a natural isomorphism

$$(E^G)' \simeq (E')_G,$$

i.e., the dual of the invariant space is isomorphic to the coinvariants of the contragredient. (*Hint*: Use (1) and duality.)

A well-known isomorphism in linear algebra states that for k-vector spaces E and F, with $\dim(F) < +\infty$, we have

$$\mathrm{Hom}_k(E, F) \simeq E' \otimes F, \tag{2.14}$$

where the isomorphism is induced by mapping a pure tensor $\lambda \otimes v$, with $v \in F$ and $\lambda : E \longrightarrow k$, to the rank 1 linear map

$$A_{\lambda,v} : \begin{cases} E & \longrightarrow & E \\ w & \mapsto & \lambda(w)v = \langle \lambda, w\rangle v \end{cases}$$

(because the image of this map lies in the space of finite-rank homomorphisms $E \to F$, we must assume that F has finite dimension to have an isomorphism).

This isomorphism shows that, if

$$\varrho : G \longrightarrow \mathrm{GL}(E), \qquad \tau : G \longrightarrow \mathrm{GL}(F)$$

are k-representations of G, the endomorphism space $\mathrm{Hom}_k(E, F)$ carries a natural representation of G. Indeed, we can define an action simply by asking that the isomorphism (2.14) be an isomorphism of representations.

[8] Note that it is only because a one-dimensional representation takes values in the abelian group $k^\times$ that $g \mapsto \varrho(g^{-1})$ is a homomorphism!

It is useful to have a more direct description of this action, and this leads to a definition which does not require the representations to be finite dimensional. We state this as a proposition:

Proposition 2.2.21 (Action on homomorphisms). *Let k be a field, let G be a group, and let*

$$\varrho : G \longrightarrow \mathrm{GL}(E), \qquad \tau : G \longrightarrow \mathrm{GL}(F)$$

be k-representations of G. Then G acts on $\mathrm{Hom}_k(E, F)$ by

$$(g \cdot \Phi) = \tau(g)\Phi\varrho(g)^{-1} : E \longrightarrow F \tag{2.15}$$

for $g \in G$ and $\Phi : E \longrightarrow F$.

If π denotes this representation and if $\dim(\tau) < +\infty$, then we have an isomorphism

$$\pi \simeq \check{\varrho} \otimes \tau. \tag{2.16}$$

Furthermore, for any ϱ and τ, we have

$$\mathrm{Hom}_k(\varrho, \tau)^G = \mathrm{Hom}_G(\varrho, \tau), \tag{2.17}$$

the space of intertwiners between ϱ and τ.

Note that the definition of the representation on $\mathrm{Hom}_k(E, F)$ is such that the diagrams

$$\begin{array}{ccc} E & \xrightarrow{\Phi} & F \\ \varrho(g) \downarrow & & \downarrow \tau(g) \\ E & \xrightarrow{\pi(g)\Phi} & F \end{array}$$

commute for all $g \in G$. Concretely, we thus have

$$(g \cdot \Phi)(v) = g \cdot \Phi(g^{-1} \cdot v) \tag{2.18}$$

for all $v \in E$. Yet another way to remember this is to write the formula in the form

$$(g \cdot \Phi)(g \cdot w) = g \cdot \Phi(w) \tag{2.19}$$

for $g \in G$ and $v \in E$.

Proof. We leave it to the reader to check that (2.15) defines a representation of G. If we grant this, we note that the important relation (2.17) is an immediate consequence of the definition.

We now check the isomorphism (2.16), which is in fact the same as (2.14). This means that we must check that this linear isomorphism is an intertwiner between π and $\check{\varrho} \otimes \tau$. This is a simple computation, which (again) the reader should attempt before reading on. Let $\lambda \otimes v$ be a pure tensor in $E' \otimes F$, and let $A = A_{\lambda,v}$ be the associated homomorphism. Then the rank 1 map associated to

$$g \cdot (v \otimes \lambda) = g \cdot v \otimes g \cdot \lambda$$

is given by

$$w \mapsto \langle g \cdot \lambda, w\rangle(g \cdot v) = \langle \lambda, g^{-1}w\rangle(g \cdot v) = g \cdot (\langle \lambda, g^{-1}w\rangle v) = g \cdot A(g^{-1}w).$$

This property exactly states that the linear isomorphism (2.14) is an isomorphism of representations. □

Remark 2.2.22. The simplest example of the action (2.15) comes from the natural representation of $\mathrm{GL}_n(k)$ on k^n. It is then the same as the action of $\mathrm{GL}_n(k)$ on the space $\mathrm{M}_n(k)$ of square matrices of size n by conjugation: $g \cdot A = gAg^{-1}$ for any $g \in \mathrm{GL}_n(k)$ and any matrix A.

These representations on homomorphism spaces are extremely useful and are used in many contexts to compare two representations. This arises from the relation (2.17) which identifies the space $\mathrm{Hom}_G(\varrho_1, \varrho_2)$ of G-homomorphisms between ϱ_1 and ϱ_2 with the invariant space in $\mathrm{Hom}_k(\varrho_1, \varrho_2)$. This interpretation makes it possible to understand and study intertwining operators from within representation theory. For instance, from one of the parts of Schur's Lemma 2.2.6, we see that $\mathrm{Hom}_k(\pi, \varrho)^G = 0$ if π and ϱ are non-isomorphic irreducible representations. We suggest looking at the proof of Proposition 2.8.2 below for another good illustration of the use of the homomorphism representation to compare two representations.

Remark 2.2.23 (Other actions on homomorphism spaces)**.** Given representations ϱ_1 and ϱ_2 of G on E and F, there is another action, say τ, on $\mathrm{Hom}_k(E, F)$ that may come to mind: for $A \in \mathrm{Hom}_k(E, F)$, simply putting

$$(\tau(g)A)(w) = \varrho_2(g)(A(w)), \tag{2.20}$$

for $g \in G$ and $w \in E$, one defines also an action of G on $\mathrm{Hom}_k(E, F)$. This will turn out to be useful below in the proof of Burnside's irreducibility criterion, but it is usually less important than the one previously described. One can guess why—the formula shows that this representation really only involves the representation ϱ_2 and does not *mix* intelligently ϱ_1 and ϱ_2 (a fact that might be obscured from writing the definition in shorthand such as $(g \cdot A)w = g \cdot Aw$, and which is also less clear if $\varrho_1 = \varrho_2$, and we consider representations on $\mathrm{End}_k(\varrho_1)$).

Exercise 2.2.24. (1) Show that the representation τ just described is isomorphic to a direct sum of $\dim E$ copies of ϱ_2 (i.e., to a direct sum of $\dim E$ representations, each of which is isomorphic to ϱ_2; in particular, if ϱ_2 is irreducible, the representation τ is ϱ_2-isotypic). (*Hint*: For a basis $(w_j)_{j\in J}$ of

E, show that the map

$$\bigoplus_{j\in J} \varrho_2 \longrightarrow \tau \tag{2.21}$$

given by mapping a family (v_j) of vectors in F to the unique linear map such that

$$A(w_j) = v_j$$

is an isomorphism.)

(2) Define similarly a representation $\tilde{\tau}$ on $\mathrm{Hom}_k(E, F)$ by putting

$$(\tilde{\tau}(g)A)(w) = A(\varrho_1(g^{-1})w).$$

Show that this is a k-representation of G. Show that if $\dim(\varrho_1)$ is finite, then $\tilde{\tau}$ is isomorphic to a direct sum of $\dim(F)$ copies of $\check{\varrho}_1$.

2.3. Formalism: changing the group

Because composites of homomorphisms are homomorphisms, we see that whenever there exists a group homomorphism

$$H \xrightarrow{\phi} G,$$

it provides a way to associate a k-representation of H to any k-representation

$$\varrho : G \to \mathrm{GL}(E)$$

of G, simply by composition

$$H \xrightarrow{\varrho\circ\phi} \mathrm{GL}(E).$$

The underlying vector space is therefore unchanged, and the dimension of $\varrho \circ \phi$ is also the same as that of ϱ. Moreover, this operation is compatible with intertwining operators of representations of G (in the language of category theory, it is a *functor*): whenever

$$\Phi : E_1 \longrightarrow E_2$$

is a morphism between k-representations ϱ_1 and ϱ_2 of G on E_1 and E_2, respectively, the linear map Φ is *also* a morphism between $\varrho_1 \circ \phi$ and $\varrho_2 \circ \phi$. Since the morphism of representations of H attached to a composite $\Phi_1 \circ \Phi_2$ is the corresponding composition, one can say that this operation from representations of G to those of H is also *functorial*. In general, this correspondence has no reason to be injective or surjective: some representations of H may not "come from" G in this way, and non-isomorphic representations of G may become isomorphic when "pulled back" to H. The reader is invited to look for (easy!) examples of both phenomena.

When H is a subgroup of G and ϕ is the inclusion, the operation is called, naturally enough, the *restriction* of representations of G to H. Because of this, one uses the standard notation $\mathrm{Res}_H^G(\varrho)$, which we will use even when ϕ is not of this type (note the ambiguity due to the fact that this representation depends on ϕ, which is not present in the notation).

Example 2.3.1 (Representations of quotients)**.** There is one very common type of *restriction* associated to a non-injective morphism: if $\phi : G \to H$ is surjective or, in other words, if $H \simeq G/K$ for some normal subgroup $K \subset G$. One can then describe precisely the representations of G obtained by restriction (using ϕ) of those of H:

Proposition 2.3.2 (Representations of a quotient)**.** *Let G be a group, let K be a normal subgroup, and let $H = G/K$ with quotient map*

$$\phi : G \longrightarrow H.$$

For any field k, the map

$$\varrho \mapsto \varrho \circ \phi$$

is a bijection between k-representations ϱ of H and k-representations π of G which are trivial on K, i.e., such that $K \subset \mathrm{Ker}(\pi)$.

This is simply a special case of the fact that, for any group Γ, a homomorphism $G \to \Gamma$ factors through K (i.e., is of the form $f \circ \phi$ for some $f : G/K \to \Gamma$) if and only if it is trivial on K.

There is a very important special case of this discussion. Recall that the derived subgroup $G' = [G, G]$ of G, generated by commutators, has the property that for any group G and abelian group A, there is a canonical bijection between homomorphisms $G \to A$ and homomorphisms $G/[G, G] \to A$. Since $\mathrm{GL}_1(k) = k^\times$ is abelian, applying this to $A = k^\times$, we obtain

Proposition 2.3.3 (One-dimensional representations)**.** *Let G be a group, and let $G^{ab} = G/[G, G]$ be the* abelianization *of G. For any field k, the one-dimensional representations of G correspond with the homomorphisms*

$$G^{ab} \longrightarrow k^\times.$$

In particular, if G is perfect, *i.e., if $[G, G] = G$, then any non-trivial representation of G, over any field k, has dimension at least* 2.

The last part of this proposition applies in many cases. For instance, if $d \geqslant 2$ and k is any field, $\mathrm{SL}_d(k)$ is known to be perfect except when $d = 2$ and $k = \mathbf{F}_2$ or $k = \mathbf{F}_3$ (see, e.g., [**40**, Th. 8.3, Th. 9.2]). Thus, no such group has a non-trivial one-dimensional representation.

Exercise 2.3.4 (Semisimplicity of restriction)**.** Let G be a group, let k be a field, and let $\varrho : G \longrightarrow \mathrm{GL}(E)$ be a semisimple k-representation of G. Let $H \subset G$ be a finite-index normal subgroup of G. Show that $\mathrm{Res}_H^G(\varrho)$ is also semisimple as a representation of H. (*Hint*: One can assume that ϱ is irreducible—show that there exists a maximal semisimple subrepresentation of $\mathrm{Res}_H^G(\varrho)$.)

The converse of this statement is not true without restrictions, but it is valid when k has characteristic 0; see Exercise 4.1.2.

One of the most basic and important constructions of representation theory, and in some sense the first notion that may not be immediately related to notions of linear algebra,[9] is the operation of *induction*. We will now define it and spend a fair amount of time discussing its basic properties, and it will reappear throughout the book.

This operation proceeds in the direction opposite to restriction: given a homomorphism

$$\phi : H \longrightarrow G,$$

it associates—in a functorial way, i.e., in a way that is natural enough to be compatible with intertwining operators—a k-representation of G to a k-representation of H. When ϕ is the inclusion of a subgroup H of G, this means going from a representation of a subgroup to one of a larger group, which may seem surprising at first. Once more, a reader who has not seen the definition before might want to stop for a few minutes to think about possible ways of doing this. We also recommend reading what follows first by assuming ϕ to be an inclusion map, and removing it from the notation.

One defines the induced[10] representation as follows: given

$$\varrho : H \longrightarrow \mathrm{GL}(E),$$

we define first the k-vector space

$$F = \{f : G \to E \mid f(\phi(h)x) = \varrho(h)f(x) \text{ for all } h \in H,\ x \in G\} \tag{2.22}$$

(which is a vector subspace of the space of functions on G with values in E). In other words, F is the space of E-valued functions on G which happen to transform *like the representation ϱ under H acting on the left.* On this vector space F, we now have an action π of G (namely the restriction π to F of the analogue of the regular representation) defined by

$$(\pi(g))f(x) = f(xg)$$

for $f \in F$, $g \in G$, and $x \in G$. Indeed, we need only check that F is stable under the regular representation of G, which is true, because F is defined using conditions relating to multiplication on the left by elements of H. Formally, if $f_1 = \pi(g)f$, we find that

$$f_1(\phi(h)x) = f(\phi(h)xg) = \varrho(h)f(xg) = \varrho(h)f_1(x),$$

[9] It is, however, related to certain tensor products.

[10] It is unfortunate that the terminology "induced" may clash with the use of the adjective "induced" in less formal senses, and that "induction" conflicts with, e.g., proofs by induction.

for all $h \in H$ and $x \in G$, which means that—as desired—we have $f_1 \in F$ again.

Especially when ϕ is an inclusion, one writes

$$\pi = \mathrm{Ind}_H^G(\varrho)$$

for this induced representation, but as for restriction, we will use it in the general case (keeping in mind the ambiguity that comes from not indicating ϕ explicitly). One may even drop H and G from the notation when they are clear from the context.

Remark 2.3.5. If we take $h \in \mathrm{Ker}(\phi)$, the transformation formula in (2.22) for elements of F gives

$$f(x) = \varrho(h)f(x)$$

so that, in fact, any function $f \in F$ takes values in the space $E^{\mathrm{Ker}(\phi)}$ of invariants of E under the action of the subgroup $\mathrm{Ker}(\phi)$ through ϱ. However, we do not need to state it explicitly in the definition of F, and this avoids complicating the notation. It will reappear in the computation of the dimension of F (Proposition 2.3.11 below). Of course, when ϕ is an inclusion (the most important case), the target space is genuinely E anyway. It is worth observing, however, that as a consequence of Lemma 2.1.12, this subspace $E^{\mathrm{Ker}(\phi)}$ is in fact a subrepresentation of E, so that in the condition

$$f(\phi(h)x) = \varrho(h)f(x),$$

the right-hand side also is always in $E^{\mathrm{Ker}(\phi)}$.

Example 2.3.6 (Elementary examples of induction)**.** (1) By the definition of F and comparison with the definition of the regular representation, we see that the latter can be expressed as

$$C_k(G) = \mathrm{Ind}_1^G(\mathbf{1}), \tag{2.23}$$

the result of inducing to G the one-dimensional trivial k-representation of the trivial subgroup $1 \hookrightarrow G$.

(2) For further simple orientation, suppose first that $\phi : G \to G$ is the identity. We then have

$$\mathrm{Ind}_G^G(\varrho) \simeq \varrho$$

for any k-representation $\varrho : G \longrightarrow \mathrm{GL}(E)$ of G, the map $F \to E$ giving this isomorphism being simply

$$f \mapsto f(1) \in E,$$

as the reader should make sure to check. The inverse maps sends $v \in E$ to the function defined by $f(g) = \varrho(g)v$.

(3) More generally, consider the canonical projection $\phi : G \to G/K$ (the context of Example 2.3.1). For a representation

$$\varrho : G \longrightarrow \mathrm{GL}(E),$$

we then claim that we have

$$\mathrm{Ind}_G^H(\varrho) \simeq E^K$$

with the action of G/K induced by ϱ (note that by Lemma 2.1.12, the subspace E^K is a subrepresentation of E). This isomorphism is again given by $f \mapsto f(1)$, which—as we have remarked—is a vector in $E^{\mathrm{Ker}(\phi)} = E^K$, as the reader is again invited to check.

(4) Suppose now that $\phi : G \to G$ is an automorphism. Then, for a representation ϱ of the *source* G, acting on E, the induced representation $\mathrm{Ind}_G^G(\varrho)$ is *not* in general isomorphic to ϱ; rather it is isomorphic to

$$\phi_* \varrho = \varrho \circ \phi^{-1} : G \longrightarrow \mathrm{GL}(E).$$

Indeed, the k-linear isomorphism

$$\Phi \begin{cases} F & \longrightarrow & E \\ f & \mapsto & f(1) \end{cases}$$

satisfies

$$\Phi(g \cdot f) = f(g) = f(\varrho(\varrho^{-1}(g))) = \varrho(\phi^{-1}(g))f(1) = \phi_* \varrho(f),$$

i.e., it intertwines the induced representation with the representation $\varrho \circ \phi^{-1}$. Incidentally, using again ϕ and seeing ϱ as a representation of the target G, one has of course

$$\mathrm{Res}_G^G(\varrho) = \phi^* \varrho = \varrho \circ \phi.$$

Although this looks like a quick way to produce many "new" representations from one, it is not so efficient in practice because if ϕ is an *inner* automorphism (i.e., if $\phi(g) = xgx^{-1}$ for some fixed $x \in G$), then we do have $\phi_* \varrho \simeq \varrho$: by definition, the linear isomorphism $\Phi = \varrho(x)$ satisfies

$$\Phi \circ \phi_* \varrho(g) = \varrho(x)\varrho(x^{-1}gx) = \varrho(g)\Phi$$

for all $g \in G$, so that it is an isomorphism $\phi_* \varrho \longrightarrow \varrho$.

(5) Finally, one can see from the above how to essentially reduce a general induction to one computed using an inclusion homomorphism. Indeed, we always have an isomorphism

$$\mathrm{Ind}_H^G(\varrho) \simeq \mathrm{Ind}_{\mathrm{Im}(\phi)}^G(\phi_*(\varrho^{\mathrm{Ker}(\phi)})),$$

where the right-hand side is computed using the inclusion homomorphism $\mathrm{Im}(\phi) \subset G$. This isomorphism is a combination of the previous cases using the factorization

$$H \xrightarrow{\phi_1} H/\mathrm{Ker}(\phi) \simeq \mathrm{Im}(\phi) \hookrightarrow G,$$

where the first map is a quotient map, the second is the isomorphism induced by ϕ, and the third is an injection. (This is also a special case of *induction in stages*; see Proposition 2.3.20 below.)

(6) Another important special case of induction occurs when the representation ϱ is one-dimensional, i.e., it is a homomorphism

$$H \longrightarrow k^{\times}.$$

In that case, the space F of $\mathrm{Ind}_H^G(\varrho)$ is a subspace of the space $C_k(G)$ of k-valued functions on G, and since G acts on this space by the regular representation, the induced representation is a subrepresentation of $C_k(G)$, characterized as those functions which transform like ϱ under H:

$$f(\phi(h)x) = \varrho(h)f(x),$$

where now $\varrho(h)$ is just a (non-zero) scalar in k.

This type of example is significant because of the crucial importance of the regular representation. Indeed, it is often a good strategy to (attempt to) determine the irreducible k-representations of a group by trying to find them as being either induced by one-dimensional representations of suitable subgroups or subrepresentations of such induced representations. We will see this in effect in Chapter 4 in the special case of the groups $\mathrm{GL}_2(\mathbf{F}_q)$, where $\mathbf{F}_q$ is a finite field.

Remark 2.3.7. Although we have given a specific model of the induced representation by writing down a concrete vector space on which G acts, one should attempt to think of it in a more abstract way. As we will see in the remainder of the book, many representations constructed differently—or even those given by nature—turn out to be isomorphic to induced representations, even if the vector space does not look like the one above.

Note also that we have defined induction purely algebraically. As one may expect, in cases where G is an infinite topological group, this definition requires some changes to behave reasonably. The model (2.22) is then a good definition as it can immediately suggest that we consider restricted classes of functions on G instead of all of them (see Example 5.2.10 and Section 7.4.)

The following two propositions are the most important facts to remember about induction.

Proposition 2.3.8 (Functoriality of induced representations). *Let k be a field, and let $\phi : H \to G$ be a group homomorphism. For any ho-*

momorphism $\varrho_1 \xrightarrow{\Phi} \varrho_2$ *of k-representations of H, there is a corresponding homomorphism*

$$\mathrm{Ind}(\Phi) : \mathrm{Ind}_H^G(\varrho_1) \longrightarrow \mathrm{Ind}_H^G(\varrho_2),$$

and this is functorial*: the identity maps to the identity, and composites map to composites.*

Proposition 2.3.9 (Frobenius reciprocity; adjointness of induction and restriction)**.** *Let k be a field, and let $\phi : H \to G$ be a group homomorphism. For any k-representation ϱ_1 of G and ϱ_2 of H, there is a natural isomorphism of k-vector spaces*

$$\mathrm{Hom}_G(\varrho_1, \mathrm{Ind}_H^G(\varrho_2)) \simeq \mathrm{Hom}_H(\mathrm{Res}_H^G(\varrho_1), \varrho_2), \tag{2.24}$$

where we recall that $\mathrm{Hom}_G(\cdot,\cdot)$ *denotes the k-vector space of a homomorphism between two representations of G.*

The last isomorphism, or its immediate corollary

$$\dim \mathrm{Hom}_G(\varrho_1, \mathrm{Ind}_H^G(\varrho_2)) = \dim \mathrm{Hom}_H(\mathrm{Res}_H^G(\varrho_1), \varrho_2) \tag{2.25}$$

is called the *Frobenius reciprocity formula.* As we will see many times, it is an extremely important result. In fact, in some (precise) sense, it *characterizes* the induced representation, and can almost be said to define it (see Remark 2.3.21 for an explanation). We will use induction and Frobenius reciprocity extensively to analyze the properties and the decomposition of induced representations.

We also remark that the definition of the induced representation that we chose is the best for handling situations in which $[G : H]$ can be infinite. If $[G : H]$ is finite, there is another natural model (say $\underline{\mathrm{Ind}}_H^G$) which leads to isomorphisms

$$\mathrm{Hom}_G(\underline{\mathrm{Ind}}_H^G(\varrho_1), \varrho_2) \simeq \mathrm{Hom}_H(\varrho_1, \mathrm{Res}_H^G(\varrho_2)), \tag{2.26}$$

and those are sometimes considered to be the incarnation of Frobenius reciprocity (see Exercise 2.3.16 and, e.g., [**28**, Ch. 5]).

Proof of Proposition 2.3.8. The induced homomorphism $\Phi_* = \mathrm{Ind}(\Phi)$ is easy to define using the model of the induced representation given above: denoting by F_1, F_2 the spaces corresponding to $\mathrm{Ind}_H^G(\varrho_1)$ and $\mathrm{Ind}_H^G(\varrho_2)$, respectively, we define $\Phi_*(f)$ for $f \in F_1$ to be given by

$$\Phi_*(f)(x) = \Phi(f(x))$$

for $x \in G$. This is a function from G to E_2, by definition, and the relation

$$\Phi_*(f)(\phi(h)x) = \Phi(f(\phi(h)x)) = \Phi(\varrho_1(h)f(x)) = \varrho_2(h)\Phi(f(x))$$

for all $h \in H$ shows that $\Phi_*(f)$ is in the space F_2 of the induced representation of ϱ_2. We leave it to the reader to check that Φ_* is indeed a homomorphism between the representations F_1 and F_2. □

Proof of Proposition 2.3.9. Here also there is little that is difficult, except maybe a certain bewildering accumulation of notation, especially parentheses, when checking the details—the reader should however make sure that these checks are done.

Assume that G acts on the space F_1 through ϱ_1 and that H acts on E_2 through ϱ_2. Then the *restriction* of ϱ_1 acts on F_1 through $\varrho_1 \circ \phi$, while the induced representation of ϱ_2 acts on the space F_2 as defined in (2.22).

We will describe how to associate to

$$\Phi : F_1 \longrightarrow F_2,$$

which intertwines ϱ_1 and $\mathrm{Ind}_H^G(\varrho_2)$, a map

$$T(\Phi) : F_1 \longrightarrow E_2$$

intertwining the restriction of ϱ_1 and ϱ_2. We will then describe, conversely, how to start with an intertwiner

$$\Psi : F_1 \longrightarrow E_2$$

and construct another one

$$\tilde{T}(\Psi) : F_1 \longrightarrow F_2,$$

and then it will be seen that $T \circ \tilde{T}$ and $\tilde{T} \circ T$ are the identity morphism, so that T and $\tilde{T}$ give the claimed isomorphisms.

The main point to get from this is that both T and $\tilde{T}$ more or less write themselves: they express the simplest way (except for putting zeros everywhere!) to move between the desired spaces. One must then check various things (e.g., that functions on G with values in E_2 actually lie in F_2, that the maps are actually intertwiners, that they are reciprocal), but at least once this is done, it is quite easy to recover the definitions.

To begin, given Φ as above and a vector $v \in F_1$, we must define a map $F_1 \longrightarrow E_2$. Since $\Phi(v)$ is in F_2, it is a function on G with values in E_2; hence, it seems natural to evaluate it somewhere, and the most natural guess is to try to evaluate at the identity element. In other words, we define $T(\Phi)$ to be the map

$$T(\Phi) : \begin{cases} F_1 & \longrightarrow & E_2 \\ v & \mapsto & \Phi(v)(1). \end{cases} \tag{2.27}$$

We can already easily check that $\tilde{\Phi} = T(\Phi)$ is an H-homomorphism (between the restriction of ϱ_1 and ϱ_2): indeed, we have

$$\tilde{\Phi}(h \cdot v) = \tilde{\Phi}(\phi(h)v) = \Phi(\phi(h)v)(1) = \Phi(v)(\phi(h)),$$

where the last equality reflects the fact that Φ intertwines ϱ_1 and the induced representation of ϱ_2, the latter acting like the regular representation on F_2. Now because $\Phi(v) \in F_2$, we get

$$\Phi(v)(\phi(h)) = \varrho_2(h)\Phi(v)(1) = \varrho_2(h)\tilde{\Phi}(v),$$

which is what we wanted.

In the other direction, given an H-homomorphism

$$\Psi : F_1 \to E_2,$$

we must construct a map $\tilde{\Psi} = \tilde{T}(\Psi)$ from F_1 to F_2. Given $v \in F_1$, we need to build a function on G with values in E_2. The function

$$x \mapsto \Psi(\varrho_1(x)v) \tag{2.28}$$

is the most natural that comes to mind, since the values of Ψ are elements of E_2. Thus $\tilde{\Psi}(v)$ is defined to be this function.

We now finish checking that these constructions give the Frobenius reciprocity isomorphisms. First, we check that $f = \tilde{\Psi}(v)$ is indeed in F_2: for all $x \in G$ and $h \in H$, we have

$$f(\phi(h)x) = \Psi(\varrho_1(\phi(h)x)v) = \varrho_2(h)\Psi(\varrho_1(x)v) = \varrho_2(h)f(x)$$

(using the fact that Ψ is a homomorphism from $\mathrm{Res}_H^G(\varrho_1)$ to ϱ_2). Next, $\tilde{\Psi}$ intertwines ϱ_1 and $\mathrm{Ind}_H^G(\varrho_2)$: for $g \in G$, the function $\tilde{\Psi}(\varrho_1(g)v)$ is

$$x \mapsto \Psi(\varrho_1(xg)v),$$

and this coincides with

$$g \cdot \tilde{\Psi}(v) = (x \mapsto \tilde{\Psi}(v)(xg)).$$

The remaining property we need is that the two constructions are inverse of each other. If we start with $\Psi \in \mathrm{Hom}_H(F_1, E_2)$, then construct $\tilde{\Psi} = \tilde{T}\Psi$, the definitions (2.27) and (2.28) show that

$$T\tilde{T}\Psi(v) = \tilde{\Psi}(v)(1) = \Psi(v)$$

for all v, i.e., $T \circ \tilde{T}$ is the identity. If we start with $\Phi \in \mathrm{Hom}_G(F_1, F_2)$, define $\Psi = T\Phi$ and $\tilde{\Phi} = \tilde{T}\Psi = \tilde{T}T\Phi$, and unravel the definitions again, we obtain the inescapable conclusion that, given $v \in F_1$, the function $\tilde{\Phi}(v)$ is given by

$$(x \mapsto \Psi(\varrho_1(x)v) = \Phi(\varrho_1(x)v)(1)),$$

and this function of x does coincide with $\Phi(v)$ because

$$\Phi(\varrho_1(x)v) = x \cdot \Phi(v) = (y \mapsto \Phi(v)(yx)).$$

Thus $\tilde{T} \circ T$ is also the identity, and the proof is finished. □

Example 2.3.10. Let $\varrho_1 = \mathbf{1}$ be the trivial (one-dimensional) representation of G. Then its restriction to H is the trivial representation $\mathbf{1}_H$ of H. By Frobenius reciprocity, we derive

$$\mathrm{Hom}_G(\mathbf{1}_G, \mathrm{Ind}(\varrho_2)) \simeq \mathrm{Hom}_H(\mathbf{1}_H, \varrho_2).$$

Comparing with (2.7), we deduce that there is a (canonical) isomorphism

$$\mathrm{Ind}_H^G(\varrho_2)^G \simeq \varrho_2^H$$

of the invariant subspaces of ϱ_2 and its induced representation.

We now wish to compute the dimension of an induced representation.

Proposition 2.3.11. *Let k be a field, and let $\phi : H \longrightarrow G$ be a group homomorphism. For a k-representation ϱ of H, acting on a space E, we have*

$$\dim(\mathrm{Ind}_H^G(\varrho)) = [G : \mathrm{Im}(\phi)] \dim(E^{\mathrm{Ker}(\phi)}).$$

In particular, if H is a subgroup of G, we have

$$\dim(\mathrm{Ind}_H^G(\varrho)) = [G : H] \dim(\varrho).$$

Remark 2.3.12. Note that this formula is one case where one must be careful in the infinite-dimensional case. We mentioned in Section 1.3 that we do not distinguish between infinite cardinals, and with this convention the formula is valid (i.e., both sides are infinite if and only if one of them is). However, the formula is *not* true if one interprets the left-hand side as the cardinal (say c_1) of a basis of $\mathrm{Ind}_H^G(\varrho)$ and the right-hand side as the product (say c_2) of the cardinals of $[G : \mathrm{Im}(\phi)]$ and that of a basis of $E^{\ker \phi}$, if the index $[G : \mathrm{Im}(\phi)]$ is infinite (and $E^{\ker \phi}$ is non-zero): we then have $c_1 > c_2$.

Proof. The idea is very simple: the definition of the space F on which the induced representation acts shows that the value of $f \in F$ at a point x determines the values at all other points of the form $\phi(h)x$, i.e., at all points which are in the same left-coset of G modulo the image of ϕ. Thus there should be $[G : \mathrm{Im}(\phi)]$ independent values of f; each seems to belong to the space E, but as we observed in Remark 2.3.5, it is in fact constrained to lie in the possibly smaller space $E^{\mathrm{Ker}(\phi)}$.

To check this precisely, we select a set R of representatives of $\mathrm{Im}(\phi)\backslash G$, we let $\tilde{F}$ denote the space of all functions

$$\tilde{f} : R \longrightarrow E^{\mathrm{Ker}(\phi)},$$

and we consider the obvious k-linear map

$$F \longrightarrow \tilde{F}$$

defined by restricting functions on G to R (using Remark 2.3.5 to see that this is well defined). Now we claim that this is an isomorphism of vector spaces. This implies the formula for the dimension of F (distinguishing the infinite-dimensional case from the finite-dimensional one).

To check the injectivity, we simply observe that if $f \in F$ is identically zero on R, we have

$$f(\phi(h)x) = \varrho(h)f(x) = 0$$

for all $x \in R$ and $h \in H$. Since these elements, by definition, cover all of G, we get $f = 0$. This is really the content of the observation at the beginning of the proof.

For surjectivity, for any $x \in G$, we denote by $r(x)$ the element of R equivalent to x, and we select one $h(x) \in H$ such that

$$x = \phi(h(x))r(x),$$

with $h(x) = 1$ if $x \in R$.

Given an arbitrary function $\tilde{f} : R \to E^{\mathrm{Ker}(\phi)}$, we then *define*

$$f(x) = f(\phi(h(x))r(x)) = \varrho(h(x))\tilde{f}(r(x)),$$

which is a well-defined E-valued function on G. Thus f is equal to $\tilde{f}$ on R. By definition of F, this is in fact the only possible definition for such a function, but we must check that $f \in F$ to conclude. Consider $x \in G$ and $h \in H$; let $y = \phi(h)x$, so that we have the two expressions

$$y = \phi(hh(x))r(x), \qquad y = \phi(h(y))r(y) = \phi(h(y))r(x)$$

since y and x are left-equivalent under $\mathrm{Im}(\phi)$. It follows that $hh(x)$ and $h(y)$ differ by an element (say κ) in $\mathrm{Ker}(\phi)$. Thus we get

$$\begin{aligned} f(y) = f(\phi(h(y))r(x)) &= \varrho(h(y))\tilde{f}(r(x)) \\ &= \varrho(\kappa)\varrho(hh(x))\tilde{f}(r(x)) \\ &= \varrho(h)\varrho(h(x))\tilde{f}(r(x)) \end{aligned}$$

since κ acts trivially on the space $E^{\mathrm{Ker}(\phi)}$, and (as in Lemma 2.1.12) the vector

$$\varrho(hh(x))\tilde{f}(r(x))$$

does belong to it. We are now done because

$$f(\phi(h)x) = f(y) = \varrho(h)\varrho(h(x))\tilde{f}(r(x)) = \varrho(h)f(x)$$

finishes the proof that $f \in F$. ☐

Remark 2.3.13. (1) From the proof we see that one could have defined the induced representation as the k-vector space of all functions

$$\mathrm{Im}(\phi)\backslash G \longrightarrow E^{\mathrm{Ker}(\phi)}$$

together with a suitable action of G. However, this *restriction model* of $\mathrm{Ind}_H^G(\varrho)$ is not very convenient because the action of G, by *transport of structure*, is not very explicit.

(2) See Exercise 4.2.8 for an application to proving a lower bound for the minimal index of a proper subgroup of a finite group.

Exercise 2.3.14. Let G be a group, let $H \subset G$ be a subgroup, and let $\varrho : G \longrightarrow \mathrm{GL}(E)$ be a k-representation of H.

(1) Show that if F is a subrepresentation of E, then $\mathrm{Ind}_H^G(F)$ is naturally isomorphic to a subrepresentation of $\mathrm{Ind}_H^G(\varrho)$.

(2) Show that if $\mathrm{Ind}_H^G(\varrho)$ is irreducible, then so is ϱ. Is the converse true?

The proof of Proposition 2.3.11 implicitly reveals more information than the dimension of the induced representation. In particular, we can use it to give one answer to the question of recognizing when a given representation of a group G is induced from a subgroup. Not only is this useful in practice (see Proposition 2.8.1, for instance), but it certainly helps in visualizing what the operation of induction is.

Proposition 2.3.15. *Let G be a group, and let k be a field. Let $\varrho : G \longrightarrow \mathrm{GL}(E)$ be a finite-dimensional k-representation of G. Assume that there exists a finite-index subgroup $H \subset G$ and a direct sum decomposition*

$$E = \bigoplus_{i=1}^{[G:H]} E_i,$$

where each $E_i \subset E$ is H-stable, such that for any i, we have $E_i = \varrho(g_i^{-1})E_1$ for some $g_i \in G$. Then the representation ϱ is isomorphic to the induced representation $\mathrm{Ind}_H^G(E_1)$.

Proof. Note that the assumption implies that the dimension of E is $[G : H] \dim E_1$, which is the dimension of $\mathrm{Ind}_H^G(E_1)$, as we have just seen, so the result is certainly plausible. We will construct an intertwining map

$$\Phi : \mathrm{Ind}_H^G(E_1) \longrightarrow E$$

and show that it is injective. Since we also assume that $\dim E < +\infty$, this will be enough to finish the proof.

The definition of Φ is easy. For f in the space F of the induced representation $\mathrm{Ind}_H^G(E_1)$, we define

$$\Phi(f) = \sum_{i=1}^{[G:H]} \varrho(g_i^{-1})f(g_i).$$

This defines a k-linear map to E, since $f(g_i) \in E_1$ by definition of F. In fact the assumption shows that $\varrho(g_i^{-1})f(g_i) \in E_i$ for each i, and since these spaces are in direct sum, we also immediately see that

$$\operatorname{Ker}\Phi = \{f \in F \mid f(g_i) = 0 \text{ for all } i\}.$$

To deduce that Φ is injective, and indeed also to show that Φ is an intertwiner, we now claim that the (g_i) form a set of representatives of left H-cosets in G, i.e., that

$$G = \bigcup_i Hg_i$$

with the union being disjoint. The number of g_i is the right one, so it suffices to prove that the g_i are in distinct H-cosets. But if $hg_i = g_j$ for some $h \in H$, we deduce that

$$\varrho(g_i^{-1})\varrho(h)E_1 = \varrho(g_j^{-1})E_1,$$

and since E_1 is H-stable, this means that $E_i = E_j$, which means that $i = j$.

We first apply this to prove injectivity of Φ: $f \in \operatorname{Ker}\Phi$ means that f is zero on a set of representatives of $H\backslash G$, and (as in the proof of Proposition 2.3.11), this means that $f = 0$.

We conclude by checking that Φ is an intertwiner. Consider $g \in G$ and $f \in F$. Then the definition gives

$$\Phi(g \cdot f) = \sum_i \varrho(g_i^{-1})f(g_ig).$$

Multiplication on the right by g permutes the left H-cosets: there exists a permutation σ of the indices such that for each i, we have

$$g_ig = h_{\sigma(i)}g_{\sigma(i)} \tag{2.29}$$

for some $h_{\sigma(i)} \in H$. Since $f \in F$, we deduce that

$$\Phi(g \cdot f) = \sum_i \varrho(g_i^{-1})f(g_ig) = \sum_i \varrho(g_i^{-1}h_{\sigma(i)})f(g_{\sigma_i}).$$

But (2.29) gives

$$gg_{\sigma(i)}^{-1} = g_i^{-1}h_{\sigma(i)}$$

so that, rearranging the sum, this this becomes

$$\Phi(g \cdot f) = \sum_i \varrho(gg_{\sigma(i)}^{-1})f(g_{\sigma(i)}) = \varrho(g)(\Phi(f)),$$

concluding the proof. □

Exercise 2.3.16. Let G be a group, let $H \subset G$ be a finite-index subgroup, and let k be a field. Let $\{g_1, g_2, \ldots, g_k\}$ be a set of representatives for right

H-cosets in G with $g_1 = 1$. For a k-representation $\varrho : H \longrightarrow \mathrm{GL}(E)$, define a representation π of G as follows. The space F of π is

$$F = \bigoplus_{i=1}^{k} g_i E,$$

where $g_i E$ denotes a vector space isomorphic to E, and the action is obtained by formally using the action of H on E (given by ϱ) and the requirement that $g_i E$ is the translate of $g_1 E = E$ by g_i, i.e. if $g \in G$, $v \in E$ and

$$gg_i = g_j h$$

for some j and $h \in H$, we put

$$g \cdot (g_i v) = g_j(\varrho(h)v) \in g_j E.$$

(1) Show that π is indeed a representation of G on F, and that it is isomorphic to $\mathrm{Ind}_H^G(\varrho)$.

(2) Show that for any k-representation $\varrho_2 : G \longrightarrow \mathrm{GL}(F_2)$, there exists a canonical isomorphism

$$\mathrm{Hom}_G(\pi, \varrho_2) \simeq \mathrm{Hom}_H(\varrho, \mathrm{Res}_H^G(\varrho_2))$$

(this is the alternate formula (2.26) for Frobenius reciprocity).

The degree relation makes it clear, if needed, that the operations of restriction and induction are *not* inverse to each other (as the dimensions of the underlying vector spaces change). In fact, there is no inverse of restriction in general:

Exercise 2.3.17. Show that there is no operation inverse of restriction: there exist subgroups $H \subset G$ and representations of H which are *not* the restriction of any representation of G. (*Hint*: Even very simple examples will do, and Proposition 2.3.3 can help.)

Nevertheless, there are relations between restriction and induction, as we have seen with the Frobenius reciprocity formula. Here is another one.

Proposition 2.3.18 (Projection formula). *Let k be a field, and let $\phi : H \to G$ be a group homomorphism. For a k-representation ϱ_1 of G and a k-representation ϱ_2 of H, we have a natural isomorphism*

$$\mathrm{Ind}_H^G(\varrho_2 \otimes \mathrm{Res}_H^G(\varrho_1)) \simeq \mathrm{Ind}_H^G(\varrho_2) \otimes \varrho_1$$

of representations of G.

As in the case of the Frobenius reciprocity isomorphism (2.24), the proof is not very difficult as the isomorphism can be described explicitly, but the full details are a bit tedious. The reader should attempt to guess a homomorphism between the two representations (it is easier to go from right to left here), and after checking that the guess is right, then try to verify that it satisfies the required properties.[11]

Proof. We denote by F_1 the space of ϱ_1, by E_2 that of ϱ_2, and by F_2 the space (2.22) of the induced representation $\mathrm{Ind}_H^G(\varrho_2)$. Moreover, we denote by τ the representation

$$\tau = \varrho_2 \otimes \mathrm{Res}_H^G(\varrho_1)$$

of H and by $\tilde{F}_2$ the space of

$$\mathrm{Ind}_H^G(\tau) = \mathrm{Ind}_H^G(\varrho_2 \otimes \mathrm{Res}_H^G(\varrho_1)),$$

also defined using (2.22).

The isomorphism of representations of G that is claimed to exist is defined as the k-linear map

$$F_2 \otimes F_1 \xrightarrow{\Phi} \tilde{F}_2,$$

which extends by linearity the definition

$$\Phi(f \otimes v) = (x \mapsto f(x) \otimes x \cdot v)$$

for $f \in F_2$ and $v \in F_1$.

Note that the right-hand side is indeed a function $G \longrightarrow E_2 \otimes F_1$, and that $E_2 \otimes F_1$ is the space of τ (in this proof, we write $x \cdot v$ for the action of ϱ_1 on F_1). It is also a bilinear expression of the arguments f and v. Hence, to see that Φ is well defined, it is enough to check that its image does lie in $\tilde{F}_2$. But if $\tilde{f} = \Phi(f \otimes v)$, then (using the fact that $f \in F_2$), we obtain

$$\begin{aligned} \tilde{f}(\phi(h)x) &= f(\phi(h)x) \otimes (\phi(h)x) \cdot v \\ &= \varrho_2(h) f(x) \otimes \phi(h)(x \cdot v) \\ &= \tau(h)\{f(x) \otimes x \cdot v\} \end{aligned}$$

for all $x \in G$, $h \in H$, which is the property required for a function $G \longrightarrow E_2 \otimes F_1$ to be in $\tilde{F}_2$.

We will now check that Φ is a G-isomorphism. First, the fact that it is a homomorphism is straightforward, as it can be checked on the generating tensors $f \otimes v$. Let $\tilde{f} = \Phi(f \otimes v)$ and $g \in G$; then we have

$$(g \cdot \tilde{f})(x) = \tilde{f}(xg) = f(xg) \otimes (xg) \cdot v,$$

which we can also write as

$$f_1(x) \otimes x \cdot w,$$

where $f_1(x) = f(xg) = g \cdot f(x)$ and $w = g \cdot v$, or in other words as

$$\Phi(f_1 \otimes w)(x) = \Phi(g \cdot (f \otimes v))(x),$$

as desired.

[11] In fact, the details of this and similar proofs are probably not worth trying to read without attempting such a process of self-discovery of the arguments.

To conclude, it remains to prove that Φ is a k-linear isomorphism. Here a little trick is needed, since pure tensors are not enough. We fix a basis (v_j) of F_1 (it could be infinite, of course). Then, for any $x \in G$, a vector w of $E_2 \otimes F_1$ can be written *uniquely* as a linear combination

$$w = \sum_j w_j(x) \otimes (x \cdot v_j) \tag{2.30}$$

for some $w_j(x) \in E_2$. This is simply because, for every x, the vectors $(x \cdot v_j)_j$ also form a basis of F_1.

We first show the injectivity of Φ: any element of $F_2 \otimes F_1$ can be expressed as

$$\sum_j f_j \otimes v_j$$

for some functions $f_j \in F_2$. Let us assume such an element is in $\mathrm{Ker}(\Phi)$. This means that for all $x \in G$, we have

$$\sum_j f_j(x) \otimes (x \cdot v_j) = 0 \in E_2 \otimes F_1.$$

Thus by the uniqueness of the representations (2.30), we get

$$f_j(x) = 0$$

for all j, or in other words $f_j = 0$ for all j, and this gives $\mathrm{Ker}(\Phi) = 0$.

We now come to surjectivity. Let $\tilde{f} \in \tilde{F}_2$ be given. Again by the observation above, for any $x \in G$, we can write uniquely[12]

$$\tilde{f}(x) = \sum_j \tilde{f}_j(x) \otimes (x \cdot v_j),$$

thus defining coefficient functions $\tilde{f}_j : G \to E_2$. We next show that, because $\tilde{f} \in \tilde{F}_2$, each $\tilde{f}_j$ is in fact in F_2, which will ensure that

$$\tilde{f} = \sum_j \Phi(\tilde{f}_j \otimes v_j)$$

is in the image of Φ, which is therefore surjective.

The condition $\tilde{f} \in \tilde{F}_2$ means that

$$\tilde{f}(\phi(h)x) = \tau(h)\tilde{f}(x)$$

for all $h \in H$ and $x \in G$. The left-hand side is

$$\sum_j \tilde{f}_j(\phi(h)x) \otimes (\phi(h)x \cdot v_j)$$

[12] This is the trick: use (2.30) for a varying x, not for a single fixed basis.

by definition, while the right-hand side is

$$(\varrho_2 \otimes \operatorname{Res} \varrho_1)(h)\tilde{f}(x) = \sum_j \varrho_2(h)\tilde{f}_j(x) \otimes \{\phi(h) \cdot (x \cdot v_j)\}$$
$$= \sum_j \varrho_2(h)\tilde{f}_j(x) \otimes (\phi(h)x \cdot v_j).$$

Comparing using the uniqueness of (2.30) with x replaced by $\phi(h)x$, we find that, for all j, we have

$$\tilde{f}_j(\phi(h)x) = \varrho_2(h)\tilde{f}_j(x),$$

and this does state that each coefficient function $\tilde{f}_j$ is in F_2. □

Remark 2.3.19. If ϕ is an injective homomorphism and the groups G and H are finite, then all spaces involved are finite dimensional. Since Proposition 2.3.11 shows that both sides of the projection formula are of degree $[G : H] \dim(\varrho_1) \dim(\varrho_2)$, the injectivity of Φ is sufficient to finish the proof.

Yet another important property of induction (and restriction) is the following, which is called *induction in stages* in the case of induction:

Proposition 2.3.20 (Transitivity). *Let k be a field, let*

$$H_2 \xrightarrow{\phi_2} H_1 \xrightarrow{\phi_1} G$$

be group homomorphisms, and let $\phi = \phi_1 \circ \phi_2$. For any k-representations ϱ_2 of H_2 and ϱ of G, we have canonical isomorphisms

$$\operatorname{Res}^{H_1}_{H_2}(\operatorname{Res}^G_{H_1} \varrho) \simeq \operatorname{Res}^G_{H_2}(\varrho), \qquad \operatorname{Ind}^G_{H_1}(\operatorname{Ind}^{H_1}_{H_2} \varrho_2) \simeq \operatorname{Ind}^G_{H_2}(\varrho_2).$$

Proof. As far as the restriction is concerned, this is immediate from the definition. For induction, the argument is pretty much of the same kind as the ones we used before: defining maps both ways is quite simple and hard to miss, and then one merely needs to make various checks to make sure that everything works out. We will simplify those by mostly omitting the homomorphisms ϕ, ϕ_1, ϕ_2 in the notation.

So here we go again: let E, F_1, F_2, F denote, respectively, the spaces of the representations

$$\varrho_2, \quad \operatorname{Ind}^{H_1}_{H_2}(\varrho_2), \quad \operatorname{Ind}^G_{H_1}(\operatorname{Ind}^{H_1}_{H_2}(\varrho_2)), \quad \operatorname{Ind}^G_{H_2}(\varrho_2),$$

so that we must define a G-isomorphism

$$T : F \longrightarrow F_2.$$

Note that F is a space of functions from G to E, and F_2 a space of functions from G to F_1. We define T as follows: given $f \in F$, a function from G to E, it is natural to consider

$$g \cdot f = (x \mapsto f(xg)),$$

the image of f under the regular representation on E-valued functions. Then $g \mapsto g \cdot f$ is a function from G to F, so its values are themselves functions from G to E. We want $T(f) \in F_2$, so it must be an F_1-valued function on G, i.e., $T(f)(g)$ must be a function from H_1 to E. Hence it seems plausible to define

$$T(f)(g) = (g \cdot f) \circ \phi_1.$$

If H_1 is a subgroup of G, this is the restriction to H_1—in the sense of restricting a function on G to one on H_1—of the function $g \mapsto g \cdot f$ on G.

We can then check that $T(f)$ is, in fact, F_1-valued; if we omit the group homomorphisms involved, this amounts to letting $\psi = T(f)$ and writing

$$\psi(h_2 h_1) = g \cdot f(h_2 h_1) = f(h_2 h_1 g) = \varrho(h_2) f(h_1 g) = \varrho(h_2)\psi(h_1),$$

for $h_i \in H_i$, using of course in the middle the assumption that f is in F. Again, this is unlikely to make much sense until the reader has tried and succeeded independently to follow the computation.

Now we should check that $T(f)$ is not only F_1-valued, but also lies in F_2, i.e., transforms under H_1 like the induced representation $\mathrm{Ind}_{H_2}^{H_1}(\varrho)$. We leave this to the reader; this is much helped by the fact that the action of H_1 on this induced representation is also the regular representation.

Next we must check that T is an intertwining operator; but again, both F and F_2 carry actions which are variants of the regular representation, and this should not be surprising. We therefore omit it.

The final step is the construction of the inverse $\tilde{T}$ of T.[13] We now start with $\psi \in F_2$ and must define a function from G to E. Unraveling in two steps, we set

$$\tilde{T}(\psi)(g) = \psi(g)(1)$$

($\psi(g)$ is an element of F_1, i.e., a function from H_1 to E, and we evaluate that at the unit of $H_1 \ldots$). Taking $g \in G$ and $h_2 \in H_2$, denoting $f = \tilde{T}(\psi)$, we again let the reader check that

$$f(h_2 g) = \psi(h_2 g)(1) = (h_2 \cdot \psi(g))(1) = \psi(g)(h_2) = \varrho(h_2)\psi(g)(1) = \varrho(h_2) f(g)$$

makes sense and means that $\tilde{T}(\psi)$ is in F.

Now we see that $\tilde{T}T(f)$ is the function which maps $g \in G$ to

$$(g \cdot f)(1) = f(g),$$

[13] If the vector spaces are finite-dimensional and the homomorphisms are inclusions, note that it is quite easy to check that T is injective, and since the dimensions of F and F_2 are both $[G : H_2] \dim \varrho$, this last step can be shortened.

in other words $\tilde{T} \circ T$ is the identity. Rather more abstrusely, if $\psi \in F_2$, $f = \tilde{T}(\psi)$, and $\tilde{\psi} = T(f)$, we find for $g \in G$ and $h_1 \in H_1$ that

$$\begin{aligned}\tilde{\psi}(g)(h_1) = (g \cdot f)(h_1) = f(h_1 g) \\ = \psi(h_1 g)(1) = (h_1 \cdot \psi(g))(1) = \psi(g)(h_1)\end{aligned}$$

(where we use the fact that, on F_2, H_1 acts through the regular representation), which indicates that $T \circ \tilde{T}$ is also the identity (since ψ and $\tilde{\psi}$ are functions on G whose values are functions from H_1 to $E \ldots$). Thus T and $\tilde{T}$ are reciprocal isomorphisms. □

Remark 2.3.21 (Functoriality saves time). At this point, conscientious readers may well have become bored and annoyed at this "death of a thousand checks". And there are indeed at least two ways to avoid much (if not all) of the computations we have done. One uses character theory; it is restricted to special sitations, and will be discussed later. We sketch the second now, since the reader is presumably well motivated to hear about abstract nonsense if it cuts down on the calculations.

The keyword is the adjective "natural" (or "canonical") that we attributed to the isomorphisms (2.24) of Frobenius reciprocity. In one sense, this is intuitive enough: the linear isomorphism

$$\operatorname{Hom}_G(\varrho_1, \operatorname{Ind}_H^G(\varrho_2)) \longrightarrow \operatorname{Hom}_H(\operatorname{Res}_H^G(\varrho_1), \varrho_2),$$

defined in the proof of Proposition 2.3.9 certainly feels natural. But we now take this more seriously and try to give rigorous sense to this sentence.

The point is the following fact: a representation ϱ of G is determined, up to isomorphism, by the data of all homomorphism spaces

$$V(\pi) = \operatorname{Hom}_G(\pi, \varrho),$$

where π runs over k-representations of G *together* with the data of the maps

$$V(\pi) \xrightarrow{V(\Phi)} V(\pi')$$

associated to any (reversed!) G-homomorphism $\pi' \xrightarrow{\Phi} \pi$ by mapping

$$(\Psi : \pi \to \varrho) \in V(\pi)$$

to

$$V(\Phi)(\Psi) = \Psi \circ \Phi.$$

To be precise:

Fact. Suppose that ϱ_1 and ϱ_2 are k-representations of G, and that for any representation π, there is given a k-linear isomorphism

$$I(\pi) : \operatorname{Hom}_G(\pi, \varrho_1) \longrightarrow \operatorname{Hom}_G(\pi, \varrho_2),$$

in such a way that all diagrams

$$\begin{array}{ccc} \mathrm{Hom}_G(\pi,\varrho_1) & \xrightarrow{I(\pi)} & \mathrm{Hom}_G(\pi,\varrho_2) \\ \downarrow & & \downarrow \\ \mathrm{Hom}_G(\pi',\varrho_1) & \xrightarrow{I(\pi')} & \mathrm{Hom}_G(\pi',\varrho_2) \end{array}$$

commute for any $\Phi : \pi' \to \pi$, where the vertical arrows, as above, are given by $\Psi \mapsto \Psi \circ \Phi$. Then ϱ_1 and ϱ_2 are isomorphic, and in fact there exists a unique isomorphism

$$\varrho_1 \xrightarrow{I} \varrho_2$$

such that $I(\pi)$ is given by $\Psi \mapsto I \circ \Psi$ for all π.

Let us first see why this is useful. When dealing with induction, the point is that it tells us that an induced representation $\mathrm{Ind}_H^G(\varrho)$ is characterized, up to isomorphism, by the Frobenius reciprocity isomorphisms (2.24). Indeed, the latter tells us, simply from the data of ϱ, what any G-homomorphism space

$$\mathrm{Hom}_G(\pi, \mathrm{Ind}_H^G(\varrho))$$

is supposed to be. And the fact above says that there can be only one representation $\tilde{\varrho}$ with given homomorphism groups $\mathrm{Hom}_G(\pi,\tilde{\varrho})$ which behave naturally. Precisely, the behavior under morphisms must be compatible: we get

Fact bis. Let $\phi : H \to G$ be a group-homomorphism, and let ϱ be a k-representation of H. There exists, up to isomorphism of representations of G, *at most* one k-representation ϱ' of G with k-linear isomorphisms

$$i(\pi) : \mathrm{Hom}_G(\pi,\varrho') \longrightarrow \mathrm{Hom}_H(\mathrm{Res}(\pi),\varrho)$$

such that the diagrams

$$\begin{array}{ccc} \mathrm{Hom}_G(\pi,\varrho') & \xrightarrow{i(\pi)} & \mathrm{Hom}_H(\mathrm{Res}(\pi),\varrho) \\ \downarrow & & \downarrow \\ \mathrm{Hom}_G(\pi',\varrho') & \xrightarrow{i(\pi')} & \mathrm{Hom}_H(\mathrm{Res}(\pi'),\varrho) \end{array}$$

commute for $\pi' \xrightarrow{\Phi} \pi$ a G-homomorphism, where the vertical arrows are again $\Psi \mapsto \Psi \circ \Phi$, on the left, and $\Psi \mapsto \Psi \circ \mathrm{Res}(\Phi)$ on the right (restriction of Φ to H).

Readers are invited to check that the (explicit) isomorphisms

$$i(\pi) : \mathrm{Hom}_G(\pi, \mathrm{Ind}_H^G(\varrho)) \longrightarrow \mathrm{Hom}_H(\mathrm{Res}_H^G(\pi),\varrho)$$

that we constructed (based on the explicit model (2.22)) are such that the diagrams

$$(2.31)\qquad \begin{array}{ccc} \mathrm{Hom}_G(\pi, \mathrm{Ind}_H^G(\varrho)) & \xrightarrow{i(\pi)} & \mathrm{Hom}_H(\mathrm{Res}_H^G(\pi), \varrho) \\ \downarrow & & \downarrow \\ \mathrm{Hom}_G(\pi', \mathrm{Ind}_H^G(\varrho)) & \xrightarrow{i(\pi')} & \mathrm{Hom}_H(\mathrm{Res}_H^G(\pi'), \varrho) \end{array}$$

commute. These are the same as the ones above, with $\varrho' = \mathrm{Ind}(\varrho)$. This is the real content of the observation that the Frobenius reciprocity isomorphisms are natural. Thus the *construction* of (2.22) proved the *existence* of the induced representation characterized by the abstract property of Frobenius reciprocity.

We can now see that the transitivity of induction is just a reflection of the—clearly valid—transitivity of restriction. Consider

$$H_2 \xrightarrow{\phi_2} H_1 \xrightarrow{\phi_1} G$$

as in the transitivity formula, and consider a representation ϱ of H_2, as well as

$$\varrho_1 = \mathrm{Ind}_{H_1}^G(\mathrm{Ind}_{H_2}^{H_1}(\varrho)), \qquad \varrho_2 = \mathrm{Ind}_{H_2}^G(\varrho).$$

According to Frobenius reciprocity applied twice or once, respectively, we have, for all representations π of G, k-linear isomorphisms

$$\mathrm{Hom}_G(\pi, \varrho_1) \simeq \mathrm{Hom}_{H_1}(\mathrm{Res}_{H_1}^G(\pi), \mathrm{Ind}_{H_2}^{H_1}(\varrho)) \simeq \mathrm{Hom}_{H_2}(\mathrm{Res}_{H_2}^{H_1}(\mathrm{Res}_{H_1}^G(\pi)), \varrho)$$

and

$$\mathrm{Hom}_G(\pi, \varrho_2) \simeq \mathrm{Hom}_{H_2}(\mathrm{Res}_{H_2}^G(\pi), \varrho);$$

hence, by comparison and the obvious transitivity of restriction, we obtain isomorphisms

$$I(\pi) : \mathrm{Hom}_G(\pi, \varrho_1) \simeq \mathrm{Hom}_G(\pi, \varrho_2).$$

The reader should be easily convinced (and then check!) that these isomorphisms satisfy the compatibility required in the claim to deduce that ϱ_1 and ϱ_2 are isomorphic; indeed, this is a *composition* or *tiling* of the corresponding facts for the diagrams (2.31).

At first sight, this may not seem much simpler than what we did earlier, but a second look reveals that we did not use *anything* relating to k-representations of G except the existence of morphisms, the identity maps, and the composition operations! In particular, there is no need whatsoever to know an explicit model for the induced representation.

We now prove the Fact above, using the notation in that statement. Take $\pi = \varrho_1$, so that we have

$$\mathrm{Hom}_G(\pi, \varrho_1) = \mathrm{Hom}_G(\varrho_1, \varrho_1)$$

and $I(\pi) = I(\varrho_1)$ is an isomorphism $\mathrm{Hom}_G(\varrho_1, \varrho_1) \longrightarrow \mathrm{Hom}_G(\varrho_1, \varrho_2)$.

We may not know much about the general existence of homomorphisms, but certainly this space contains the identity of ϱ_1. Hence we obtain an element

$$I = I(\varrho_1)(\mathrm{Id}_{\varrho_1}) \in \mathrm{Hom}_G(\varrho_1, \varrho_2).$$

Then (this looks like a cheat) this I *is* the desired isomorphism! To see this—but first try it!—we check first that $I(\pi)$ is given, as claimed, by precomposition with I for any π. Indeed, $I(\pi)$ is an isomorphism

$$\mathrm{Hom}_G(\pi, \varrho_1) \longrightarrow \mathrm{Hom}_G(\pi, \varrho_2).$$

Take an element $\Phi : \pi \to \varrho_1$; we can then build the associated commutative square

$$\begin{array}{ccc} \mathrm{Hom}_G(\varrho_1, \varrho_1) & \xrightarrow{I(\varrho_1)} & \mathrm{Hom}_G(\varrho_1, \varrho_2) \\ \downarrow & & \downarrow \\ \mathrm{Hom}_G(\pi, \varrho_1) & \xrightarrow{I(\pi)} & \mathrm{Hom}_G(\pi, \varrho_2). \end{array}$$

Take the element Id_{ϱ_1} in the top-left corner. If we follow the right-then-down route, we get, by definition, the element

$$I(\varrho_1)(\mathrm{Id}_{\varrho_1}) \circ \Phi = I \circ \Phi \in \mathrm{Hom}_G(\pi, \varrho_2).$$

But if we follow the down-then-right route, we get $I(\pi)(\mathrm{Id}_{\varrho_1} \circ \Phi) = I(\pi)(\Phi)$, and hence the commutativity of these diagrams says that, for all Φ, we have

$$I(\pi)(\Phi) = I \circ \Phi, \tag{2.32}$$

which is what we had claimed.

We now check that I is, indeed, an isomorphism, by exhibiting an inverse. The construction we used strongly suggests that

$$J = I(\varrho_2)^{-1}(\mathrm{Id}_{\varrho_2}) \in \mathrm{Hom}_G(\varrho_2, \varrho_1),$$

should be what we need (where we use that $I(\varrho_2)$ is an isomorphism, by assumption). Indeed, tautologically, we have

$$I(\varrho_2)(J) = \mathrm{Id}_{\varrho_2},$$

which translates, from the formula (2.32) we have just seen (applied with $\pi = \varrho_2$) to

$$I \circ J = \mathrm{Id}_{\varrho_2}.$$

Now we simply exchange the role of ϱ_1 and ϱ_2 and replace $I(\pi)$ by its inverse; then I and J are exchanged, and we get also

$$J \circ I = \mathrm{Id}_{\varrho_1}.$$

Why did we not start with this "functorial" language? Partly this is a matter of personal taste and partly of wanting to show very concretely how—especially if the reader does (or has done) all computations independently—part of the spirit of the game will have seeped in. Moreover, in some of the more down-to-earth applications of these games with induction and its variants, it may be quite important to know what the *canonical maps* actually are. The functorial language does usually give a way to compute them, but it may be more direct to have written them down as directly as we did.

To conclude with the general properties of induction, we leave the proof of the following lemma to the reader:

Lemma 2.3.22. *Let k be a field, and let $\phi : H \longrightarrow G$ be a group homomorphism with $\phi(H)$ of finite index in G. For any finite-dimensional representations ϱ and ϱ_i of H, we have natural isomorphisms*

$$(\mathrm{Ind}_H^G(\varrho))\check{} \simeq \mathrm{Ind}_H^G(\check{\varrho})$$

and

$$\mathrm{Ind}_H^G\Bigl(\bigoplus_{i\in I}\varrho_i\Bigr) \simeq \bigoplus_{i\in I}\mathrm{Ind}_H^G(\varrho_i)$$

for I finite.

The corresponding statements for the restriction are also valid and equally easy to check. On the other hand, although the isomorphism

$$\mathrm{Res}_H^G(\varrho_1\otimes\varrho_2) \simeq \mathrm{Res}_H^G(\varrho_1)\otimes\mathrm{Res}_H^G(\varrho_2)$$

is immediate, it is usually definitely false (say when ϕ is injective but is not an isomorphism) that

$$\mathrm{Ind}_H^G(\varrho_1\otimes\varrho_2), \quad \mathrm{Ind}_H^G(\varrho_1)\otimes\mathrm{Ind}_H^G(\varrho_2)$$

are isomorphic, for instance, because the degrees do not match (from left to right, they are given by

$$[G:H]\dim(\varrho_1)\dim(\varrho_2), \qquad [G:H]^2(\dim\varrho_1)(\dim\varrho_2),$$

respectively).

We conclude this longish section with another type of "change of groups". Fix a field k and two groups G_1 and G_2. Given k-representations ϱ_1 and ϱ_2 of G_1 and G_2, acting on E_1 and E_2, respectively, we can define a representation of the direct product $G_1\times G_2$ on the tensor product $E_1\otimes E_2$. For pure tensors $v_1\otimes v_2$ in $E_1\otimes E_2$, we let

$$(\varrho_1\boxtimes\varrho_2)(g_1,g_2)(v_1\otimes v_2) = \varrho_1(g_1)v_1\otimes\varrho_2(g_2)v_2,$$

which extends by linearity to the desired action, sometimes called the *external tensor product* of ϱ_1 and ϱ_2,

$$\varrho_1\boxtimes\varrho_2 : G_1\times G_2 \longrightarrow \mathrm{GL}(E_1\otimes E_2).$$

Of course, the dimension of this representation is again $(\dim \varrho_1)(\dim \varrho_2)$. In particular, it is clear that not all representations of $G_1 \times G_2$ can be of this type, simply because their dimensions might not factor non-trivially. However, in some cases, *irreducible* representations must be external tensor products of irreducible representations of the factors.

Proposition 2.3.23 (Irreducible representations of direct products)**.** *Let k be an* algebraically closed field, *and let G_1, G_2 be two groups. If ϱ is a finite-dimensional irreducible k-representation of $G = G_1 \times G_2$, then there exist irreducible k-representations ϱ_1 of G_1 and ϱ_2 of G_2, respectively, such that*

$$\varrho \simeq \varrho_1 \boxtimes \varrho_2.$$

Moreover, ϱ_1 and ϱ_2 are unique, up to isomorphism of representations of their respective groups.

Conversely, if ϱ_1 and ϱ_2 are irreducible finite-dimensional k-representations of G_1 and G_2, respectively, the external tensor product $\varrho_1 \boxtimes \varrho_2$ is an irreducible representation of $G_1 \times G_2$.

The proof of this requires some preliminary results, so we defer it to the end of Section 2.7. The statement is false, in general, over non-algebraically closed fields (see Example 2.7.33).

Remark 2.3.24 (Relation with the ordinary tensor product)**.** Consider a group G. There is an injective *diagonal* homomorphism

$$\phi \begin{cases} G & \longrightarrow & G \times G \\ g & \mapsto & (g, g). \end{cases}$$

If ϱ_1 and ϱ_2 are k-representations of G, the definitions show that

$$\mathrm{Res}_G^{G\times G}(\varrho_1 \boxtimes \varrho_2) = \varrho_1 \otimes \varrho_2.$$

2.4. Formalism: changing the field

We will not say much about changing the field. Clearly, whenever K is an extension of k, we can turn a k-representation

$$G \longrightarrow \mathrm{GL}(E)$$

into a representation (which we denote $\varrho \otimes K$) over K, by composing with the group homomorphism

$$\mathrm{GL}(E) \longrightarrow \mathrm{GL}(E \otimes_k K)$$

which, concretely (see the next section also), can be interpreted simply by saying that a matrix with coefficients in the subfield k of K can be seen as a matrix with coefficients in K, i.e., by looking at the inclusion

$$\mathrm{GL}_n(k) \hookrightarrow \mathrm{GL}_n(K).$$

If ϱ is a representation of G over a field K, and it is *isomorphic* to a representation arising in this manner from a k-representation, for some subfield k of K, one customarily says that ϱ can be *defined over* k.

Sometimes, given a field extension K/k (for instance, with K algebraically closed) and a certain property $\mathcal{P}(\varrho)$ of a representation ϱ, it may happen that $\mathcal{P}(\varrho)$ does not hold for a k-representation ϱ, but that $\mathcal{P}(\varrho \otimes K)$ does (or conversely). If K is an algebraic closure of k, and if $\varrho \otimes K$ has the desired property, one then says that the k-representation ϱ "has $\mathcal{P}$ absolutely".

Example 2.4.1. We give here an example of a representation which is irreducible but not *absolutely irreducible*. Consider the (infinite) abelian group $G = \mathbf{R}/\mathbf{Z}$, and the two-dimensional real representation given by

$$\varrho : \begin{cases} G \longrightarrow \mathrm{GL}_2(\mathbf{R}) \\ \theta \mapsto \begin{pmatrix} \cos(2\pi\theta) & \sin(2\pi\theta) \\ -\sin(2\pi\theta) & \cos(2\pi\theta) \end{pmatrix} \end{cases}$$

(which corresponds to the action of $\mathbf{R}/\mathbf{Z}$ on the real plane by the rotation with angle $2\pi\theta$). This makes it clear that this is a homomorphism (which is otherwise easy to check using trigonometric identities), and it also makes it clear that ϱ is irreducible (there is no non-zero real subspace of $\mathbf{R}^2$ which is stable under all rotations $\varrho(\theta)$, except $\mathbf{R}^2$ itself).

However, the irreducibility breaks down when extending the base field to $\mathbf{C}$. Indeed, on $\mathbf{C}^2$, we have

$$\varrho(\theta)\begin{pmatrix} 1 \\ i \end{pmatrix} = \begin{pmatrix} \cos\theta + i\sin\theta \\ -\sin\theta + i\cos\theta \end{pmatrix} = (\cos\theta + i\sin\theta)\begin{pmatrix} 1 \\ i \end{pmatrix}$$

and

$$\varrho(\theta)\begin{pmatrix} 1 \\ -i \end{pmatrix} = \begin{pmatrix} \cos\theta - i\sin\theta \\ \sin\theta - i\cos\theta \end{pmatrix} = (\cos\theta - i\sin\theta)\begin{pmatrix} 1 \\ -i \end{pmatrix}.$$

This means that $\mathbf{C}^2$, under the action of G through ϱ, splits as a direct sum

$$\mathbf{C}^2 = \begin{pmatrix} 1 \\ i \end{pmatrix}\mathbf{C} \oplus \begin{pmatrix} 1 \\ -i \end{pmatrix}\mathbf{C}$$

of two complex lines which are both subrepresentations, one of them isomorphic to the one-dimensional complex representation

$$\begin{cases} G \to \mathrm{GL}(\mathbf{C}) \simeq \mathbf{C}^\times \\ \theta \mapsto e^{i\theta} \end{cases}$$

and the other to

$$\begin{cases} G \to \mathbf{C}^\times \\ \theta \mapsto e^{-i\theta} \end{cases}$$

(its *complex conjugate*, in the sense of Example 2.4.2 below). Thus ϱ is *not absolutely irreducible.*

Another way to change the field, which may be more confusing, is to apply *automorphisms* of k. Formally, this is not different: we have an automorphism $\sigma : k \longrightarrow k$, and we define a representation $\varrho^{(\sigma)}$ by the composition

$$\varrho^{(\sigma)} : G \xrightarrow{\varrho} \mathrm{GL}(E) \longrightarrow \mathrm{GL}(E \otimes_k k),$$

where we have to be careful to see k, in the second argument of the tensor product, as given with the k-algebra structure σ. Concretely, $E_\sigma = E \otimes_k k$ is the k-vector space with the same underlying abelian group as E, but with scalar multiplication given by

$$\alpha \cdot v = \sigma(\alpha) v \in E.$$

Here again matrix representations may help understand what happens. A basis (v_i) of E is still a basis of E_σ but, for any $g \in G$, the matrix representing $\varrho(g)$ in the basis (v_i) of E_σ is obtained by applying σ^{-1} to all coefficients of the matrix that represents $\varrho(g)$. Indeed, for any i, we can write

$$\varrho(g) v_i = \sum_j \alpha_j v_j = \sum_j \sigma^{-1}(\alpha_j) \cdot v_j$$

for some coefficients α_j, so that the (j,i)-th coefficient of the matrix for $\varrho(g)$ is α_j, while it is $\sigma^{-1}(\alpha_j)$ for $\varrho^{(\sigma)}(g)$.

This operation on representations can be interesting because ϱ and $\varrho^{(\sigma)}$ are usually *not* isomorphic as representations, despite the fact that they are closely related. In particular, there is a bijection between the subrepresentations of E and those of E_σ (given by $F \mapsto F_\sigma$), and hence ϱ and $\varrho^{(\sigma)}$ are simultaneously irreducible or not irreducible, semisimple or not semisimple.

Example 2.4.2 (Complex conjugate). Consider $k = \mathbf{C}$. Although $\mathbf{C}$, considered as an abstract field, has many automorphisms, the only continuous ones, and therefore the most important, are the identity and the complex conjugation $\sigma : z \mapsto \bar{z}$. It follows therefore that any time we have a complex representation $G \longrightarrow \mathrm{GL}(E)$, where E is a $\mathbf{C}$-vector space, there is a naturally associated *conjugate* representation $\bar{\varrho}$ obtained by applying the construction above to the complex conjugation. If E is finite dimensional, then from the basic theory of characters (Proposition 2.7.38 below and Example 2.7.41) one derives the fact that $\bar{\varrho}$ is isomorphic to ϱ if and only if the function $g \mapsto \operatorname{Tr} \varrho(g)$ is real valued. (This can already be checked when ϱ is one-dimensional, since $\bar{\varrho}$ is then the conjugate function $G \to \mathbf{C}$, which equals ϱ if and only if ϱ is real valued.) In particular, the examples in (2.4), (2.5), or (2.6) lead to many cases of representations where ϱ and $\bar{\varrho}$ are not isomorphic.

Field extensions (including automorphisms) are the only morphisms for fields. However, there are sometimes other possibilities for changing fields, which are more subtle. Suppose for instance that

$$\varrho : G \longrightarrow \mathrm{GL}(E)$$

is a complex representation of degree $d \geqslant 1$ of some group G, and that *with respect to some chosen basis of* E, the image of ϱ is given by matrices with integral coefficients. If we then fix a prime number p, we may consider the reduction modulo p (say $\bar{\varrho}(g)$) of these matrices, which will be elements in $\mathrm{GL}_d(\mathbf{Z}/p\mathbf{Z})$. The properties of the reduction modulo p imply that $g \mapsto \bar{\varrho}(g)$ is a homomorphism from G to $\mathrm{GL}_d(\mathbf{Z}/p\mathbf{Z})$, or in other words, an $\mathbf{F}_p$-representation of G.

More abstractly, this definition corresponds to the existence of a *G-stable lattice M of E*, which is an abelian group $M \subset E$ such that $M \otimes_{\mathbf{Z}} \mathbf{C} \simeq E$ and such that $\varrho(g)m \in M$ for all $g \in G$ and $m \in M$. We can then define an $\mathbf{F}_p$-representation of G on M/pM, which is a d-dimensional $\mathbf{F}_p$-vector space, simply because pM is also G-stable (by linearity).

This construction can be extremely useful and important. However, it is delicate: first of all, it is not always defined (the G-stable lattice M may not exist), and also it may not be *well defined*, in the sense that taking a different G-stable lattice (there is no uniqueness, since for instance pM works just as well as M) might lead to a non-isomorphic $\mathbf{F}_p$-representation of G. We refer to [**53**, §15.2] for further discussion of this theory.

Exercise 2.4.3. Let $G = \mathbf{Z}/2\mathbf{Z}$, and let ϱ be the two-dimensional regular representation of G on $\mathbf{C}^2$, with canonical basis e_1, e_2. Let $f_1 = e_1 + e_2$, $f_2 = e_1 - e_2$. Let $k = \mathbf{Z}/2\mathbf{Z}$.

Show that $M = \mathbf{Z}^2$ is a G-stable lattice and that the k-representation of G on $M/2M$ is not semisimple. On the other hand, show that $M' = f_1\mathbf{Z} \oplus f_2\mathbf{Z}$ is another G-stable lattice, and that the k-representation of G on $M'/2M'$ is trivial, in particular semisimple.

In Exercise 2.6.6, we will see an example of an irreducible representation that reduces modulo a prime to one which is not.

2.5. Matrix representations

We have emphasized in Definition 2.1.1 the abstract view where a representation is seen as a linear action of G on a k-vector space E. However, in practice, if one wishes to *compute* with representations, one will select a fixed basis of E and express ϱ as the homomorphism

$$\varrho^m : G \longrightarrow \mathrm{GL}_n(k), \qquad n = \dim(E),$$

that maps g to the matrix representing $\varrho(g)$ in the chosen basis. Indeed, this is what we already did in the cases of the example in (2.12) and in Example 2.4.1.

Although such matrix representations can be awkward when used exclusively (especially because of the choice of a basis), it is useful and important to know how to express in these terms the various operations on representations that we have described previously. These concrete descriptions may also help clarify these operations, especially for readers less familiar with abstract algebra. We will explain this here fairly quickly.

For a direct sum $\varrho_1 \oplus \varrho_2$, we may concatenate bases $(e_1, \ldots, e_n)$ of E_1 and $(f_1, \ldots, f_m)$ of E_2 to obtain a basis

$$(e_1, \ldots, e_n, f_1, \ldots, f_m)$$

in which the representation $\varrho_1 \oplus \varrho_2$ takes the form of block-diagonal matrices

$$g \mapsto \left(\begin{array}{c|c} \varrho_1^m(g) & 0 \\ \hline 0 & \varrho_2^m(g) \end{array}\right)$$

of size $m + n$. Corresponding to a short exact sequence

$$0 \to E_1 \longrightarrow E \xrightarrow{\Phi} E_2 \to 0$$

of representations, which may not be split, we select a basis $(e_1, \ldots, e_n)$ of the subspace E_1 of E, and we extend it to a basis $(e_1, \ldots, e_n, f_1, \ldots, f_m)$, $m = \dim(E_2)$, of E. Then

$$(f_1', \ldots, f_m') = (\Phi(f_1), \ldots, \Phi(f_m))$$

is a basis of E_2 and we get in these bases a block-triangular matrix representation of ϱ acting on E:

$$g \mapsto \varrho^m(g) = \left(\begin{array}{c|c} \varrho_1^m(g) & \star \\ \hline 0 & \varrho_2^m(g) \end{array}\right), \tag{2.33}$$

where ϱ_1^m is the matrix representation in $(e_1, \ldots, e_n)$ and ϱ_2^m the one in $(f_1', \ldots, f_m')$. The block denoted $\star$ is an important invariant of the short exact sequence. If we view it as a map

$$c : G \longrightarrow \mathrm{M}_{n,m}(k)$$

from G to rectangular $n \times m$ matrices with coefficients in k, then c is not a homomorphism, but writing down the relation

$$\varrho^m(gh) = \varrho^m(g)\varrho^m(h),$$

we see that it satisfies instead

$$c(gh) = \varrho_1^m(g)c(h) + c(g)\varrho_2^m(h)$$

for $g, h \in G$.

In the case of a tensor product $\varrho = \varrho_1 \otimes \varrho_2$, one usually represents it in the basis of pure tensors $\delta_{i,j} = e_i \otimes f_j$. If this basis is ordered as

$$(\delta_{1,1}, \ldots, \delta_{1,m}, \delta_{2,1}, \ldots, \delta_{2,m}, \ldots, \delta_{n,1}, \ldots, \delta_{n,m}),$$

and we denote by $A = (a_{i,j})_{1 \leqslant i,j \leqslant n}$ the matrix $\varrho_1^{\boldsymbol{m}}(g)$ and by B the matrix $\varrho_2^{\boldsymbol{m}}(g)$, then $\varrho^{\boldsymbol{m}}(g)$ is a block matrix with n rows and columns of square blocks of size m, given by

$$\left(\begin{array}{c|c|c|c} a_{1,1}B & a_{1,2}B & \ldots & a_{1,n}B \\ \hline \vdots & \vdots & & \vdots \\ \hline a_{n,1}B & \ldots & \ldots & a_{n,n}B \end{array}\right).$$

It should be noted however that this explicit form is very rarely useful.

The matrix representation of the contragredient of a representation ϱ is also easy to describe. We have

$$\check{\varrho}^{\boldsymbol{m}}(g) = {}^t\varrho^{\boldsymbol{m}}(g)^{-1},$$

the inverse-transpose homomorphism.

The case of the restriction to a subgroup is immediate: the matrices of the restriction do not change. For induction, the situation is more involved, but we will see some examples in the next chapters.

2.6. Examples

We collect here some more examples of representations. The first one, in particular, is very important, and it will reappear frequently in various ways in the rest of the book.

2.6.1. Binary forms and invariants. Let k be any field. For an integer $m \geqslant 0$, we denote by V_m the vector space of polynomials in $k[X, Y]$ which are homogeneous of degree m, i.e., the k-subspace of $k[X, Y]$ generated by the monomials

$$X^i Y^{m-i}, \qquad 0 \leqslant i \leqslant m.$$

In fact, these monomials are independent, and therefore form a basis of V_m. In particular $\dim V_m = m + 1$.

If we take $G = \mathrm{SL}_2(k)$, we can let G act on V_m by

$$(\varrho_m(g)f)(X, Y) = f((X, Y) \cdot g),$$

where $(X, Y) \cdot g$ denotes the multiplication of the row vector (X, Y) by the matrix g. In other words, if

$$g = \begin{pmatrix} a & b \\ c & d \end{pmatrix}, \tag{2.34}$$

we have

$$(g \cdot f)(X, Y) = f(aX + cY, bX + dY)$$

(one just says that G acts on V_m by linear change of variables).

We then have

Theorem 2.6.1 (Irreducible representations of SL_2). (1) *For $k = \mathbf{C}$, the representations ϱ_m, for $m \geqslant 0$, are irreducible representations of* $\mathrm{SL}_2(\mathbf{C})$. *In fact, ϱ_m is an irreducible representation of the subgroup* $\mathrm{SU}_2(\mathbf{C}) \subset \mathrm{SL}_2(\mathbf{C})$.

(2) *Any finite-dimensional, continuous, irreducible representation of the subgroup* $\mathrm{SU}_2(\mathbf{C})$ *is isomorphic to one of the ϱ_m.*

(3) *On the other hand, if k is a field of non-zero characteristic p, the representation ϱ_p is not irreducible.*

The first assertion will be proved in Example 2.7.11 and Exercise 2.7.13, and the second in Theorem 5.6.3 (the notion of continuous representation is explained in Section 3.3). The analogous statement for $\mathrm{SL}_2(\mathbf{C})$ is more complicated: it is not true that the representations ϱ_m exhaust (up to isomorphism) all finite-dimensional irreducible continuous representations of $\mathrm{SL}_2(\mathbf{C})$. This can be seen quickly by noting that the complex conjugate $\bar{\varrho}_n$ of ϱ_m (as defined in Example 2.4.2), which is irreducible, is not isomorphic to any ϱ_n. Indeed, just by comparing dimensions, the only possibility would be that $\bar{\varrho}_n \simeq \varrho_n$, but from simple computations, one can see that the character $g \mapsto \mathrm{Tr}(\varrho_m(g))$ is not real valued, which is a necessary condition for ϱ_m to be isomorphic to its complex conjugate (see (2.50)). One can show that an irreducible finite-dimensional continuous representation of $\mathrm{SL}_2(\mathbf{C})$ is of the form

$$\varrho_m \otimes \bar{\varrho}_n$$

for some integers m, $n \geqslant 0$ (see, e.g., [**36**, end of II.3]).

We can however already explain the statement (3). If k has characteristic p, consider the subspace $W \subset V_p$ spanned by the monomials X^p and Y^p. Then $W \neq V_p$ (since $\dim V_p = p + 1 \geqslant 3$) and V_p is a subrepresentation. Indeed, for g given by (2.34), we have

$$(g \cdot X^p) = (aX + cY)^p = a^p X^p + c^p Y^p \in W,$$
$$(g \cdot Y^p) = (aX + cY)^p = b^p X^p + d^p Y^p,$$

by the usual properties of the p-th power operation in characteristic p (i.e., the fact that the binomial coefficients $\binom{p}{j}$ are divisible by p for $1 \leqslant j \leqslant p-1$). One can also show that $W \subset V_p$ does not have a stable complementary subspace, so V_p is not semisimple in characteristic p.

We now consider only the case $k = \mathbf{C}$. It is elementary that ϱ_m is isomorphic to the m-th symmetric power of ϱ_1 for all $m \geqslant 0$. Hence we see here a case where, using multilinear operations, all irreducible (finite-dimensional) representations of a group are obtained from a "fundamental" one. We also see here an elementary example of a group which has irreducible finite-dimensional representations of arbitrarily large dimension. (In fact, $\mathrm{SL}_2(\mathbf{C})$ also has many infinite-dimensional representations which are irreducible, in the sense of representations of topological groups.)

Exercise 2.6.2 (Matrix representation). (1) Compute the matrix representation for ϱ_2 and ϱ_3, in the bases (X^2, XY, Y^2) and (X^3, X^2Y, XY^2, Y^3) of V_2 and V_3, respectively.

(2) Compute the kernel of ϱ_2 and ϱ_3, and recover the result without using matrix representations.

A very nice property of these representations—which turns out to be crucial in quantum mechanics—illustrates another important type of results in representation theory:

Theorem 2.6.3 (Clebsch–Gordan formula). *For any integers $m \geqslant n \geqslant 0$, the tensor product $\varrho_m \otimes \varrho_n$ is semisimple*[14] *and decomposes as*

$$\varrho_m \otimes \varrho_n \simeq \varrho_{m+n} \oplus \varrho_{m+n-2} \oplus \cdots \oplus \varrho_{m-n}. \tag{2.35}$$

One point of this formula is to illustrate that, if one knows some irreducible representations of a group, one may well hope to be able to construct or identify others by trying to decompose the tensor products of these representations into irreducible components (if possible). Here, supposing one knew only the "obvious" representations $\varrho_0 = \mathbf{1}$ and ϱ_1 (which is just the tautological inclusion $\mathrm{SL}_2(\mathbf{C}) \longrightarrow \mathrm{GL}_2(\mathbf{C})$), we see that all other representations ϱ_m arise by taking tensor products iteratively and decomposing them, e.g.,

$$\varrho_1 \otimes \varrho_1 = \varrho_2 \oplus \mathbf{1}, \qquad \varrho_2 \otimes \varrho_1 = \varrho_3 \oplus \varrho_1, \qquad \text{etc.}$$

Proof. Both sides of the Clebsch–Gordan formula are trivial when $m = 0$. Using induction on m, we then see that it is enough to prove that

$$\varrho_m \otimes \varrho_n \simeq \varrho_{m+n} \oplus (\varrho_{m-1} \otimes \varrho_{n-1}) \tag{2.36}$$

for $m \geqslant n \geqslant 1$.

At least a subrepresentation isomorphic to $\varrho_{m-1} \otimes \varrho_{n-1}$ is not too difficult to find. Indeed, first of all, the tensor product $\varrho_m \otimes \varrho_n$ can be interpreted concretely as a representation on the space $V_{m,n}$ of polynomials in four variables X_1, Y_1, X_2, Y_2, which are homogeneous of degree m with respect to (X_1, Y_1), and of degree n with respect to the other variables, where the

[14] In fact, any tensor product of finite-dimensional semisimple complex representations is semisimple, but this result of Chevalley is by no means easy to prove (see Theorem 7.1.11).

group $SL_2(\mathbf{C})$ acts by simultaneous linear change of variable on the two sets of variables, i.e.,

$$(g \cdot f)(X_1, Y_1, X_2, Y_2) = f((X_1, Y_1)g, (X_2, Y_2)g)$$

for $f \in V_{m,n}$. This G-isomorphism

$$V_m \otimes V_n \longrightarrow V_{m,n}$$

is induced by

$$(X^i Y^{m-i}) \otimes (X^j Y^{n-j}) \mapsto X_1^i Y_1^{m-i} X_2^j Y_2^{n-j}$$

for the standard basis vectors.

Using this description, we have a linear map

$$\Delta \quad \begin{cases} V_{m-1,n-1} & \longrightarrow & V_{m,n} \\ f & \mapsto & (X_1 Y_2 - X_2 Y_1) f \end{cases}$$

which is a G-homomorphism. If we view the factor $X_1 Y_2 - X_2 Y_1$ as a determinant

$$\delta = \begin{vmatrix} X_1 & X_2 \\ Y_1 & Y_2 \end{vmatrix},$$

it follows that

$$\delta((X_1, Y_1)g, (X_2, Y_2)g) = \delta(X_1, X_2, Y_1, Y_2) \det(g) = \delta(X_1, X_2, Y_1, Y_2)$$

for $g \in SL_2(\mathbf{C})$. Moreover, it should be intuitively obvious that Δ is injective, but we check this rigorously: if $f \neq 0$, it has degree $d \geqslant 0$ with respect to some variable, say X_1, and then $X_1 Y_2 f$ has degree $d+1$ with respect to X_1, while $X_2 Y_1 f$ remains of degree d, and therefore $X_1 Y_2 f \neq X_2 Y_1 f$.

Now we describe a stable complement to the image of Δ. To justify the solution a little bit, note that $\mathrm{Im}(\Delta)$ only contains polynomials f such that $f(X, Y, X, Y) = 0$. Those for which this property fails must be recovered. We do this by defining W to be the representation generated by the single vector

$$e = X_1^m X_2^n,$$

i.e., the linear span in $V_{m,n}$ of the translates $g \cdot e$. To check that it has the required property, we look at the linear map "evaluating f when both sets of variables are equal" suggested by the remark above, restricted to W. This map is given by

$$T \quad \begin{cases} W \longrightarrow V_{m+n} \\ f \mapsto f(X, Y, X, Y) \end{cases}$$

(since a polynomial of the type $f(X, Y, X, Y)$ with $f \in V_{m,n}$ is homogeneous of degree $m+n$), and we notice that it is an intertwiner with ϱ_{m+n}. Since e maps to X^{m+n} which is nonzero, and ϱ_{m+n} is irreducible (Theorem 2.6.1; although the proof of this will be given only later, the reader will have no problem checking that there is no circularity), Schur's Lemma (Lemma 2.2.6) proves that T is surjective.

We now examine W more closely. Writing $g = \begin{pmatrix} a & b \\ c & d \end{pmatrix}$, we have

$$g \cdot e = (aX_1 + bY_1)^m (aX_2 + bY_2)^n = \sum_{0 \leqslant j \leqslant m+n} a^j b^{m+n-j} \varphi_j(X_1, Y_1, X_2, Y_2)$$

for some $\varphi_j \in V_{m,n}$. We deduce that the space W, spanned by the vectors $g \cdot e$, is contained in the span of the φ_j, and hence that $\dim W \leqslant m + n + 1$. But since $\dim \varrho_{m+n} = m + n + 1$, we must have equality, and in particular T is an isomorphism.

Since $\dim V_{m-1,n-1} + \dim W = mn + m + n + 1 = \dim V_{m,n}$, it only remains to check that $V_{m-1,n-1} \oplus W = V_{m,n}$ to conclude that (2.36) holds. But the intersection $V_{m-1,n-1} \cap W$ is zero, since $f(X, Y, X, Y) = 0$ for $f \in V_{m-1,n-1}$, while $f(X, Y, X, Y) = Tf \neq 0$ for a non-zero $f \in W$.... □

In Corollary 5.6.2 in Chapter 5, we will see that the Clebsch–Gordan formula for the subgroup $\mathrm{SU}_2(\mathbf{C})$ (i.e., seeing each ϱ_m as restricted to $\mathrm{SU}_2(\mathbf{C})$) can be proved—at least at the level of *existence* of an isomorphism!—in a few lines using character theory. However, the proof above has the advantage that it "explains" the decomposition, and can be used to describe concretely the subspaces of $V_m \otimes V_n$ corresponding to the subrepresentations of $\varrho_m \otimes \varrho_n$.

Now, in a slightly different direction, during the late 19th and early 20th centuries, a great amount of work was done on the topic called *invariant theory*, which in the (important) case of the invariants of $\mathrm{SL}_2(\mathbf{C})$ can be described as follows. One considers, for some $m \geqslant 0$, the algebra $S(V_m)$ of all polynomial functions on V_m. The group G acts on $S(V_m)$ according to

$$(g \cdot \phi)(f) = \phi(\varrho_m(g^{-1})f),$$

and hence $S(V_m)$ is also a representation of G (it is infinite dimensional, but splits as a direct sum of the homogeneous components of degree $d \geqslant 0$, which are finite dimensional). Then one tries to understand the subalgebra $S(V_m)^G$ of all G-invariant functions on V_m, in particular, to understand the (finite) dimensions of the homogeneous pieces $S(V_m)^G_d$ of invariant functions of degree d.

For instance, if $m = 2$, so that V_2 is the space of binary quadratic forms, one can write any $f \in V_2$ as

$$f = a_0 X^2 + 2a_1 XY + a_2 Y^2,$$

and then $S(V_2) \simeq \mathbf{C}[a_0, a_1, a_2]$ is the polynomial algebra in these coordinates. One invariant springs to mind: the *discriminant*

$$\Delta(a_0, a_1, a_2) = a_1^2 - a_0 a_2$$

of a binary quadratic form. One can then show that $S(V_2)^G \simeq \mathbf{C}[\Delta]$ is a polynomial algebra in the discriminant. For $m = 3$, with $S(V_3)$ the space of binary cubic forms, with elements

$$f = a_0 X^3 + 3a_1 X^2 Y + 3a_2 XY^2 + a_3 Y^3 \in S(V_3),$$

one can prove that $S(V_3)^G \simeq \mathbf{C}[\Delta_3]$, where

$$\Delta_3 = a_0^2 a_3^2 - 6a_0 a_1 a_2 a_3 + 4a_0 a_2^3 - 3a_1^2 a_2^2 + 4a_1^3 a_3$$

(see, e.g., [**59**, 3.4.2 and 3.4.3]). We note that this reference computes the larger rings C_2 and C_3 of the *covariants* of quadratic and cubic binary forms, which have a double grading by integers $e \geqslant 0$ and $i \geqslant 0$, and that the invariants $S(V_2)^G$ and $S(V_3)^G$ are the $i = 0$ subrings of C_2 and C_3, respectively.

The search for explicit descriptions of the invariant spaces $S(V_m)^G$—and similar questions for other linear actions of groups such as $\mathrm{SL}_m(\mathbf{C})$ acting on homogeneous polynomials in more variables—was one of main topics of the classical theory of invariants, which was extremely popular during the 19th century (see, e.g., [**59**, Ch. 3] for a modern presentation). These questions are very hard if one wishes to give concrete answers. Currently, explicit generators of $S(V_m)^G$ (as an algebra) seem to be known only for $m \leqslant 10$. For $m = 9$, one needs 92 invariants to generate $S(V_m)^G$ as an algebra (see [**10**]); these generators are *not* algebraically independent.

2.6.2. Permutation representations. At the origin of group theory, a group G was often seen as a *permutation group*, or in other words, as a subgroup of the group $\mathfrak{S}_X$ of all bijections of some set X (often finite). Indeed, any group G can be identified with a subgroup of $\mathfrak{S}_G$ by mapping $g \in G$ to the permutation $h \mapsto gh$ of the underlying set G (i.e., mapping g to the g-th row of the "multiplication table" of the group law on G).

More generally, one may consider any action of G on a set X, i.e., any homomorphism

$$\begin{cases} G \longrightarrow \mathfrak{S}_X \\ g \mapsto (x \mapsto g \cdot x) \end{cases}$$

as a permutation group analogue of a linear representation. Such actions, even if X is not a vector space, are often very useful means of investigating the properties of a group. There is always an associated linear representation which encapsulates the action by *linearizing* it: given any field k, denote by $k(X)$ the k-vector space generated by basis vectors e_x indexed by the elements of the set X, and define

$$\varrho : G \longrightarrow \mathrm{GL}(k(X))$$

by linearity using the rule

$$\varrho(g)e_x = e_{g\cdot x},$$

which exploits the action of G on X. Since $g \cdot (h \cdot x) = (gh) \cdot x$ (the crucial defining condition for an action!), we see that ϱ is, indeed, a representation of G, which is called the *permutation representation* associated to the action of G on X. It has dimension $\dim \varrho = |X|$, by construction.

Example 2.6.4. (1) As the choice of notation suggests, the representation π_G of G on the space $k(G)$ spanned by G, defined in (2.3), is simply the permutation representation associated to the left-action of G on itself by multiplication.

(2) If $H \subset G$ is a subgroup of G with finite index and $X = G/H$ is the finite set of right cosets of G modulo H with the action given by

$$g \cdot (xH) = gxH \in G/H,$$

the corresponding permutation representation ϱ is isomorphic to the induced representation

$$\mathrm{Ind}_H^G(\mathbf{1}).$$

Indeed, the space for this induced representation is given by

$$F = \{f : G \longrightarrow k \mid f(hg) = f(g) \text{ for all } h \in H\},$$

with the action of G given by the regular representation. This space has a basis given by the functions f_x which are the characteristic functions of the *left* cosets Hx. Moreover,

$$g \cdot f_x = f_{xg^{-1}}$$

(the left-hand side is non-zero at those y where $yg \in Hx$, i.e., $y \in Hxg^{-1}$), which means that mapping

$$f_x \mapsto e_{x^{-1}}$$

gives a linear isomorphism $F \longrightarrow k(X)$, which is now an intertwiner.

A feature of all permutation representations is that they are never irreducible if X is finite and $|X| \neq 1$: the element

$$\sum_{x\in X} e_x \in k(X)$$

is an invariant vector.

Exercise 2.6.5. (1) Let ϱ be the permutation representation over a field k associated to the action on G/H, for $H \subset G$ of finite index. Show that ϱ^G is spanned by this invariant vector, and explain how to recover it as the image of an explicit element

$$\Phi \in \mathrm{Hom}_G(\mathbf{1}, \varrho)$$

constructed using Frobenius reciprocity.

(2) Let X be any finite set with an action of G, and let ϱ be the associated permutation representation over k. Show that $\dim \varrho^G$ is equal to the number of orbits of the action of G on X.

Exercise 2.6.6. Let $G = \mathfrak{S}_3$, and consider the permutation representation of dimension 3 associated to the natural action of G on $X = \{1, 2, 3\}$.

(1) Show that the subspace

$$E = \{v = (x, y, z) \in \mathbf{C}^3 \mid x + y + z = 0\}$$

is a stable complement of the line spanned by the invariant vector $(1, 1, 1) \in \mathbf{C}^3$ and that the representation ϱ of G on E is irreducible.

(2) Show that $M = \mathbf{Z}^3 \subset \mathbf{C}^3$ and $N = E \cap M$ are stable lattices for their respective representations. Prove that the reduction of ϱ modulo 3 (i.e., the representation of G over the field $\mathbf{Z}/3\mathbf{Z}$ induced from $N/3N$, as in the end of Section 2.4 and Exercise 2.4.3) is reducible and not semisimple.

2.6.3. Generators and relations. From an abstract point of view, one may try to describe representations of a group G by writing down a presentation of G, i.e., a set $\underline{g} \subset G$ of generators, together with the set $\underline{r}$ describing all relations between the elements of $\underline{g}$, relations being seen as (finite) words involving the $g \in \underline{g}$, a situation which one summarizes by writing

$$G \simeq \langle \underline{g} \mid \underline{r} \rangle$$

(the relations are complete in the sense that any relation between the generators is a product of *conjugates* of some of the given words.)

Then one can see that for a given field k and dimension $d \geqslant 1$, it is equivalent to give a d-dimensional (matrix) representation

$$G \longrightarrow \mathrm{GL}_d(k)$$

or to give a family

$$(x_g)_{g \in \underline{g}}$$

of invertible matrices in $\mathrm{GL}_d(k)$, such that "all relations in $\underline{r}$ hold", i.e., if a given $r \in \underline{r}$ is given by a word

$$r = g_1 \cdots g_\ell$$

(with g_i in the free group generated by $\underline{g}$), we should ensure that

$$x_{g_1} \cdots x_{g_\ell} = 1$$

in the matrix group $\mathrm{GL}_d(k)$.

This description is usually not very useful for practical purposes if the group G is given, since it is often the case that there is no particularly natural choice of generators and relations to use, and since furthermore it might be very difficult to determine when representations built in this manner are isomorphic or not.

One very interesting well-known case of the use of generators and relations to define a representation is the construction of the *Weil representation* of $SL_2(F)$, where F is a field which is either a finite field or $\mathbf{R}$. We refer to [**41**, Ch. XI] (for $F = \mathbf{R}$) or to [**11**, Prop. 4.1.3] (for finite fields) for full details, as well as to Exercise 4.6.21.

One can also make use of this approach to provide examples of groups with "a lot" of representations. Indeed, if G is a group where there are no relations at all between a set $\underline{g}$ of generators (i.e., a free group), it is equivalent to give a homomorphism $G \longrightarrow GL(E)$ as to give elements x_g in $GL(E)$ indexed by the generators $g \in \underline{g}$. Moreover, two such representations given by $x_g \in GL(E)$ and $y_g \in GL(F)$ are isomorphic if and only if these elements are (globally) conjugate, i.e., if there exists a linear isomorphism $\Phi : E \to F$ such that

$$x_g = \Phi^{-1} y_g \Phi$$

for all $g \in \underline{g}$.

Here is a slight variant that makes this even more concrete. Consider the group $G = PSL_2(\mathbf{Z})$ of matrices of size 2 with integral coefficients and determinant 1, modulo the subgroup $\{\pm 1\}$. Then G is not free, but it is known to be generated by the (image modulo $\{\pm 1\}$ of the) two elements

$$g_1 = \begin{pmatrix} 0 & -1 \\ 1 & 0 \end{pmatrix}, \qquad g_2 = \begin{pmatrix} 0 & -1 \\ 1 & 1 \end{pmatrix}$$

in such a way that the only relations between the generators are

$$g_1^2 = 1, \qquad g_2^3 = 1$$

(i.e., G is a *free product* of $\mathbf{Z}/2\mathbf{Z}$ and $\mathbf{Z}/3\mathbf{Z}$; see [**26**, II.A, II.B.28] for a proof and more information).

Hence it is equivalent to give a representation $PSL_2(\mathbf{Z}) \longrightarrow GL(E)$ or to give two elements $x, y \in GL(E)$ such that $x^2 = 1$ and $y^3 = 1$.

Yet another use of generators and relations is in showing that there exist groups for which certain representations *do not exist*. In that case, it is enough to find some abstract presentation where the relations are incompatible with matrix groups. Here is a concrete example:

Theorem 2.6.7 (Higman and Baumslag; an example of a non-linear finitely generated group). *Let G be the group with two generators a, b subject to the* *relation*

$$a^{-1}b^2a = b^3.$$

Then, whatever the field k, there exists no faithful linear representation

$$G \longrightarrow \mathrm{GL}(E),$$

where E is a finite-dimensional k-vector space.

The first example of such a group was constructed by Higman; the example here is due to Baumslag (see [**45**]) and is an example of a family of groups called the Baumslag–Solitar groups which have similar presentations with the exponents 2 and 3 replaced by arbitrary integers.

We will only give a sketch, dependent on some fairly deep facts of group theory.

Sketch of proof. We appeal to the following two results:

- *Malcev's Theorem.* If k is a field and $G \subset \mathrm{GL}_d(k)$ is a finitely generated group, then for any $g \in G$ there exists a finite quotient $G \longrightarrow G/H$ such that g is non-trivial modulo H (one says that G is *residually finite*; for a proof, see [**46**]).
- *The Identitätssatz of Magnus or Britton's Lemma* (see, e.g., [**51**, Th. 11.81]). Let G be a finitely presented group with a *single relation* (a one-relator group). Then one can decide algorithmically if a word in the generators represents or not the identity element of G.

Now we are going to check that G fails to satisfy the conclusion of Malcev's Theorem and, therefore, has no faithful finite-dimensional representation over any field.

To begin with, iterating the single relation leads to

$$a^{-k}b^{2^k}a^k = b^{3^k}$$

for all $k \geqslant 1$. Now assume $G \xrightarrow{\pi} G/H$ is a finite quotient of G, and let $\alpha = \pi(a)$, $\beta = \pi(b)$. Taking k to be the order of α in the finite group G/H, we see that

$$\beta^{2^k - 3^k} = 1,$$

i.e., the order of β divides $2^k - 3^k$. In particular, this order is coprime with 2, and this implies that the map $\gamma \mapsto \gamma^2$ is surjective on the finite cyclic group generated by β. Thus β is a power of β^2. Similarly, after conjugation by a, the element $b_1 = a^{-1}ba$ is such that $\beta_1 = \pi(b_1)$ is a power of β_1^2.

But now we observe that $\beta_1^2 = \pi(a^{-1}b^2a) = \pi(b^3) = \beta^3$. Hence β_1 is a power of β^3, and in particular it commutes with β, so that

$$\pi([b_1, b]) = \beta_1\beta\beta_1^{-1}\beta^{-1} = 1,$$

and this relation is valid in any finite quotient.

Now Britton's Lemma [**51**, Th. 11.81] implies that the word

$$c = [b_1, b] = b_1bb_1^{-1}b^{-1} = a^{-1}baba^{-1}b^{-1}ab^{-1}$$

is non-trivial in G.[15] Thus $c \in G$ is an element which is non-trivial, but becomes so in any finite quotient of G. This is the desired conclusion. □

Remark 2.6.8. Concerning Malcev's Theorem, a good example to have in mind is the following. A group like $\mathrm{SL}_d(\mathbf{Z}) \subset \mathrm{GL}_d(\mathbf{C})$ is finitely generated, and one can check that it satisfies the desired condition simply by using the reduction maps

$$\mathrm{SL}_d(\mathbf{Z}) \longrightarrow \mathrm{SL}_d(\mathbf{Z}/p\mathbf{Z})$$

modulo primes. Indeed, for any fixed $g \in \mathrm{SL}_d(\mathbf{Z})$, if $g \neq 1$, we can find some prime p larger than the absolute values of all coefficients of g, and then g is certainly non-trivial modulo p. The proof of Malcev's Theorem is based on similar ideas (though of course one has to use more general rings than $\mathbf{Z}$).

Note that if one does not insist on finitely generated counterexamples, it is easier to find non-linear groups—for instance, "sufficiently big" abelian groups will work.

2.7. Some general results

In this section, we will prove some of the basic facts about representations. Some of them will, for the first time, require that some restrictions be imposed on the representations, namely, either we consider finite-dimensional representations or the base field k be algebraically closed.

2.7.1. The Jordan–Hölder–Noether Theorem. We first discuss a generalization of the classical Jordan–Hölder Theorem of group theory, which explains in which sense irreducible representations are in fact "building blocks" of all representations, at least for the finite-dimensional case.

Theorem 2.7.1 (Jordan–Hölder–Noether Theorem). *Let G be a group, let k be a field, and let*

$$\varrho : G \longrightarrow \mathrm{GL}(E)$$

be a k-representation of G.

[15] In the language explained in Rotman's book [**51**], G is an HNN extension for $A = 2\mathbf{Z}$, $B = 3\mathbf{Z}$, isomorphic subgroups of $\mathbf{Z} = \langle b\rangle$, with stable letter a. Thus the expression for c contains no "pinch" $a^{-1}b^2a$ or ab^3a^{-1} as a subword, and Britton's Lemma deduces from this that $c \neq 1$.

(1) *If $E \neq 0$ and E is finite dimensional, there exists a finite increasing sequence of subrepresentations*

$$0 = E_0 \subset E_1 \subset E_2 \subset \cdots \subset E_{n-1} \subset E_n = E$$

of E such that, for all i, $1 \leqslant i \leqslant n$, the quotient representations E_i/E_{i-1} are irreducible. Such sequences are called composition series, *and the irreducible representations E_i/E_{i-1} are called the* composition factors.

(2) *If ϱ admits any finite composition series,*[16] *then any two such sequences are equivalent in the following sense: the number of terms are the same, and the irreducible composition factors are isomorphic, up to a permutation. In other words, for any irreducible k-representation π of G, the integer*

$$n_\pi(\varrho) = |\{i \mid E_i/E_{i-1} \simeq \pi\}|$$

is independent of the choice of a composition series (E_i).

The uniqueness part of the statement may be considered, to some extent, as an analogue of the fundamental theorem of arithmetic: a factorization of an integer into prime powers is unique, but only up to permutation of the primes. The integer $n_\pi(\varrho)$ is called *the multiplicity of the composition factor* π in ϱ.

Remark 2.7.2. (1) The result is often simply called the Jordan–Hölder Theorem, but according to H. Weyl [**64**], the extension to representations is due to E. Noether.

(2) By definition, any composition factor of a representation ϱ is isomorphic to a quotient ϱ_1/ϱ_2 for some subrepresentations $\varrho_2 \subset \varrho_1$ of ϱ. More generally, any representation of this type, not necessarily irreducible, is called a *subquotient* of ϱ.

Proof. The existence part (1) is easy, by dimension arguments. Since $E \neq 0$, we can select an irreducible subrepresentation E_1 (for instance, a subrepresentation of minimal non-zero dimension), then—if $E_1 \neq E$—a subrepresentation $E_2 \supsetneq E_1$ of minimal dimension, etc. For dimension reasons, each quotient E_i/E_{i-1} is then irreducible, and the process terminates in finitely many steps because $\dim(E) < +\infty$.

The uniqueness is more important. Assume that we have two sequences

$$0 = E_0 \subset E_1 \subset E_2 \subset \cdots \subset E_{n-1} \subset E_n = E,$$
$$0 = F_0 \subset F_1 \subset F_2 \subset \cdots \subset F_{m-1} \subset F_m = E$$

with irreducible quotients F_j/F_{j-1} and E_i/E_{i-1}. We proceed to use the second one to insert (apparent) steps between the successive subspaces of

[16] This may happen even if $\dim(E)$ is infinite!

the first sequence, and vice versa. Precisely, for $0 \leqslant i \leqslant n-1$, let

$$E_{i,j} = E_i + (E_{i+1} \cap F_j), \qquad 0 \leqslant j \leqslant m,$$

and for $0 \leqslant j \leqslant m-1$, let

$$F_{j,i} = F_j + (F_{j+1} \cap E_i), \qquad 0 \leqslant i \leqslant n.$$

Then we have, e.g,

$$E_i = E_{i,0} \subset E_{i,1} \subset \cdots \subset E_{i,m-1} \subset E_{i,m} = E_{i+1},$$

and a similar refinement between F_j and F_{j+1}.

By construction, each $E_{i,j}$ and $F_{i,j}$ is a subrepresentation of E. Now, for each i, $0 \leqslant i \leqslant n-1$, observe that since there is no proper intermediate subrepresentation between E_i and E_{i+1} (this would contradict the fact that E_{i+1}/E_i is irreducible), there exists a *unique* index j, $0 \leqslant j \leqslant m-1$, for which

$$E_{i,j} = E_i, \qquad E_{i,j+1} = E_{i+1},$$

hence with

$$E_{i+1}/E_i = (E_i + (E_{i+1} \cap F_{j+1}))/(E_i + (E_{i+1} \cap F_j)).$$

There is a certain symmetry in this between i and j; in fact, by a standard isomorphism theorem, there is a canonical isomorphism

$$\begin{aligned}(2.37)\quad (E_i + (E_{i+1} \cap F_{j+1}))/(E_i + (E_{i+1} \cap F_j))\\ \simeq (F_j + (E_{i+1} \cap F_{j+1}))/(F_j + (E_i \cap F_{j+1})).\end{aligned}$$

(see below for a reminder on this).

The right-hand side is none other than $F_{j,i+1}/F_{j,i}$. The latter is therefore non-zero, but for the same reason as before, there is a single step of the interpolated sequence between F_j and F_{j+1} that can be non-zero, and it must satisfy

$$F_{j,i+1}/F_{j,i} \simeq F_{j+1}/F_j.$$

In other words, for each successive irreducible quotient of the *first* sequence, we have associated (in a unique way) a well-defined irreducible quotient of the second sequence. This gives a map

$$\begin{cases} \{1, \ldots, n\} & \longrightarrow & \{1, \ldots, m\} \\ \quad i & \mapsto & j, \end{cases}$$

which is injective, because j is characterized by the isomorphism

$$F_{j+1}/F_j \simeq F_{j,i+1}/F_{j,i},$$

which holds for a single index i.

Reversing the role of the two sequences, we obtain the equality $n = m$, and then a bijection between the irreducible quotients in the first sequence and those of the second. □

Remark 2.7.3 (About the standard isomorphism). The isomorphism (2.37) can be expressed as

$$(E + (\tilde{E} \cap \tilde{F}))/(E + (\tilde{E} \cap F)) \simeq (F + (\tilde{E} \cap \tilde{F}))/(F + (E \cap \tilde{F}))$$

for subrepresentations $E \subset \tilde{E}$, $F \subset \tilde{F}$ of some ambient space. This is induced simply by quotienting the reciprocal linear maps

$$e + \tilde{g} \mapsto \tilde{g},$$
$$f + \tilde{g} \mapsto \tilde{g}$$

for $e \in E$, $f \in F$, and $\tilde{g} \in \tilde{E} \cap \tilde{F}$. Indeed, formally at least, these maps are inverse to each other: losing e from left to right is no problem because the E-component is zero modulo $E + (\tilde{E} \cap F)$ anyway, and similarly for losing f. Thus the only thing one must check is that the maps are well defined, since once it is done, we see equally well that these isomorphisms are intertwining operators.

To check that the maps are well defined, it is enough to deal with the first one, because of the symmetry. First of all, the map

$$\Phi : E + (\tilde{E} \cap \tilde{F}) \longrightarrow (F + (\tilde{E} \cap \tilde{F}))/(F + (E \cap \tilde{F}))$$

mapping $e + \tilde{g}$ to $\tilde{g}$ is well defined because if

$$e_1 + \tilde{g}_1 = e_2 + \tilde{g}_2$$

(with obvious notation), we get

$$\tilde{g}_1 - \tilde{g}_2 = e_2 - e_1 \in E \cap (\tilde{E} \cap \tilde{F}) = E \cap \tilde{F},$$

and hence $\tilde{g}_1 - \tilde{g}_2$ maps to zero in the right-hand quotient modulo $F + (E \cap \tilde{F})$. It is then enough to check that $E + (\tilde{E} \cap F) \subset \mathrm{Ker}(\Phi)$ to see that Φ induces the linear map we want. But for an element of the type $e + g$ with $g \in \tilde{E} \cap F$, Φ maps it to g modulo $F + (\tilde{F} \cap E)$, which is 0 since $g \in F$!

Example 2.7.4. (1) Let ϱ be semisimple and finite dimensional, say

$$\varrho \simeq \varrho_1 \oplus \varrho_2 \oplus \cdots \oplus \varrho_n,$$

with each ϱ_i irreducible. Then a composition series of ϱ is provided by

$$E_0 = 0, \quad E_i = \varrho_1 \oplus \cdots \oplus \varrho_i, \quad 1 \leqslant i \leqslant n,$$

with $E_i/E_{i-1} \simeq \varrho_i$. It is clear that if we permute the labels i of the ϱ_i, this does not change ϱ, but the sequence changes; however, the quotients are indeed merely permuted.

In that situation, $n_\pi(\varrho)$ is the number of components ϱ_i which are isomorphic to π, and it is called the *multiplicity of π in ϱ*.

(2) Consider $k = \mathbf{C}$ and the group

$$G = \left\{ \begin{pmatrix} a & b \\ 0 & 1 \end{pmatrix} \mid a \in \mathbf{C}^\times, \quad b \in \mathbf{C} \right\}$$

with its two-dimensional representation given by the inclusion in $\mathrm{GL}_2(\mathbf{C}) = \mathrm{GL}(\mathbf{C}^2)$. This representation is not semisimple. With (e_1, e_2) the canonical basis of $\mathbf{C}^2$, one can take $E_1 = \mathbf{C}e_1$, $E_2 = \mathbf{C}^2$; indeed, E_1 is a subrepresentation because

$$\begin{pmatrix} a & b \\ 0 & 1 \end{pmatrix} e_1 = ae_1,$$

and $E_1/E_0 = E_1$ and E_2/E_1 are one dimensional, hence irreducible. In abstract terms, E_1 is the representation

$$\begin{pmatrix} a & b \\ 0 & 1 \end{pmatrix} \mapsto a \in \mathbf{C}^\times = \mathrm{GL}_1(\mathbf{C})$$

while E_2/E_1 is in fact the trivial representation of G.

Exercise 2.7.5. Find the composition factors of the representation of $\mathfrak{S}_3$ over the field $\mathbf{Z}/3\mathbf{Z}$ arising in Exercise 2.6.6(2).

Exercise 2.7.6. Let k be a field, let G be a group, and let ϱ_1, ϱ_2 be two irreducible k-representations of G acting on E_1 and E_2, respectively. Show that a proper subrepresentation π of $\varrho_1 \oplus \varrho_2$ such that the two projections $\pi \longrightarrow \varrho_1$ and $\pi \longrightarrow \varrho_2$ are surjective is the graph of an isomorphism $\varrho_1 \longrightarrow \varrho_2$.

Determining the Jordan–Hölder–Noether irreducible composition factor of a representation ϱ can be delicate. At least the following holds:

Lemma 2.7.7. *Let k be a field, let G be a group, and let ϱ be a finite-dimensional k-representation of G. If ϱ_1 is a finite-dimensional irreducible representation of G and*

$$\mathrm{Hom}_G(\varrho, \varrho_1) \neq 0 \qquad \text{or} \qquad \mathrm{Hom}_G(\varrho_1, \varrho) \neq 0,$$

then ϱ_1 is among the Jordan–Hölder–Noether composition factors of ϱ.

Proof. Both are similar and very intuitive, so we consider here only the case of a non-zero intertwiner $\varrho \xrightarrow{\Phi} \varrho_1$. Let E be the space on which ϱ acts. Because the image of Φ is a non-zero subrepresentation of ϱ_1, which is irreducible, we see that Φ is surjective. Thus $F = \mathrm{Ker}(\Phi) \subset E$ is a proper subrepresentation with

$$E/F \simeq \varrho_1.$$

Considering a composition series of F, say

$$0 = E_0 \subset E_1 \subset \cdots \subset E_n = F,$$

and defining $E_{n+1} = E$, we obtain a composition series $(E_i)_{0\leqslant i\leqslant n+1}$ of E, in which ϱ_1 is one of the composition factors. By uniqueness, this means ϱ_1 is indeed one of the composition factors of ϱ. □

Example 2.7.8. If a representation $\varrho : G \longrightarrow \mathrm{GL}(E)$ has, for a given basis, a matrix representation which is block-triangular

$$\varrho^{m}(g) = \left(\begin{array}{c|c|c|c} \varrho_1^{m}(g) & \star & \ldots & \star \\ \hline 0 & \varrho_2^{m}(g) & \star & \ldots \\ \hline \vdots & \vdots & & \vdots \\ \hline 0 & 0 & \ldots & \varrho_n^{m}(g) \end{array}\right)$$

with square blocks of size $d_1, \ldots, d_n$ on the diagonal, then the ϱ_i^{m} are matrix representations of the composition factors of ϱ. Indeed, multiplication shows that $g \mapsto \varrho_i^{m}(g)$ is a homomorphism to $\mathrm{GL}_{d_i}(k)$, and the subspaces E_i can be defined as those spanned by the

$$d_1 + \cdots + d_i$$

first basis vectors; as in (2.33), one sees that ϱ acts on E_i/E_{i-1} in the basis formed of the vectors in the i-th block of the given one, like the matrix representation ϱ_i^{m}.

If ϱ is semisimple, we can find a decomposition as above with *block-diagonal* matrices, and indeed, a block-diagonal decomposition (with irreducible blocks) corresponds to a semisimple representation. However, if the decomposition turns out to be merely block triangular (with some non-zero off-diagonal blocks), this does *not* mean that the representation is not semisimple! It might just be that the choice of basis was not the best possible.

Here is an example: Consider $G = \mathfrak{S}_3$ and the representation

$$\mathfrak{S}_3 \longrightarrow \mathrm{GL}(\mathbf{C}^3)$$

by permutation of the coordinates. The subspace

$$F = \{(x, y, z) \in \mathbf{C}^3 \mid x + y + z = 0\}$$

is a subrepresentation. In terms of the basis $(1, -1, 0)$, $(1, 0, -1)$, $(1, 0, 0)$ of $\mathbf{C}^3$, we see for instance that the action of the cycle (123) is

$$\begin{pmatrix} 0 & 1 & 0 \\ 1 & -1 & -1 \\ 0 & 0 & 1 \end{pmatrix},$$

where the third column shows that it is not block diagonal. This column is given by

$$\varrho((123))(1,0,0) = (0,0,1) = (1,0,0) - (1,0,-1).$$

However, the representation is semisimple here (taking the third basis vector $(1,1,1)$ will lead to a block-diagonal decomposition).

If

$$\varrho : G \longrightarrow \mathrm{GL}(E)$$

is a semisimple representation, it is natural to ask to classify its irreducible subspaces and, more generally, to determine what its subrepresentations look like. Here again there are possible traps. If

$$E = E_1 \oplus \cdots \oplus E_n \tag{2.38}$$

is *a* decomposition into irreducible subspaces, then we know that the isomorphism classes of the E_i, and their multiplicities, are determined by ϱ, up to isomorphism. The actual subspaces E_i, in general, are *not* determined; a related fact is that there are usually many more subrepresentations of E than the obvious ones of the form

$$F = \bigoplus_{i \in S} E_i$$

for some subset $S \subset \{1, \ldots, n\}$. Indeed, this is already clear in the case of the trivial group G, where any representation is semisimple, and a decomposition (2.38) is obtained by any choice of a basis of E: if $\dim(E) \geqslant 2$, the vector space E contains many more subspaces than those spanned by finitely many *axes* spanned by basis vectors.

A weaker uniqueness is still valid: in any decomposition (2.38), the direct sum $\mathrm{M}(\pi)$ of *all* subspaces E_i on which the action of G is isomorphic to a given irreducible representation π, is independent of the decomposition.

Proposition 2.7.9 (Isotypic components). *Let G be a group, and let k be a field. Let $\varrho : G \longrightarrow \mathrm{GL}(E)$ be a semisimple k-representation of G.*

(1) *Fix an irreducible k-representation π of G. Let*

$$E = \bigoplus_{i \in I} E_i \tag{2.39}$$

be a decomposition of E into irreducible subrepresentations. Then the subspace of E defined by

$$\bigoplus_{E_i \simeq \pi} E_i \subset E,$$

where i runs over all indices such that the subrepresentation E_i is isomorphic to π, is independent *of the decomposition. Indeed, it is equal to the* sum *of all subrepresentations of E isomorphic to π.*

This space, which is the largest subrepresentation of ϱ which is π-isotypic, is called the π-isotypic component *of E and is denoted* $\mathrm{M}(\pi)$ *or* $\mathrm{M}_E(\pi)$.

(2) *In particular, if all irreducible components ϱ_i of ϱ occur with multiplicity 1, the corresponding subspaces $E_i \subset E$ isomorphic to ϱ_i are unique, and any subrepresentation of E is equal to*

$$\bigoplus_{i \in S} E_i$$

for some subset $S \subset I$.

(3) *If ϱ_1, ϱ_2 are semisimple k-representations of G, acting on E_1 and E_2, respectively, and if $\Phi : E_1 \to E_2$ is a G-homomorphism, then the restriction of Φ to the isotypic component $\mathrm{M}_{E_1}(\pi)$ is a linear map*

$$\mathrm{M}_{E_1}(\pi) \longrightarrow \mathrm{M}_{E_2}(\pi),$$

i.e., the image of $\mathrm{M}_{E_1}(\pi)$ is contained in $\mathrm{M}_{E_2}(\pi)$.

In Proposition 2.7.21 and Exercise 2.7.23, we will find a formula for the π-isotypic component of π in ϱ, using a suitable tensor product.

Proof. In order to prove (1), we first denote by $\mathrm{M}(\pi)$ the sum (not necessarily direct, of course) of the subrepresentations of E which are isomorphic to π. This is a well-defined subspace of E, and it is clear that, for any decomposition (2.39), we have

$$\bigoplus_{E_i \simeq \pi} E_i \subset \mathrm{M}(\pi). \tag{2.40}$$

To finish proving assertion (1), it is therefore enough to show that the converse inclusion holds. But if $F \subset E$ is a subrepresentation isomorphic to π, and j is such that the representation on E_j is *not* isomorphic to π, the projection map

$$p_j : F \longrightarrow E_j$$

(defined using (2.39)) is in $\mathrm{Hom}_G(F, E_j)$, and it is therefore zero by Schur's Lemma 2.2.6. Thus the component along E_j of any vector in F is zero. This means precisely that F is a subspace of the left-hand side of (2.40). From the definition of $\mathrm{M}(\pi)$ and the fact that F was arbitrary, we get (1).

The first part of (2) follows from (1), since in the absence of multiplicity $\geqslant 2$, the isotypic components are reduced to a single E_i in the decomposition (2.39). And if $F \subset E$ is a subrepresentation, we know (Lemma 2.2.10) that F is also semisimple, and then any of its own irreducible subspaces is an irreducible subspace of E, and hence is equal to some E_i. Thus F becomes equal to the direct sum of those subspaces E_i which are in F.

Finally, (3) is due to the fact that there exists (from (1)) an "intrinsic" definition of $\mathrm{M}_{E_1}(\pi)$ (i.e., one that requires no choice to construct it). This definition must naturally be "transported" under an intertwining map to E_2. Precisely, $\mathrm{M}_{E_1}(\pi)$ is generated by the vectors v which belong to the image of some homomorphism $\Psi : \pi \longrightarrow \varrho_1$. For any such map, the composite

$$\Phi \circ \Psi : \pi \longrightarrow \varrho_2$$

has image in $\mathrm{M}_{E_2}(\pi)$, for the same reason. Hence $\mathrm{M}_{E_2}(\pi)$ contains the image under Φ of generators of $\mathrm{M}_{E_1}(\pi)$. This means that the π-isotypic component of E_2 contains the image of $\mathrm{M}_{E_1}(\pi)$, which is statement (3).[17] □

Example 2.7.10 (Isotypic components for the trivial and one-dimensional representations)**.** The simplest example concerns the trivial representation $\mathbf{1}_G$. This is always irreducible, and for any representation $\varrho : G \longrightarrow \mathrm{GL}(E)$, we have

$$\mathrm{M}_{\varrho}(\mathbf{1}) = \varrho^G,$$

the subspace of invariant vectors. This is something that was stated, without using the same words, in Example 2.1.9.

For instance, if ϱ is the regular representation of G, we find (from the same example) that $\mathrm{M}_{\varrho_G}(\mathbf{1})$ is the one-dimensional subspace of *constant* k-valued functions on G.

More generally, if $\dim(\pi) = 1$, so that it is automatically irreducible, and $\pi(g) \in k^\times$ is just a scalar, we have

$$\mathrm{M}_{\varrho}(\pi) = \{v \in E \mid \varrho(g)v = \pi(g)v \text{ for all } g \in G\},$$

the space of vectors in E that *transform like* π.

Applied to the regular representation, the reader will easily check that $\mathrm{M}_{\varrho_G}(\pi)$ is again one dimensional and is generated by π itself, seen as a k-valued function.

Example 2.7.11. Here is an application of these ideas, leading to the proof of Theorem 2.6.1. We consider the representation ϱ_m of $G = \mathrm{SL}_2(\mathbf{C})$ on the space V_m of homogeneous polynomials of degree m in $\mathbf{C}[X, Y]$ (see Section 2.6.1).

Proposition 2.7.12. *For each $m \geqslant 0$, the representation ϱ_m of $\mathrm{SL}_2(\mathbf{C})$ is irreducible.*

Proof. We use a fairly common strategy, which we will see again later, to attempt to analyze ϱ_m (which, at this point, we do not even know to be semisimple). We first consider its restriction to some subgroup of G for

[17] The underlying principle of compatability of intrinsic subspaces (or other structures) under homomorphisms is a very useful general fact of algebra.

which we understand well (or better, at least) the representation theory. Here we consider

$$T = \left\{ t(\lambda) = \begin{pmatrix} \lambda & 0 \\ 0 & \lambda^{-1} \end{pmatrix} \mid \lambda \in \mathbf{C}^{\times} \right\} \simeq \mathbf{C}^{\times}$$

(a choice justified partly by the fact that T is abelian). We can see easily how T acts on the basis vectors $e_i = X^i Y^{m-i}$, $0 \leqslant i \leqslant m$: for $\lambda \in \mathbf{C}^{\times}$, we have by definition

$$\varrho_m(t(\lambda))e_i = (\lambda X)^i (\lambda^{-1} Y)^{m-i} = \lambda^{2i-m} e_i.$$

This means that the lines $\mathbf{C}e_i$ are all stable under the action of T, and that $\operatorname{Res}_T^G \varrho_m$ acts on $\mathbf{C}e_i$ according to the representation

$$\chi_{2i-m} \begin{cases} T & \longrightarrow & \mathrm{GL}_1(\mathbf{C}) \\ t(\lambda) & \mapsto & \lambda^{2i-m}. \end{cases}$$

Since (e_i) is a basis of V_m, this means that we have proved that

$$\operatorname{Res}_T^G(\varrho_m) \simeq \bigoplus_{0 \leqslant i \leqslant m} \chi_{2i-m} = \chi_{-m} \oplus \chi_{-m+2} \oplus \cdots \oplus \chi_m. \tag{2.41}$$

Thus, $\operatorname{Res}_T^G(\varrho_m)$ is semisimple, and its irreducible components (the χ_{2i-m}, which are irreducible since they are one dimensional) occur with multiplicity 1.

Now consider any non-zero G-stable subspace $F \subset V_m$. It is also a subrepresentation of the restriction of ϱ_m to T, obviously, and from what we just observed, Proposition 2.7.9(2), implies that the subspace F is a direct sum of some of the lines $\mathbf{C}e_i$ corresponding to the representations of T occurring in F. Thus, there exists some non-empty subset

$$I \subset \{0, \ldots, m\}$$

such that

$$F = \bigoplus_{i \in I} \mathbf{C}e_i. \tag{2.42}$$

Now we *bootstrap* this information using the action of other elements of G than those in T. Namely, fix some $i \in I$, and consider the action of

$$u = \begin{pmatrix} 1 & 1 \\ 0 & 1 \end{pmatrix} \tag{2.43}$$

(a *unipotent* element); since F is a stable under G, we know that $\varrho_m(u)e_i \in F$, and this means

$$X^i(X+Y)^{m-i} \in F.$$

Expanding by the binomial theorem, we get

$$X^m + (m-i)X^{m-1}Y + \cdots + (m-i)X^iY^{m-i-1} + X^iY^{m-i} = \sum_{j=i}^{m} \binom{m-i}{j-i} e_j \in F,$$

and comparison with (2.42) leads to the conclusion that all $j \geqslant i$ are also in I. Similarly, considering the action of

$$\begin{pmatrix} 1 & 0 \\ 1 & 1 \end{pmatrix},$$

we conclude that $j \in I$ if $j \leqslant i$. Hence, we must have $I = \{0, \ldots, m\}$, which means that $F = V_m$. This shows that V_m is irreducible. □

The argument in this proof is related to the *highest weight theory* of representations of $\mathrm{SL}_2(\mathbf{C})$ and of similar groups like $\mathrm{SL}_n(\mathbf{C})$ for all $n \geqslant 2$. We will see other hints of this theory (e.g., in Section 3.2), and it is described in detail in many books; see, for instance, [**20**] or [**35**].

Exercise 2.7.13 (Irreducibility of ϱ_m restricted to a smaller group)**.** Consider again the representation ϱ_m of $\mathrm{SL}_2(\mathbf{C})$ for $m \geqslant 0$. We restrict it now to the subgroup $\mathrm{SU}_2(\mathbf{C})$ of unitary matrices of size 2. The proof of irreducibility of ϱ_m in the previous example used the element (2.43) and its transpose, which do not belong to $\mathrm{SU}_2(\mathbf{C})$. However, ϱ_m restricted to $\mathrm{SU}_2(\mathbf{C})$ is still irreducible, as claimed in Theorem 2.6.1. Of course, this gives another proof of the irreducibility of ϱ_m as a representation of $\mathrm{SL}_2(\mathbf{C})$.

(1) Show that the decomposition (2.42) still holds with I not empty for a non-zero subspace F stable under $\mathrm{SU}_2(\mathbf{C})$.

(2) Let j be such that $e_j = X^jY^{n-j}$ is in F. Show that for

$$r(\theta) = \begin{pmatrix} \cos\theta & \sin\theta \\ -\sin\theta & \cos\theta \end{pmatrix} \in \mathrm{SU}_2(\mathbf{C}), \qquad \theta \in \mathbf{R},$$

we have

$$\varrho_m(r(\theta))e_j = \sum_{0 \leqslant i \leqslant m} f_i(\theta)e_i,$$

where the f_i are functions on $[0, 2\pi]$ which are not identically zero. Deduce that $F = V_m$.

Exercise 2.7.14 (Another example of an irreducible representation)**.** The following example will be used in Chapter 6. We consider $k = \mathbf{C}$, and we let $V = \mathbf{C}^n$ for $n \geqslant 1$ and $E = \mathrm{End}(V)$. The group $G = \mathrm{GL}_n(\mathbf{C})$ acts on V (by matrix multiplication!) and therefore there is an associated representation on E, as in (2.15) (see also Remark 2.2.22).

(1) Show that this representation on E is the conjugation action

$$g \cdot A = gAg^{-1}$$

and that the space E_0 of endomorphisms $A \in E$ with trace 0 is a subrepresentation of E. (This representation is also called the *adjoint representation* of $\mathrm{SL}_n(\mathbf{C})$.)

We will now prove that E_0 is an irreducible representation of G and, in fact, that it is already an irreducible representation of the subgroup $\mathrm{SU}_n(\mathbf{C})$ of unitary matrices with determinant 1.

(2) Let $T \subset \mathrm{SU}_n(\mathbf{C})$ be the diagonal subgroup of $\mathrm{SU}_n(\mathbf{C})$. Show that the restriction of E_0 to T decomposes as the direct sum of T-subrepresentations

$$E_0 = H \oplus \bigoplus_{i \neq j} \mathbf{C}E_{i,j},$$

where H is the subspace of diagonal matrices in E_0 and $E_{i,j} \in E_0$ is, for $1 \leqslant i \neq j \leqslant n$, the rank 1 matrix with a single coefficient equal to 1 on the (i,j)-th entry. Moreover show that the subspaces $\mathbf{C}E_{i,j}$ each carry distinct non-trivial irreducible representations of T and that $H = (E_0)^T$ is the space of T-invariants.

(3) Let $F \subset E_0$ be a non-zero subspace stable under $\mathrm{SU}_n(\mathbf{C})$. Show that F cannot be contained in H.

(4) Deduce that F contains all $E_{i,j}$ for $i \neq j$. Then conclude that $F = E_0$. (*Hint*: Show that suitable combinations of vectors generating H are $\mathrm{SU}_n(\mathbf{C})$-conjugate of combinations of some $E_{i,j}$, $i \neq j$.)

2.7.2. Schur's Lemma. The next result is fundamental. It is usually called *Schur's Lemma*, but it was known and used by others such as Frobenius and Burnside, independently of Schur. In fact, it is a refinement of Lemma 2.2.6; the version we state is valid for finite-dimensional representations, but there are variants when infinite-dimensional representations are considered with some topological restrictions (see for instance Proposition 3.4.17).

Proposition 2.7.15 (Schur's Lemma, version II). *Let G be a group, and let k be an algebraically closed field, for instance $k = \mathbf{C}$.*

(1) *If π_1 and π_2 are irreducible k-representations of G which are non-isomorphic, we have*

$$\mathrm{Hom}_G(\pi_1, \pi_2) = 0.$$

(2) *If π_1 and π_2 are isomorphic finite-dimensional irreducible k-representations of G, then*

$$\dim \mathrm{Hom}_G(\pi_1, \pi_2) = 1,$$

and in fact if π is an irreducible k-representation of G of finite dimension, *we have*

$$\mathrm{Hom}_G(\pi, \pi) = \mathrm{Hom}_k(\pi, \pi)^G = k\mathrm{Id}_\pi.$$

(3) *Conversely, if* π *is a finite-dimensional,* semisimple k*-representation of* G *such that* $\dim \mathrm{Hom}_G(\pi,\pi) = 1$*, it follows that* π *is irreducible.*

Note that here we used here the natural representation of G on homomorphism spaces, in which the G-homomorphisms are the G-invariants. The statement gives a very strong expression of the fact that non-isomorphic irreducible representations of G are independent of each other; it is frequently used in the form of the formula

$$\dim \mathrm{Hom}_G(\pi_1,\pi_2) = \delta(\pi_1,\pi_2) = \begin{cases} 1 & \text{if } \pi_1 \simeq \pi_2 \\ 0 & \text{otherwise,} \end{cases} \tag{2.44}$$

for irreducible finite-dimensional representations of a group G over an algebraically closed field.

We will see other incarnations of this independence later (e.g., Theorem 2.7.28).

Proof. The first part is a consequence of the earlier version of Schur's Lemma (Lemma 2.2.6). For the second, it is enough to consider the case in which $\pi_1 = \pi_2$ (since $\mathrm{Hom}_G(\pi_1,\pi_2) \simeq \mathrm{Hom}_G(\pi_1,\pi_1)$ if $\pi_2 \simeq \pi_1$). Denote $\pi = \pi_1$, and consider a G-homomorphism Φ from π to itself. The fact that k is algebraically closed and π is finite dimensional implies that Φ, as a linear map, has an eigenvalue, say $\lambda \in k$. But $\Phi - \lambda\mathrm{Id}$ is then a G-homomorphism of π which is *not injective*. By Lemma 2.2.6, the only possibility is that $\Phi - \lambda\mathrm{Id}$ be identically zero, which is the desired conclusion.

Finally we prove the converse when π is semisimple. Let E be the k-vector space on which π acts. We can assume that $\dim E \geqslant 1$, and then we let $F \subset E$ be an irreducible subrepresentation and F_1 a complementary subrepresentation, so that $E = F \oplus F_1$. The projection $\Phi : E \longrightarrow E$ onto F with kernel F_1 is an element of $\mathrm{Hom}_G(\pi,\pi)$ (we saw this explicitly in Lemma 2.2.4), and our assumption on π implies that it is a multiple of the identity. Since it is non-zero, it is therefore equal to the identity, which means that $F_1 = 0$ and $E = F$ is irreducible. □

Exercise 2.7.16 (Schur's Lemma and semisimplicity)**.** The last statement in Schur's Lemma can be a very useful irreducibility criterion; however, one should not forget the semisimplicity condition! Consider the representation ϱ of

$$G = \left\{ \begin{pmatrix} a & b \\ 0 & d \end{pmatrix} \right\} \subset \mathrm{GL}_2(\mathbf{C})$$

on $\mathbf{C}^2$ by left-multiplication.

(1) What are its composition factors? Is it semisimple?

(2) Compute $\mathrm{Hom}_G(\varrho, \varrho)$, and conclude that the converse of Schur's Lemma (part (3)) does not always hold when π is not assumed to be semisimple. What happens if, instead of G, one uses its subgroup where $a = d = 1$?

A simple, but important, corollary of Schur's Lemma is the following:

Corollary 2.7.17 (Abelian groups and central character). *Let G be an abelian group, and let k be an algebraically closed field. Then any finite-dimensional irreducible representation of G is of dimension 1, i.e., the finite-dimensional irreducible representations of G coincide with the homomorphisms $G \longrightarrow k^{\times}$.*

More generally, if G is any group and $\varrho : G \longrightarrow \mathrm{GL}(E)$ is a finite-dimensional irreducible k-representation of G, there exists a one-dimensional representation ω of the center $Z(G)$ of G such that

$$\varrho(z) = \omega(z)\mathrm{Id}_E$$

for all $z \in Z(G)$. This representation ω is called the central character *of ϱ.*

Proof. (1) Let ϱ be a finite-dimensional irreducible representation of G, acting on E. *Because G is abelian*, any $\Phi = \varrho(g) : E \to E$ is in fact a homomorphism in $\mathrm{Hom}_G(\varrho, \varrho)$. By Schur's Lemma in Proposition 2.7.15, there exists therefore $\lambda(g) \in k$ such that $\varrho(g) = \lambda(g)\mathrm{Id}$ is a scalar. Then any one-dimensional subspace of E is invariant under all operators $\varrho(g)$, and by irreducibility, this means that E is equal to any such subspace.

(2) Similarly, for G arbitrary, if z is an element of the center of G, we see that $\varrho(z)$ commutes with all $\varrho(g)$, for any representation of G, i.e., $\varrho(z) \in \mathrm{End}_G(\varrho)$. If ϱ is irreducible, Schur's Lemma implies that $\varrho(z)$ is multiplication by a scalar, and of course the latter is a one-dimensional representation of $Z(G)$. □

Remark 2.7.18 (Division algebras). Example 2.4.1 shows that this result does not hold in general if the field is not necessarily algebraically closed.

If ϱ is an irreducible (finite-dimensional) k-representation of G, the earlier version of Schur's Lemma already shows that $A = \mathrm{End}_G(\varrho)$, the space of G-endomorphisms of ϱ, has a remarkable structure: it is a subalgebra of the matrix algebra $\mathrm{End}_k(\varrho)$ which is a *division algebra*, i.e., any non-zero element of A has an inverse in A.

In the case of Example 2.4.1, the reader is invited to show explicitly that A is isomorphic to $\mathbf{C}$, as an $\mathbf{R}$-algebra.

Another easy and useful corollary is the following algebraic characterization of multiplicities of irreducible representations in a semisimple representation:

Corollary 2.7.19 (Multiplicities). *Let G be a group, and let k be an algebraically closed field. If ϱ is a finite-dimensional semisimple k-representation of G, then for any irreducible k-representation π of G, we have*

$$n_\pi(\varrho) = \dim \mathrm{Hom}_G(\pi, \varrho) = \dim \mathrm{Hom}_G(\varrho, \pi),$$

where $n_\pi(\varrho)$ is the multiplicity of π as a summand in ϱ.

Proof. If we express

$$\varrho \simeq \bigoplus_i \varrho_i,$$

where ϱ_i are (necessarily finite-dimensional) irreducible representations of G, then we have

$$\mathrm{Hom}_G(\pi, \varrho) \simeq \bigoplus_i \mathrm{Hom}_G(\pi, \varrho_i)$$

for all irreducible representation π. This space has dimension equal to the number of indices i for which $\varrho_i \simeq \pi$, by Schur's Lemma (i.e., by (2.44)), which is of course $n_\pi(\varrho)$. A similar argument applies to $\mathrm{Hom}_G(\varrho, \pi)$. □

Example 2.7.20 (Multiplicity version of Frobenius reciprocity). Let G be a group, let H be a subgroup of G, and let $\varrho_1 : G \longrightarrow \mathrm{GL}(V_1)$ and $\varrho_2 : H \longrightarrow \mathrm{GL}(V_2)$ be two irreducible k-representations of G and H respectively, where k is algebraically closed. Assuming that

$$\mathrm{Ind}_H^G(\varrho_2) \quad \text{and} \quad \mathrm{Res}_H^G(\varrho_1)$$

are semisimple, we obtain from (2.25) and Corollary 2.7.19 the formula

$$n_{\varrho_1}(\mathrm{Ind}_H^G(\varrho_2)) = n_{\varrho_2}(\mathrm{Res}_H^G(\varrho_1)),$$

which is a very useful form of Frobenius reciprocity. (For cases where the semisimplicity assumption holds, see Chapters 4 and 5.)

If k is algebraically closed, we can also use Schur's Lemma to give a nice description of the isotypic component $\mathrm{M}(\pi)$ of a finite-dimensional semisimple representation ϱ of G (acting on E). To describe this, let E_π be the space on which π acts; then there is a natural k-linear map

$$\Theta \begin{cases} \mathrm{Hom}_G(E_\pi, E) \otimes E_\pi & \longrightarrow \quad E \\ \Phi \otimes v & \mapsto \quad \Phi(v). \end{cases}$$

The image of this map is, almost by definition, equal to the isotypic component $\mathrm{M}(\pi) \subset E$ (because any non-zero $\Phi \in \mathrm{Hom}_G(E_\pi, E)$ is injective by Schur's Lemma, so that $\Phi(v)$ is in the subrepresentation $\mathrm{Im}(\Phi)$ isomorphic to ϱ.)

If E is finite dimensional, we then see (by the previous corollary) that the dimensions of $\mathrm{M}(\pi)$ and of the source

$$\mathrm{Hom}_G(E_\pi, E) \otimes E_\pi$$

coincide, and we conclude that Θ gives an isomorphism

$$\mathrm{Hom}_G(E_\pi, E) \otimes E_\pi \simeq \mathrm{M}(\pi) \subset E.$$

Moreover, Θ is a G-homomorphism, if we let G act trivially on the space $\mathrm{Hom}_G(E_\pi, E)$ (which is natural by (2.17)) and through π on E_π. From this (picking, if needed, a basis of $\mathrm{Hom}_G(E_\pi, E)$), we see that $\mathrm{M}(\pi)$ is isomorphic, as representation of G, to a direct sum of d copies of π, where

$$d = \dim \mathrm{Hom}_G(E_\pi, E).$$

As it happens, the injectivity of Θ can also be proved directly, and this leads to the following useful result, where ϱ is not assumed to be semisimple, but we can still characterize the π-isotypic component using the same construction:

Proposition 2.7.21 (A formula for isotypic components)**.** *Let G be a group, and let k be an algebraically closed field. If*

$$\varrho : G \longrightarrow \mathrm{GL}(E)$$

is a finite-dimensional k-representation of G and

$$\pi : G \longrightarrow \mathrm{GL}(F)$$

is any irreducible k-representation, the map

$$\Theta \begin{cases} \mathrm{Hom}_G(F, E) \otimes F & \longrightarrow \quad E \\ \Phi \otimes v & \mapsto \quad \Phi(v) \end{cases}$$

is injective and its image is equal to the π-isotypic component $\mathrm{M}(\pi) = \mathrm{M}_E(\pi)$ of E, the sum of all subrepresentations of E isomorphic to π. In particular,

$$\mathrm{M}(\pi) \simeq (\dim \mathrm{Hom}_G(F, E))\pi$$

as representation.[18]

Note that, if ϱ is not semisimple, π might also appear as a composition factor outside of $\mathrm{M}(\pi)$ (i.e., as a genuine quotient or subquotient).

Proof. As before, it is clear that the image of Θ is equal to the subrepresentation $\mathrm{M}(\pi) \subset E$. It remains thus to show that Θ is injective, as this leads to the isomorphism of representations

$$\mathrm{M}(\pi) \simeq \mathrm{Hom}_G(F, E) \otimes F,$$

from which the last step follows.

[18] We use here a relatively standard notation $n\pi$, for $n \geqslant 0$ and a representation π, to denote a direct sum of n copies of π.

Let (Φ_j) be a basis of the space $\mathrm{Hom}_G(F, E)$, so that any element of the tensor product is of the form

$$\sum_j \Phi_j \otimes v_j$$

for some $v_j \in F$. Then we have

$$\Theta\Big(\sum_j \Phi_j \otimes v_j\Big) = \sum_j \Phi_j(v_j) \in E,$$

and the injectivity of Θ is seen to be equivalent to saying that the spaces $F_j = \mathrm{Im}(\Phi_j)$ are in direct sum in E. We prove this in a standard manner: Assume the contrary is true, and let $J \subset I$ be a set of smallest order for which there is a relation

$$\sum_{j \in J} \Phi_j(v_j) = 0$$

with $\Phi_j(v_j) \neq 0$ for $j \in J$. Consider any $\ell \in J$. We find that

$$0 \neq \Phi_\ell(v_\ell) \in \mathrm{Im}(\Phi_\ell) \cap \bigoplus_{j \neq \ell} \mathrm{Im}(\Phi_j),$$

so that, by irreducibility, this intersection is in fact equal to $\mathrm{Im}(\Phi_\ell)$ (note that we wrote that the Φ_j, $j \neq \ell$, are in direct sum because otherwise we could replace the set J by $J - \{\ell\}$). This means that Φ_ℓ belongs, in an obvious sense, to the space

$$\mathrm{Hom}_G(F, \bigoplus_{j \neq \ell} \mathrm{Im}(\Phi_j)),$$

which is spanned by the homomorphisms $(\Phi_j)_{j \neq \ell}$. This is impossible however, since all the Φ_j's are linearly independent by assumption. □

The following addition is also useful:

Lemma 2.7.22. *With assumptions and notation as in Proposition* 2.7.21, *if* (π_i) *is a family of pairwise non-isomorphic irreducible representations of* G, *the isotypic subspaces* $\mathrm{M}(\pi_i) \subset E$ *are in direct sum.*

Proof. Indeed, for any fixed π in this family, the intersection of $\mathrm{M}(\pi)$ with the sum of all $\mathrm{M}(\pi_i)$, $\pi_1 \neq \pi$, is necessarily zero: it cannot contain any irreducible subrepresentation, since the possibilities coming from π are incompatible with those coming from the other π_i. □

Exercise 2.7.23 (Isotypic components when k is not algebraically closed). Let A be a ring with unit, which may not be commutative. Given a *right* A-module M (which means that ma is defined for $m \in M$ and $a \in A$, with the expected rules, such that $(m_1 + m_2)a = m_1 a + m_2 a$), a left A-module

N, and an abelian group V, a $\mathbf{Z}$-linear map $b : M \times N \longrightarrow V$ is said to be *A-bilinear* if we have

$$b(ma, n) = b(m, an)$$

for all $m \in M$, $n \in N$ and $a \in A$. Then one can show (imitating the usual construction of the tensor product over commutative rings) that there exists an abelian group $M \otimes_A N$, called the tensor product of M and N over A, and an A-bilinear map $\otimes : M \times N \longrightarrow M \otimes_A N$ which *represents* all A-bilinear maps, in the sense that for any abelian group V and for any A-bilinear map $b : M \times N \longrightarrow V$, there exists a unique homomorphism $M \otimes_A N \xrightarrow{\tilde{b}} V$ of abelian groups such that

$$b(m, n) = \tilde{b}(m \otimes n)$$

for all m in M and n in N.

Let G be a group, let k be a field, let $\varrho : G \longrightarrow \mathrm{GL}(E)$ be a finite-dimensional k-representation of G, and let $\pi : G \longrightarrow \mathrm{GL}(F)$ be a finite-dimensional irreducible k-representation of G.

(1) Let $A = \mathrm{End}_G(\pi)$. Show that $\mathrm{Hom}_G(F, E)$ is naturally a right A-module and that F is naturally a left A-module.

(2) Let $\mathrm{M}_E(\pi)$ be the π-isotypic component of ϱ. Show that there exists a natural isomorphism of abelian groups

$$\mathrm{M}_E(\pi) \simeq \mathrm{Hom}_G(F, E) \otimes_A F.$$

(3) Show that $\mathrm{Hom}_G(F, E) \otimes_A F$ has a natural structure of k-vector space such that the isomorphism in (3) is k-linear.

(4) Show that there is a k-linear action of G on $\mathrm{Hom}_G(F, E) \otimes_A F$ determined by

$$g \cdot \Phi \otimes v = \Phi \otimes \varrho(g)v,$$

and that the isomorphism in (2) and (3) is then an intertwiner.

If k is algebraically closed, the algebra A is isomorphic to k, and hence is commutative, and these results are then the same as those of Proposition 2.7.21.

2.7.3. Burnside's irreducibility criterion and its generalizations, 1. We now show how to use Schur's Lemma to prove a result of Burnside which provides a frequently useful irreducibility criterion for finite-dimensional representations, and we derive further consequences along the same lines. In fact, we will prove this twice (and a third time in Chapter 4 in the case of finite groups); in this section, we argue in the style of Burnside, and below in Section 2.7.4, we will recover the same results in the style of Frobenius.

We will motivate the first result, Burnside's criterion, from the following point of view: given a finite-dimensional k-representation ϱ of a group G

acting on the vector space E, the image of ϱ is a subset of the vector space $\mathrm{End}_k(E)$. We then ask, What are the linear relations satisfied by these elements? For instance, the block-diagonal shape (2.33) of a representation which is is not irreducible shows clearly *some* relations: those that express that the matrices in the bases indicated have lower-left corner(s) equal to 0, for instance. These are obvious. Are there others?

Theorem 2.7.24 (Burnside's irreducibility criterion). *Let k be an algebraically closed field, and let G be a group. A* finite-dimensional *k-representation*

$$\varrho : G \longrightarrow \mathrm{GL}(E)$$

is irreducible *if and only if the image of ϱ satisfies no non-trivial linear relation in* $\mathrm{End}_k(E)$. *Equivalently, ϱ is irreducible if and only if the linear span of the image of ϱ in* $\mathrm{End}_k(E)$ *is equal to* $\mathrm{End}_k(E)$.

The proof we give is a modern version of Burnside's original argument. One can give much shorter proofs (the one in the next section is an example), but this one has the advantage of "exercising" the basic formalism of representation theory and of being easy to motivate.

Proof. First of all, the two statements we give are equivalent by duality of finite-dimensional k-vector spaces. More precisely, let $V = \mathrm{End}_k(E)$. Then by "relations satisfied by the image of ϱ", we mean the k-linear subspace

$$R = \{\phi \in V' \mid \langle \phi, \varrho(g)\rangle = 0 \text{ for all } g \in G\},$$

of V', the linear dual of $\mathrm{End}_k(E)$. Then we are saying that $R = 0$ if and only if the image of G spans V, which is part of duality theory.

The strategy of the proof is going to be the following:

(1) For some natural representation of G on V', we show that R is a subrepresentation.

(2) We find an explicit decomposition of V' (with its G-action) as a direct sum of irreducible representations, embedded in V' in a specific manner.

(3) Using this description, we can see what the possibilities for R are, and especially that $R = 0$ if ϱ is irreducible.

This strategy will also be used afterward to give a more general result of comparison of distinct irreducible representations.

We let G act on V' by the contragredient of the representation of G on V given by

$$g \cdot T = \varrho(g) \circ T$$

for $g \in G$ and $T : E \to E$. Note that this corresponds to the action (2.20), and not the action (2.18). To check that $R \subset V'$ is a subrepresentation, we need simply note that if $\phi \in R$ and $g \in G$, then we have

$$\langle g \cdot \phi, \varrho(h) \rangle = \langle \phi, g^{-1} \cdot \varrho(h) \rangle = \langle \phi, \varrho(g^{-1}h) \rangle = 0$$

for all $h \in G$, which means that $g \cdot \phi$ is also in R.

From (2.21), we know that V—with the above action—is isomorphic to a direct sum of $\dim \varrho$ copies of ϱ; hence V' is isomorphic to a direct sum of the same number of copies of the contragredient $\check{\varrho}$ of ϱ, which we know to be irreducible (Proposition 2.2.18). It follows that any irreducible subrepresentation π of R (which exists if $R \neq 0$) must be itself isomorphic to $\check{\varrho}$. (This fact is clear from the Jordan–Hölder–Noether Theorem, but can also be seen directly by considering the composites

$$p_i : \pi \hookrightarrow R \hookrightarrow V' \simeq \bigoplus_{i \leqslant \dim \varrho} W_i \longrightarrow W_i \simeq \check{\varrho},$$

where W_i are subrepresentations of V' isomorphic to $\check{\varrho}$. Each of these composites is in $\mathrm{Hom}_G(\pi, \check{\varrho})$, and hence is either 0 or an isomorphism, by Schur's Lemma. Since $\pi \neq 0$, not all p_i can be zero, hence $\pi \simeq \check{\varrho}$.)

However, we claim that there is an isomorphism (of k-vector spaces)

$$\begin{cases} E & \longrightarrow & \mathrm{Hom}_G(E', V') \\ v & \longmapsto & \alpha_v, \end{cases}$$

where $\alpha_v : E' \to V'$ is defined by

$$\langle \alpha_v(\lambda), T \rangle = \langle \lambda, T(v) \rangle$$

for $\lambda \in E'$ and $T : E \to E$. If this is the case, then assuming that $R \neq 0$, and hence that R contains a copy of $\check{\varrho}$, means that, for some $v \neq 0$, the image of α_v is in R. But this implies that for all $\lambda \in E'$, and $g \in G$, we have

$$0 = \langle \alpha_v(\lambda), \varrho(g) \rangle = \langle \lambda, \varrho(g)v \rangle,$$

which is impossible even for a single g, since $\varrho(g)v \neq 0$.

Checking the claim is not very difficult; we leave it to the reader to verify that each α_v is indeed a G-homomorphism from E' to V' and that $v \mapsto \alpha_v$ is k-linear. We then observe that

$$\dim \mathrm{Hom}_G(E', V') = \dim \mathrm{Hom}_G(\check{\varrho}, (\dim E)\check{\varrho}) = \dim(E) \dim_G(\check{\varrho}, \check{\varrho}) = \dim E$$

by Schur's Lemma again (using the fact that k is algebraically closed). So the map will be an isomorphism as soon as it is injective. However, $\alpha_v = 0$ means that

$$\langle \lambda, T(v) \rangle = 0$$

for all $\lambda \in E'$ and $T \in V$, and that only happens when $v = 0$ (take T to be the identity). □

Example 2.7.25. Consider again the representation ϱ of $\mathbf{R}/\mathbf{Z}$ by rotations in the plane of Example 2.4.1, which is irreducible over $\mathbf{R}$ but not absolutely irreducible. We can see easily that the linear span of the image of ϱ is the proper subalgebra

$$\left\{\begin{pmatrix} a & b \\ -b & a \end{pmatrix} \mid a,\ b \in \mathbf{R}\right\}$$

in $\mathrm{M}_2(\mathbf{R})$. In particular, it does satisfy non-trivial relations, such as the diagonal coefficients are equal.

We emphasize again the strategy we used, because it is a common pattern in applications of representation theory. One wishes to analyze a certain vector space (here, the relation space R). This space is seen to be a subspace of a bigger one, on which a group G acts, and then the space is seen to be a *subrepresentation* of this bigger space. Independent analysis of the latter is performed, e.g., a decomposition in irreducible summands, and then one deduces *a priori* restrictions on the possibilities for the space of original interest. (For another concrete application, see Section 4.7.3.)

We now implement this principle again in a generalization (due to Frobenius and Schur) of Burnside's Theorem. To motivate it, consider a finite-dimensional k-representation of G

$$\varrho : G \longrightarrow \mathrm{GL}(E),$$

which is irreducible, with k algebraically closed. Burnside's Theorem means, in particular, that if we fix a basis (v_i) of E and express ϱ in the basis (v_i) producing the homomorphism

$$\varrho^{\boldsymbol{m}} : G \longrightarrow \mathrm{GL}_{\dim(E)}(k),$$

the resulting *matrix coefficients* $(\varrho^{\boldsymbol{m}}_{i,j}(g))$, seen as functions on G, are k-linearly independent. Indeed, if we denote by (λ_i) the dual basis of E', we have

$$\varrho^{\boldsymbol{m}}_{i,j}(g) = \langle \lambda_i, \varrho(g)v_j\rangle,$$

so that a relation

$$\sum_{i,j} \alpha_{i,j}\varrho^{\boldsymbol{m}}_{i,j}(g) = 0$$

valid for all g, for some fixed $\alpha_{i,j} \in k$, means that the element ϕ of $\mathrm{End}_k(E)'$ defined by

$$\langle \phi, T\rangle = \sum_{i,j} \alpha_{i,j}\langle \lambda_i, T(v_j)\rangle$$

is in the relation space R of the proof above, hence is identically zero, which means that $\alpha_{i,j} = 0$ for all i and j.

The interest of these matrix coefficients is that they are functions on G (with values in k). As such, they might be written down without mentioning the representation at all and, in particular, without knowing the representation space. However, the choice of basis is annoying, so the following definition is more convenient:

Definition 2.7.26 (Matrix coefficient)**.** Let G be a group, let k be a field, and let

$$\varrho : G \longrightarrow \mathrm{GL}(E)$$

be a k-representation of G. A *matrix coefficient* of ϱ is any function on G of the type

$$f_{v,\lambda} \begin{cases} G \to k \\ g \mapsto \lambda(\varrho(g)v) = \langle \lambda, \varrho(g)v \rangle, \end{cases}$$

for some fixed $v \in E$ and $\lambda \in E'$.

Remark 2.7.27. Note that these functions are typically *not* multiplicative. An exception is when $\varrho : G \longrightarrow \mathrm{GL}_1(k) = k^\times$ is a one-dimensional representation. In that case one can take $v = 1 \in k$ and λ the identity of k, so that the matrix coefficient is equal to ϱ as a function $G \longrightarrow k$.

Now we come back to the discussion: matrix coefficients of a fixed irreducible representation are k-linearly independent. What is more natural than to ask, What about different representations? Is it possible that their matrix coefficients satisfy non-trivial linear relations? The answer is very satisfactory: No! This is another expression of the fact that distinct (i.e., non-isomorphic) irreducible representations of a group are independent.

Theorem 2.7.28 (Linear independence of matrix coefficients)**.** *Let G be a group, and let k be an algebraically closed field.*

(1) *For any finite collection (ϱ_i) of pairwise non-isomorphic, finite-dimensional, irreducible k-representations of G, acting on E_i, let*

$$\varrho = \bigoplus_i \varrho_i \text{ acting on } E = \bigoplus_i E_i.$$

Then the k-linear span of the elements $\varrho(g)$, for $g \in G$, in $\mathrm{End}_k(E)$ is equal to

$$\bigoplus_i \mathrm{End}_k(E_i). \tag{2.45}$$

(2) *The matrix coefficients of finite-dimensional irreducible k-representations of G are linearly independent, in the following sense. For any finite collection (ϱ_i) of pairwise non-isomorphic, finite-dimensional, irreducible k-representations of G, acting on E_i, for any choice $(v_{i,j})_{1 \leqslant j \leqslant \dim E_i}$ of bases of E_i, and for the dual bases $(\lambda_{i,j})$, the family of functions*

$$(f_{v_{i,j},\lambda_{i,k}})_{i,j,k}$$

on G are k-linearly independent as elements of the vector space $C_k(G)$ of k-valued functions on G.

Note that there are

$$\sum_i (\dim E_i)^2$$

functions on G in this family, given by

$$\begin{cases} G \to k \\ g \mapsto \langle \lambda_{i,j}, \varrho_i(g) v_{i,k} \rangle_{E_i}, \end{cases}$$

for $1 \leqslant j, k \leqslant \dim E_i$.

Proof. It is easy to see first that (1) implies (2): writing down, as above, a matrix representation for $\varrho = \bigoplus \varrho_i$ in the direct sum of the given bases of E_i, a k-linear relation on the matrix coefficients implies one on the k-linear span of the image of ϱ, but there is no non-trivial such relation on

$$\bigoplus_i \mathrm{End}_k(E_i).$$

To prove (1), we use the same strategy as in the proof of Burnside's Theorem 2.7.24, for the representation ϱ of G on E: let $V = \mathrm{End}_k(E)$, and let

$$R = \{\phi \in V' \mid \langle \phi, \varrho(g) \rangle = 0 \text{ for all } g \in G\}$$

be the relation space. We will compute R and show that $R = S$, where

$$S = \{\phi \in V' \mid \langle \phi, T \rangle = 0 \text{ for all } T \in \bigoplus_i \mathrm{End}_k(E_i)\}, \tag{2.46}$$

so that by duality, the linear span of $\varrho(g)$ is equal to (2.45), as claimed.

As before, we consider the representation of G on V' by the contragredient of the action $g \cdot T = \varrho(g) \circ T$ on V, and we see that $R \subset V'$ is a subrepresentation and that $S \subset V'$ is also one (because ϱ leaves each E_i stable).

We now show that V' is semisimple by exhibiting a decomposition into irreducibles. For this purpose, denoting

$$V_{i,j} = \mathrm{Hom}_k(E_j, E_i),$$

we have also (as in Remark 2.2.23) the similar action of G on each $V_{i,j}$, and we obtain first, as k-vector spaces, the direct sum decompositions

$$V = \bigoplus_{i,j} V_{i,j}, \qquad V' = \bigoplus_{i,j} V'_{i,j}.$$

Note that, more precisely, we identify $V_{i,j}$ with the subspace of V consisting of all $T : E \to E$ that map the summand E_j to E_i and that map the other summands E_ℓ with $\ell \neq j$ to 0. This identification is, in fact, also implicit in the statement of the theorem involving (2.45). These subspaces $V'_{i,j}$ are not irreducible in general: by (2.21), we get isomorphisms of representations

$$V_{i,j} \simeq (\dim E_j)\varrho_i,$$

and hence

$$V'_{i,j} \simeq (\dim E_j)\check{\varrho}_i,$$

for the contragredient, leading to the decomposition

$$V' \simeq \bigoplus_{i,j}(\dim E_j)\check{\varrho}_i \simeq \bigoplus_{i}(\dim E)\check{\varrho}_i,$$

of V' as direct sum of irreducible representations.

Since R and $S \subset V'$ are subrepresentations, they are therefore also semisimple and have irreducible components among the $\check{\varrho}_i$. We now determine which subspaces of V', isomorphic to some $\check{\varrho}_i$, can be in R.

Fix an index i. As in the proof of Burnside's Theorem, we first claim that there is an isomorphism of k-vector spaces

$$\begin{cases} E \longrightarrow \mathrm{Hom}_G(E'_i, V') \\ v \mapsto \alpha_v \end{cases}$$

defined by the formula

$$\langle \alpha_v(\lambda), T\rangle = \langle \lambda, T(v)\rangle$$

for $\lambda \in E'_i$ and $T \in V$, where $\lambda \in E'_i$ is extended to E by being 0 on the other summands E_j, $j \neq i$.

Indeed, the α_v are G-morphisms, and this map is injective (the arguments are the same here as in the case of Burnside's Theorem). Then we find that

$$\begin{aligned} \dim \mathrm{Hom}_G(E'_i, V') &= \sum_j \dim \mathrm{Hom}_G(\check{\varrho}_i, (\dim E)\check{\varrho}_j) \\ &= \dim(E)\sum_j \dim \mathrm{Hom}_G(\check{\varrho}_i, \check{\varrho}_j) = \dim(E) \end{aligned}$$

since, using Schur's Lemma,[19] only the term $j = i$ contributes a non-zero factor 1 to the sum.

Any subspace of V' isomorphic to the fixed E'_i is therefore of the form $\mathrm{Im}(\alpha_v)$ for some $v \in E$. Now $\mathrm{Im}(\alpha_v) \subset R$ is equivalent with

$$\langle \alpha_v(\lambda), \varrho(g)\rangle = 0$$

for all $g \in G$ and $\lambda \in E'_i$. But since ϱ is the direct sum of the ϱ_i and $\lambda \in E'_i$, we have

$$\langle \alpha_v(\lambda), \varrho(g)\rangle = \langle \lambda, \varrho(g)v\rangle = \langle \lambda, \varrho_i(g)v_i\rangle,$$

[19] This is the crucial point, where the "independence" of distinct irreducible representations comes into play.

where v_i is the component of v in E_i. Hence (putting $g = 1$) the condition $\mathrm{Im}(\alpha_v) \subset R$ is equivalent with $v_i = 0$.

But a similar computation shows that $\mathrm{Im}(\alpha_v) \subset S$ is *also* equivalent with $v_i = 0$ (see (2.46)). Varying i, we see that R and S contain exactly the same irreducible subrepresentations. Hence $R = S$, and we saw at the beginning that this implies the conclusion (2.45). $\square$

Example 2.7.29 (Linear independence of one-dimensional representations). Following Remark 2.7.27, since any one-dimensional representation is irreducible and two of them are isomorphic if and only if they coincide as functions $G \longrightarrow k^\times$, the theorem shows that any family of homomorphisms

$$\chi_i : G \longrightarrow k^\times \subset k$$

is k-linearly independent in $C_k(G)$ when k is algebraically closed. But in fact, this last assumption is not needed, because we can see the representations χ_i as taking values in an arbitrary algebraic closure of k, and they remain irreducible when seen in this manner, like all one-dimensional representations.

This result is important in Galois theory. As one might expect, it is possible to prove it more directly and elementarily, and the reader should attempt to do it (see, e.g., [**40**, Th. VI.4.1]).

The linear independence of matrix coefficients turns out to have many important applications. In particular, it gives quite precise information on the structure of the regular representation of G acting on the space $C_k(G)$ of k-valued functions on the group.

Corollary 2.7.30 (Matrix coefficients as subrepresentations of the regular representation). *Let G be a group, let k be an algebraically closed field, and let ϱ be a finite-dimensional irreducible k-representation of G. Let* $\mathrm{M}(\varrho)$ *be the subspace of $C_k(G)$* spanned *by all matrix coefficients $f_{v,\lambda}$ of ϱ.*

(1) *The space* $\mathrm{M}(\varrho)$ *depends only on ϱ up to isomorphism.*

(2) *It is a subrepresentation of the regular representation of G acting on $C_k(G)$; moreover,* $\mathrm{M}(\varrho)$ *is semisimple and isomorphic to a direct sum of* $\dim(\varrho)$ *copies of ϱ.*

(3) *Any subrepresentation of $C_k(G)$ isomorphic to ϱ is contained in the subspace* $\mathrm{M}(\varrho)$, *i.e.,* $\mathrm{M}(\varrho)$ *is the ϱ-isotypic component of $C_k(G)$, as defined in Proposition* 2.7.9.

For one-dimensional representations ϱ (e.g., $\varrho = \mathbf{1}$), we already computed $\mathrm{M}(\varrho)$ in Example 2.1.9: it is a one-dimensional space, spanned by ϱ seen as a k-valued function. This verifies (3) directly in these simple cases.

Proof. We first check (1), which states that $\mathrm{M}(\varrho)$ is a canonical subspace of $C_k(G)$. Let E be the space on which ϱ acts, and let $\tau : G \longrightarrow \mathrm{GL}(F)$ be a k-representation isomorphic to ϱ with the linear map

$$\Phi : E \longrightarrow F$$

giving this isomorphism. Then for any $w \in F$ and $\lambda \in F'$, writing $w = \Phi(v)$ for some $v \in E$, we have

$$\begin{aligned} f_{w,\lambda}(g) &= \langle \lambda, \tau(g)w \rangle_F \\ &= \langle \lambda, \tau(g)\Phi(v) \rangle_F \\ &= \langle \lambda, \Phi(\varrho(g)v) \rangle_F \\ &= \langle {}^t\Phi(\lambda), \varrho(g)v \rangle_E \\ &= f_{v,{}^t\Phi(\lambda)}(g), \end{aligned}$$

for all $g \in G$, showing that any matrix coefficient for τ is also one for ϱ. By symmetry, we see that $\mathrm{M}(\varrho)$ and $\mathrm{M}(\tau)$ are *equal* subspaces of $C_k(G)$.

We next check that $\mathrm{M}(\varrho) \subset C_k(G)$ is indeed a subrepresentation: for $v \in E$ (the space on which G acts), $\lambda \in E'$, and $g \in G$, we have

$$\varrho_G(g) f_{v,\lambda}(x) = f_{v,\lambda}(xg) = \langle \lambda, \varrho(xg)v \rangle = f_{\varrho(g)v,\lambda}(x) \in \mathrm{M}(\varrho).$$

In fact, this formula says more: it shows that, for a *fixed* $\lambda \in E'$, the linear map

$$\Phi_\lambda : v \mapsto f_{v,\lambda}$$

is an intertwining operator between ϱ and $\mathrm{M}(\varrho)$.

Fix a basis (v_j) of the space E on which ϱ acts, and let (λ_j) be the dual basis. By construction, $\mathrm{M}(\varrho)$ is spanned by the matrix-coefficients

$$f_{i,j} = f_{v_i,\lambda_j}, \qquad 1 \leqslant i, j \leqslant \dim(\varrho),$$

and from the linear independence of matrix coefficients, these functions form a *basis* of the space $\mathrm{M}(\varrho)$. In particular, we have $\dim \mathrm{M}(\varrho) = \dim(\varrho)^2$. Now the intertwining operator

$$\Phi = \bigoplus_i \Phi_{\lambda_i} : \bigoplus_i \varrho \longrightarrow \mathrm{M}(\varrho)$$

is surjective (since its image contains each basis vector $f_{i,j}$), and both sides have the same dimension. Hence it must be an isomorphism, which shows that $\mathrm{M}(\varrho)$ is isomorphic to $(\dim(\varrho))\varrho$ as a representation of G.

There only remains to check the last part. Let $E \subset C_k(G)$ be a subrepresentation of the regular representation that is isomorphic to ϱ. To show that $E \subset \mathrm{M}(\varrho)$, we will check that the elements $f \in E$, which are functions on G, are all matrix coefficients of E. Let $\delta \in C_k(G)'$ be the linear form defined by

$$\delta(f) = f(1)$$

for $f \in C_k(G)$. Consider then the linear form $\delta_E \in E'$ that is the restriction of δ to E. Then, for any function $f \in E$ and $x \in G$, we have

$$\langle \delta_E, \varrho_G(x) f \rangle = \varrho_G(x) f(1) = f(x)$$

(by definition of the regular representation).

The left-hand side (seen as a function of x) is a matrix coefficient for ϱ, since ϱ_G on E is isomorphic to ϱ, and hence we see that $f \in \mathrm{M}(\varrho)$. □

The next corollary will be improved in the chapter on representations of finite groups. We state it here because it is the first a priori restriction we have found on irreducible representations for certain groups:

Corollary 2.7.31. *Let G be a finite group, and let k be an algebraically closed field. There are only finitely many irreducible k-representations of G up to isomorphism, and they satisfy*

$$\sum_{\varrho} (\dim \varrho)^2 \leqslant |G|,$$

where the sum is over isomorphism classes of irreducible k-representations of G.

Proof. First of all, since the space of an irreducible representation of a finite group is spanned by finitely many vectors $\varrho(g)v$, $g \in G$ (for any vector $v \neq 0$), any irreducible representation of G is finite dimensional.

Then by the previous corollary, for any such irreducible representation ϱ, the regular representation ϱ_G contains a subspace isomorphic to $\dim(\varrho)$ copies of ϱ. By the linear independence of matrix coefficients of non-isomorphic irreducible representations (Theorem 2.7.28), the sum over ϱ of these representations is a direct sum and has dimension

$$\sum_{\varrho} (\dim \varrho)^2,$$

hence the result since the dimension of the regular representation is $|G|$. □

Remark 2.7.32. If G is finite, there is equality in this formula if and only if the regular representation is semisimple. If there is equality, this means that

$$C_k(G) = \bigoplus_{\varrho} \mathrm{M}(\varrho) \simeq \bigoplus_{\varrho} (\dim \varrho) \varrho$$

is semisimple; conversely, if $C_k(G)$ is semisimple, by Lemma 2.2.11 there exists a stable subspace F such that

$$C_k(G) = F \oplus \Big(\bigoplus_{\varrho} \mathrm{M}(\varrho) \Big),$$

but F cannot contain any irreducible subrepresentation π, as it would be isomorphic to some ϱ and hence contained in $\mathrm{M}(\varrho)$, so that $F = 0$. In Chapter 4, we will see that this semisimplicity occurs if and only if the characteristic of the field k does not divide the order of G. Readers may enjoy trying to think about it beforehand and should also write down explicitly a matrix representation of (say) the regular representation of $G = \mathbf{Z}/2\mathbf{Z}$ over a field of characteristic 2, to check that the latter is not semisimple (see Exercise 2.4.3).

In the next section, we will derive further consequences of the linear independence of matrix coefficients, related to *characters* of finite-dimensional representations. Before this, as another application of Schur's Lemma and its corollaries, we can now prove Proposition 2.3.23 about irreducible representations of a direct product $G = G_1 \times G_2$.

Proof of Proposition 2.3.23. Recall that we are considering an algebraically closed field k and two groups G_1 and G_2, and we want to prove that all finite-dimensional irreducible k-representations of $G = G_1 \times G_2$ are external tensor products of the form $\varrho \simeq \varrho_1 \boxtimes \varrho_2$ for some irreducible representations ϱ_i of G_i (unique up to isomorphism).

To begin with, if ϱ_1 and ϱ_2 are finite-dimensional irreducible representations of G_1 and G_2, the irreducibility of $\varrho = \varrho_1 \boxtimes \varrho_2$ follows from Burnside's irreducibility criterion: since the $\varrho_1(g_1)$ and $\varrho_2(g_2)$, for $g_i \in G_i$, span the k-linear endomorphism spaces of their respective spaces, it follows by elementary linear algebra that the $\varrho_1(g_1) \otimes \varrho_2(g_2)$ also span the endomorphism space of the tensor product. (Here we used the fact that k is algebraically closed.)

Thus what matters is to prove the converse. Therefore, let

$$\varrho : G_1 \times G_2 \longrightarrow \mathrm{GL}(E)$$

be an irreducible k-representation. We restrict ϱ to the subgroup $G_1 = G_1 \times \{1\} \subset G$, and we let $E_1 \subset E$ be an irreducible subrepresentation of E seen as representation of G_1. We denote by ϱ_1 the corresponding abstract representation of G_1. We now proceed to find a representation ϱ_2 of G_2 such that

$$\varrho_1 \boxtimes \varrho_2 \simeq \varrho.$$

For this purpose, define the k-vector space

$$E_2 = \mathrm{Hom}_{G_1}(\varrho_1, \mathrm{Res}^G_{G_1} \varrho)$$

of intertwiners between ϱ_1 and the restriction of ϱ to G_1; note that it is non-zero by definition of ϱ_1. We claim that the definition

$$(\varrho_2(g_2)\Phi)(v) = \varrho(1, g_2)\Phi(v),$$

for $g_2 \in G_2$, $\Phi \in E_2$, and $v \in E_1$, defines a representation ϱ_2 of $G_2 = \{1\} \times G_2 \subset G$ on E_2. The point is that because G_1 and G_2, seen as subgroups of G, commute with each other, the k-linear map $\varrho_2(g_2)\Phi$ is still a homomorphism $\varrho_1 \to \varrho$ (and not merely a linear map). Indeed, denoting $\Psi = \varrho_2(g_2)\Phi$, we compute

$$\begin{aligned}\Psi(\varrho_1(g_1)v) &= \varrho(1, g_2)\Phi(\varrho_1(g_1)v)\\ &= \varrho(1, g_2)(\varrho(g_1, 1)\Phi(v)) \qquad (\text{since } \Phi \in E_2)\\ &= \varrho(g_1, 1)(\varrho(1, g_2)\Phi(v)) = \varrho(g_1, 1)\Psi(v),\end{aligned}$$

for all $v \in E_1$, which is to say, $\Psi \in E_2$.

Now we define a k-linear map

$$\Theta \begin{cases} E_1 \otimes E_2 & \longrightarrow & E \\ v \otimes \Phi & \mapsto & \Phi(v), \end{cases}$$

and we claim that Θ is an intertwiner between $\varrho_1 \boxtimes \varrho_2$ and ϱ. Indeed, we can check this on pure tensors. For $g_i \in G_i$, $v \otimes \Phi \in E_1 \otimes E_2$, we have

$$\begin{aligned}\Theta(\varrho_1 \boxtimes \varrho_2(g_1, g_2)(v \otimes \Phi)) &= \Theta(\varrho_1(g_1)v \otimes \varrho_2(g_2)\Phi)\\ &= (\varrho_2(g_2)\Phi)(\varrho_1(g_1)v)\\ &= \varrho(1, g_2)\Phi(\varrho_1(g_1)v)\\ &= \varrho(1, g_2)\varrho(g_1, 1)\Phi(v)\\ &= \varrho(g_1, g_2)\Phi(v) = \varrho(g_1, g_2)\Theta(v \otimes \Phi).\end{aligned}$$

Note that this really must be written down by hand to not look like gibberish.

We will now show that Θ is bijective. Let $F_2 \subset E_2$ be any irreducible subrepresentation (of G_2) with action denoted $\tilde{\varrho}_2$; restricting Θ to $E_1 \otimes F_2$ gives an intertwiner

$$\varrho_1 \boxtimes \tilde{\varrho}_2 \longrightarrow E.$$

By the first part of Proposition 2.3.23, which was proved above, the representation $\varrho_1 \boxtimes \tilde{\varrho}_2$ of G is irreducible, and so is E by assumption; moreover, if $v_0 \neq 0$ is a vector in E_1 and $0 \neq \Phi_0 \in F_2$, then $\Theta(v_0 \otimes \Phi_0) = \Phi_0(v_0)$ is non-zero (it is injective, by Schur's Lemma, because Φ_0 is an embedding of the irreducible representation ϱ_1 in ϱ). Thus Θ restricted to $\varrho_1 \boxtimes \tilde{\varrho}_2$ is non-zero, and again by Schur's Lemma, it must be an isomorphism.

We are thus already done proving that ϱ is an external tensor product, but we will continue with some (minor) additional work that shows that, in fact, Θ itself is bijective. For this, we just need to show that

$$\dim(E_1 \otimes E_2) = \dim E = \dim(E_1)\dim(F_2)$$

(since we now know that Θ is surjective) or equivalently that $\dim F_2 = \dim E_2$. But

$$E_2 = \mathrm{Hom}_{G_1}(\varrho_1, \mathrm{Res}^G_{G_1} \varrho),$$

and, by fixing a basis (v_j) of the space F_2 of $\tilde{\varrho}_2$, we see that the restriction to G_1 of $\varrho = \varrho_1 \boxtimes \tilde{\varrho}_2$ is the direct sum

$$\bigoplus_j E_1 \otimes k v_j \simeq (\dim \tilde{\varrho}_2) E_1,$$

so that the dimension of E_2 is given by

$$\dim E_2 = (\dim \tilde{\varrho}_2) \dim \mathrm{Hom}_{G_1}(\varrho_1, \varrho_1) = \dim \tilde{\varrho}_2$$

by the last part of Schur's Lemma (we use again the fact that k is algebraically closed).

Finally, coming back to the general situation, note that this last observation on the restriction of $\varrho_1 \boxtimes \varrho_2$ to G_1 (and the analogue for G_2) show that ϱ_1 and ϱ_2 are indeed unique up to isomorphism by the Jordan–Hölder–Noether Theorem. □

Example 2.7.33. As we have mentioned before, Proposition 2.3.23 is not valid in general for k-representations of a group $G_1 \times G_2$ if k is not algebraically closed. We give here a simple example. Let $k = \mathbf{R}$, $G_1 = G_2 = \mathbf{R}/\mathbf{Z}$, and $G = G_1 \times G_2$, and consider the composite

$$\pi : G \xrightarrow{a} \mathbf{R}/\mathbf{Z} \xrightarrow{\varrho} \mathrm{GL}_2(\mathbf{R})$$

where $a(\theta_1, \theta_2) = \theta_1 + \theta_2$ and ϱ is the $\mathbf{R}$-irreducible representation of Example 2.4.1. Since $\mathbf{R}/\mathbf{Z}$ is abelian, π is indeed a representation of G over $\mathbf{R}$. It is irreducible, because the addition map a is surjective, and ϱ is $\mathbf{R}$-irreducible, as we saw. But we claim that it is *not* isomorphic to an external tensor product of representations of G_1 and G_2.

One way to see this, anticipating a bit some elementary parts of Section 2.7.5 below, is to note that an exterior tensor product $\tau = \varrho_1 \boxtimes \varrho_2$ of two finite-dimensional representations ϱ_1 and ϱ_2 satisfies

$$\mathrm{Tr}(\tau(g_1, g_2)) = \mathrm{Tr}(\varrho_1(g)) \,\mathrm{Tr}(\varrho_2(g))$$

for all $(g_1, g_2) \in G_1 \times G_2$. Since the traces of isomorphic representations are equal functions on G (an easy property, see Proposition 2.7.38 for a formal proof), the claim will be established if we can show that the trace of π is not a product of a function of g_1 and a function of g_2. But for $(\theta_1, \theta_2) \in G_1 \times G_2$, we have

$$\mathrm{Tr}(\pi(\theta_1, \theta_2)) = 2\cos(\theta_1 + \theta_2),$$

and we leave it as an exercise to check that this is not a product function.

2.7.4. Burnside's theorem and its generalizations, 2. As promised, we now explain how to recover the results of the previous section in a style closer (maybe) to that of Frobenius. Even for readers who have fully understood the arguments already used, this may be useful. In fact, the proofs are simpler, but not so well motivated. The viewpoint is to start this time by determining directly the isotypic components of the regular representation.

Proposition 2.7.34 (Isotypic component of the regular representation). *Let k be an algebraically closed field, and let G be a group. For any finite-dimensional irreducible k-representation*

$$\varrho : G \longrightarrow \mathrm{GL}(E)$$

of G, the ϱ-isotypic component $\mathrm{M}(\varrho) \subset C_k(G)$ of the regular representation is isomorphic to a direct sum of $\dim(\varrho)$ *copies of ϱ, and is spanned by the matrix coefficients of ϱ.*

Proof. We start with the realization (2.23) of $C_k(G)$ as the induced representation $\mathrm{Ind}_1^G(\mathbf{1})$ and then apply Frobenius reciprocity (2.24), which gives us linear isomorphisms

$$\begin{aligned}\mathrm{Hom}_G(\varrho, C_k(G)) &= \mathrm{Hom}_G(\varrho, \mathrm{Ind}_1^G(\mathbf{1}))\\ &\simeq \mathrm{Hom}_1(\mathrm{Res}_1^G(\varrho), \mathbf{1}) = \mathrm{Hom}_k(E, k) = E'.\end{aligned}$$

Thus the isotypic component $\mathrm{M}(\varrho)$, which is equal to the image of the injective G-homomorphism

$$\begin{cases} \mathrm{Hom}_G(\varrho, C_k(G)) \otimes E & \longrightarrow & C_k(G) \\ \Phi \otimes v & \mapsto & \Phi(v) \end{cases}$$

(see Proposition 2.7.21), is isomorphic to $E' \otimes E$ as a representation of G, where E' has the trivial action of G, and E the action under ϱ. This is the same as the direct sum of $\dim(E)$ copies of ϱ.

We now show, also directly, that the image of the linear map above is spanned by matrix coefficients. Indeed, given $\lambda \in E'$, the corresponding homomorphism

$$\Phi_\lambda : E \longrightarrow C_k(G)$$

in $\mathrm{Hom}_G(\varrho, C_k(G))$ is given, according to the recipe in (2.28), by

$$\Phi_\lambda(v) = (x \mapsto \lambda(\varrho(x)v)) = f_{v,\lambda},$$

i.e., its values are indeed matrix coefficients. □

Re-proof of Theorem 2.7.28. First of all, we recover Burnside's irreducibility criterion. Consider an irreducible finite-dimensional representation $\varrho : G \longrightarrow \mathrm{GL}(E)$. We know from the proposition that

$$\dim \mathrm{M}(\varrho) = \dim(\varrho)^2.$$

Since $\mathrm{M}(\varrho)$ is spanned by the $\dim(\varrho)^2$ matrix coefficients

$$f_{v_i,\lambda_j}, \quad 1 \leqslant i, j \leqslant \dim \varrho,$$

associated with any basis (v_i) of E and the dual basis (λ_j) of E', these must be independent. But then if we consider the matrix representation ϱ^m of ϱ in the basis (v_i), it follows that there is no non-trivial linear relation between the coefficients of the $\varrho^m(g)$, and hence, by duality, the span of those matrices must be the space of all matrices of size $\dim \varrho$, which means that the linear span of the $\varrho(g)$ in $\mathrm{End}(E)$ is equal to $\mathrm{End}(E)$. This recovers Burnside's criterion (since the converse direction was immediate).

Now consider finitely many irreducible representations (ϱ_i) which are pairwise non-isomorphic. The subspaces $\mathrm{M}(\varrho_i) \subset C_k(G)$ are in direct sum (see Lemma 2.7.22: the intersection of any one with the sum of the others is a subrepresentation where no composition factor is permitted, hence it is zero), and this means that the matrix coefficients of the ϱ_i must be linearly independent—in the sense of the statement of Theorem 2.7.28. □

Exercise 2.7.35. Let G be a group, and let k be an algebraically closed field. Consider the representation of $G \times G$ on $C_k(G)$ by

$$\pi(g,h)f(x) = f(g^{-1}xh) \qquad g,h,x \in G,\ f \in C_k(G),$$

of Exercise 2.1.3. Let $\varrho : G \to \mathrm{GL}(E)$ be a finite-dimensional irreducible k-representation, and let $\mathrm{M}(\varrho)$ denote as usual the span of the matrix coefficients of ϱ. Show that $\mathrm{M}(\varrho)$ is an irreducible subrepresentation of π, and describe its isomorphism class as representation of $G \times G$ as an external tensor product.

2.7.5. Characters of finite-dimensional representations. As another consequence of Theorem 2.7.28, we see that if we are given one matrix coefficient of each of ϱ_1 and ϱ_2, some irreducible k-representations of G, both finite-dimensional, we are certain that they will be distinct functions if ϱ_1 and ϱ_2 are not isomorphic. The converse is not true, since even a single representation has typically many matrix coefficients. However, one can combine some of them in such a way that one obtains a function which only depends on the representation up to isomorphism, and which *characterizes* (finite-dimensional) irreducible representations, up to isomorphism.

Definition 2.7.36 (Characters). Let G be a group, and let k be a field.

(1) A *character* of G over k, or k-character of G, is any function $\chi : G \longrightarrow k$ of the type

$$\chi(g) = \mathrm{Tr}\, \varrho(g),$$

where ϱ is a finite-dimensional k-representation of G. One also says that χ is the character of G.

(2) An *irreducible character* of G over k is a character associated to an irreducible k-representation of G.

Remark 2.7.37. One can, in certain contexts, define characters even for infinite-dimensional representations (see, e.g., [**36**, Ch. X], which discusses the important work of Harish-Chandra in this direction) and obtain a formalism with properties similar to the finite-dimensional case. This requires many tools from functional analysis and is much deeper than what we are discussing here.

We will typically write χ_ϱ for the character of a given representation ϱ. Then we have

Proposition 2.7.38. *Let G be a group, and let k be a field.*

(1) *Two isomorphic finite-dimensional representations ϱ_1 and ϱ_2 of G have the same character as function on G.*

(2) *If k is algebraically closed, then two irreducible finite-dimensional representations ϱ_1 and ϱ_2 of G are isomorphic if and only if their characters are equal as functions on G. More generally, the characters of the irreducible finite-dimensional representations of G, up to isomorphism, are linearly independent in $C_k(G)$.*

(3) *If k is algebraically closed and of characteristic zero, then any two finite-dimensional semisimple representations of G are isomorphic if and only if their characters are equal.*

Proof. For (1), if two representations (irreducible or not, but assumed to be finite dimensional) ϱ_1 and ϱ_2 are isomorphic, with $\Phi : E_1 \to E_2$ giving this isomorphism, we have

$$\Phi \circ \varrho_1(g) \circ \Phi^{-1} = \varrho_2(g)$$

for all $g \in G$, and hence, by the invariance of the trace of endomorphisms under conjugation, we obtain

$$\operatorname{Tr}(\varrho_1(g)) = \operatorname{Tr}(\varrho_2(g))$$

so that their characters are equal.

Now, to prove the converse assertion in (2), we note simply that the character $\operatorname{Tr} \varrho(g)$ is a sum of (diagonal) matrix coefficients: if (v_i) is a basis of the space of ϱ, with dual basis (λ_i), we have

$$\chi_\varrho = \sum_i f_{v_i,\lambda_i}$$

(i.e,

$$\operatorname{Tr} \varrho(g) = \sum_i \langle \lambda_i, \varrho(g) v_i \rangle$$

for all $g \in G$). Hence an equality

$$\chi_{\varrho_1} = \chi_{\varrho_2}$$

is a linear relation between certain matrix coefficients of ϱ_1 and ϱ_2, respectively. If ϱ_1 and ϱ_2 are irreducible but not isomorphic, it follows that such a relation is impossible.

Similarly, expanding in terms of matrix coefficients, we see that any linear relation

$$\sum_{\pi} \alpha(\pi)\chi_\pi = 0$$

among characters of the irreducible finite-dimensional k-representations of G (taken up to isomorphism) must have $\alpha(\pi) = 0$ for all π.

Finally, for (3), let ϱ be a semisimple k-representation. If we write the decomposition

$$\varrho \simeq \bigoplus_{\pi} n_\pi(\varrho)\pi$$

in terms of the irreducible finite-dimensional k-representations π of G (up to isomorphism), with $n_\pi(\varrho) \geqslant 0$ the multiplicity of π in ϱ, we find a corresponding decomposition of the character

$$\chi_\varrho = \sum_{\pi} n_\varrho(\pi)\chi_\pi.$$

By (1), if $\chi_{\varrho_1} = \chi_{\varrho_2}$ for two (semisimple finite-dimensional) representations, we must have $n_\pi(\varrho_1) = n_\pi(\varrho_2)$ for all π. *But* this equality is an equality in k (the characters are k-valued functions); if k has characteristic 0, this implies the corresponding equality of integers, from which we see that ϱ_1 and ϱ_2 are indeed isomorphic. □

Example 2.7.39 (A zero character)**.** Continuing with the notation of the last proof, if k has positive characteristic p, it is possible that the integer $n_\pi(\varrho_1) - n_\pi(\varrho_2)$ is a multiple of p for all π, and then the characters of ϱ_1 and ϱ_2 are the same. Here is an easy example showing that this indeed happens. Consider any subgroup G of the group $U_n(\mathbf{Z}/p\mathbf{Z})$ of unipotent upper-triangular $n \times n$-matrices with coefficients in $\mathbf{Z}/p\mathbf{Z}$, for instance, a subgroup of

$$U_3(\mathbf{Z}/p\mathbf{Z}) = \left\{ \begin{pmatrix} 1 & a & b \\ 0 & 1 & c \\ 0 & 0 & 1 \end{pmatrix} \mid a,\ b,\ c \in \mathbf{Z}/p\mathbf{Z} \right\}.$$

With $k = \mathbf{Z}/p\mathbf{Z}$, the inclusion in $\mathrm{GL}_n(k)$ gives a k-representation

$$\varrho : U_n(\mathbf{Z}/p\mathbf{Z}) \longrightarrow \mathrm{GL}(k^n).$$

Then, if n is a multiple of p, we have

$$\chi_\varrho(g) = 0$$

for all g, despite the fact that ϱ is not trivial. (Here, of course, ϱ is not semisimple and has n trivial composition factors, but the dimension n is equal to 0 in the field k.)

Example 2.7.40 (The character of the regular representation)**.** Let k be any field, and let G be a finite group, so that the space $C_k(G)$ of the regular representation of G is finite dimensional. Then its character is given by

$$\chi(g) = \begin{cases} |G| & \text{if } g = 1 \\ 0 & \text{if } g \neq 1. \end{cases} \tag{2.47}$$

Indeed, we can take as a basis of $C_k(G)$ the family of functions $(\delta_x)_{x \in G}$ equal to 1 at $g = x$ and 0 elsewhere. Then

$$\varrho_G(g)\delta_x = \delta_{xg^{-1}}$$

for all g and $x \in G$. This means that $\varrho_G(g)$ acts on the basis vectors by permuting them, so that the corresponding matrix is a permutation matrix. The trace of $\varrho_G(g)$ is the number of fixed points of this permutation, but we see that $\varrho_G(g)\delta_x = \delta_x$ if and only if $g = 1$, and then x is arbitrary. This gives the formula we claimed. (Note that here also, if the order of the group G is zero in k, the character becomes identically zero.)

Example 2.7.41 (Character of the complex conjugate)**.** We can now justify the remark in Example 2.4.2 concerning the complex conjugate $\bar{\varrho}$ of a finite-dimensional complex representation ϱ of a group G. We have

$$\chi_{\bar{\varrho}}(g) = \overline{\chi_{\varrho}(g)},$$

since (as explained in the paragraph before that example) the matrix representing $\bar{\varrho}(g)$ in a fixed basis is the conjugate of the matrix representing $\varrho(g)$. Since $\bar{\varrho}$ is semisimple if and only if ϱ is (again, this was explained before Example 2.4.2), it follows that if ϱ is semisimple, we have $\varrho \simeq \bar{\varrho}$ if and only if the character of ϱ is real valued.

Proposition 2.7.38 is quite remarkable. It gives us a tool to study representations of a group using only functions on G, and—especially for finite and compact groups—it is so successful that in some cases, one can know all the characters of irreducible representations of a group—as explicit functions—*without* knowing explicit descriptions of the corresponding representations!

Exercise 2.7.42 (Characters are unique)**.** The character of a representation can be interpreted as a linear combination of its matrix coefficients, which turns out to be independent of the specific choice of model of the representation, in the sense of depending only on its isomorphism class. Can we construct any other similar function?

(1) Let k be a field. Explain why finding a linear combination of matrix coefficients for k-representations of dimension n (of any group) which is an invariant under isomorphism amounts to finding a linear form $\lambda : \mathrm{M}_n(k) \longrightarrow k$ such that $\lambda(gxg^{-1}) = \lambda(x)$ for all $x \in \mathrm{GL}_n(k)$ and $g \in \mathrm{M}_n(k)$.

(2) Prove that the trace is, up to multiplication by a scalar, the only such linear form. (*Hint*: The problem is that of finding the $\mathrm{GL}_n(k)$-invariants for the contragredient of the natural conjugation action of $\mathrm{GL}_n(k)$ on $\mathrm{M}_n(k)$.)

Part of the appeal of characters is that they are quite manageable in various computations. The following summarizes some of their formal properties.

Proposition 2.7.43 (Formalism of characters). *Let G be a group, and let k be a field. For any finite-dimensional k-representation ϱ of G, the character χ_ϱ satisfies*

$$\chi_\varrho(gxg^{-1}) = \chi_\varrho(x), \qquad \chi_\varrho(xg) = \chi_\varrho(gx) \tag{2.48}$$

for all g, $x \in G$. Moreover, we have the identities

$$\begin{aligned} \chi_\varrho(1) &= \dim \varrho \quad \text{(seen as an element of } k\text{)},\\ \chi_{\varrho_1 \oplus \varrho_2} &= \chi_{\varrho_1} + \chi_{\varrho_2},\\ \chi_{\varrho_1 \otimes \varrho_2} &= \chi_{\varrho_1}\chi_{\varrho_2},\\ \chi_{\check{\varrho}}(g) &= \chi_\varrho(g^{-1}). \end{aligned}$$

If ϱ is finite-dimensional and has distinct composition factors ϱ_i with multiplicities $n_i \geqslant 1$, then

$$\chi_\varrho = \sum_i n_i \chi_{\varrho_i}.$$

Moreover if $H \subset G$ is a subgroup, we have

$$\chi_{\mathrm{Res}_H^G(\varrho)}(h) = \chi_\varrho(h) \text{ for all } h \in H,$$

and if H is a finite-index subgroup, we have

$$\chi_{\mathrm{Ind}_H^G(\varrho)}(g) = \sum_{\substack{s \in H\backslash G \\ sgs^{-1} \in H}} \chi_\varrho(sgs^{-1}). \tag{2.49}$$

The two statements in (2.48) are equivalent, and they state that the value of a character at some $x \in G$ only depends on the conjugacy class of x in G. Functions with this property are usually called *class functions* on G.

Note that we restrict to a finite-index subgroup for induction because otherwise the dimension of the induced representation is not finite.

The character formula (2.49) makes sense because the property that sgs^{-1} be in H, and the value of the trace of $\varrho(sgs^{-1})$, are both unchanged if s is replaced by any other element of the coset Hs. At least if k has characteristic 0, one can also use this remark to rewrite this formula as

$$\chi_{\mathrm{Ind}_H^G(\varrho)}(g) = \frac{1}{|H|} \sum_{\substack{s \in G \\ sgs^{-1} \in H}} \chi_\varrho(sgs^{-1}),$$

since the sum on the right-hand side is equal to $|H|$ times the expression in (2.49).

It may also be useful to observe that one *cannot* use the invariance of χ_ϱ under conjugation to remove the s in $\chi_\varrho(sgs^{-1})$, since ϱ is only a representation of H, and usually $s \notin H$.

Note also that by taking $g = 1$, this formula implies

$$\dim \mathrm{Ind}_H^G(\varrho) = [G : H] \dim \varrho$$

which recovers Proposition 2.3.11 (as it should), if the field k has characteristic zero (in which case the equality in k gives the same in $\mathbf{Z}$).

Proof. The first formulas are direct consequences of the definitions and the properties of the trace of linear maps. Similarly, the formula for the restriction is clear, and only the case of induction requires proof. The argument will be an elaboration of the proof of the dimension formula (Proposition 2.3.11, or of Proposition 2.3.15).

Let E be the space on which ϱ acts, and let F be the space

$$F = \{f : G \to E \mid f(hx) = \varrho(h)f(x) \text{ for } h \in H,\ x \in G\}$$

of the induced representation. We will compute the trace by decomposing F (as a linear space) conveniently, much as was done in the proof of Proposition 2.3.11 when computing the dimension of F (which is also, of course, the value of the character at $g = 1$). First of all, for any $s \in G$, let $F_s \subset F$ be the subspace of those $f \in F$ which vanish for all $x \notin Hs$; thus F_s only depends on the coset $Hs \in H\backslash G$. We have a direct sum decomposition

$$F = \bigoplus_{s \in H\backslash G} F_s,$$

where the components of a given f are just obtained by taking the restrictions of f to the cosets Hs and extending this by zero outside Hs.

Now, for a fixed $g \in G$, the action of $\mathrm{Ind}(g)$ on F is given by the formula

$$\mathrm{Ind}(g)f(x) = f(xg),$$

which is analogous to that of the regular representation.

It follows from this that $\mathrm{Ind}(g)$ *permutes* the subspaces F_s, and more precisely that $\mathrm{Ind}(g)$ sends F_s to $F_{sg^{-1}}$. In other words, in terms of the direct sum decomposition above, the action of $\mathrm{Ind}(g)$ is a *block-permutation matrix*. Taking the trace (it helps to visualize this as a permutation matrix), we see that it is the sum of the trace of the maps on F_s induced by $\mathrm{Ind}(g)$ where s runs over those cosets $s \in H\backslash G$ for which $Hsg^{-1} = Hs$, i.e., over those s for which $sgs^{-1} \in H$.

Now for any $s \in G$ with $sgs^{-1} \in H$, we compute the trace of the linear map

$$\pi_{s,g} : F_s \longrightarrow F_s$$

induced by $\mathrm{Ind}(g)$. To do this, we use the fact (already used in Proposition 2.3.11) that F_s is isomorphic to E, as a k-vector space. More precisely, there are reciprocal k-linear isomorphisms

$$E \xrightarrow{\alpha} F_s \xrightarrow{\beta} E$$

such that

$$\beta(f) = f(s)$$

on the one hand, and $\alpha(v)$ is the element of F_s mapping hs to $\varrho(h)v$ (and all $x \notin Hs$ to 0). The fact that α and β are inverses of each other is left for the reader to prove (it is contained in the proof of Proposition 2.3.11).

Thus the trace of $\mathrm{Ind}(g)$ is the sum, over those s with $sgs^{-1} \in H$, of the trace of the linear map on E given by $\beta \circ \pi_{s,g} \circ \alpha$. But—and this shouldn't be much of a surprise—this map is simply given by

$$\varrho(sgs^{-1}) : E \longrightarrow E,$$

with trace $\chi_\varrho(sgs^{-1})$ (which is defined because $sgs^{-1} \in H$, of course). The stated formula follows by summing over the relevant s.

We check the claim: given $v \in E$, and $f = \alpha(v)$, we have

$$\begin{aligned}(\beta \circ \pi_{s,g} \circ \alpha)(v) &= \mathrm{Ind}(g)f(s) = f(sg)\\ &= f((sgs^{-1})s) = \varrho(sgs^{-1})f(s) = \varrho(sgs^{-1})v,\end{aligned}$$

using the definitions of α and β. □

The formula for an induced character may look strange or complicated at first. In particular, it is probably not clear just by looking at the right-hand side that it *is* the character of a representation of G! Nevertheless, we will see, here and especially in Chapter 4, that the formula is quite flexible and much nicer than it may seem.

Example 2.7.44. (1) Example 2.7.40 is also a special case of the formula (2.49) for the character of an induced representation. Indeed, we know that the regular representation ϱ_G of a group G (over a field k) is the same as the induced representation $\mathrm{Ind}_1^G(\mathbf{1})$ of the trivial representation of the trivial subgroup $\{1\}$ of G. Hence

$$\chi_{\varrho_G}(g) = \sum_{\substack{s \in G \\ sgs^{-1}=1}} 1,$$

which leads to the formula (2.47) (since the sum is empty, except when $g = 1$, and the conjugacy class of 1 is reduced to $\{1\}$).

(2) Generalizing this, let ϱ be the permutation representation (Section 2.6.2) associated with the action of G on a finite set X. Then we have

$$\chi_\varrho(g) = |\{x \in X \mid g \cdot x = x\}|,$$

i.e., the character value at g is the number of fixed points of g acting on X. Indeed, in the basis (e_x) of the space of ϱ, each $\varrho^m(g)$ is a permutation matrix, and its trace is the number of non-zero diagonal entries (which are equal to 1). These correspond to those $x \in X$ where $\varrho(g)x = e_{gx}$ is equal to e_x, i.e., to the fixed points of g.

(3) Let $H \lhd G$ be a normal subgroup of G of finite index, and let ϱ be a finite-dimensional representation of H. Let $\pi = \mathrm{Ind}_H^G(\varrho)$. Then

$$\chi_\pi(g) = 0$$

for $g \notin H$, since the condition $sgs^{-1} \in H$ means $g \in s^{-1}Hs = H$. For $h \in H$, on the other hand, we have $shs^{-1} \in H$ for all s, and thus

$$\chi_\pi(h) = \sum_{s \in G/H} \chi_\varrho(shs^{-1}).$$

Exercise 2.7.45. Show directly using the character formula that if H is a subgroup of finite index in G, the characters on both sides of the projection formula

$$\mathrm{Ind}_H^G(\varrho_2 \otimes \mathrm{Res}_H^G(\varrho_1)) \simeq \mathrm{Ind}_H^G(\varrho_2) \otimes \varrho_1$$

are identical functions on G.

Example 2.7.46 (Characters of $\mathrm{SL}_2(\mathbf{C})$)**.** Consider $G = \mathrm{SL}_2(\mathbf{C})$ and the representations V_m on homogeneous polynomials of degree m defined in Section 2.6.1 for $m \geqslant 0$. We can compute the character of V_m, to some extent, by using the basis of monomials $e_i = X^i Y^{m-i}$ of the space V_m: by definition, if

$$g = \begin{pmatrix} a & b \\ c & d \end{pmatrix},$$

we have

$$\begin{aligned}\varrho_m(g)(e_i) &= (aX+cY)^i(bX+dY)^{m-i}\\ &= \sum_{k=0}^{i}\sum_{l=0}^{m-i}\binom{i}{k}\binom{m-i}{l}a^kc^{i-k}b^ld^{m-i-l}X^{k+l}Y^{m-k-l}\\ &= \sum_{j=0}^{m}\Bigl(\sum_{k+l=j}\binom{i}{k}\binom{m-i}{l}a^kc^{i-k}b^ld^{m-i-l}\Bigr)e_j\end{aligned}$$

by binomial expansion. The diagonal coefficient here is

$$\sum_{k+l=i}\binom{i}{k}\binom{m-i}{l}a^kc^{i-k}b^ld^{m-i-l},$$

and hence

$$\chi_{\varrho_m}(g) = \sum_{i=0}^{m}\sum_{k+l=i}\binom{i}{k}\binom{m-i}{l}a^kc^{i-k}b^ld^{m-i-l}.$$

This may—or may not—look forbidding. However, if one remembers that the value of the character at g depends only on the conjugacy class of g, one can simplify this, at least for certain elements. Suppose for instance that g is diagonalizable; hence, it is conjugate to a matrix

$$t(\lambda) = \begin{pmatrix}\lambda & 0\\ 0 & \lambda^{-1}\end{pmatrix}$$

for some $\lambda \in \mathbf{C}^\times$ (this will be true very often, e.g., whenever the eigenvalues of g are distinct). The computation of $\chi_{\varrho_m}(g)$ is then much easier: we have indeed

$$\varrho_m(t(\lambda))e_i = (\lambda X)^j(\lambda^{-1}Y)^{m-i} = \lambda^{2i-m}e_i$$

for $0 \leqslant i \leqslant m$ (as in the proof of Proposition 2.7.12). Hence, we obtain the formula

$$\chi_{\varrho_m}(t(\lambda)) = \lambda^{-m} + \lambda^{-m+2} + \cdots + \lambda^{m-2} + \lambda^m = \frac{\lambda^{m+1}-\lambda^{-m-1}}{\lambda-\lambda^{-1}}. \tag{2.50}$$

(This computation corresponds to the fact that the restriction of V_m to the diagonal subgroup T is the direct sum (2.41), as seen in Example 2.7.11. Note in particular that, as a function of $\lambda \in \mathbf{C}$, this expression is not real-valued, and hence by Example 2.7.41, the complex conjugate of ϱ_m is not isomorphic to ϱ_m, as we stated after Theorem 2.6.1.)

If we specialize even further to $\lambda = e^{i\theta}$ with $\theta \in \mathbf{R}$ (i.e., to $t(\lambda)$ being a unitary matrix), we obtain

$$\chi_{\varrho_m}\left(\begin{pmatrix}e^{i\theta} & 0\\ 0 & e^{-i\theta}\end{pmatrix}\right) = \frac{\sin((m+1)\theta)}{\sin(\theta)} \tag{2.51}$$

(with $\theta = 0$ mapping of course to $m + 1$); these character values are simple, and fundamental, trigonometric functions.

Suppose on the other hand that $g = u(t)$ is conjugate to an upper-triangular unipotent matrix

$$u(t) = \begin{pmatrix} 1 & t \\ 0 & 1 \end{pmatrix}$$

with $t \in \mathbf{C}$. Then we have

$$\varrho_m(u(t))e_i = X^i(tX + Y)^{m-i},$$

and if we expand the second term, we see quickly that this is of the form

$$X^iY^{m-i} + (\text{combination of } e_j \text{ with } j > i),$$

leading in particular to

$$\chi_{\varrho_m}(u(t)) = m + 1$$

for all t.

Exercise 2.7.47. Prove that if $m \geqslant n \geqslant 0$, the characters of the two sides of the Clebsch–Gordan formula (2.35) coincide for as large a set of (conjugacy classes of) $g \in \mathrm{SL}_2(\mathbf{C})$ as you can.

The first part of Proposition 2.7.43 shows that the subset of $C_k(G)$, whose elements are characters of finite-dimensional k-representations of G, is stable under addition and multiplication. It is therefore quite natural to consider the abelian group generated by all characters, as a subgroup of the additive group of $C_k(G)$. Indeed, the tensor product formula shows that this group is in fact a ring.

Definition 2.7.48 (Generalized, or virtual, characters)**.** Let G be a group, and let k be a field of characteristic 0. The *character ring* $R_k(G)$ of *generalized characters* of G over k is the ring generated, as an abelian group, by the characters of finite-dimensional k-representations of G. The elements of $R_k(G)$ are also called *virtual characters*. The *dimension* of a virtual character χ is defined to be $\chi(1) \in \mathbf{Z} \subset k$.

Note that $R_k(G)$ is *not* a k-vector space: we do not allow linear combinations of characters with coefficients which are not integers. Concretely, a virtual character $\chi \in R_k(G)$ is a function

$$\chi : G \to k$$

of the form

$$\chi = \chi_{\varrho_1} - \chi_{\varrho_2}$$

for some finite-dimensional k-representations ϱ_1 and ϱ_2 of G. Note that ϱ_1 and ϱ_2 are by no means unique (e.g., one may replace ϱ_1 by $\varrho_1 \oplus \varrho_3$, and ϱ_2 by $\varrho_2 \oplus \varrho_3$, for an arbitrary finite-dimensional representation ϱ_3).

Example 2.7.49 (Induced representations as an ideal in $R_k(G)$). Consider the ring $R_k(G)$ of a group G and a homomorphism $\phi : H \to G$ with image $\mathrm{Im}(\phi)$ of finite index in G. Now consider the subgroup I_ϕ of $R_k(G)$ generated—as an abelian group—by all characters of induced representations $\mathrm{Ind}_H^G(\varrho)$, where ϱ is a finite-dimensional representation of H. Then the projection formula (Proposition 2.3.18) shows that $I_\phi \subset R_k(G)$ is *an ideal*, i.e., $\chi_1\chi_2 \in I_\phi$ if χ_1 in I_ϕ and χ_2 is arbitrary.

We will present some applications of the character ring in Section 4.8.1. We will also say quite a bit more about characters, for finite and compact groups, in Chapters 4 and 5.

2.8. Some Clifford theory

We have already observed that, in general, there are no obvious relations between irreducible representations of a group G and those of a subgroup H. We consider here a special case of this question as an illustration of many aspects of the general formalism. This is the case when we have a group G with an exact sequence

$$1 \longrightarrow H \longrightarrow G \longrightarrow A \longrightarrow 1$$

and the quotient group A is abelian. We can then find some precise links between irreducible representations of G and those of H. This situation, although it seems rather special, does have important applications, in particular in number theory. In these, it is usually the case that the representation theory for H is considered to be simpler: this indicates why we attempt to understand representations of G by restricting them to H.

The first result does not require that A be abelian.

Proposition 2.8.1. *Let k be an algebraically closed field, let G be a group, and let $H \lhd G$ be a normal subgroup. Let ϱ be a finite-dimensional irreducible representation of G. Then either $\mathrm{Res}_H^G(\varrho)$ is an isotypic representation of H or else there exists a proper subgroup $H_1 \supset H$ of G and an irreducible representation π of H_1 such that $\varrho \simeq \mathrm{Ind}_{H_1}^G(\pi)$. In that second case, the character of ϱ is zero outside of the subgroup H_1.*

Proof. First, by Exercise 2.3.4,[20] the restriction of ϱ to H is semisimple. Let E be the space on which ϱ acts, and let

$$E = \bigoplus_{\pi} \mathrm{M}(\pi)$$

[20] If one does not wish to use this exercise, one might just add this as an assumption, and note that it holds in the cases where all k-representations of G are semisimple, e.g., when G is finite.

be the canonical decomposition of E, as a representation of H, into a direct sum of isotypic components, each of which is an isotypic representation attached to an irreducible representation π of H. If there is only a single term in this direct sum, then by definition we are in the case where $\mathrm{Res}_H^G(\varrho)$ is isotypic. Thus we assume that this is not the case.

Because H is normal in G, the group G acts on the H-subrepresentations of E: if $F \subset E$ is H-stable, then $\varrho(g)F$ is also an H-subrepresentation since

$$\varrho(h)(\varrho(g)v) = \varrho(g)\varrho(g^{-1}hg)v \in \varrho(g)F.$$

This formula also indicates that in this action, any H-irreducible subrepresentation F isomorphic to some given representation π is sent by $\varrho(g)$ to a subrepresentation isomorphic to the representation

$$g \mapsto \pi(g^{-1}hg)$$

of H, which is also irreducible (the H-subrepresentations of F and $\varrho(g)F$ correspond under $\varrho(g)$ for the same reasons). In other words, G acts on the finite set of H-isotypic components $\mathrm{M}(\pi)$.

This action is transitive, because the direct sum of the isotypic components corresponding to any non-empty G-orbit is a G-subrepresentation of E, hence equal to E since we assumed that ϱ is irreducible.

Now fix some isotypic component π_1 occurring in E, and let H_1 denote the stabilizer of $F = \mathrm{M}(\pi_1)$ for this action of G. We note that H_1 contains H, since $\mathrm{M}(\pi_1)$ is an H-subrepresentation of E, and is not equal to G, since otherwise E would be H-isotypic, which we assumed here is not the case. Also the index of H_1 in G is finite since $\dim E < +\infty$.

By definition of the stabilizer, the isotypic space F is an H_1-subrepresentation of E. Similarly, all other isotypic spaces are stable under the action of H_1 (since they are of the form $\varrho(g)F$ for some $g \in G$). We therefore find a decomposition

$$E = \bigoplus_{\pi} \mathrm{M}(\pi)$$

into H_1-subrepresentations, where each summand is of the form $\varrho(g)F$ for some $g \in G$. The number of these summands is $|G|/|H_1| = [G : H_1]$ (since G acts transitively on the set of summands and H_1 is the stabilizer of F). Thus Proposition 2.3.15 allows us to recognize that $E \simeq \mathrm{Ind}_{H_1}^G(F)$. Now we conclude that F is irreducible as a representation of H_1 because ϱ was assumed to be irreducible (see Exercise 2.3.14).

The final statement concerning the character values was already mentioned in Example 2.7.44(3). □

Now we consider some special properties of the case where the quotient A is abelian.

Proposition 2.8.2. *Let k be an algebraically closed field, let G be a group, and let $H \lhd G$ be a normal subgroup such that $A = G/H$ is abelian. Let ϱ_1 and ϱ_2 be finite-dimensional irreducible k-representations of G acting on E_1 and E_2, respectively. Then*

$$\mathrm{Res}_H^G(\varrho_1) \simeq \mathrm{Res}_H^G(\varrho_2)$$

if and only if there exists a one-dimensional representation χ of A such that

$$\varrho_2 \simeq \varrho_1 \otimes \chi,$$

where χ is viewed as a character of G via the quotient map $G \longrightarrow A$.

Proof. One implication is clear: since a character χ of A is extended to G in such a way its kernel contains H, the restriction to H of $\varrho \otimes \chi$ is equal to that of ϱ, for any representation ϱ.

The converse seems intuitive enough, but the argument, easy as it might look, is a rather good illustration of the formalism of representation theory, and the reader might want to try to solve the problem individually first (see also the following Remark 2.8.3).

We consider the space $\mathrm{Hom}_k(E_1, E_2)$ with its natural action (2.15) of G and the subspace $E = \mathrm{Hom}_H(E_1, E_2) = \mathrm{Hom}_k(E_1, E_2)^H$. The space E is non-zero, since the restrictions of ϱ_1 and ϱ_2 to H are isomorphic. Since H is normal in G, E is a G-subrepresentation of $\mathrm{Hom}_k(E_1, E_2)$ (see Lemma 2.1.12), and since H acts trivially, this means that E is a representation of the abelian quotient A. Since $E \neq 0$, this representation contains some irreducible subrepresentation, which is one dimensional (because A is abelian and k is algebraically closed, see Corollary 2.7.17). Let χ be such an irreducible representation of A in E. Now one can either deduce from the G-intertwiner

$$\chi \hookrightarrow \mathrm{Hom}_k(\varrho_1, \varrho_2) \simeq \check{\varrho}_1 \otimes \varrho_2$$

that, by *twisting* by $\chi^{-1} = \check{\chi}$, we have

$$\mathbf{1} \hookrightarrow (\varrho_1 \otimes \chi)\check{} \otimes \varrho_2$$

and hence $\varrho_1 \otimes \chi \simeq \varrho_2$ by Schur's Lemma (both are irreducible), or more concretely, one can take a non-zero element $\Phi \in E$ in the space of the subrepresentation isomorphic to χ. Spelling out what this means, one finds that

$$\varrho_2(g)\Phi(\varrho_1(g^{-1})v) = \chi(g)\Phi(v)$$

for all $g \in G$ and $v \in E_1$, which means that $\Phi \in \mathrm{Hom}_G(\varrho_1, \varrho_2 \otimes \chi^{-1})$. Since it is non-zero, Schur's Lemma again proves that $\varrho_1 \simeq \varrho_2 \otimes \chi^{-1}$. ☐

Remark 2.8.3. We used the representation on $\mathrm{Hom}_k(\varrho_1, \varrho_2)$ to compare the two representations ϱ_1 and ϱ_2. It is worth contrasting this argument with the following cruder version: since both representations are isomorphic on H, they have in particular the same dimension, and composing with an H-isomorphism, we are reduced to the case where ϱ_1 and ϱ_2 act on the same space E, and satisfy $\varrho_1(h) = \varrho_2(h)$ for all $h \in H$. Now one can check directly that, for any fixed $g \in G$, the linear map $\alpha(g) = \varrho_2(g)^{-1} \circ \varrho_1(g)$ on E is an H-intertwiner. If we assume that $\mathrm{Res}_H^G(\varrho_1)$ is irreducible, then Schur's Lemma shows that $\alpha(g) = \chi(g)\mathrm{Id}$ for some scalar $\chi(g) \in k^\times$, and of course it is then easy to check that χ is a character, and hence that $\varrho_1 = \varrho_2 \otimes \chi$. The problem with this down-to-earth reasoning is that it is rather messy to try to weaken the assumption of irreducibility of the restriction to H! Philosophically, it is not representation-theoretic enough....

Exercise 2.8.4. Give a different proof of Proposition 2.8.2 by showing directly that (with the same notation) if ϱ_1 and ϱ_2 have isomorphic restrictions to H, then

$$\varrho_2 \simeq \varrho_1 \otimes \mathrm{Hom}_H(\varrho_1, \varrho_2)$$

as representation of G, where $\mathrm{Hom}_H(\varrho_1, \varrho_2)$ is given the structure of a G-representation as in the above proof.

Exercise 2.8.5. We consider in this exercise the situation and notation of Proposition 2.8.2, and we assume that A is finite.

(1) If ϱ_1 and ϱ_2 are irreducible representations of G, which are both H-isotypic and have the same H-irreducible component, show that $\varrho_1 \simeq \varrho_2 \otimes \chi$ for some character χ of A.

(2) For a finite-dimensional representation ϱ of G, show that

$$\mathrm{Ind}_H^G(\mathrm{Res}_H^G(\varrho)) \simeq \bigoplus_{\chi} \varrho \otimes \chi,$$

where χ runs over all one-dimensional representations of A.

2.9. Conclusion

We have now built, in some sense, the basic foundations of representation theory. Interestingly, there are other basic results, also valid in great generality, for which no proof is known using only the methods of this chapter (which consist, essentially, of elementary abstract algebra). These statements require the theory of *algebraic groups*. A very short introduction with a discussion of some of the results it leads to can be found in the beginning of Chapter 7.

Before closing this chapter, we can use the vocabulary we have accumulated to ask, What are the fundamental questions of representation theory? The following are maybe the two most obvious ones:

- *Classification*. For a group G, one may want to classify all its irreducible representations (say over the complex numbers) or all representations of a certain type. This is possible in a number of very important cases and is often of tremendous importance for applications. One should think here of a group G which is fairly well understood from a group-theoretic point of view; knowing its representations is a natural way to deepen the understanding of the group.

- *Decomposition.* Given a group G again and a representation E of G, which arises naturally in some application, one may ask to find explicitly the irreducible components of E, either as summands if E is semisimple or simply as composition factors. If the group is sufficiently complicated, this decomposition might involve a *direct integral* (the simplest example comes from the theory of Fourier integrals, see Section 7.3) and not merely a direct sum.

In Chapters 4 and 5, we will see a number of examples of the first problem for specific cases of finite groups, and for some compact topological groups. Indeed, this problem is rather well understood for some of these groups, including very important ones for applications. We will also see a few cases of the second problem, one of them being responsible for some of the basic properties of the hydrogen atom (see Section 6.4).

Of course, these few examples are only sample illustrations. Among the areas of modern mathematics where one finds many others, we will just mention that representation theory is absolutely crucial to the so-called Langlands Program in number theory. We refer to [**3**] for a general survey of many aspects of this theory.

Chapter 3

Variants

We present here some of the variants of the notions of representation theory that were discussed in the previous chapter. They are all very important topics in their own right, and we will encounter them intermittently in later chapters.

3.1. Representations of algebras

If, instead of a group G, we consider an *algebra* A over a field k, i.e., a (not necessarily commutative) ring which has a compatible structure of k-vector space, the analogue of a representation of A is an algebra homomorphism

$$A \xrightarrow{\varrho} \mathrm{End}_k(E).$$

This may be called a *representation of* A, but it is more usual to focus on E and to note that such a map defines a structure of A-module on the vector space E by

$$a \cdot v = \varrho(a)v$$

for $a \in A$ and $v \in E$.

It is important for certain aspects of representation theory that the k-representations of a group G can be understood in this language. Associated to G is the vector space $k(G)$ freely generated by G (which already appeared at the beginning of the previous chapter). Using temporarily the notation $[g]$ to indicate the g-th basis vector of $k(G)$, this vector space has a structure of k-algebra if one defines the product on $k(G)$ by

$$[g] \cdot [h] = [gh]$$

and extends it by linearity. Thus, given elements ℓ, $m \in k(G)$ that we express as linear combinations

$$\ell = \sum_{g \in G} \lambda_g [g], \qquad m = \sum_{h \in G} \mu_h [h],$$

with only finitely many non-zero coefficients $\lambda_g \in k$ and $\mu_h \in k$, the product ℓm is given by

$$\ell m = \sum\sum_{g,h \in G} \lambda_g \mu_h [g][h] = \sum_{x \in G} \Big(\sum_{gh=x} \lambda_g \mu_h \Big) [x] = \sum_{x \in G} \Big(\sum_{y \in G} \lambda_{xy^{-1}} \mu_y \Big) [x].$$

Note that if G is not abelian, this algebra is not commutative.

Example 3.1.1. Let G be a finite group, let k be a field, and consider the element

$$s = \sum_{g \in G} [g] \in k(G).$$

Then by expanding the square, we find in $k(G)$ that

$$s^2 = \sum\sum_{g,h \in G} [gh] = \sum_{x \in G} \Big(\sum_{[g][h]=x} 1 \Big) x = |G| s. \tag{3.1}$$

Multiplying on the left or the right with $x \in k(G)$, we get

$$s[x] = \sum_{g \in G} [g][x] = \sum_{g \in G} [gx] = \sum_{h \in G} [h] = s \tag{3.2}$$

and similarly

$$[x]s = s.$$

By linearity, we see that $sa = as$ for all $a \in k(G)$; thus the element s is in the *center* of the algebra $k(G)$.

Definition 3.1.2 (Group algebra)**.** Let k be a field, and let G be a group. The algebra $k(G)$ defined above is called the *group algebra* of G over k.

The characteristic algebraic property of the group algebra is the following.

Proposition 3.1.3 (Representations as modules over the group algebra)**.** *Let G be a group, and let k be a field. If*

$$G \xrightarrow{\varrho} \mathrm{GL}(E)$$

is a k-representation of E, then E has a structure of $k(G)$-module defined by extending ϱ by linearity, i.e.,

$$\Big(\sum_{g \in G} \lambda_g [g] \Big) \cdot v = \sum_{g \in G} \lambda_g \varrho(g) v.$$

Conversely, any $k(G)$-module E inherits a representation of G by restricting the multiplication-by-a *maps to the basis vectors $[g]$, i.e., by putting*

$$\varrho(g) = (v \mapsto [g] \cdot v),$$

and furthermore, a k-linear map $\Phi : \varrho_1 \longrightarrow \varrho_2$ is a G-morphism if and only if Φ is $k(G)$-linear.

Sometimes, one denotes by $\varrho(a)$ the map $v \mapsto a \cdot v$. Thus $\varrho(a)$ becomes an element in $\mathrm{Hom}_k(E, E)$.

Here are some reasons why this approach can be very fruitful:

- It gives access to all the terminology and results of the theory of algebras; in particular, for any ring A, the notions of sums, intersections, direct sums, etc., of A-modules, are well defined and well known. For $A = k(G)$, they correspond to the definitions already given in Chapter 2 for linear representations. Other constructions are however more natural in the context of algebras. For instance, one may consider the ideals of $k(G)$ and their properties, which is not something so natural at the level of the representations themselves.
- The $k(G)$-modules parametrize, in the almost tautological way we have described, the k-representations of G. It may happen that one wishes to concentrate attention on a special class $\mathcal{C}$ of representations, characterized by some property. It happens sometimes that these representations correspond in a natural way to all (or some of) the representations of another group $G_{\mathcal{C}}$. An example is to consider for $\mathcal{C}$ the class of one-dimensional representations, which correspond to those of the abelianized group $G/[G, G]$, as in Proposition 2.3.3. This phenomenon is rather rare, but it may happen more often that there is a natural k-algebra $A_{\mathcal{C}}$ such that *its* modules correspond precisely (and "naturally") to the representations in $\mathcal{C}$.

Partly for reasons of personal habit (or taste), we will not exploit the group algebra systematically in this book. This can be justified by the fact that it is not absolutely necessary at this level of work. But we will say a bit more, e.g., in Section 4.3.6 and Exercise 4.3.31 describes a property of representations with cyclic vectors which is much more natural from the point of view of the group algebra.

Exercise 3.1.4 (Representations with a fixed vector under a subroup)**.** Let G be a finite group, and let $H \subset G$ be a subgroup. We will describe a subalgebra $\mathcal{H}$ of $\mathbf{C}(H)$ such that the irreducible complex representations of

G which contain non-zero H-invariant vectors are in correspondance with certain $\mathcal{H}$-modules.

(1) Show that

$$\mathcal{H} = \{a \in \mathbf{C}(G) \mid h_1 a h_2 = a \text{ for all } h_1,\ h_2 \in H\}$$

is a subalgebra of $\mathbf{C}(G)$, and that it is generated as $\mathbf{C}$-vector space by the elements

$$\eta_x = \sum_{y \in HxH} [y] \in \mathbf{C}(G), \qquad x \in G,$$

where HxH denotes a *double coset* of H in G, i.e.,

$$HxH = \{h_1 x h_2 \mid h_1,\ h_2 \in H\} \subset G.$$

(2) Show that if $\varrho : G \longrightarrow \mathrm{GL}(E)$ is a representation of G, the subspace E^H of H-invariant vectors is stable under multiplication by $\mathcal{H}$, i.e., that E^H is an $\mathcal{H}$-module.

(3) Show that if $\Phi \in \mathrm{Hom}_G(E, F)$, the restriction of Φ to E^H is an $\mathcal{H}$-linear map from E^H to F^H.

(4) Let $\varrho : G \longrightarrow \mathrm{GL}(E)$ be a complex representation of G. If $F \subset E^H$ is an $\mathcal{H}$-submodule of E^H, and $\tilde{F} = \mathbf{C}(G)F$ is the subrepresentation of E generated by F, show that $\tilde{F}^H = F$. (*Hint*: Show that a vector $v \in \tilde{F}^H$ is of the form $v = \frac{1}{|HxH|}\eta_x w$ for some $x \in G$ and $w \in F$.)

(5) Let $\varrho : G \longrightarrow \mathrm{GL}(E)$ be a semisimple representation of G such that $E^H \neq 0$. Show that ϱ is irreducible if and only if E^H is simple as an $\mathcal{H}$-module, i.e., if and only if E^H contains no proper non-zero submodule.

(6) Show that if ϱ_1, ϱ_2 are irreducible representations of G on E_1, E_2, respectively, with $E_i^H \neq 0$, the $\mathcal{H}$-modules E_1^H and E_2^H are isomorphic if and only if $\varrho_1 \simeq \varrho_2$. (*Hint*: When $E_1^H \simeq E_2^H$, consider the subrepresentation of $E_1 \oplus E_2$ generated by the graph of an $\mathcal{H}$-isomorphism.)

We will use the $k(G)$-module structure corresponding to representations a number of times in the remainder of the book. Usually, the notation $[g]$ will be abandoned in doing so, and we will write

$$\sum_i a_i g_i$$

for elements in $k(G)$, for instance we will write

$$s = \sum_{g \in G} g \in k(G)$$

for the element of Example 3.1.1.

Exercise 3.1.5 (The group ring as *universal* endomorphisms)**.** A fixed $a \in k(G)$ has the feature that, for any representation $\varrho : G \longrightarrow \mathrm{GL}(E)$, there is a corresponding k-linear endomorphism given by its action on E, i.e., the linear map

$$\varepsilon_\varrho^{(a)} \begin{cases} E \longrightarrow E \\ v \mapsto \varrho(a)v \end{cases}$$

in $\mathrm{Hom}_k(E, E)$.

These maps are *functorial* in ϱ, in the sense that for any other representation $\pi : G \longrightarrow \mathrm{GL}(F)$ and any morphism of representations $\Phi \in \mathrm{Hom}_G(\varrho, \pi)$, the square diagram

$$\begin{array}{ccc} E & \xrightarrow{\varepsilon_\varrho^{(a)}} & E \\ \Phi \downarrow & & \downarrow \Phi \\ F & \xrightarrow{\varepsilon_\pi^{(a)}} & F \end{array}$$

is commutative. This exercise shows that this property characterizes the group algebra. Namely consider now any map

$$\varrho \mapsto \varepsilon_\varrho$$

that associates a linear map $\varepsilon_\varrho \in \mathrm{Hom}_k(E, E)$ to any k-representation ϱ of G, in such a way that the analogue of the property above is valid, i.e., $\Phi \circ \varepsilon_\varrho = \varepsilon_\pi \circ \Phi$ for $\Phi \in \mathrm{Hom}_G(\varrho, \pi)$. We will show that there exists a unique $a \in k(G)$, such that we have $\varepsilon_\varrho = \varepsilon_\varrho^{(a)}$ for all representations ϱ.

(1) Show that there exists $a \in k(G)$ such that $\varepsilon_{k(G)}$ is the linear map

$$\varepsilon_{k(G)}^{(a)} : x \mapsto ax$$

on $k(G)$, seen as a $k(G)$-module by multiplication on the left. (*Hint*: Consider the maps $\Phi : x \mapsto xb$ on $k(G)$.)

(2) With a as in (1), show that $\varepsilon_\varrho = \varepsilon_\varrho^{(a)}$ for any representation ϱ with a cyclic vector v_0.

(3) Conclude that $\varepsilon_\varrho = \varepsilon_\varrho^{(a)}$ for all representations.

(4) Show that the universal endomorphisms associated to a and $b \in k(G)$ are the same if and only if $a = b$.

(5) Show that $a \in k(G)$ is such that $\varepsilon_\varrho^{(a)}$ is in the subspace $\mathrm{End}_G(\varrho)$ of self-intertwiners of ϱ, for all representations ϱ of G, if and only if a is in the center of the group algebra, i.e., if and only if $ax = xa$ for all $x \in k(G)$.

A motivating application of this exercise appears in Section 4.3.6.

3.2. Representations of Lie algebras

This section can be safely skipped in a first reading, since it will only be mentioned again in Chapters 6 and 7. However, it should become clear then (especially from the analysis of representations of $\mathrm{SL}_2(\mathbf{R})$ in Section 7.4) that the concepts we quickly discuss here are of great importance and usefulness.

Besides associative algebras, as in the previous section, there exist other important and natural algebraic structures which lead to corresponding representation theories. A particularly appealing one is that of a *quiver*, for which an excellent introduction is found in [**18**, §5]. But the most important case, besides that of groups and algebras, is probably that of Lie algebras, because of the intimate connection between Lie groups and Lie algebras.

We present here only the simplest definitions and one of the most basic and crucial results in the representation theory of Lie algebras: the classification of the finite-dimensional irreducible representations of the three-dimensional simple complex Lie algebra $\mathfrak{sl}_2$.

Definition 3.2.1 (Lie algebra)**.** Let k be a field. A *Lie algebra* over k is a k-vector space L given with a k-bilinear map

$$[\cdot,\cdot]\begin{cases} L \times L & \longrightarrow & L \\ (x,y) & \mapsto & [x,y] \end{cases}$$

often called the *(Lie) bracket* of L, which satisfies the following rules:

$$[x,y] = -[y,x] \text{ for all } x,y \in L,$$
$$[[x,y],z] + [[y,z],x] + [[z,x],y] = 0 \text{ for all } x,y,z \in L.$$

The second condition is called the *Jacobi identity*.[1]

There are two major natural sources of Lie algebras. One is the tangent space at the identity of a Lie group, with its Lie bracket (see Section 6.1), or the Lie algebra of smooth vector fields on an arbitrary smooth manifold, which are real Lie algebras. The other is any associative k-algebra A, which has an associated Lie algebra structure obtained by defining

$$[x,y] = xy - yx$$

[1] A mnemonic: a sum of three iterated brackets, where the bracket positions are fixed, and the arguments are permuted cyclically.

for $(x, y) \in A \times A$. Indeed, this clearly defines a k-bilinear operation on A which is antisymmetric, and the Jacobi identity is obtained by a straightforward computation:

$$\begin{aligned}[][x,y],z] + [[y,z],x] &= (xy - yx)z - z(xy - yx) + (yz - zy)x - x(yz - zy) \\ &= -yxz - zxy + yzx + xzy \\ &= y(zx - xz) - (zx - xz)y \\ &= [y,[z,x]] = -[[z,x],y].\end{aligned}$$

In particular, if E is a k-vector space, we obtain a Lie algebra structure, denoted $\mathfrak{gl}(E)$, on the k-algebra $\mathrm{End}_k(E)$.

Definition 3.2.2 (Homomorphisms, representations). Let k be a field.

(1) A morphism between Lie algebras L_1 and L_2 over k is a k-linear map $\phi : L_1 \longrightarrow L_2$ such that $\phi([x, y]) = [\phi(x), \phi(y)]$ for all x, $y \in L_1$. A subspace M of a Lie algebra L is a subalgebra if the restriction of the bracket to M gives it a Lie algebra structure, i.e., if $[x, y] \in M$ for all x, y in M.

(2) A k-representation of a k-Lie algebra L is a morphism

$$\phi : L \longrightarrow \mathfrak{gl}(E)$$

for some k-vector space E. One says that L "acts on E by ϕ", and one writes $x \cdot v$ instead of $\phi(x)(v)$ for $x \in L$ and $v \in E$.

(3) A homomorphism $\phi_1 \longrightarrow \phi_2$ of representations of L, with ϕ_1 and ϕ_2 acting on E_1 and E_2, respectively, is a k-linear map $E_1 \xrightarrow{\Phi} E_2$ such that

$$\Phi(x \cdot v) = \Phi(x) \cdot v$$

for all $x \in L$ and $v \in E_1$.

Example 3.2.3. (The statements in these examples are simple exercises that the reader should work out individually.)

(1) Let E be a k-vector space. The subspace

$$\mathfrak{sl}(E) = \{x \in \mathfrak{gl}(E) = \mathrm{End}_k(E) \mid \mathrm{Tr}(x) = 0\}$$

is a subalgebra of $\mathfrak{gl}(E)$. Similarly, if B is a bilinear form defined on E, the subspace

$$\mathfrak{o}(E, B) = \{x \in \mathfrak{gl}(E) \mid B(x \cdot v, w) + B(v, x \cdot w) = 0\}$$

of $\mathfrak{gl}(E)$ is a subalgebra of E. (To check this, it may be simpler to express B in the form $B(v, w) = {}^t vMw$ for some matrix M.)

(2) Let A be an associative k-algebra. A derivation ∂ of A is a k-linear map $A \xrightarrow{\partial} A$ such that

$$\partial(xy) = x\partial(y) + \partial(x)y$$

for all x, $y \in A$. The set $\mathrm{Der}_k(A)$ of all derivations of A is a vector subspace of $\mathrm{End}_k(A)$, and a simple computation (left as exercise) shows that $\mathrm{Der}_k(A)$ is a subalgebra of $\mathfrak{gl}(A)$. The name of derivation is justified, of course, by the example where A is the algebra of smooth functions defined on $\mathbf{R}$, and $\partial f = f'$ is the derivative of a function $f \in A$.

(3) Let L be a k-Lie algebra. The adjoint representation of L is the representation

$$\mathrm{ad} : L \longrightarrow \mathfrak{gl}(L)$$

of L on its underlying vector space L that is defined by

$$\mathrm{ad}(x)(y) = [x, y]$$

for all x, y in L. It is indeed a representation, and in fact this property is equivalent with the Jacobi identity. Furthermore, another easy computation using the Jacobi identity shows that $\mathrm{ad}(x)$ is a derivation of $\mathrm{End}_k(E)$.

It is clear that the kernel of ad is the *center*

$$\mathfrak{z}(L) = \{x \in L \mid [x, y] = 0 \text{ for all } y \in L\}$$

of L. For instance, one deduces easily that the adjoint representation of $\mathfrak{sl}(E)$ is injective.

There is a formalism of representations of Lie algebras that closely parallels the formalism for groups, and which is indeed directly related when specialized to Lie groups. For instance, an irreducible representation is one of dimension $\geqslant 1$, where there is no non-zero subspace that is stable under the action of the Lie algebra.

Exercise 3.2.4. Let k be an algebraically closed field, and let L be a k-Lie algebra. We denote by $\mathrm{Hom}_L(\phi_1, \phi_2)$ the vector space of homomorphisms of Lie algebra representations between two such representations ϕ_1 and ϕ_2.

(1) Show that Schur's Lemma holds for finite-dimensional irreducible k-representations of L: for two irreducible Lie algebra representations ϕ_1 and ϕ_2, we have

$$\dim \mathrm{Hom}_L(\phi_1, \phi_2) = \begin{cases} 0 & \text{if } \phi_1 \text{ and } \phi_2 \text{ are not isomorphic,} \\ 1 & \text{if } \phi_1 \text{ and } \phi_2 \text{ are isomorphic.} \end{cases}$$

(2) What is the analogue of a trivial representation for L? What is the analogue of the space of invariants of a representation? (*Hint*: It might be useful here to think of the example of a Lie group and of how representations of Lie groups give rise to representations of the Lie algebra; see for instance (3.4).)

(3) Given a representation ϕ of L acting on E, can you define a representation of L on $\mathrm{End}_k(E)$ such that the space of invariants of this representation is equal to $\mathrm{Hom}_L(\phi, \phi)$?

We will present only one gem from the vast field of Lie algebras: we will determine the finite-dimensional irreducible representations of the complex Lie algebra $\mathfrak{sl}(\mathbf{C}^2)$, or in other words, of the space of complex matrices of size 2 with trace 0.

Theorem 3.2.5. *For any integer $n \geqslant 0$, there exists a unique irreducible representation ϕ_n of dimension $n+1$ of $\mathfrak{sl}(\mathbf{C}^2)$. It is characterized among irreducible representations, up to isomorphism, by the fact that there exists a non-zero vector v such that*

$$\phi_n\left(\begin{pmatrix} 0 & 1 \\ 0 & 0 \end{pmatrix}\right)v = 0, \qquad \phi_n\left(\begin{pmatrix} 1 & 0 \\ 0 & -1 \end{pmatrix}\right)v = nv.$$

This vector is unique up to multiplication by a non-zero scalar and is called a highest weight vector *of ϕ_n.*

Sketch of proof. First, we denote $L = \mathfrak{sl}(\mathbf{C}^2)$ for simplicity. Then we observe that the elements

$$e = \begin{pmatrix} 0 & 1 \\ 0 & 0 \end{pmatrix}, \qquad h = \begin{pmatrix} 1 & 0 \\ 0 & -1 \end{pmatrix}, \qquad f = \begin{pmatrix} 0 & 0 \\ 1 & 0 \end{pmatrix}$$

form a basis of L (as a vector space). A straightforward computation shows that they satisfy the relations

$$[h, e] = 2e, \qquad [h, f] = -2f, \qquad [e, f] = h.$$

We see from these in particular that both h and $\mathrm{ad}(h)$ are diagonalizable with *integral* eigenvalues (in the basis (e, h, f) of L, the matrix of $\mathrm{ad}(h)$ is diagonal with eigenvalues $(2, 0, -2)$). This fact will generalize to all irreducible representations of L, and it is a crucial point.

Let $\phi : L \longrightarrow \mathfrak{gl}(E)$ be a finite-dimensional irreducible representation of L. We attempt to understand the decomposition of $\phi(h)$ in eigenspaces in order to determine the structure of E. One first easy observation is that $\mathrm{Tr}(\phi(h)) = \mathrm{Tr}(\phi([e, f])) = \mathrm{Tr}([\phi(e), \phi(f)])$ is equal to zero.

Let λ be an eigenvalue of $\phi(h)$, selected so that $\mathrm{Re}(\lambda)$ is as large as possible, and let v be a non-zero λ-eigenvector of $\phi(h)$. (This exists because E is non-zero and finite dimensional, and $\mathbf{C}$ is algebraically closed.)

We first observe that $\phi(e)v = 0$. Indeed, using the relations above and the fact that ϕ is a representation, we have

$$\phi(h)\phi(e)v = \phi(e)\phi(h)v + \phi([h, e])v = \lambda\phi(e)v + 2\phi(e)v = (\lambda + 2)\phi(e)v.$$

Thus $\phi(e)v$ is in the $(\lambda + 2)$-eigenspace of $\phi(v)$; but the maximality condition imposed on λ shows that this eigenspace must be 0, and hence $\phi(e)v = 0$.

Next, consider the vectors $v_i = \phi(f)^i v$ for $i \geqslant 0$ (the power refers to repeated compositions in $\mathrm{End}_k(E)$). By induction, using the relation $[h, f] = -2f$, one sees that v_i satisfies

$$\phi(h)v_i = (\lambda - 2i)v_i.$$

Since $\dim E$ is finite, all these v_i must be zero from some point on; otherwise, we would get infinitely many distinct eigenvectors of $\phi(h)$. Hence there exists $n \geqslant 0$ such that $v_n \neq 0$ but $v_i = 0$ for all $i \geqslant n + 1$.

We denote $w = v_n$ and $\mu = \lambda - 2n$ its eigenvalue. Now, we go back "upwards", defining $w_i = \phi(e)v_{i+1}$ for $0 \leqslant i < n$. The point is that

$$w_i = (i + 1)(\lambda - i)v_i$$

for all i. This is proved by induction on i, using the relation $[e, f] = h$. For instance, for $i = 0$, we have

$$w_0 = \phi(e)\phi(f)v = \phi(f)\phi(e)v + \phi(h)v = \lambda v = \lambda v_0$$

(since $\phi(e)v = 0$), and for $i = 1$ we get

$$\begin{aligned} w_1 = \phi(e)v_2 &= \phi(e)\phi(f)v_1 \\ &= \phi(f)\phi(e)v_1 + \phi(h)v_1 = \phi(f)w_0 + (\lambda - 2)v_1, \end{aligned}$$

which (using $w_0 = \lambda v_0$ and $\phi(f)v_0 = v_1$) gives

$$w_1 = \lambda v_1 + (\lambda - 2)v_1 = 2(\lambda - 1)v_1.$$

This is the case $i = 1$, and the induction is completely similar.

A direct consequence of this formula is that the span of $(v_0, \ldots, v_n)$ is a non-zero subrepresentation of E. Thus, by irreducibility, we have

$$E = \bigoplus_{0 \leqslant j \leqslant n} \mathbf{C} v_n,$$

and in particular $\dim E = n+1$. This decomposition gives a diagonalization of $\phi(h)$, and we finally determine λ by recalling that $\mathrm{Tr}(\phi(h)) = 0$: this gives

$$0 = \sum_{j=0}^{n} (\lambda - 2j) = (n + 1)\lambda - 2\sum_{j=0}^{n} j = (n + 1)(\lambda - n),$$

so that $\lambda = n$. The eigenvalues of $\phi(h)$ are therefore

$$n,\ n - 2,\ \ldots,\ -n + 2,\ -n,$$

each with multiplicity 1 (this gives the generalization of the property we observed for h and $\mathrm{ad}(h)$). This shows that E satisfies the properties of the representation of dimension $n + 1$ whose existence we claim.

To conclude the proof, we therefore only need to prove the existence and uniqueness of a representation of dimension $n+1$.

We leave these as exercises with some hints. For uniqueness, we can start from a non-zero vector v with $\phi(h)v = nv$ and $\phi(e)v = 0$, and the only subtlety is to check that $w = \phi(f)^n v$ is non-zero, to deduce that the vectors

$$(v, \phi(f)v, \ldots, \phi(f)^n v) \tag{3.3}$$

form a basis of E on which the action is completely determined by the computations above.

The existence can then also be checked by hand. For any fixed basis $(v_0, \ldots, v_n)$ of an $(n+1)$-dimensional vector space E, we define linear maps $\tilde{h}$, $\tilde{e}$ and $\tilde{f}$ on E by

$$\tilde{h}v_i = (n-2i)v_i, \qquad \tilde{f}v_i = v_{i+1},$$

and by following the rules for $\tilde{e}$ found in expressing $\phi(e)v_i$ in terms of the basis (3.3) during the uniqueness step. We may then check by brute force that these linear maps satisfy the commutation relations

$$[\tilde{h}, \tilde{e}] = 2\tilde{e}, \qquad [\tilde{h}, \tilde{f}] = -2\tilde{f}, \qquad [\tilde{e}, \tilde{f}] = \tilde{h},$$

and this implies that the map ϕ given by

$$\phi(\alpha e + \beta f + \gamma h) = \alpha\tilde{e} + \beta\tilde{f} + \gamma\tilde{h},$$

for $(\alpha, \beta, \gamma) \in \mathbf{C}^3$, is a representation of L. See also, for instance, [**35**, §4.8] for a slightly different approach. □

This classification should remind the reader of the existence of the representations ϱ_n of $\mathrm{SL}_2(\mathbf{C})$ of any integral dimension $n \geqslant 1$ (Theorem 2.6.1). Indeed, in a certain precise sense, these are "the same". Namely, the irreducible representation ϕ_n of dimension $n+1$ of $\mathfrak{sl}(\mathbf{C}^2)$ can be constructed in a different way as a Lie algebra representation acting on the same space V_n of homogeneous polynomials of degree n in $\mathbf{C}[X, Y]$ on which the irreducible representation ϱ_n acts.

To explain this, we first note that if $x \in \mathfrak{sl}(\mathbf{C}^2)$, we have $\exp(x) \in \mathrm{SL}_2(\mathbf{C})$ (where we compute the exponential of a matrix by the usual power series expansion), since $\det(\exp(x)) = \exp(\mathrm{Tr}(x))$ for any matrix x. This gives a connection between the Lie algebra and the group $\mathrm{SL}_2(\mathbf{C})$. Now let

$$\tilde{\phi}_n(x) = \frac{d}{dt}\varrho_n(\exp(tx))\Big|_{t=0} \tag{3.4}$$

for $x \in \mathfrak{sl}(\mathbf{C}^2)$. Then $\tilde{\phi}_n$ is a representation of $\mathfrak{sl}(\mathbf{C}^2)$, and it is irreducible (hence it is isomorphic to the representation of dimension $n+1$ given by the theorem).

We leave it as an exercise (see below) to check that, because ϱ_n is a group homomorphism, the map $\tilde{\phi}_n : \mathfrak{sl}(\mathbf{C}^2) \longrightarrow \mathfrak{gl}(V_n)$ is indeed a representation of the Lie algebra. Now let $H_n = \tilde{\phi}_n(h)$, where h is the element of $\mathfrak{sl}(\mathbf{C}^2)$ occurring in the proof of Theorem 3.2.5. Then

$$H_n = \frac{d}{dt}\varrho_n\left(\begin{pmatrix} e^t & 0 \\ 0 & e^{-t} \end{pmatrix}\right)\Big|_{t=0}$$

and in particular

$$H_n(X^jY^{n-j}) = \frac{d}{dt}e^{jt-(n-j)t}X^jY^{n-j}\Big|_{t=0} = (2j-n)X^jY^{n-j}$$

for $0 \leqslant j \leqslant n$. Thus the element $\tilde{v} = X^n$ is an eigenvector of H_n with eigenvalue n. In addition, $E_n = \tilde{\phi}_n(e)$ satisfies

$$\begin{aligned} E_n(X^jY^{n-j}) &= \frac{d}{dt}\varrho_n\left(\begin{pmatrix} 1 & t \\ 0 & 1 \end{pmatrix}\right)\Big|_{t=0}(X^jY^{n-j}) \\ &= \frac{d}{dt}X^j(tX+Y)^{n-j}\Big|_{t=0} = (n-j)X^{j+1}Y^{n-j-1}, \end{aligned}$$

and especially $E_n(\tilde{v}) = 0$. Thus the vector $\tilde{v} \neq 0$ has both properties of a highest weight vector, as in Theorem 3.2.5, and it is then easy to conclude that $\tilde{\phi}_n$ is isomorphic to ϕ_n, and that such an isomorphism can be chosen to map $\tilde{v}$ to a highest weight vector v. (What remains to check is that $\tilde{\phi}_n$ is irreducible. The point is that we have diagonalized the action of H_n, and that the eigenvectors are seen to be obtained from $\tilde{v}$ by repeated application of $\tilde{\phi}_n(f)$. Comparison with the proof of the theorem shows that the span of these eigenvectors of H_n is irreducible.)

We conclude by observing that it is no coincidence that the argument above has parallels with the proof of irreducibility of ϱ_n in Example 2.7.11: Lie algebras and Lie groups, which are groups with a differentiable structure that allows us to write expressions like (3.4) above, are intimately related. We refer for instance to the book of Fulton and Harris [**20**] for abundant illustrations and insights.

Exercise 3.2.6. (1) For matrices x and y of size n, show that

$$\exp(-tx)\exp(-ty)\exp(tx)\exp(ty) = \exp(t^2[x,y] + O(t^3)),$$

where $O(t^3)$ represents a matrix-valued function of the form $t^3f(t)$ with f bounded for t sufficiently close to 0. (*Hint*: This can be done by brute force using the power series expansion.)

(2) Let ϱ be a finite-dimensional representation of $\mathrm{SL}_n(\mathbf{C})$. For $x \in \mathfrak{sl}(\mathbf{C}^n)$, let

$$\phi(x) = \frac{d}{dt}\varrho(\exp(tx))\Big|_{t=0}.$$

Show that ϕ is a representation of the Lie algebra $\mathfrak{sl}(\mathbf{C}^n)$. (*Hint*: Use the previous result and differentiate carefully....)

3.3. Topological groups

In many applications, it is particularly important to restrict the representations to respect some additional structure. Among these, topological conditions are the most common.

The corresponding structure is that of a *topological group*, i.e., a group G equipped with a topology such that the product map

$$G \times G \to G$$

and the inverse map

$$G \to G$$

are both continuous (with $G \times G$ being given the product topology). There are many examples: for instance, any finite group can be seen as a topological group with the discrete topology; the additive group $\mathbf{R}$ or the multiplicative group $\mathbf{R}^\times$, with the usual euclidean topology, are also topological groups; similarly, for any $n \geqslant 1$, $\mathrm{GL}_n(\mathbf{C})$ is a group with the topology coming from its inclusion in $\mathrm{M}_n(\mathbf{C}) \simeq \mathbf{C}^{n^2}$; and moreover, any subgroup H of a topological group G in G inherits from G a topology and is then a topological group. This includes, for instance, groups like $\mathbf{Z} \subset \mathbf{R}$, $\mathrm{SL}_n(\mathbf{R}) \subset \mathrm{GL}_n(\mathbf{R})$ or $\mathrm{SL}_n(\mathbf{Z}) \subset \mathrm{SL}_n(\mathbf{R})$.

When dealing with a topological group, one typically wishes to restrict the representations which are considered to include some continuity property. This usually means taking the base field to be either $k = \mathbf{C}$ or $\mathbf{R}$, and ensuring that the vector space carries a suitable topology.[2] We will restrict our attention to Banach spaces, i.e., k-vector spaces E with a norm $\|\cdot\|$ on E such that E is complete for this norm. Of special interest, in fact, will be Hilbert spaces, when the norm derives from an inner product $\langle \cdot, \cdot \rangle$. As a special case, it is important to recall that if E is a finite-dimensional real or complex vector space, it carries a unique topology of Banach space, which can be defined using an inner product if desired. (Though, as is well known, there are many equivalent norms which can be used to define this topology.)

Given a topological group G and a Banach space E, the first restriction concerning representations ϱ of G on E is that the operators $\varrho(g)$, $g \in G$, should be *continuous*. Precisely, we denote by $\mathrm{L}(E)$ the space of continuous linear maps

$$T : E \longrightarrow E,$$

[2] Although other fields with a topology do arise; for instance, in number theory and algebraic geometry, one often encounters p-adic fields.

(which are also called bounded linear maps) and by $\mathrm{BGL}(E)$ the group of those $T \in \mathrm{L}(E)$ that are invertible, i.e., bijective with a continuous inverse. In fact, this last continuity condition is automatic when E is a Banach space by the Banach isomorphism theorem (see, e.g., [**49**, Th. III.11].) Of course, if $\dim(E) < +\infty$, we have $\mathrm{BGL}(E) = \mathrm{GL}(E)$.

If $\mathrm{L}(E)$ is given any topology, the group $\mathrm{BGL}(E)$ inherits, as a subset of $\mathrm{L}(E)$, a topology from the latter. It is then natural to think of considering representations

$$\varrho : G \longrightarrow \mathrm{BGL}(E)$$

that are continuous. However, as readers familiar with functional analysis will already know, quite a few different topologies are commonly encountered on $\mathrm{L}(E)$. The most natural[3] is the *norm topology*, associated to the norm

$$\|T\|_{\mathrm{L}(E)|} = \sup_{\substack{v\in E\\ v\neq 0}} \frac{\|Tv\|}{\|v\|} = \sup_{\substack{v\in E\\ \|v\|=1}} \|T(v)\| \tag{3.5}$$

defined for $T \in \mathrm{L}(E)$. In other words, it is the topology of uniform convergence on bounded subsets of E; the two expressions coincide because of linearity. However, it turns out that asking for homomorphisms to $\mathrm{BGL}(E)$ which are continuous for this topology does not lead to a good theory: there are *too few* representations in that case, as we will illustrate below. The "correct" definition is the following:

Definition 3.3.1 (Continuous representation)**.** Let G be a topological group, and let $k = \mathbf{R}$ or $\mathbf{C}$. Let E be a k-Banach space.

(1) A *continuous representation*, often simply called a representation, of G on E is a homomorphism

$$\varrho : G \longrightarrow \mathrm{BGL}(E)$$

such that the corresponding action map

$$\begin{cases} G \times E \longrightarrow E \\ (g, v) \mapsto \varrho(g)v \end{cases}$$

is continuous, where $G \times E$ has the product topology.

(2) If ϱ_1 and ϱ_2 are continuous representations of G, acting on E_1 and E_2, respectively, a homomorphism

$$\varrho_1 \xrightarrow{\Phi} \varrho_2$$

is a *continuous* linear map $E_1 \longrightarrow E_2$ such that

$$\Phi(\varrho_1(g)v) = \varrho_2(g)\Phi(v)$$

for all $g \in G$ and $v \in E_1$.

[3] It is natural, for instance, because $\mathrm{L}(E)$ becomes a Banach space for this norm.

If $\dim(E) < +\infty$, note that this is indeed equivalent to asking that ϱ be continuous, where $\mathrm{BGL}(E)$ has its usual (euclidean) topology.

Example 3.3.2 (Too many representations). Consider $G = \mathbf{R}$ and $k = \mathbf{C}$. If we consider simply one-dimensional representations

$$\mathbf{R} \to \mathbf{C}^{\times}$$

with no continuity assumptions at all, there are *too many* for most purposes. Indeed, as an abelian group, $\mathbf{R}$ has infinite rank, and it is a relatively direct application of Zorn's Lemma that any homomorphism $\chi : H \longrightarrow \mathbf{C}^{\times}$, where $H \subset \mathbf{R}$ is any subgroup, can be extended to a one-dimensional representation of $\mathbf{R}$. Since $\mathbf{R}$ is torsion free, it follows that for *any* finite subset $J \subset \mathbf{R}$ which generates a free abelian group of rank $|J|$ and *any* complex numbers z_j for $j \in J$, we can find a one-dimensional representation χ such that $\chi(j) = z_j$ for $j \in J$.

But as soon as we impose some regularity on the homomorphisms $\mathbf{R} \to \mathbf{C}^{\times}$, we obtain a much better understanding:

Proposition 3.3.3 (Continuous characters of $\mathbf{R}$). *Let $\chi : \mathbf{R} \to \mathbf{C}^{\times}$ be a* continuous *group homomorphism. Then there exists a unique $s \in \mathbf{C}$ such that*

$$\chi(x) = e^{sx}$$

for all $x \in \mathbf{R}$.

In fact, one can show that it is enough to ask that the homomorphism be *measurable*. The intuitive meaning of this is that, if one can "write down" a homomorphism $\mathbf{R} \to \mathbf{C}^{\times}$ in any concrete way or by using standard limiting processes, it will automatically be continuous, and hence it will be one of the ones above.

The proof we give uses differential calculus, but one can give completely elementary arguments (as in [**2**, Example A.2.5]).

Proof. If χ is differentiable and not merely continuous, we can use differential equations. We have

$$\chi'(x) = \lim_{h \to 0} \frac{\chi(x+h) - \chi(x)}{h} = \chi(x) \lim_{h \to 0} \frac{\chi(h) - 1}{h} = \chi(x)\chi'(0)$$

for all $x \in \mathbf{R}$. Denoting $s = \chi'(0)$, any solution of the differential equation $y' = sy$ is given by

$$y(x) = \alpha e^{sx}$$

for some parameter $\alpha \in \mathbf{C}$. In our case, putting $x = 0$ leads to $1 = \chi(0) = \alpha$, and hence we get the result.

We now use a trick to show that any continuous homomorphism $\chi : \mathbf{R} \longrightarrow \mathbf{C}^\times$ is in fact differentiable. We define

$$\Psi(x) = \int_0^x \chi(u)du$$

(note that this would be $s^{-1}(e^{sx} - 1)$ if $\chi(x) = e^{sx}$; this formula explains the manipulations to come), which is a differentiable function on $\mathbf{R}$. Then we write

$$\Psi(x+t) = \int_0^{x+t} \chi(u)du = \int_0^x \chi(u)du + \int_x^{x+t} \chi(u)du$$
$$= \Psi(x) + \int_0^t \chi(x+u)du = \Psi(x) + \chi(x)\Psi(t)$$

for all real numbers x and t. The function Ψ cannot be identically zero (its derivative χ would then also be zero, which is not the case), so picking a fixed $t_0 \in \mathbf{R}$ with $\Psi(t_0) \neq 0$, we obtain

$$\chi(x) = \frac{\Psi(x+t_0) - \Psi(x)}{\Psi(t_0)},$$

which is a differentiable function! Thus our first argument shows that $\chi(x) = e^{sx}$ for all x, with $s = \chi'(0)$. □

Example 3.3.4 (Too few representations)**.** Let G be the compact group $\mathbf{R}/\mathbf{Z}$. We now show that, if one insisted on considering only homomorphisms as representations

$$\varrho : \mathbf{R}/\mathbf{Z} \longrightarrow \mathrm{BGL}(E)$$

which are continuous with respect to the norm topology on $\mathrm{BGL}(E)$, there would be *too few* (similar examples hold for many other groups). In particular, there would be no good analogue of the regular representation. Indeed, if f is a complex-valued function on $\mathbf{R}/\mathbf{Z}$ and $t \in \mathbf{R}/\mathbf{Z}$, it is natural to try to define the latter with

$$\varrho(t)f(x) = f(x+t).$$

To have a Banach space of functions, we must impose some regularity condition. Although the most natural spaces turn out to be the L^p-spaces, with respect to Lebesgue measure, we start with $E = C(\mathbf{R}/\mathbf{Z})$, the space of continuous functions $f : \mathbf{R}/\mathbf{Z} \to \mathbf{C}$. If we use the supremum norm

$$\|f\| = \sup_{t \in \mathbf{R}/\mathbf{Z}} |f(t)|,$$

this is a Banach space. Moreover, the definition above clearly maps a function $f \in E$ to $\varrho(t)f \in E$; indeed, we have

$$\|\varrho(t)f\| = \|f\|$$

(since the graph of $\varrho(t)f$, seen as a periodic function on $\mathbf{R}$, is just obtained from that of f by translating it to the left by t units), and this further says that $\varrho(t)$ is a continuous linear map on E. Hence, ϱ certainly defines a homomorphism

$$\varrho : \mathbf{R}/\mathbf{Z} \longrightarrow \mathrm{BGL}(E).$$

But now we claim that ϱ is *not* continuous for the topology on $\mathrm{BGL}(E)$ coming from the norm (3.5). This is quite easy to see: in fact, for any $t \neq 0$ with $0 < t < 1/2$, we have

$$\|\varrho(t) - \mathrm{Id}_E\|_{\mathrm{L}(E)} = \|\varrho(t) - \varrho(0)\|_{\mathrm{L}(E)} \geqslant 1,$$

which shows that ϱ is very far from being continuous at 0. To see this, we take as a test function any $f_t \in E$ which is zero outside $[0, t]$, non-negative, always $\leqslant 1$, and equal to 1 at $t/2$, for instance (where we view $\mathbf{R}/\mathbf{Z}$ as obtained from $[0, 1]$ by identifying the end points).

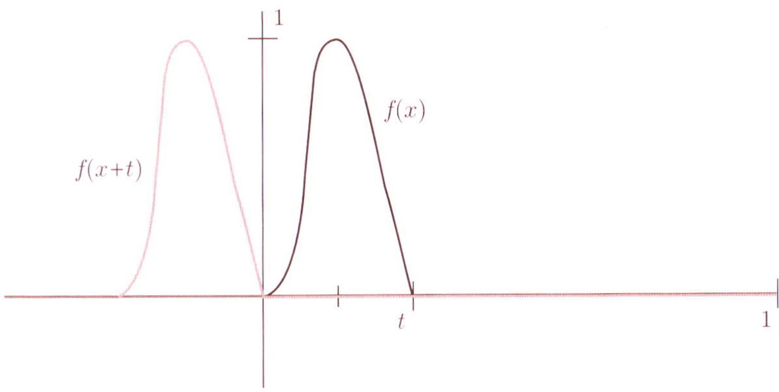

Then $\|f_t\| = 1$, and therefore

$$\|\varrho(t) - \mathrm{Id}\|_{\mathrm{L}(E)} \geqslant \|\varrho(t)f_t - f_t\| = \sup_{x \in \mathbf{R}/\mathbf{Z}} |f_t(t + x) - f_t(x)| = 1$$

since $\varrho(t)f_t$ is supported on the image modulo $\mathbf{Z}$ of the interval $[1 - t, t]$ which is *disjoint* from $[0, t]$, apart from the common endpoint t.

However, the point of this is that we had to use a different test function for each t: for a fixed f, the problem disappears, since by uniform continuity of $f \in E$, we have

$$\lim_{t \to 0} \|\varrho(t)f - f\| = 0$$

for any fixed $f \in E$.

Exercise 3.3.5. Show that ϱ defined above on $E = C(\mathbf{R}/\mathbf{Z})$ is a continuous representation in the sense of Definition 3.3.1.

The general formalism of representation theory can, to a large extent, be transferred or adapted to the setting of representations of topological groups. In particular, for finite-dimensional representations, since all operations considered are *obviously continuous*, every construction goes through. This applies to direct sums, tensor products, the contragredient, symmetric and exterior powers, subrepresentations and quotients, etc.

Some care is required when considering infinite-dimensional representations. For instance, the definition of subrepresentations and quotient representations, as well as that of irreducible representations, should be adjusted to take the topology into account:

Definition 3.3.6 (Subrepresentations and irreducibility). Let G be a topological group, and let

$$\varrho : G \longrightarrow \mathrm{BGL}(E)$$

be a representation of G on a Banach space E.

(1) A representation π of G is a subrepresentation of ϱ if π acts on a *closed* subspace $F \subset E$ which is invariant under ϱ and $\pi(g)$ is the restriction of $\varrho(g)$ to F. Given such a subrepresentation π, the quotient ϱ/π is the induced representation on the quotient Banach space E/F with the norm

$$\|v\|_{E/F} = \min\{\|w\|_E \mid w \,(\mathrm{mod}\, F) = v\}.$$

(2) The representation ϱ is irreducible (sometimes called *topologically irreducible*) if E is non-zero and E has no non-zero proper subrepresentation, i.e., there is no non-zero proper *closed* subspace F of E which is stable under all $\varrho(g)$, $g \in G$.

If $\dim(E) < +\infty$, since any subspace of F is closed, there is no difference between these definitions and the previous one. But in general the distinction between closed subspaces and general subspaces is necessary to obtain a good formalism. There exist indeed infinite-dimensional representations which are topologically irreducible but have many non-zero proper stable subspaces (these are necessarily dense in the space E, but they are not closed).

The second example we give of adapting the formalism is that of the contragredient representation. Given a Banach space E, the dual Banach space is the vector space E' of continuous linear maps $E \longrightarrow \mathbf{C}$ with the norm

$$\|\lambda\|_{E'} = \sup_{\|v\|=1} |\lambda(v)| = \sup_{v \neq 0} \frac{|\lambda(v)|}{\|v\|}.$$

Given a representation

$$\varrho : G \longrightarrow \mathrm{BGL}(E)$$

of a topological group G on E, the *topological contragredient* $\check{\varrho}$ acts on E' by the usual formula

$$\langle g \cdot \lambda, v\rangle = \langle \lambda, \varrho(g^{-1})v\rangle.$$

If E is finite-dimensional, this is obviously continuous. Otherwise, the following gives a simple condition under which this representation is continuous.

Lemma 3.3.7 (Continuity of the contragredient)**.** *Let G be a topological group, and let ϱ be a representation of G on the Banach space E such that*

$$\sup_{g \in G} \|\varrho(g)\|_{\mathrm{L}(E)} < +\infty.$$

Then the topological contragredient representation on E' is a continuous representation.

Proof. It is enough to check continuity of the action map at the pair $(1, 0) \in G \times E'$ (we leave this reduction as an exercise). But we have

$$\|\check{\varrho}(g)\lambda\|_{E'} = \sup_{\|v\|=1} |\lambda(\varrho(g^{-1})v)| \leqslant M\|\lambda\|$$

with $M = \sup \|\varrho(g)\|$. Thus if λ is close enough to 0 (and g is arbitrary), then so is $\check{\varrho}(g)\lambda$, which means that $\check{\varrho}$ is continuous. □

Although it is an abuse of notation, we will sometimes omit the adjective "topological" when considering the contragredient of continuous representations. Unless otherwise specificed, whenever we consider continuous representations of topological groups, the "contragredient" will mean the topological contragredient.

3.4. Unitary representations

When the representation space (for a topological group) is a Hilbert space H with an inner product[4] $\langle \cdot, \cdot \rangle$, it is natural to consider particularly closely the *unitary representations*, which are those for which the operators $\varrho(g)$ are *unitary*, i.e., preserve the inner product

$$\langle \varrho(g)v, \varrho(g)w\rangle = \langle v, w\rangle$$

for all $g \in G$, v and $w \in \mathrm{H}$.

We present here the most basic facts about such representations. They will be considered in more detail first for finite groups (see Chapter 4) and

[4] Recall from the introduction that our inner products are linear in the first variable, and conjugate linear in the second: $\langle v, \lambda w\rangle = \bar{\lambda}\langle v, w\rangle$ for $\lambda \in \mathbf{C}$.

for compact and locally compact groups (see Chapters 5 and 6 and Sections 7.2, 7.3, 7.4). For additional information on the general theory of unitary representations, especially with respect to infinite-dimensional cases, we recommend the Appendices to [**2**] as well as [**44**].

Definition 3.4.1 (Unitary representations, unitarizable representations). Let G be a topological group.

(1) A *unitary representation* of G is a continuous representation of G on a Hilbert space H where $\varrho(g) \in \mathrm{U}(\mathrm{H})$ for all $g \in G$, where $\mathrm{U}(\mathrm{H})$ is the group of unitary operators of G. A morphism $\varrho_1 \to \varrho_2$ of unitary representations, acting on H_1 and H_2, respectively, is a morphism of continuous representations $\varrho_1 \xrightarrow{\Phi} \varrho_2$ (one does *not* require that Φ preserve the inner product).

(2) An arbitrary representation

$$G \longrightarrow \mathrm{GL}(E)$$

of G on a complex vector space E is *unitarizable* if there exists an inner product $\langle \cdot, \cdot \rangle$ on E, defining a structure of Hilbert space, such that the values of ϱ are unitary for this inner product, and the resulting map $G \longrightarrow \mathrm{U}(E)$ is a unitary representation.

Remark 3.4.2. (1) If E is a finite-dimensional complex vector space and G carries the discrete topology (e.g., if G is finite), it is enough to construct an inner product on the vector space E such that ϱ takes value in $\mathrm{U}(E)$ in order to check that a representation is unitarizable (the continuity is automatic).

(2) If $\varrho_1 : G \longrightarrow \mathrm{U}(\mathrm{H}_1)$ and $\varrho_2 : G \longrightarrow \mathrm{U}(\mathrm{H}_2)$ are unitary representations and $\Phi : \mathrm{H}_1 \longrightarrow \mathrm{H}_2$ is a G-homomorphism, the adjoint $\Phi^* : \mathrm{H}_2 \longrightarrow \mathrm{H}_1$ of Φ is also an intertwiner. Indeed, given $g \in G$ and $v \in \mathrm{H}_2$, we have (with obvious notation)

$$\begin{aligned}
\langle \varrho_1(g)(\Phi^*(v)), w\rangle_1 &= \langle \Phi^*(v), \varrho_1(g^{-1})w\rangle_1 \\
&= \langle v, \Phi(\varrho_1(g^{-1})w)\rangle_2 \\
&= \langle v, \varrho_2(g^{-1})\Phi(w)\rangle_2 \\
&= \langle \varrho_2(g)v, \Phi(w)\rangle_2 = \langle \Phi^*(\varrho_2(g)v), w\rangle_1,
\end{aligned}$$

for all $w \in \mathrm{H}_1$, which implies that $\Phi^*(\varrho_2(g)v) = \varrho_1(g)(\Phi^*(v))$, and hence that Φ^* intertwines ϱ_2 and ϱ_1.

The continuity requirement for unitary representations can be rephrased in a way that is easier to check:

Proposition 3.4.3 (Strong continuity criterion for unitary representations). *Let $\varrho : G \longrightarrow \mathrm{U}(\mathrm{H})$ be a homomorphism of a topological group G to a unitary group of some Hilbert space. Then ϱ is a unitary representation, i.e., the* *action map is continuous, if and only if* ϱ *is* strongly continuous*: for any fixed* $v \in \mathrm{H}$*, the map*

$$\begin{cases} G \longrightarrow \mathrm{H} \\ g \mapsto \varrho(g)v \end{cases}$$

is continuous on G*. Equivalently, this holds when these maps are continuous at* $g = 1$*.*

Proof. The joint continuity of the two-variable action map implies that of the maps above, which are one-variable specializations. For the converse, we use the unitarity and a splitting of epsilons.

Let (g, v) be given in $G \times \mathrm{H}$. For any $(h, w) \in G \times \mathrm{H}$, we have

$$\|\varrho(g)v - \varrho(h)w\| = \|\varrho(h)(\varrho(h^{-1}g)v - w)\| = \|\varrho(h^{-1}g)v - w\|$$

by unitarity. Then, by the triangle inequality, we get

$$\|\varrho(g)v - \varrho(h)w\| \leqslant \|\varrho(h^{-1}g)v - v\| + \|v - w\|.$$

Under the assumption of the continuity of $x \mapsto \varrho(x)v$ when $x \to 1$, this shows that when w is close to v and h is close to g, the difference becomes arbitrarily small, which is the continuity of the action map at (g, v). To be precise, given $\varepsilon > 0$, we can first find an open neighborhood U_v of v in H such that $\|v - w\| < \varepsilon$ for $w \in U_v$, and we can use the continuity assumption to find an open neighborhood U_1 of $1 \in G$ such that $\|\varrho(x)v - v\| < \varepsilon$ for $x \in U_1$. Then, when $(h, w) \in gU_1^{-1} \times U_v$, we have

$$\|\varrho(g)v - \varrho(h)w\| < 2\varepsilon.$$

□

Remark 3.4.4 (The strong topology)**.** The name *strong continuity* refers to the fact that the condition above is equivalent with the assertion that the homomorphism

$$\varrho : G \longrightarrow \mathrm{U}(\mathrm{H})$$

is continuous, with respect to the topology induced on U(H) by the so-called *strong operator topology* on L(H). The latter is defined as weakest topology such that all linear maps

$$\begin{cases} \mathrm{L}(\mathrm{H}) & \longrightarrow & \mathbf{C} \\ T & \mapsto & Tv \end{cases}$$

(for $v \in \mathrm{H}$) are continuous. This means that a basis of neighborhoods of $T_0 \in \mathrm{L}(\mathrm{H})$ for this topology is given by the finite intersections

$$\bigcap_{1 \leqslant i \leqslant m} V_i,$$

where

$$V_i = \{T \in \mathrm{L}(\mathrm{H}) \mid \|Tv_i - T_0 v_i\| < \varepsilon\}$$

for some unit vectors $v_i \in \mathrm{H}$ and some fixed $\varepsilon > 0$ (see, e.g., [**49**, p. 183]).

Example 3.4.5 (The regular representation on $L^2(\mathbf{R})$)**.** Here is a first example of infinite-dimensional unitary representation. We consider the additive group $\mathbf{R}$ of real numbers, with the usual topology, and the space $\mathrm{H} = L^2(\mathbf{R})$ of square-integrable functions on $\mathbf{R}$, defined using Lebesgue measure. Defining

$$\varrho(t)f(x) = f(x+t),$$

we claim that we obtain a unitary representation. This is formally obvious, but as we have seen in Example 3.3.4, the continuity requirement needs some care. We will see this in a greater context in Proposition 5.2.6 in Chapter 5, but we sketch the argument here.

First, one must check that the definition of $\varrho(t)f$ makes sense (since elements of $L^2(\mathbf{R})$ are not really functions): this is indeed so, because if we change a measurable function f on a set of measure zero, we only change $x \mapsto f(x+t)$ on a translate of this set, which still has Lebesgue measure zero.

Then the unitarity of the action is clear: we have

$$\int_{\mathbf{R}} |\varrho(t)f(x)|^2 dx = \int_{\mathbf{R}} |f(x+t)|^2 dx = \int_{\mathbf{R}} |f(x)|^2 dx,$$

and only continuity remains to be checked. This is done in two steps using Proposition 3.4.3 (the reader can fill the outline or look at Proposition 5.2.6). Fix $f \in L^2(\mathbf{R})$, and assume first that it is continuous and compactly supported. Then the continuity of $t \mapsto \varrho(t)f$ amounts to the limit formula

$$\lim_{h\to 0} \int_{\mathbf{R}} |f(x+t+h) - f(x+t)|^2 dx = 0$$

for all $t \in \mathbf{R}$, which follows from the dominated convergence theorem. Next, one uses the fact that continuous functions with compact support form a dense subspace of $L^2(\mathbf{R})$ (for the L^2-norm) in order to extend this continuity statement to all $f \in L^2(\mathbf{R})$.

The operations of representation theory, when applied to unitary representations, lead most of the time to other unitary representations. Here are the simplest cases:

Proposition 3.4.6 (Operations on unitary representations)**.** *Let G be a topological group.*

(1) *If ϱ is a unitary representation of G, then any subrepresentation and any quotient representation are naturally unitary. Similarly, the restriction of ϱ to any subgroup H is a unitary representation with the same inner product.*

(2) *Direct sums of unitary representations are unitary with inner product*

$$\langle v_1 \oplus w_1, v_2 \oplus w_2 \rangle_{\varrho_1 \oplus \varrho_2} = \langle v_1, v_2 \rangle_{\varrho_1} + \langle w_1, w_2 \rangle_{\varrho_2}$$

for $\varrho_1 \oplus \varrho_2$*, so that the subrepresentations* ϱ_1 *and* ϱ_2 *in* $\varrho_1 \oplus \varrho_2$ *are orthogonal complements of each other.*

(3) *The tensor product* $\varrho_1 \otimes \varrho_2$ *of finite-dimensional unitary representations* ϱ_1 *and* ϱ_2 *is unitary, with respect to the inner product defined for pure tensors by*

$$\langle v_1 \otimes w_1, v_2 \otimes w_2 \rangle_{\varrho_1 \otimes \varrho_2} = \langle v_1, v_2 \rangle_{\varrho_1} \langle w_1, w_2 \rangle_{\varrho_2}. \tag{3.6}$$

Similarly, external tensor products of finite-dimensional unitary representations, are unitary, with the same inner product on the underlying tensor-product space.

We leave to the reader the simple (and standard) argument that checks that the definition of (3.6) does extend to a well-defined inner product on the tensor product of two finite-dimensional Hilbert spaces. In fact, the following exercise shows how to extend this definition to (possibly) infinite-dimensional unitary representations.

Exercise 3.4.7. Let H_1 and H_2 be Hilbert spaces, one or both possibly infinite dimensional.

(1) Let $H = H_1 \otimes H_2$, defined as a **C**-vector space. Show that one can define a unique hermitian form on H such that

$$\langle v_1 \otimes w_1, v_2 \otimes w_2 \rangle_H = \langle v_1, v_2 \rangle_{H_1} \langle w_1, w_2 \rangle_{H_2}$$

for all $v_1, v_2 \in H_1$ and $w_1, w_2 \in H_2$.

(2) Show that $\langle \cdot, \cdot \rangle_H$ is a positive-definite inner-product on H.

(3) Define $H_1 \hat{\otimes} H_2$ to be the completion of H with respect to $\langle \cdot, \cdot \rangle_H$, seen as a Hilbert space. If G is a topological group and ϱ_1 (resp. ϱ_2) is a unitary representation of G on H_1 (resp. on H_2), show that there exists a unique unitary representation $\varrho_1 \hat{\otimes} \varrho_2$ of G on $H_1 \hat{\otimes} H_2$ extending the representation $\varrho_1 \otimes \varrho_2$ of G on H.

One can also extend, to a large extent, the notion of induction to unitary representations of topological groups. However, it is not obvious to do this, and we will consider this in Chapter 5 for compact groups, and in Section 7.4 in another special case. In both cases, the reader may look at [**2**, App. A, App. E] for some more results and general facts.

Example 3.4.8. A special case of Proposition 3.4.6(3) concerns the tensor product (*twist*) of an arbitrary unitary representation $\varrho : G \longrightarrow U(H)$ with a one-dimensional unitary representation $\chi : G \longrightarrow U(\mathbf{C}) = \mathbf{S}^1$. In that case, one can define $\varrho \otimes \chi$ on the same underlying Hilbert space H, by

$$(\varrho \otimes \chi)(g)v = \chi(g)\varrho(g)v$$

with the same inner product, which is still invariant for $\varrho \otimes \chi$: for any v, $w \in \mathrm{H}$, we have

$$\begin{aligned}\langle (\varrho \otimes \chi)(g)v, (\varrho \otimes \chi)(g)w \rangle_{\mathrm{H}} &= \langle \chi(g)\varrho(g)v, \chi(g)\varrho(g)w \rangle_{\mathrm{H}} \\ &= |\chi(g)|^2 \langle \varrho(g)v, \varrho(g)w \rangle_{\mathrm{H}} = \langle v, w \rangle_{\mathrm{H}}.\end{aligned}$$

Now we discuss the contragredient in the context of unitary representations. There are special features here, which arise from the canonical duality properties of Hilbert spaces. For a Hilbert space H, the *conjugate space* $\bar{\mathrm{H}}$ is defined to be the Hilbert space with the same underlying abelian group (i.e., addition) but with scalar multiplication and inner products defined by

$$\alpha \cdot v = \bar{\alpha} v, \qquad \langle v, w \rangle_{\bar{\mathrm{H}}} = \langle w, v \rangle_{\mathrm{H}}.$$

The point of the conjugate Hilbert space is that it is canonically isometric[5] (as Banach space) to the dual of H, by the map

$$\Phi \begin{cases} \bar{\mathrm{H}} & \longrightarrow & \mathrm{H}' \\ w & \mapsto & (\lambda_w : v \mapsto \langle v, w \rangle). \end{cases}$$

Vectors in w are "the same" as vectors in H, but because $\lambda_{\alpha w} = \bar{\alpha}\lambda_w$, this map is only linear when the conjugate Hilbert space structure is used as the source.

If ϱ is a unitary representation of a group G on H, this allows us to rewrite the basic matrix coefficients $f_{v,\lambda}$ of ϱ using inner products on H only: given $\lambda = \lambda_w \in \mathrm{H}'$ and $v \in \mathrm{H}$, we have

$$f_{v,\lambda}(g) = \lambda_w(\varrho(g)v) = \langle \varrho(g)v, w \rangle.$$

These are now parametrized by the two vectors v and w in H; though it is formally better to see w as being an element of $\bar{\mathrm{H}}$, one can usually dispense with this extra formalism without much loss.

Using the map Φ, we can also transport the contragredient representation of ϱ to an isomorphic representation $\bar{\varrho}$ acting on $\bar{\mathrm{H}}$. Its character, when ϱ is finite-dimensional, is given by

$$\chi_{\bar{\varrho}}(g) = \overline{\chi_{\varrho}(g)},$$

since the eigenvalues of a unitary matrix are complex numbers of modulus 1, hence their inverse is the same as their conjugate.

Exercise 3.4.9 (Matrix coefficients of the conjugate representation)**.** Let ϱ be a unitary representation of G on a Hilbert space H. Show that the functions

$$g \mapsto \overline{\langle \varrho(g)v, w \rangle},$$

for v, $w \in \mathrm{H}$, are matrix coefficients of $\bar{\varrho}$.

[5] This is the Riesz representation theorem for Hilbert spaces; see e.g., [**49**, Th. II.4].

Example 3.4.10 (A unitarizability criterion)**.** Let $G = \mathbf{R}$ with its usual topology. We have seen that there are many one-dimensional representations of G as a topological group, given by

$$\omega_s : \begin{cases} \mathbf{R} & \longrightarrow & \mathbf{C}^\times \\ x & \mapsto & e^{sx} \end{cases}$$

for $s \in \mathbf{C}$ (these are different functions, hence non-isomorphic representations).

However, these are only unitary (or, more properly speaking, unitarizable) when $s = it$ is purely imaginary. Indeed, when $\mathrm{Re}(s) = 0$, we have $|\omega_s(x)| = 1$ for all $x \in \mathbf{R}$, and the unit circle is the unitary group of the one-dimensional Hilbert space $\mathbf{C}$ with inner product $z\bar{w}$. Conversely, the following lemma is a quick and convenient necessary condition for unitarity or unitarizability because it gives a property which does not depend on any information on the inner product, and it implies that ω_s is not unitarizable when $\mathrm{Re}(s) \neq 0$.

Lemma 3.4.11. *Let ϱ be a finite-dimensional unitary representation of a group G. Then any eigenvalue of $\varrho(g)$, for any g, is a complex number of modulus* 1.

Proof. This is linear algebra for unitary matrices, of which the $\varrho(g)$ are examples.... □

This lemma applies also to the representations ϱ_m of $\mathrm{SL}_2(\mathbf{C})$ of Section 2.6.1 and Example 2.7.46: from (2.41), for instance, we see that if $m \geqslant 1$, ϱ_m is not unitarizable (since the restriction to the subgroup T is diagonal with explicit eigenvalues which are not of modulus 1). However, the restriction of ϱ_m to the compact subgroup $\mathrm{SU}_2(\mathbf{C})$ is unitarizable (see Theorem 5.2.11 for the general fact concerning compact groups which implies this).

Along the same lines, observe that if we consider the compact group $\mathbf{R}/\mathbf{Z}$, its one-dimensional representations induce, by composition, some representations of $\mathbf{R}$

$$\mathbf{R} \longrightarrow \mathbf{R}/\mathbf{Z} \longrightarrow \mathbf{C}^\times,$$

which must be among the ω_s. Which ones occur in this manner is easy to see: we must have $\mathbf{Z} \subset \mathrm{Ker}(\omega_s)$, and from $\omega_s(1) = 1$, it follows that $s = 2ik\pi$ for some *integer* $k \in \mathbf{Z}$. In particular, we observe a feature which will turn out to be a general feature of compact groups: all these irreducible representations of $\mathbf{R}/\mathbf{Z}$ are unitary!

Example 3.4.12 (Regular representation of a finite group)**.** Let G be a finite group, and let $C(G)$ be the space of complex-valued functions on G, with the regular representation of G. A natural inner product on the vector space $C(G)$ is

$$\langle \varphi_1, \varphi_2 \rangle = \frac{1}{|G|} \sum_{x \in G} \varphi_1(x)\overline{\varphi_2(x)}.$$

One could omit the normalizing factor $1/|G|$, but it has the advantage that $\|1\| = 1$ for the constant function[6] 1 on G, independently of the order of G.

It is quite natural that, with respect to this inner product, the regular representation ϱ_G on $C(G)$ is unitary. Indeed, we have

$$\|\varrho_G(g)\varphi\|^2 = \frac{1}{|G|} \sum_{x \in G} |\varphi(xg)|^2 = \frac{1}{|G|} \sum_{y \in G} |\varphi(y)|^2 = \|\varphi\|^2$$

for all $g \in G$, using the change of variable $y = xg$.

A similar property holds for all locally compact groups, but if G is infinite, the inner product must be defined using integration on G with respect to a natural invariant measure μ, and the space $C(G)$ must be replaced by the Hilbert space $L^2(G, \mu)$. We will come back to this later (see Proposition 5.2.6).

In addition to the usual formalism of direct sums, the extra structure of Hilbert spaces leads to a definition of infinite orthogonal direct sums. If G is a topological group and $(\varrho_i)_{i \in I}$ is any family of unitary representations of G, acting on the Hilbert spaces H_i, we can define the Hilbert space orthogonal direct sum

$$\mathrm{H} = \widehat{\bigoplus_{i \in I}} \mathrm{H}_i, \tag{3.7}$$

and a corresponding representation $\varrho = \hat{\bigoplus}_i \varrho_i$ acting on H_i. Precisely, recall that H is defined to be the space of families $v = (v_i)_{i \in I}$ such that

$$\|v\|_{\mathrm{H}}^2 = \sum_{i \in I} \|v_i\|_i^2 < +\infty,$$

and we define

$$\varrho(g)v = (\varrho_i(g)v_i)_{i \in I},$$

which is easily checked to be a unitary representation acting on H (see Exercise 3.4.13 below). Of course, for each i, the subspace $\mathrm{H}_i \subset \mathrm{H}$ is a subrepresentation of H isomorphic to ϱ_i. Moreover, the algebraic direct sum $\bigoplus \mathrm{H}_i$ of the H_i (the space of families (v_i) where v_i is zero for all but finitely many i) is a dense subspace of H. It is stable under the action of ϱ, but since it is not closed in general, it is usually not a subrepresentation in the topological sense.

[6] This should not be confused with the neutral element in the group.

Exercise 3.4.13 (Pre-unitary representation)**.** It is often convenient to define a unitary representation by first considering an action of G on a pre-Hilbert space, which *extends by continuity* to a proper unitary representation (e.g., when defining a representation of a space of functions, it may be easier to work with a dense subspace of regular functions; for a concrete example, see the construction of the regular representation of a compact topological group in Proposition 5.2.6).

We consider a fixed topological group G. A pre-unitary representation of G is a strongly continuous homomorphism

$$\varrho : G \longrightarrow \mathrm{U}(\mathrm{H}_0),$$

where H_0 is a pre-Hilbert space, i.e., a complex vector space given with a (positive-definite) inner product, which is not necessarily complete.

(1) Show that if ϱ is a pre-unitary representation, the operators $\varrho(g)$ extend by continuity to unitary operators of the completion H of H_0, and that the resulting map is a unitary representation of G, such that $\mathrm{H}_0 \subset \mathrm{H}$ is a stable subspace. (*Hint*: To check the strong continuity, use the fact that H_0 is dense in H.)

(2) Suppose H is a Hilbert space and $\mathrm{H}_0 \subset \mathrm{H}$ a dense subspace. If ϱ is a pre-unitary representation on H_0, explain why the resulting unitary representation is a representation of G on H.

(3) Use this to check that the Hilbert direct sum construction of (3.7) above leads to unitary representations.

Unitary representations are better behaved than general (complex) representations. One of the main reasons is the following fact.

Proposition 3.4.14 (Reducibility of unitary representations)**.** *Let $G \xrightarrow{\varrho} \mathrm{U}(\mathrm{H})$ be a unitary representation of a topological group G. Then any closed subspace $F \subset \mathrm{H}$ invariant under ϱ has a closed stable complement given by $F^{\perp} \subset \mathrm{H}$. In particular, any* finite-dimensional *unitary representation is semisimple.*

Proof. Since any finite-dimensional subspace of a Hilbert space is closed, the second part follows from the first using the criterion of Lemma 2.2.11.

Thus we consider a subrepresentation $F \subset \mathrm{H}$. Taking a hint from the statement, we just check that the orthogonal complement

$$F^{\perp} = \{v \in \mathrm{H} \mid \langle v, w\rangle = 0 \text{ for all } w \in F\} \subset \mathrm{H}$$

is a closed stable complement of F.

Indeed, the theory of Hilbert spaces[7] shows that $F^{\perp}$ is closed in H and that $F \oplus F^{\perp} = \mathrm{H}$. From the fact that ϱ preserves the inner product, the same property follows for its orthogonal complement: if $v \in F^{\perp}$, we have

$$\langle \varrho(g)v, w \rangle = \langle v, \varrho^{-1}(g)w \rangle = 0$$

for all $g \in G$ and $w \in F$, i.e., $F^{\perp}$ is indeed a subrepresentation of H. □

Example 3.4.15 (Failure of semisimplicity for unitary representations). Although this property is related to semisimplicity, it is *not* the case that any unitary representation

$$\varrho : G \longrightarrow \mathrm{U}(\mathrm{H})$$

is semisimple, even in the sense that there exists a family $(\mathrm{H}_i)_{i \in I}$ of stable subspaces of H such that we have a Hilbert-space orthogonal direct sum

$$\mathrm{H} = \widehat{\bigoplus}_{i \in I} \mathrm{H}_i$$

as described above. The reader can of course look back at the proof of Lemma 2.2.11 where the equivalence of semisimplicity and complete reducibility was proved for the algebraic case: the problem is that the first step, the *existence* of an irreducible subrepresentation of H (which is Exercise 2.2.13) may fail in this context. To see this, take the representation ϱ of $\mathbf{R}$ on $L^2(\mathbf{R})$ described in Example 3.4.5. This is an infinite-dimensional unitary representation, *and it contains no irreducible subrepresentation*!

To see this, we need to know that (as in Corollary 2.7.17) all irreducible unitary representations of $\mathbf{R}$ are one dimensional. This requires a version of Schur's Lemma for unitary representations, possibly infinite dimensional, which is proved in Proposition 3.4.17 below. Then, by Proposition 3.3.3 and the unitarizability criterion, it follows that all irreducible unitary representations of $\mathbf{R}$ are given by

$$\chi_x \begin{cases} \mathbf{R} & \longrightarrow & \mathbf{C}^{\times} \\ t & \mapsto & e^{itx} \end{cases}$$

for some $x \in \mathbf{R}$. Now, a non-zero function $f \in L^2(\mathbf{R})$ spans an irreducible subrepresentation of ϱ isomorphic to χ_x if and only if we have

$$f(x+t) = \varrho(t)f(x) = \chi_x(t)f(x) = e^{itx}f(x)$$

for all x and $t \in \mathbf{R}$. But this means that $|f(t)| = |f(0)|$ is constant for (almost) all $t \in \mathbf{R}$, and this constant is non-zero since we started with $f \neq 0$. However, we get

$$\int_{\mathbf{R}} |f(x)|^2 dx = |f(0)|^2 \int_{\mathbf{R}} dx = +\infty,$$

which contradicts the assumption $f \in L^2(\mathbf{R})$. . . .

[7] If H is finite-dimensional, of course, this is mere linear algebra.

In Chapter 5, we will see that this type of behavior does not occur for *compact* topological groups. But this shows, obviously, that the study of unitary representations of non-compact groups is fraught with new difficulties. This is illustrated in Section 7.4 with the prototypical case of $\mathrm{SL}_2(\mathbf{R})$.

Exercise 3.4.16. Show, without using Schur's Lemma, that the regular representation ϱ of $\mathbf{R}$ on $L^2(\mathbf{R})$ is not irreducible. (*Hint*: Use the classical Fourier transform to define proper closed invariant subspaces.)

Schur's Lemma takes the following natural form for unitary representations:

Proposition 3.4.17 (Schur's Lemma). *Let G be a topological group.*

(1) *If $\varrho_1 : G \longrightarrow \mathrm{U}(\mathrm{H}_1)$ and $\varrho_2 : G \longrightarrow \mathrm{U}(\mathrm{H}_2)$ are non-isomorphic irreducible unitary representations of G, then $\mathrm{Hom}_G(\varrho_1, \varrho_2) = 0$.*

(2) *If $\varrho : G \longrightarrow \mathrm{U}(\mathrm{H})$ is an irreducible unitary representation of G, then the space of intertwiners $\mathrm{H} \longrightarrow \mathrm{H}$ is the one-dimensional space of multiples of the identity.*

Although the statement is exactly the same as that of the previous version of Schur's Lemma, the proof lies quite a bit deeper, and the reader may skip it in a first reading. Precisely, the proof requires the spectral theorem for (bounded) self-adjoint operators, which we state in a convenient version (see [**49**, Ch. VII, Th. VII.3] for the proof in the separable case and enlightening discussions).

Theorem 3.4.18 (The spectral theorem). *Let H be a Hilbert space, and let A be a continuous self-adjoint operator on H. There exists a measure space (X, μ), an isometric isomorphism*

$$T : \mathrm{H} \longrightarrow L^2(X, \mu),$$

and a real-valued function $f \in L^\infty(X)$ such that A is, by means of T, equivalent to the multiplication operator M_f defined on $L^2(X, \mu)$ by

$$M_f(\varphi) = f\varphi$$

for $\varphi \in L^2(X, \mu)$. In other words, we have

$$T(Av) = M_f(Tv) = fT(v) \quad \text{for all} \quad v \in \mathrm{H}.$$

If H is separable, one may take (X, μ) in such a way that $\mu(X) < +\infty$.

If A is a unitary operator on H, the same statement is valid with f taking values in the unit circle $\mathbf{S}^1 \subset \mathbf{C}$.

The meaning of this theorem is that any property of an individual self-adjoint (resp. unitary) operator may be studied by reducing to the case of a multiplication operator by a bounded real-valued function (resp. by a measurable function with values of modulus 1).

Exercise 3.4.19. Let (X, μ) be a measure space, let $\mathrm{H} = L^2(X, \mu)$, and let $A \in \mathrm{BGL}(\mathrm{H})$ be the multiplication operator by a real-valued function $f \in L^\infty(X, \mu)$. Furthermore, let $B \in \mathrm{BGL}(\mathrm{H})$ be an operator that commutes with A, so that $AB = BA$.

Let $I \subset \mathbf{R}$ be a bounded interval, and define C to be the multiplication operator by $\mathbf{1}_I \circ f = \mathbf{1}_{f^{-1}(I)}$. Show that B commutes with C. (*Hint*: Let $M = \|f\|_\infty$; there exist polynomials P_n on $[-M, M]$ such that $P_n(f(x)) \to \mathbf{1}_I(f(x))$ for all x and $\|P_n\|_\infty$ is absolutely bounded. Then use the fact that $P_n(A)B = BP_n(A)$.)

Proof of Schur's Lemma. We begin this time with the proof of Proposition 3.4.17(2). Let $\Phi : \mathrm{H} \longrightarrow \mathrm{H}$ be an intertwiner. Then the adjoint Φ^* is also an intertwiner (see Remark 3.4.2(2)), and we can write

$$\Phi = \frac{\Phi + \Phi^*}{2} + i\frac{\Phi - \Phi^*}{2i} = A + iB,$$

where both A and B are self-adjoint intertwiners. Thus, if we prove that a self-adjoint intertwiner is of the form $\lambda\mathrm{Id}$, with $\lambda \in \mathbf{R}$, then the result will follow for all Φ. We can therefore assume that Φ itself is self-adjoint.

Now we apply the spectral theorem to Φ. This means that we may assume that $\mathrm{H} = L^2(X, \mu)$ for some measure space (X, μ) and that Φ is the multiplication operator M_f by some real-valued bounded function f on X. We claim that f is almost everywhere constant, say equal to λ. It is then clear that $\Phi = M_f = \lambda\mathrm{Id}$.

For the proof of the claim, let C be the support of the image measure $\nu = f_*\mu$, which is defined on $\mathbf{R}$ by $\nu(B) = \mu(f^{-1}(B))$ for measurable subsets $B \subset \mathbf{R}$. This set is a non-empty compact (closed and bounded) subset of $\mathbf{R}$ (in fact, it is the spectrum of the multiplication operator, in the sense of spectral theory [**49**, VI.3]). Let $a = \min(C)$, $b = \max(C)$. If $a = b$, we deduce that $C = \{a\}$ is a single point, and then f is almost everywhere equal to a, and we are done. Otherwise, let $c = (a + b)/2$, $U = [a, c[$ and $V = [c, b]$. By definition of the support of a measure, we then have $\nu(U) > 0$ and $\nu(V) > 0$.

Now let finally A be the operator of multiplication by the characteristic function of $f^{-1}(U)$, and let H_1 be its kernel. This is a closed subspace of H, and the conditions $\nu(U) > 0$ and $\nu(V) > 0$ allow us to deduce that $0 \neq \mathrm{H}_1 \neq \mathrm{H}$. But by Exercise 3.4.19 (applied with $B = \varrho(g)$ for all $g \in G$), A is an intertwiner of ϱ, and therefore H_1 is a non-trivial subrepresentation of the representation on H. This contradicts the irreducibility assumption and finishes the proof.

We use this to prove (1). Let $\Phi : \mathrm{H}_1 \longrightarrow \mathrm{H}_2$ be a non-zero intertwiner between two given irreducible unitary representations of G. Then $\Phi^* \circ \Phi : \mathrm{H}_1 \longrightarrow \mathrm{H}_1$ is a self-intertwiner of H_1. According to (2), we have therefore $\Phi^* \circ \Phi = \lambda \mathrm{Id}$ for some constant $\lambda \in \mathbf{C}$. Then from the formula

$$\lambda\langle v, w\rangle = \langle \Phi^*\Phi v, w\rangle = \langle \Phi v, \Phi w\rangle$$

for all v and w in H_1, we deduce that $\lambda > 0$ (take $v = w$ and use the fact that Φ is assumed to be non-zero) and then that $\lambda^{-1/2}\Phi$ is an isometric intertwiner from H_1 to H_2.

But now $\mathrm{Ker}\,\Phi$ is a (closed) subrepresentation of H_1, and it is not H_1, so it must be 0, i.e., Φ is injective. Similarly, the image of Φ must be dense in H_2, but the fact that Φ is isometric (up to a scalar) implies that $\Phi(\mathrm{H}_1)$ is closed. Hence, in fact, Φ is an isomorphism and, by contraposition, we derive (1). □

Exercise 3.4.20 (Uniqueness of invariant inner product)**.** Let G be a topological group, and let $\varrho : G \longrightarrow \mathrm{U}(\mathrm{H})$ be an irreducible unitary representation of G. Show that the inner product on H for which ϱ is unitary is unique up to multiplication by a positive scalar, i.e., if $\langle\cdot,\cdot\rangle_1$ is a positive-definite inner product on H, defining the same topology on H as the original inner product $\langle\cdot,\cdot\rangle$, and if

$$\langle \varrho(g)v, \varrho(g)w\rangle_1 = \langle v, w\rangle_1$$

for all v, $w \in \mathrm{H}$, then there exists $\lambda > 0$ such that

$$\langle v, w\rangle_1 = \lambda\langle v, w\rangle$$

for all v, $w \in \mathrm{H}$.

The following result is also very useful:

Lemma 3.4.21 (Unrelated unitary subrepresentations are orthogonal)**.** *Let G be a topological group, and let $G \xrightarrow{\varrho} \mathrm{U}(\mathrm{H})$ be a unitary representation. If H_1 and H_2 are subrepresentations of* H *such that there is no non-zero G-intertwiner $\mathrm{H}_1 \to \mathrm{H}_2$, then H_1 and H_2 are orthogonal. In particular, isotypic components in* H *of non-isomorphic irreducible representations of G are pairwise orthogonal.*

Proof. Consider the orthogonal projection

$$\Phi : \mathrm{H} \longrightarrow \mathrm{H}$$

on the closed subspace H_2. This linear map Φ is also a G-homomorphism: indeed, if $v \in \mathrm{H}$, its projection $\Phi(v)$ is characterized by

$$v = \Phi(v) + (v - \Phi(v)), \qquad \Phi(v) \in \mathrm{H}_2, \quad v - \Phi(v) \in \mathrm{H}_2^{\perp},$$

and for any $g \in G$, the subsequent relation

$$\varrho(g)v = \varrho(g)(\Phi(v)) + \varrho(g)(v - \Phi(v))$$

together with the condition $\varrho(g)\Phi(v) \in \mathrm{H}_2$, $\varrho(g)(v - \Phi(v)) \in \mathrm{H}_2^{\perp}$ (which follow from the fact that H_2 is a subrepresentation) imply that

$$\Phi(\varrho(g)v) = \varrho(g)\Phi(v),$$

as claimed.

Since the image of Φ is H_2, its restriction to H_1 is a linear map

$$\mathrm{H}_1 \longrightarrow \mathrm{H}_2,$$

which is a G-intertwiner. The assumption then says that it is zero, so that $\mathrm{H}_1 \subset \mathrm{Ker}(\Phi) = \mathrm{H}_2^{\perp}$, which is the same as to say that H_1 and H_2 are orthogonal.

The last statement is of course a corollary of this fact together with Schur's Lemma. □

Chapter 4

Linear representations of finite groups

In this chapter, we take up the special case of finite groups, building on the basic results of Section 2.7. There are however still two very distinct situations: if the field k has characteristic coprime with the order of a finite group G (for instance, if $k = \mathbf{C}$, or any other field of characteristic 0), a fundamental result of Maschke shows that *any k-representation of G is semisimple.* Thus, we can use characters to characterize all (finite-dimensional) representations of G, and this leads to very powerful methods to analyze representations. Most of this chapter will be devoted to this case. However, if k is of characteristic p dividing the order of G, the semisimplicity property fails. The classification and structure of the representations of G is then much more subtle, and we refer to [**53**, Part III] for a very clear introduction.

4.1. Maschke's Theorem

As already hinted, the next result is the most important result about the representation theory of finite groups.

Theorem 4.1.1 (Maschke). *Let G be a finite group, and let k be a field with characteristic* not dividing $|G|$. *Then any k-linear representation*

$$\varrho : G \longrightarrow \mathrm{GL}(E)$$

of G is semisimple. In fact, the converse is also true: if all k-representations of G are semisimple, then the characteristic of k does not divide $|G|$.

Thus, in some sense, in the case where Maschke's Theorem applies, the classification of all representations of G is reduced to the question of classifying the irreducible ones. Note that it is not required to assume that k be algebraically closed here.

Proof. We use the criterion of Lemma 2.2.11. For a given subrepresentation $F \subset E$, the idea is to construct a stable supplement $F^{\perp}$ as the kernel of a linear projection

$$P : E \longrightarrow E$$

with image F which is a G-morphism, i.e., $P \in \mathrm{Hom}_G(E, E)$. Indeed, if $P^2 = P$ (which means P is a projection) and $\mathrm{Im}(P) = F$, we have

$$E = F \oplus \mathrm{Ker}(P),$$

and of course $\mathrm{Ker}(P)$ is a subrepresentation if $P \in \mathrm{Hom}_G(E, E)$.

From linear algebra again, we know that there exists a projection $P_0 \in \mathrm{Hom}_k(E, E)$ with $\mathrm{Im}(P_0) = F$, but a priori not one that commutes with the action of G. Note that $P \in \mathrm{Hom}_G(E, E)$ means that $P \in \mathrm{End}_k(E)^G$ (for the action of G on $\mathrm{End}_k(E)$ defined in Proposition 2.2.21; see (2.17)). The trick is to construct P using P_0 by *averaging* the action (2.18) of G on P_0 in order to make it invariant.

Let then

$$P = \frac{1}{|G|} \sum_{g \in G} g \cdot P_0 \in \mathrm{End}_k(E).$$

We claim that P is the desired projection in $\mathrm{End}_G(E)$. The first thing to notice is that it is this definition which requires that $p \nmid |G|$, since $|G|$ must be invertible in k in order to compute P.

By averaging, it is certainly the case that $P \in \mathrm{End}_G(E) = \mathrm{End}_k(E)^G$: acting on the left by some $h \in G$ just permutes the summands

$$h \cdot P = \frac{1}{|G|} \sum_{g \in G} h \cdot (g \cdot P_0) = \frac{1}{|G|} \sum_{g \in G} (hg) \cdot P_0 = \frac{1}{|G|} \sum_{x \in G} x \cdot P_0 = P$$

(in the notation of Example 3.1.1, we are using the fact that $P = e \cdot P_0$ where $e = \frac{1}{|G|} s \in k(G)$, and that $he = e$ by (3.2)).

Next, $\mathrm{Im}(P) \subset \mathrm{Im}(P_0) = F$: indeed, F is G-invariant and each term

$$(g \cdot P_0)v = \varrho(g)(P_0(\varrho(g^{-1})v)) \in F$$

in the sum is in F for any fixed v. Moreover, P is the identity on F, since P_0 is and F is stable: for $v \in F$, we have

$$P(v) = \frac{1}{|G|} \sum_{g \in G} \varrho(g) P_0(\varrho(g^{-1})v) = \frac{1}{|G|} \sum_{g \in G} \varrho(gg^{-1})v = v.$$

Thus $(P \circ P)(v) = P(P(v)) = P(v)$ since $P(v) \in F$, and hence P is indeed an intertwining projection onto F.

We now prove the converse of Maschke's Theorem. In fact, the result is stronger than what we claim: we now show that the regular representation on $C_k(G)$ is never semisimple if k has characteristic p dividing $|G|$. To do this, we consider the subspace

$$C_0 = \{\varphi \in C_k(G) \mid \sum_{g \in G} \varphi(g) = 0\} \subset C_k(G).$$

This subspace is always a subrepresentation of $C_k(G)$, as one checks immediately (as before, the values of $\varrho_G(g)\varphi$ are a permutation of the values of φ). As the kernel of the non-zero linear form

$$\lambda : \varphi \mapsto \sum_{g \in G} \varphi(g),$$

we see that C_0 is of codimension 1 in $C_k(G)$. If $p \nmid |G|$, a complementary stable subspace is easy to find: it is the space of constant functions (it is in fact the unique stable complement). But if $\varphi = c \in k$ is constant, and $p \mid |G|$, we have $\lambda(\varphi) = c|G| = 0$, so this complement does not work in characteristic p. We now check that no other will do: if φ_0 is a basis of such a complement, the action of G on $k\varphi_0$ is by a one-dimensional representation χ, so we have

$$\varrho_G(g)\varphi_0 = \chi(g)\varphi_0$$

for all $g \in G$. Evaluating at 1, we find that $\varphi_0(g) = \chi(g)\varphi_0(1)$ for all G. But then

$$\lambda(\varphi_0) = \varphi_0(1) \sum_{g \in G} \chi(g),$$

and this equal to 0, which contradicts the assumption that $\varphi_0 \notin \ker(\lambda)$. Indeed, either χ is trivial and then the value is $|G| = 0$ in k, or there exists $x \in G$ with $\chi(x) \neq 1$, and then writing

$$\sum_{g \in G} \chi(g) = \sum_{h \in G} \chi(xh) = \chi(x)\lambda(\chi),$$

we find again that the sum is zero. □

Exercise 4.1.2. This exercise shows that the idea of averaging can also be used in more general contexts than that of finite group.

Let $H \lhd G$ be a finite-index subgroup of an arbitrary group G. Let k be a field, and let ϱ be a finite-dimensional k-representation of G.

(1) Show that if $\operatorname{Res}_H^G \varrho$ is semisimple as a representation of H, and if the characteristic of k is coprime to $[G : H]$, then ϱ is semisimple. (This is a partial converse to the result of Exercise 2.3.4.)

(2) Show that this statement is false without a condition on the characteristic of k.

Remark 4.1.3 (Semisimplicity of unitary representations)**.** If $k = \mathbf{C}$, we can also prove Theorem 4.1.1 by exploiting Proposition 3.4.14, at least when dealing with finite-dimensional representations. Indeed, we have the following result.

Proposition 4.1.4 (Unitarizability for finite groups)**.** *For a finite group G, any finite-dimensional representation of G over $\mathbf{C}$ is unitarizable.*

Proof. The idea is similar to the one in the proof of Maschke's Theorem. Let ϱ be a finite-dimensional representation of G on E. What we must do is find an inner product $\langle\cdot,\cdot\rangle$ on E with respect to which ϱ is unitary, i.e., such that

$$\langle\varrho(g)v, \varrho(g)w\rangle = \langle v, w\rangle$$

for all v, $w \in E$. Since E is finite-dimensional, we can certainly find some inner product $\langle\cdot,\cdot\rangle_0$ on E, although it is not necessarily invariant. But then if we let

$$\langle v, w\rangle = \frac{1}{|G|}\sum_{g\in G}\langle\varrho(g)v, \varrho(g)w\rangle_0,$$

it is easy to check that we obtain the desired invariant inner product. □

More generally, suppose $\langle\cdot,\cdot\rangle_0$ is a non-negative, but not necessarily positive-definite, hermitian form on E, with kernel

$$F = \{v \in E \mid \langle v, v\rangle_0 = 0\}.$$

Then it is clear that the construction of $\langle\cdot,\cdot\rangle$ above still leads to an invariant non-negative hermitian form. By positivity, it will be a genuine, positive-definite, inner product if

$$\bigcap_{g\in G} g\cdot F = 0.$$

An example of this is the regular representation, where we can take

$$\langle\varphi_1, \varphi_2\rangle_0 = \varphi_1(1)\overline{\varphi_2(1)}.$$

This has a huge kernel F (all functions vanishing at 1) but since

$$\varrho_G(g)F = \{\varphi \in C(G) \mid \varrho_G(g)\varphi(1) = \varphi(g) = 0\},$$

it follows that the intersection of the translates of F is in fact 0. Quite naturally, the resulting inner product on $C(G)$ is exactly the one described in Example 3.4.12.

The meaning of Maschke's Theorem is that for any k-representation ϱ of G on a finite-dimensional vector space E, there always exists a direct sum decomposition of E in irreducible stable subspaces if $|G|$ is invertible in k. As already discussed in Chapter 2, this decomposition is not unique. However, by Proposition 2.7.9, the isotypic components of ϱ, denoted $\mathrm{M}(\pi)$ or $\mathrm{M}_E(\pi)$, are defined for any irreducible k-representation π of G, and they are intrinsic subspaces of E such that

$$E = \bigoplus_{\pi} \mathrm{M}_E(\pi), \tag{4.1}$$

where π runs over isomorphism classes of irreducible k-representations of G. Recall that because they are intrinsic, it follows that for any G-homomorphism

$$\Phi : E \longrightarrow F,$$

the restriction of Φ to $\mathrm{M}_E(\pi)$ gives a linear map

$$\mathrm{M}_E(\pi) \longrightarrow \mathrm{M}_F(\pi).$$

In order to analyze the representations concretely, one needs some way of obtaining information concerning this decomposition. We will see how to describe explicitly the projections on E mapping onto the isotypic components (when k is algebraically closed), and when k is of characteristic 0, how to use characters to compute the multiplicities of the irreducible representations (which are of course related to the dimension of the isotypic components).

4.2. Applications of Maschke's Theorem

We are now going to apply Maschke's Theorem. For most of the remainder of this chapter, we will consider algebraically closed fields k and finite groups G such that the characteristic of k does not divide $|G|$, so that Maschke's Theorem is applicable, as well as Schur's Lemma (Proposition 2.7.15).

We introduce the notation $\widehat{G}$ for the set of isomorphism classes of k-irreducible representations of G (it will always be clear which algebraically closed field k is used). This set can be identified, by character theory, with the set of characters of irreducible k-representations of G.

Applying first Maschke's Theorem to the regular representation of G on $C_k(G)$, we deduce from Corollary 2.7.30 a fundamental result:

Corollary 4.2.1 (Decomposition of the regular representation). *Let G be a finite group, and let k be an algebraically closed field of characteristic not dividing $|G|$. Then the regular representation of G on $C_k(G)$ is isomorphic to the direct sum, over all isomorphism classes of irreducible representations $\varrho \in \widehat{G}$ of G, of subrepresentations isomorphic to* $\dim(\varrho)$ *copies of ϱ.*

In particular, we have

$$\sum_{\varrho\in\hat{G}} (\dim \varrho)^2 = |G|. \tag{4.2}$$

One naturally wants to get more information about the irreducible representations than what is contained in formula (4.2). The first basic question is, What is the *number* of irreducible representations (up to isomorphism)? The general answer is known, but we start with a particularly simple case:

Proposition 4.2.2 (Irreducible representations of finite abelian groups). *Let G be a finite group, and let k be an algebraically closed field of characteristic not dividing $|G|$. Then all irreducible finite-dimensional representations of G are of dimension 1 if and only if G is abelian. In particular, if G is abelian, there are $|G|$ non-isomorphic irreducible k-representations of G.*

Proof of Proposition 4.2.2. We know that the one-dimensional representations of G are in bijection with those of the abelianized group $G/[G,G]$ (Proposition 2.3.3). Thus if all irreducible k-representations of G are of dimension 1, Corollary 4.2.1 implies that $|G| = |G/[G,G]|$ (both are equal to the number of irreducible k-representations of G), which means that $[G,G] = 1$, i.e., that G is commutative. □

Although the representations of abelian groups are quite elementary in comparison with the general case, they are of great importance in applications. We will say more about them in Section 4.5, which includes in particular a sketch of the proof of Dirichlet's Theorem on primes in arithmetic progressions. (That later section could be read right now without much difficulty.)

Note that here also the assumption on k cannot be removed: there are cases where G is non-abelian, k is algebraically closed of characteristic dividing G, and the only irreducible k-representation of G is trivial. Indeed, this is true for all non-abelian p-groups and representations over algebraically closed fields of characteristic p (see, e.g., [**12**, 27.28]); examples are given by the groups from Example 2.7.39 and in the following exercise.

Exercise 4.2.3. (1) Let G be a p-group acting on a finite set X with $p \nmid |X|$. Show that this action has at least one fixed point.

(2) Let

$$G = \left\{ \begin{pmatrix} 1 & x & y \\ & 1 & z \\ & & 1 \end{pmatrix} \mid x, y, z \in \mathbf{F}_p \right\}.$$

Let k be any field of characteristic p. Show that any irreducible k-representation of G is trivial. (*Hint*: Consider first the case where k is a finite field.)

Example 4.2.4. Consider $G = \mathfrak{S}_3$, the symmetric group on three letters. It is non-abelian of order 6, and hence the only possible values for the degrees of irreducible $\mathbf{C}$-representations of G are 1, 1 and 2 (there are no other integers with squares summing to 6, where not all are equal to 1). The two one-dimensional representations are of course the trivial one and the signature

$$\varepsilon : \mathfrak{S}_3 \longrightarrow \{\pm 1\} \subset \mathbf{C}^\times,$$

and the two-dimensional one is isomorphic to the representation by permutation of the coordinates on the vector space

$$E = \{(x, y, z) \in \mathbf{C}^3 \mid x + y + z = 0\}$$

(see Exercise 2.6.6).

More generally, the decomposition of the regular representation implies that there are at most $|G|$ irreducible representations, and Proposition 4.2.2 shows that this upper bound is reached if and only G is abelian. In fact, we have the following:

Theorem 4.2.5 (Number of irreducible characters)**.** *Let G be a finite group, and let k be an algebraically closed field of characteristic not dividing $|G|$. Then the number $|\hat{G}|$ of isomorphism classes of irreducible k-representations of G is equal to the number of conjugacy classes in G.*

This is another striking fact, which can tell us how many irreducible representations exist even when none is explicitly known (except for the trivial one...). We will give more comments on this theorem after its proof.

Proof. We have the decomposition

$$C_k(G) = \bigoplus_{\varrho \in \hat{G}} \mathrm{M}(\varrho), \tag{4.3}$$

where $\mathrm{M}(\varrho)$ is the space spanned by all matrix coefficients of the irreducible k-representation ϱ. To compute the number of summands, we try to find some invariant of $C_k(G)$ which will involve *counting* each ϱ only once. The dimension does not work since $\dim \mathrm{M}(\varrho) = \dim(\varrho)^2$; computing the intertwiners from $C_k(G)$ to itself is also tempting, but since $\mathrm{M}(\varrho)$ is a direct sum of $\dim(\varrho)$ copies of ϱ, we have (by Schur's Lemma) again

$$\begin{aligned}\dim \operatorname{Hom}_G(C_k(G), C_k(G)) &= \sum_{\varrho} \dim \operatorname{Hom}_G(\mathrm{M}(\varrho), \mathrm{M}(\varrho)) \\ &= \sum_{\varrho} (\dim \varrho)^2 = |G|.\end{aligned}$$

One way to do this turns out to be a bit tricky but enlightening (two other possibilities are discussed in Section 4.3.3 and Remark 4.3.30): we use the fact that $C_k(G)$ carries in fact a representation π of $G \times G$ defined by

$$\pi(g_1, g_2)f(x) = f(g_1^{-1} x g_2)$$

(see Exercise 2.1.3) and that the decomposition (4.3) is in fact a decomposition of $C_k(G)$ into subrepresentations of π. Indeed, if $f_{v,\lambda} \in \mathrm{M}(\varrho)$ is a matrix coefficient

$$f_{v,\lambda}(x) = \langle \lambda, \varrho(g)v \rangle$$

of an irreducible representation ϱ, we have

$$\pi(g_1, g_2) f_{v,\lambda}(x) = \langle \lambda, \varrho(g_1^{-1} x g_2)v \rangle = \langle \check{\varrho}(g_1)\lambda, \varrho(x)\varrho(g_2)v \rangle = f_{\check{\varrho}(g_1)\lambda, \varrho(g_2)v}(x)$$

so that $\mathrm{M}(\varrho)$ is indeed stable under the action of $G \times G$.

The point is that $\mathrm{M}(\varrho)$, as a subrepresentation of π in $C_k(G)$, is isomorphic to the external tensor product $\check{\varrho} \boxtimes \varrho$. Indeed, if ϱ acts on the space E, this isomorphism is the canonical one given by

$$\begin{cases} E' \otimes E & \longrightarrow & \mathrm{M}(\varrho) \\ \lambda \otimes v & \mapsto & f_{v,\lambda}, \end{cases}$$

which is a linear isomorphism by linear independence of the matrix coefficients, and an intertwiner by the computation just performed.[1]

Since the representations $\check{\varrho} \boxtimes \varrho$ of $G \times G$ are all irreducible and nonisomorphic, as ϱ runs over the irreducible representations of G (by Proposition 2.3.23), each appears with multiplicity 1 in π. As a consequence, we can use Schur's Lemma (on $G \times G$) to express the number of irreducible representations ϱ by the formula

$$\dim \operatorname{Hom}_{G\times G}(\pi, \pi) = \sum_{\varrho_1, \varrho_2} \dim \operatorname{Hom}_{G\times G}(\varrho_1 \boxtimes \check{\varrho}_1, \varrho_2 \boxtimes \check{\varrho}_2) = \sum_{\varrho} 1.$$

We now compute directly the left-hand side dimension to deduce the desired formula. In fact, consider a linear map

$$\Phi : C_k(G) \longrightarrow C_k(G),$$

[1] Note that the order of the factors is important here! If we consider $E \otimes E'$ instead, we do not obtain an intertwiner....

which commutes with the representation π. If $\delta_g \in C_k(G)$ denotes the function which is 0 except at $x = g$, where it is equal to 1, and

$$\ell_g = \Phi(\delta_g),$$

we must therefore have

$$\pi(g_1, g_2)\ell_g = \Phi(\pi(g_1, g_2)\delta_g) = \Phi(\delta_{g_1 g g_2^{-1}}) = \ell_{g_1 g g_2^{-1}}$$

for all g, g_1, $g_2 \in G$. Evaluating at $x \in G$ means that we must have

$$\ell_g(g_1^{-1} x g_2) = \ell_{g_1 g g_2^{-1}}(x), \tag{4.4}$$

and in fact, since (δ_g) is a basis of $C_k(G)$, these equations on ℓ_g are equivalent with Φ being an intertwiner of π with itself.

We can solve these equations as follows. Picking first $g_2 = 1$ and $g_1 = x^{-1}$ and then $g_1 = 1$ and $g_2 = x$, we get

$$\ell_{gx}(x) = \ell_g(1) = \ell_{xg}(x),$$

and then we deduce the two relations

$$\ell_{gx^{-1}}(1) = \ell_g(x) = \ell_{x^{-1}g}(1). \tag{4.5}$$

Thus Φ is entirely defined by the function $\psi : G \longrightarrow k$ defined by

$$\psi(g) = \ell_g(1).$$

In view of the two expressions for $\ell_g(x)$ above, this function must satisfy

$$\psi(ab) = \psi(ba) \tag{4.6}$$

for all a, $b \in G$. But conversely, given a function ψ that satisfies these relations, we may define $\ell_g(x)$ by

$$\ell_g(x) = \psi(gx^{-1}) = \psi(x^{-1}g)$$

(see (4.5)), and then obtain

$$\ell_g(g_1^{-1} x g_2) = \psi(g g_2^{-1} x^{-1} g_1)$$

and

$$\ell_{g_1 g g_2^{-1}}(x) = \psi(x^{-1} g_1 g g_2^{-1}),$$

from which (4.4) follows by putting $a = gg_2^{-1}$, $b = x^{-1}g_1$ in (4.6).

The conclusion is that $\mathrm{Hom}_{G \times G}(\pi, \pi)$ is isomorphic (by mapping Φ to ψ) to the linear space $c_k(G)$ of all functions ψ such that (4.6) holds. But this condition is equivalent with

$$\psi(x) = \psi(gxg^{-1}), \qquad \text{for all } x, g \in G,$$

or in other words, $c_k(G)$ is the space of class functions on G. Since the dimension of $c_k(G)$ is equal to the number of conjugacy classes of G, by definition (a class function is determined by the values at the conjugacy classes), we obtain the desired formula. □

Remark 4.2.6 (Sum of dimensions)**.** Thus we have *directly accessible* group-theoretic expressions for the number of irreducible k-representations of a finite group G (which is the number of conjugacy classes), and for the sum of the squares of their degrees (which is simply $|G|$). It seems natural to ask, What about the sum of the degrees themselves, or what about other powers of $\dim(\varrho)$? Although there does not seem to exist any nice expression valid for all groups, there are some special cases (including important examples like symmetric groups) where the sum of the dimensions has a nice interpretation, as explained in Lemma 6.2.6.

Remark 4.2.7 (Bijections anyone?)**.** Theorem 4.2.5 is extremely striking; since it gives an equality between the cardinality of the set of irreducible characters of G (over an algebraically closed field of characteristic not dividing $|G|$) and the cardinality of the set of conjugacy classes, it implies that there exist some bijection between the two sets. However, even for $k = \mathbf{C}$, there is no general natural definition of such a bijection, and there probably is none.

Despite this, there are many cases where one can understand both the conjugacy classes and the irreducible representations, and where some rough features of the two sets seem to correspond in tantalizing parallels (see the discussion of $\mathrm{GL}_2(\mathbf{F}_p)$ in Section 4.6.4). Additionally, in some particularly favorable circumstances, a precise correspondence can be found; the most striking and important of these cases is that of the symmetric groups $\mathfrak{S}_m$ (see Section 4.6.5).

Another remarkable aspect of the result is that the number of irreducible representations does not depend on the field k (assumed to be algebraically closed, and with characteristic coprime to $|G|$). Here again, this means that there are bijections between the sets of irreducible characters for different fields. This is of course surprising when the characteristics of the fields are distinct! One may also ask here if such a bijection can be described explicitly, and the situation is better than the previous one. Indeed, Brauer developed a theory which, among other things, does lead to bijections between irreducible characters of G over any algebraically closed fields of characteristic coprime with $|G|$; see, e.g, [**28**, Ch. 15, Th. 15.13] or [**53**, Ch. 18], for an account of this theory, from which it follows, in particular, that the family of dimensions of the irreducible characters over two such fields always coincide.

Exercise 4.2.8 (Minimal index of a proper subgroup)**.** Here is an application of Maschke's Theorem and of induction. Let G be a finite group G, and let $H \subset G$ be a proper subgroup. We want to obtain a lower bound for its index $[G : H]$ in terms of representation theory.

(1) Show that the (complex) induced representation

$$\varrho = \mathrm{Ind}_H^G(\mathbf{1}_{\mathbf{C}}),$$

contains the trivial representation of G with multiplicity one.

(2) Deduce that

$$[G : H] \geqslant 1 + \min_{\pi \neq \mathbf{1}} \dim(\pi),$$

where π runs over non-trivial irreducible $\mathbf{C}$-representations of G.

(Although sometimes this result only implies $[G : H] \geqslant 2$, we will see later in Exercises 4.6.13 and 4.7.3 that for the finite groups $\mathrm{SL}_2(\mathbf{F}_p)$, for instance, it is non-trivial.)

4.3. Decomposition of representations

4.3.1. The projection on invariant vectors. Especially when written in terms of the group algebra (using Example 3.1.1), the core argument of the proof of Maschke's Theorem can be immediately generalized:

Proposition 4.3.1 (Projection on the invariant vectors). *Let G be a finite group, and let k be an algebraically closed field of characteristic not dividing $|G|$. For any k-representation $\varrho : G \longrightarrow \mathrm{GL}(E)$, the map*

$$\frac{1}{|G|} \sum_{g \in G} \varrho(g) \; : \; E \longrightarrow E$$

is a homomorphism of representations, which is a projection with image equal to E^G, the space of invariant vectors in E. Moreover, if E is finite dimensional, we have

$$\dim(E^G) = \frac{1}{|G|} \sum_{g \in G} \chi_\varrho(g) \tag{4.7}$$

as an equality in the field k.

Proof. Let P be the indicated linear map. As in the proof of Maschke's Theorem, we see that that $\mathrm{Im}(P) \subset E^G$, and that P is the identity on E^G. Moreover, we have

$$P(\varrho(g)v) = P(v) = \varrho(g)P(v),$$

since $\mathrm{Im}(P) \subset E^G$. All this shows that $P \in \mathrm{Hom}_G(E, E)$ is a projection with image exactly equal to E^G.

Finally, the trace of a projection is equal to the dimension of the image, as seen in the field k, hence the last formula. □

We can use fruitfully Proposition 4.3.1 in many ways. One is to take any representation for which we know the invariants, and to see what form the projection takes. The next few sections give some important examples.

4.3.2. "Orthogonality" of characters and matrix coefficients. Consider G and k as before. If

$$\pi_1 : G \longrightarrow \mathrm{GL}(E_1), \qquad \pi_2 : G \longrightarrow \mathrm{GL}(E_2)$$

are irreducible k-representations, we know by Schur's Lemma that for the natural action of G on $E = \mathrm{Hom}_k(E_1, E_2)$, the space of invariants

$$\mathrm{Hom}_G(E_1, E_2) = \mathrm{Hom}_k(E_1, E_2)^G$$

has dimension 0 or 1 depending on whether π_1 and π_2 are isomorphic. Applying Proposition 4.3.1, this leads to some fundamental facts.

We assume, to begin with, that π_1 is *not* isomorphic to π_2. Then the invariant space is 0, and hence the associated projection is also zero: for any $\Phi : E_1 \longrightarrow E_2$, we have

$$\frac{1}{|G|} \sum_{g \in G} g \cdot \Phi = 0. \tag{4.8}$$

To see what this means concretely, we select some element $\Phi \in E = \mathrm{Hom}_k(E_1, E_2)$. Because there is no intrinsic relation between the spaces here, some choice must be made. We consider linear maps of rank 1: let $\lambda \in E_1'$ be a linear form, let $w \in E_2$ be a vector, and let

$$\Phi : v \mapsto \langle \lambda, v \rangle w$$

be the corresponding rank 1 map in $\mathrm{Hom}_k(E_1, E_2)$. Spelling out the identity (4.8) by applying it to a vector $v \in E_1$, we get

$$0 = \frac{1}{|G|} \sum_{g \in G} \pi_2(g)(\langle \lambda, \pi_1(g^{-1}) v \rangle w) = \frac{1}{|G|} \sum_{g \in G} f_{v,\lambda}(g) \pi_2(g^{-1}) w$$

for all v (we replaced g by g^{-1} in the sum, which merely permutes the terms).

We can make this even more concrete by applying an arbitrary linear form $\mu \in E_2'$ to obtain numerical identities:

Proposition 4.3.2 (Orthogonality). *With notation as above, let π_1, π_2 be* non-isomorphic *k-irreducible representations of G. Then for all vectors $v \in E_1$, $w \in E_2$ and linear forms $\lambda \in E_1'$, $\mu \in E_2'$, we have*

$$\frac{1}{|G|} \sum_{g \in G} f_{v,\lambda}(g) f_{w,\mu}(g^{-1}) = \frac{1}{|G|} \sum_{g \in G} \langle \lambda, g \cdot v \rangle_{E_1} \langle \mu, g^{-1} \cdot w \rangle_{E_2} = 0. \tag{4.9}$$

Because such sums will come up often, it is convenient to make the following definition:

Definition 4.3.3 (*Inner product* of functions on G)**.** Let G be a finite group, and let k be a field with $|G|$ invertible in k. For φ_1, φ_2 in $C_k(G)$, we denote

$$[\varphi_1, \varphi_2] = \frac{1}{|G|} \sum_{g \in G} \varphi_1(g)\varphi_2(g^{-1}).$$

This is a non-degenerate[2] symmetric bilinear form on $C_k(G)$, called the k-inner product.

Thus we have shown that matrix coefficients of non-isomorphic representations are orthogonal for the k-inner product on $C_k(G)$.

Before going on, we can also exploit the formula (4.7) in this situation. Since $E^G = 0$, we derive

$$0 = \sum_{g \in G} \chi_E(g).$$

On the other hand, using the isomorphism

$$E \simeq E_1' \otimes E_2,$$

which intertwines the action on E with $\check{\varrho}_1 \otimes \varrho_2$, we have (by Proposition 2.7.43)

$$\chi_E(g) = \chi_{\pi_1}(g^{-1})\chi_{\pi_2}(g).$$

So we derive the formula

$$[\chi_{\pi_2}, \chi_{\pi_1}] = \frac{1}{|G|} \sum_{g \in G} \chi_{\pi_1}(g^{-1})\chi_{\pi_2}(g) = 0, \tag{4.10}$$

i.e., the characters of distinct irreducible representations are also orthogonal.

Now we consider the case where the representations π_1 and π_2 are equal, and we then denote $\pi = \pi_1 = \pi_2$, acting on the space E. The space $\mathrm{End}_k(E)^G$ is one dimensional and consists of the scalar operators. The precise application of the projection on the invariant space $\mathrm{End}_k(E)^G$ will now require us to identify the scalar which is obtained.

But first we apply (4.7), which does not involve such computations. The argument leading to (4.10) still applies, with the sole change that the trace of the projection is now 1. Hence, we get

$$[\chi_\pi, \chi_\pi] = \frac{1}{|G|} \sum_{g \in G} \chi_\pi(g)\chi_\pi(g^{-1}) = 1. \tag{4.11}$$

We can proceed as before with rank 1 linear maps, but in the present case we should also observe that, because we deal with $\mathrm{End}_k(E)$, there are some other obvious linear maps to apply the projection to, namely the endomorphisms $\varrho(h)$, for $h \in G$.

---

[2] If $\varphi_1 \neq 0$, pick $x \in G$ with $\varphi_1(x) \neq 0$, and let φ_2 be the characteristic function of x^{-1}. Then $[\varphi_1, \varphi_2] \neq 0$.

We obtain that, for some $\lambda(h) \in k$, we have

$$\frac{1}{|G|}\sum_{g\in G} g\cdot\pi(h) = \frac{1}{|G|}\sum_{g\in G}\pi(ghg^{-1}) = \lambda(h)\mathrm{Id}_E.$$

To determine $\lambda(h)$, we take the trace (this is a standard technique): since

$$\mathrm{Tr}(\pi(ghg^{-1})) = \chi_\pi(h)$$

for all g, we get

$$\chi_\pi(h) = \lambda(h)\dim(\pi)$$

for any $h \in G$.

We have to be careful before concluding, because if k has positive characteristic, it might conceivably be the case that $\dim(\pi) = 0$ in k. However, we can see that this is not the case by noting that the formula just derived would then say that the character of π is identically 0, which contradicts the linear independence of irreducible characters, or simply the formula (4.11). Hence we have:

Proposition 4.3.4. *Let G be a finite group, and let k be an algebraically closed field of characteristic coprime to $|G|$. For any irreducible k-representation π of G and any $h \in G$, we have*

$$\frac{1}{|G|}\sum_{g\in G}\pi(ghg^{-1}) = \frac{\chi_\pi(h)}{\dim(\pi)}.$$

We now finally come back to rank 1 maps. For given $w \in E$ and $\lambda \in E'$ defining

$$\Phi : \begin{cases} E \to E \\ v \mapsto \langle\lambda, v\rangle w, \end{cases}$$

as before, the operator

$$\frac{1}{|G|}\sum_{g\in G} g\cdot\Phi$$

is now a multiplication by a scalar α. To determine the scalar in question, we compute the trace. Since

$$\mathrm{Tr}(g\cdot\Phi) = \mathrm{Tr}(\pi(g)\Phi\pi(g^{-1})) = \mathrm{Tr}(\Phi)$$

for all g, we get

$$(\dim E)\alpha = \mathrm{Tr}(\alpha\mathrm{Id}_E) = \mathrm{Tr}(\Phi) = \langle\lambda, w\rangle$$

(the trace of the rank 1 map can be computed by taking a basis where w, the generator of the image, is the first basis vector).

Since we have already seen that $\dim(E)$ is invertible in k, we obtain therefore

$$\frac{1}{|G|}\sum_{g\in G} g\cdot\Phi = \frac{\langle\lambda, w\rangle}{\dim E},$$

and applying this to a vector $v \in E$, and applying further a linear form $\mu \in E'$ to the result, we get

Proposition 4.3.5 (Orthogonality of matrix coefficients). *With notation as above, let π be an irreducible k-representation of G. Then for all vectors $v, w \in E$ and linear forms $\lambda, \mu \in E'$, we have*

$$[f_{v,\lambda}, f_{w,\mu}] = \frac{1}{|G|}\sum_{g\in G}\langle\lambda, g\cdot v\rangle\langle\mu, g^{-1}\cdot w\rangle = \frac{\langle\lambda, w\rangle\langle\mu, v\rangle}{\dim E}. \tag{4.12}$$

We also summarize the orthogonality for characters:

Corollary 4.3.6 (Orthogonality of characters). *With notation as above, for any two irreducible representations π and ϱ of G, we have*

(4.13)

$$[\chi_\pi, \chi_\varrho] = \frac{1}{|G|}\sum_{g\in G}\chi_\pi(g)\chi_\varrho(g^{-1}) = \begin{cases} 0 & \text{if } \pi \text{ is not isomorphic to } \varrho, \\ 1 & \text{otherwise,}\end{cases}$$

or equivalently

$$\frac{1}{|G|}\sum_{g\in G}\chi_{\pi(g)}\chi_{\check{\varrho}}(g) = \begin{cases} 0 & \text{if } \pi \text{ is not isomorphic to } \varrho, \\ 1 & \text{otherwise.}\end{cases} \tag{4.14}$$

Remark 4.3.7 (Invertibility of dimensions of irreducible representations). We have seen that for k algebraically closed of characteristic p not dividing $|G|$, the dimension of any irreducible representation is invertible in k. In fact, much more is known:

Theorem 4.3.8 (Divisibility). *Let G be a finite group. For any algebraically closed field of characteristic not dividing $|G|$, the family of the dimensions of irreducible k-representations of G is the same, and these dimensions divide the order of $|G|$.*

We will explain later how to show that $\dim(E) \mid |G|$ when $k = \mathbf{C}$ (Proposition 4.7.8); showing that the dimensions of the irreducible representations are the same for any algebraically closed field of characteristic $p \nmid |G|$ is more delicate, since it involves the Brauer characters mentioned in Remark 4.2.7.

4.3.3. Decomposition of class functions. As an application of the previous sections, we now give a slightly different proof of Theorem 4.2.5. The motivation is that the characters of irreducible k-representations of G are linearly independent class functions on G. The k-subspace they span in

$C_k(G)$ is a subspace of $c_k(G)$, and the number of distinct characters is therefore at most $\dim c_k(G)$, which is the number of conjugacy classes. Hence the equality—which is the claim of the theorem—amounts to the statement that the characters actually generate $c_k(G)$, i.e., that they form a basis of this space.

We now prove this fact directly (yet another argument is contained in the proof we give of the corresponding statements for compact groups in Theorem 5.5.1). Let $\varphi \in c_k(G)$ be a class function. Like any function on G it can be expanded into a linear combination of matrix coefficients. We will show that, in this decomposition, only *diagonal* coefficients appear, and that those diagonal ones are constant (for a given irreducible representation), and this means that in fact the linear combination in question is a combination of characters.

To be precise, we fix, for every distinct irreducible representation π, a basis $(e_i^{(\pi)})_i$ of the space E_π of the representation, and denote by $(\lambda_j^{(\pi)})_j$ the dual basis. Let

$$f_{i,j}^{(\pi)} \in C_k(G), \qquad 1 \leqslant i,j \leqslant \dim(\pi),$$

denote the corresponding matrix coefficients. Theorem 2.7.28 (which corresponds simply to the isotypic decomposition of the regular representation) shows that there exist coefficients $\alpha_{i,j}^{(\pi)} \in k$ such that

$$\varphi = \sum_\pi \sum_{i,j} \alpha_{i,j}^{(\pi)} f_{i,j}^{(\pi)}. \tag{4.15}$$

Our claims follow:

(1) for any π, and any distinct indices $i \neq j$, the coefficient $\alpha_{i,j}^{(\pi)}$ is zero;

(2) for any π, the diagonal coefficients $\alpha_{i,i}^{(\pi)}$ are constant as i varies. Given this, if α_π denotes this last common value, we get

$$\varphi = \sum_\pi \alpha_\pi \chi_\pi$$

by interpreting the character as a sum of diagonal matrix coefficients.

To prove the claim, we use the orthogonality of matrix coefficients. Using the choice of a basis and its dual, Propositions 4.3.2 and 4.3.5 show that

$$[f_{i,j}^{(\pi)}, f_{k,l}^{(\varrho)}] = \begin{cases} 0 & \text{if } \pi \neq \varrho, \text{ or } (i,j) \neq (l,k), \\ \dfrac{1}{\dim(\varrho)} & \text{if } \pi = \varrho,\ (i,j) = (l,k). \end{cases}$$

Hence, taking the inner product with some $f_{k,l}^{(\varrho)}$ on both sides of (4.15), we get

$$[\varphi, f_{k,l}^{(\varrho)}] = \sum_{\pi} \sum_{i,j} \alpha_{i,j}^{(\pi)} [f_{i,j}^{(\pi)}, f_{k,l}^{(\varrho)}] = \frac{\alpha_{l,k}^{(\varrho)}}{\dim(\varrho)}. \tag{4.16}$$

We now think of ϱ as fixed. If we remember that $f_{k,l}^{(\varrho)}$ is the (l,k)-th coefficient of the matrix representing $\varrho(g)$ in the basis $(e_l^{(\varrho)})_l$, we can reinterpret the left-hand side of this computation as the (l,k)-th coefficient of the matrix representing

$$A_\varphi = \frac{1}{|G|} \sum_{g \in G} \varphi(g)\varrho(g^{-1}) \in \mathrm{End}_k(\varrho).$$

Now, *because* φ *is a class function*, it follows that, in fact, A_φ is in $\mathrm{End}_G(\varrho)$. This will be presented in the context of the group algebra later, but it is easy to check. We have

$$\begin{aligned}
A_\varphi(\varrho(h)v) &= \frac{1}{|G|} \sum_{g \in G} \varphi(g)\varrho(g^{-1}h)v \\
&= \frac{1}{|G|} \sum_{g \in G} \varphi(g)\varrho(h(h^{-1}g^{-1}h))v \\
&= \frac{1}{|G|} \sum_{g \in G} \varphi(hgh^{-1})\varrho(h)\varrho(g^{-1})v = \varrho(h)A_\varphi(v).
\end{aligned}$$

Consequently, Schur's Lemma ensures—once again!—that A_φ is a scalar matrix. In particular, its off-diagonal coefficients $[\varphi, f_{k,l}^{(\varrho)}]$ with $k \neq l$ are zero, and the diagonal ones are all identical. Translating in terms of the coefficients $\alpha_{l,k}^{(\varrho)}$ using (4.16), we obtain the claim concerning the latter.

4.3.4. Orthogonality for unitary representations. We consider in this short section the case $k = \mathbf{C}$. Then we can proceed with the same arguments as before, but using the self-duality of Hilbert spaces, we may use the rank 1 linear maps

$$\mathrm{H}_1 \xrightarrow{\Phi} \mathrm{H}_2$$

between two Hilbert spaces defined by

$$\Phi(v) = \langle v, v_1 \rangle v_2,$$

where $v_1 \in \mathrm{H}_1$ and $v_2 \in \mathrm{H}_2$, and the bracket denotes the inner product between vectors of H_1. The same analysis leads, when H_1 and H_2 are not isomorphic, to the relation

$$\frac{1}{|G|}\sum_{g\in G} g\cdot\Phi = 0,$$

and spelling it out by applying this to a $v \in E_1$ and taking the inner product of the result with $w \in E_2$, we obtain[3]

$$\frac{1}{|G|}\sum_{g\in G}\langle g^{-1}\cdot v, v_1\rangle\langle g\cdot v_2, w\rangle = \frac{1}{|G|}\sum_{g\in G}\langle g\cdot v, v_1\rangle\overline{\langle g\cdot w, v_2\rangle} = 0.$$

We interpret this in terms of the invariant inner product

$$\langle \varphi_1, \varphi_2\rangle = \frac{1}{|G|}\sum_{g\in G}\varphi_1(g)\overline{\varphi_2(g)}$$

for which the regular representation on the space of complex-valued functions $C(G)$ is unitary. The formula says that

$$\langle \varphi_{v_1,v_2}, \varphi_{v_3,v_4}\rangle = 0$$

for any *unitary* matrix coefficients of non-isomorphic irreducible unitary representations of G.

Similarly, if $E = E_1 = E_2$ carries the irreducible unitary representation π, the same argument as in the previous example leads to

$$\frac{1}{|G|}\sum_{g\in G} g\cdot\Phi = \frac{\langle v_2, v_1\rangle}{\dim E} \in \mathrm{End}_{\mathbf{C}}(E).$$

Applying to v and taking inner product with $w \in E$, we get

$$\begin{aligned}\frac{1}{|G|}\sum_{g\in G}\langle g\cdot v, v_1\rangle\langle g^{-1}\cdot v_2, w\rangle &= \frac{1}{|G|}\sum_{g\in G}\langle g\cdot v, v_1\rangle\overline{\langle g\cdot w, v_2\rangle}\\ &= \frac{\langle v_2, v_1\rangle\langle v, w\rangle}{\dim E},\end{aligned}$$

i.e., renaming the vectors, we have

$$\langle \varphi_{v_1,v_2}, \varphi_{v_3,v_4}\rangle = \frac{\langle v_1, v_3\rangle\overline{\langle v_2, v_4\rangle}}{\dim E}$$

for any $v_i \in E$. Hence:

Corollary 4.3.9 (Orthonormality of unitary matrix coefficients)**.** *Let G be a finite group, and let $\pi : G \to \mathrm{U}(E)$ be an irreducible unitary representation of G. For any orthonormal basis (e_i) of E, the normalized unitary matrix coefficients*

$$\varphi_{i,j} : x \mapsto \sqrt{\dim(E)}\langle \pi(x)e_i, e_j\rangle$$

are orthonormal in $C(G)$, with respect to the invariant inner product.

---

[3] Changing again g into g^{-1}.

Indeed, we get

$$\langle \varphi_{i,j}, \varphi_{k,l} \rangle = \dim(E) \frac{\langle e_i, e_k \rangle \overline{\langle e_j, e_l \rangle}}{\dim E} = \begin{cases} 1 & \text{if } i = k \text{ and } j = l, \\ 0 & \text{otherwise.} \end{cases}$$

Similarly, for the characters themselves, we obtain the fundamental orthonormality of characters in the unitary case:

Corollary 4.3.10 (Orthonormality of characters). *Let G be a finite group and π, and let ϱ be two irreducible unitary representations of G. We then have*

$$(4.17)\quad \langle \chi_\varrho, \chi_\pi \rangle = \frac{1}{|G|} \sum_{g \in G} \chi_\pi(g) \overline{\chi_\varrho(g)} = \begin{cases} 0 & \textit{if } \pi \textit{ is not isomorphic to } \varrho, \\ 1 & \textit{otherwise.} \end{cases}$$

Hence the characters of irreducible unitary representations of G form an orthonormal basis of the space $c_k(G)$ of class functions on G with respect to the invariant inner product.

The last part is due to the fact that we know that the characters of irreducible unitary representations form an orthonormal family in $c_k(G)$ and that there are as many of them as there are conjugacy classes, i.e., as many as the dimension of $c_k(G)$, so that they must form an orthonormal basis.

4.3.5. Multiplicities. A crucial consequence of the orthogonality of characters is a formula for the multiplicities of irreducible representations in a given representation, at least in characteristic 0.

Proposition 4.3.11 (Multiplicity formula). *Let G be a finite group, and let k be an algebraically closed field of characteristic 0. For any finite-dimensional k-representation*

$$\varrho : G \longrightarrow \mathrm{GL}(E)$$

and for any irreducible k-representation π of G, the multiplicity $n_\pi(\varrho)$ of π in ϱ is given by

$$n_\pi(\varrho) = \dim \mathrm{Hom}_G(\pi, \varrho) = [\chi_\varrho, \chi_\pi] = \frac{1}{|G|} \sum_{g \in G} \chi_\varrho(g) \chi_\pi(g^{-1}).$$

If $k = \mathbf{C}$, then we can also write

$$n_\pi(\varrho) = \langle \chi_\pi, \chi_\varrho \rangle = \frac{1}{|G|} \sum_{g \in G} \chi_\varrho(g) \overline{\chi_\pi(g)}.$$

Note that we also have $n_\pi(\varrho) = \dim \mathrm{Hom}_G(\varrho, \pi)$ by the symmetry between irreducible quotient representations and subrepresentations, which is valid for a semisimple representation (Corollary 2.7.19).

Proof. Since ϱ is semisimple, we know that its character is given by

$$\chi_\varrho = \sum_{\pi \in \hat{G}} n_\pi(\varrho)\chi_\pi$$

in terms of the multiplicities. By orthogonality, we obtain

$$[\chi_\varrho, \chi_\pi] = n_\pi(\varrho)[\chi_\pi, \chi_\pi] = n_\pi(\varrho).$$

This is an equality in the field k, but since k has characteristic zero, it is also one in $\mathbf{Z}$ (in particular the left-hand side is an integer). □

More generally, we can extend this multiplicity formula by linearity:

Proposition 4.3.12 (Multiplicity formula). *Let G be a finite group, and let k be an algebraically closed field of characteristic 0. For $i = 1, 2$, let*

$$\varrho_i : G \longrightarrow \mathrm{GL}(E_i)$$

be a finite-dimensional k-representation of G. Then we have

$$[\chi_{\varrho_1}, \chi_{\varrho_2}] = \dim \mathrm{Hom}_G(\varrho_1, \varrho_2) = \dim \mathrm{Hom}_G(\varrho_2, \varrho_1).$$

If $k = \mathbf{C}$, we have

$$\langle \chi_{\varrho_1}, \chi_{\varrho_2} \rangle = \dim \mathrm{Hom}_G(\varrho_1, \varrho_2) = \dim \mathrm{Hom}_G(\varrho_2, \varrho_1).$$

Remark 4.3.13. It is customary to use (especially when $k = \mathbf{C}$) the short-hand notation

$$\langle \varrho_1, \varrho_2 \rangle = \langle \chi_{\varrho_1}, \chi_{\varrho_2} \rangle$$

for the two representations ϱ_1 and ϱ_2 of G. We will do so to simplify notation, indicating sometimes the underlying group by writing $\langle \varrho_1, \varrho_2 \rangle_G$.

The multiplicity formula also leads to the following facts, which are very useful when attempting to decompose a representation and when one does not know a priori all the irreducible representations of G. Indeed, this leads to a very convenient *numerical criterion* for irreducibility:

Corollary 4.3.14 (Irreducibility criterion). *Let G be a finite group, and let k be an algebraically closed field of characteristic 0. For any finite-dimensional k-representation ϱ of G, we have*

$$[\chi_\varrho, \chi_\varrho] = \sum_{\pi \in \hat{G}} n_\pi(\varrho)^2$$

or, if $k = \mathbf{C}$, we have the formula

$$\langle \chi_\varrho, \chi_\varrho \rangle = \sum_{\pi} n_\pi(\varrho)^2$$

for the squared norm of the character of ϱ. In particular, ϱ is irreducible if and only if

$$[\chi_\varrho, \chi_\varrho] = 1,$$

and if $k = \mathbf{C}$, if and only if

$$\langle \chi_\varrho, \chi_\varrho \rangle = \frac{1}{|G|} \sum_{g \in G} |\chi_\varrho(g)|^2 = 1.$$

Proof. By linearity and orthogonality

$$[\chi_\varrho, \chi_\varrho] = \sum_{\pi_1, \pi_2} n_{\pi_1}(\varrho) n_{\pi_2}(\varrho) [\chi_{\pi_1}, \chi_{\pi_2}] = \sum_{\pi} n_\pi(\varrho)^2,$$

and similarly for $k = \mathbf{C}$. And if this is equal to 1, as an equality in $\mathbf{Z}$, the only possibility is that one of the multiplicities $n_\pi(\varrho)$ be equal to 1, and all the others are 0, which means $\varrho \simeq \pi$ is irreducible. $\square$

Exercise 4.3.15 (Direct product)**.** Let $G = G_1 \times G_2$, where G_1 and G_2 are finite groups. Use the irreducibility criterion to prove Proposition 2.3.23 directly for complex representations: all irreducible complex representations of G are of the form $\pi_1 \boxtimes \pi_2$ for some (unique) irreducible representations π_i of G_i.

Example 4.3.16 (Permutation representations)**.** Suppose we have a complex representation ϱ of a finite group G with $\langle \chi_\varrho, \chi_\varrho \rangle = 2$. Then ϱ is necessarily a direct sum of two non-isomorphic irreducible subspaces, since 2 can only be written as $1^2 + 1^2$ as the sum of positive squares of integers.

A well-known source of examples of this is given by certain permutation representations (see Section 2.6.2 for the definition). Consider an action of G on a finite set X, and the associated permutation representation ϱ on the space $\mathbf{C}(X)$ with basis vectors (e_x), so that

$$\varrho(g) e_x = e_{g \cdot x}.$$

The character of ϱ is given in Example 2.7.44. We have

$$\chi_\varrho(g) = |\{x \in X \mid g \cdot x = x\}|.$$

We first deduce from this that

$$\langle \chi_\varrho, \mathbf{1} \rangle = \frac{1}{|G|} \sum_{g \in G} \sum_{\substack{x \in X \\ g \cdot x = x}} 1 = \frac{1}{|G|} \sum_{x \in X} |G_x|,$$

where $G_x = \{g \in G \mid g \cdot x = x\}$ is the stabilizer of x in G. The order of this subgroup depends only on the orbit of x, since we have $G_y = g G_x g^{-1}$ if $y = g \cdot x$, and hence, summing over the orbits, we get

$$\langle \chi_\varrho, \mathbf{1} \rangle = \sum_{o \in G \backslash X} \frac{|G_o||o|}{|G|} = |G \backslash X|, \tag{4.18}$$

the number of orbits (we used the standard bijection $G_o \backslash G \longrightarrow o$ induced by mapping $g \in G$ to $g \cdot x_0$ for some fixed $x_0 \in o$).

We assume now that there is a single orbit, i.e., that the action of G on X is transitive (otherwise, ϱ already contains at least two copies of the trivial representation). Then, since the character of ϱ is real valued, we have

$$\begin{aligned}\langle \varrho, \varrho \rangle &= \frac{1}{|G|} \sum_{g \in G} \Big(\sum_{\substack{x \in X \\ g \cdot x = x}} 1 \Big)^2 \\ &= \frac{1}{|G|} \sum_{x,y \in X} \sum_{g \in G_x \cap G_y} 1 \\ &= \frac{1}{|G|} \sum_{x \in G} |G_x| + \frac{1}{|G|} \sum_{x \neq y} \sum_{g \in G_x \cap G_y} 1 \\ &= 1 + \frac{1}{|G|} \sum_{g \in G} |\{(x,y) \in X \times X \mid x \neq y \text{ and } g \cdot (x,y) = (x,y)\}|.\end{aligned}$$

We recognize (from the character formula for a permutation representation again) that this gives

$$\langle \varrho, \varrho \rangle = 1 + \langle \varrho^{(2)}, \mathbf{1} \rangle,$$

where $\varrho^{(2)}$ is the permutation representation associated to the natural action of G on the set

$$Y = \{(x,y) \in X \mid x \neq y\}$$

(recall that $g \cdot x = g \cdot y$ implies that $x = y$, so the action of G on $X \times X$ leaves Y invariant). By (4.18), we see that we have $\langle \varrho, \varrho \rangle = 2$ if (and, in fact, only if) this action on Y is *transitive*. This, by definition, is saying that the original action was *doubly transitive*: not only can G bring some element $x \in X$ to any other (transitivity), but a single element can simultaneously bring some x to any x', and some other y to any y', provided the conditions $x \neq y$ and $x' \neq y'$ are satisfied.

Thus:

Proposition 4.3.17. *Let G be a finite group acting doubly transitively on a finite set X. Then the representation ϱ of G on the space*

$$E = \Big\{ \sum_{x \in X} \lambda_x e_x \mid \sum_{x \in X} \lambda_x = 0 \Big\} \subset \mathbf{C}(X)$$

defined by $\varrho(g)e_x = e_{gx}$ is an irreducible complex representation of G of dimension $|X| - 1$, with character

$$\chi_\varrho(g) = |\{x \in X \mid g \cdot x = x\}| - 1.$$

Indeed, the subspace E is stable under the action of G (it is the orthogonal of the space E_X^G for the natural inner product on E_X such that (e_x) is an orthonormal basis).

For a concrete example, consider $G = \mathfrak{S}_n$ acting on $X = \{1, \ldots, n\}$ by permutations. If $n \geqslant 2$, this action is doubly transitive (as the reader should make sure to check, if needed!), and this means that the representation of $\mathfrak{S}_n$ on the hyperplane

$$E_n = \Big\{(x_i) \in \mathbf{C}^n \mid \sum_i x_i = 0\Big\} \tag{4.19}$$

is irreducible of dimension $n - 1$.

Remark 4.3.18. A warning about the irreducibility criterion: it only applies if one knows that χ_ϱ is, indeed, the character of a representation of G. There are many class functions φ with squared norm 1 which are not characters, for instance

$$\varphi = \frac{3\chi_{\pi_1} + 4\chi_{\pi_2}}{5}$$

if π_1 and π_2 are non-isomorphic irreducible representations. If π_1 and π_2 have dimension divisible by 5, the non-integrality might not be obvious from looking simply at the character values!

However, note that if $\varphi \in R(G)$ is a *virtual character* (over $\mathbf{C}$, say), i.e., $\varphi = \chi_{\varrho_1} - \chi_{\varrho_2}$ for some actual complex representations ϱ_1 and ϱ_2, the condition

$$\langle \varphi, \varphi \rangle = 1$$

means that either ϱ_1 or ϱ_2 is irreducible and the other is zero, or in other words, either φ or $-\varphi$ is a character of an irreducible representation of G. Indeed, we can write

$$\varphi = \sum_\pi n_\pi \chi_\pi$$

as a combination of irreducible characters with *integral* coefficients n_π, and we have again

$$\langle \varphi, \varphi \rangle = \sum_\pi n_\pi^2.$$

So one, and only one, of the n_π is equal to ± 1, and the others are 0.

Example 4.3.19 (Frobenius reciprocity)**.** From the general multiplicity formula, we get a numerical version of Frobenius reciprocity for induced representations (Proposition 2.3.9 and (2.25)): given a subgroup H of a finite group G, a (complex, say) representation ϱ_1 of G and a representation ϱ_2 of H, we have[4]

$$\langle \varrho_1, \mathrm{Ind}_H^G(\varrho_2)\rangle_G = \langle \mathrm{Res}_H^G\, \varrho_1, \varrho_2\rangle_H.$$

In particular, if ϱ_1 and ϱ_2 are irreducible representations, then the multiplicity of ϱ_1 in $\mathrm{Ind}_H^G(\varrho_2)$ is equal to the multiplicity of ϱ_2 in $\mathrm{Res}_H^G(\varrho_1)$.

[4] We use the notation of Remark 4.3.13.

This numerical form of Frobenius reciprocity can easily be checked directly as an identity between characters. Denoting $\chi_i = \chi_{\varrho_i}$, we find using (2.49) that we have

$$\begin{aligned}
\langle \varrho_1, \mathrm{Ind}_H^G(\varrho_2)\rangle_G &= \frac{1}{|G|}\sum_{g\in G}\chi_1(g)\overline{\sum_{\substack{s\in H\backslash G\\ sgs^{-1}\in H}}\chi_2(sgs^{-1})}\\
&= \frac{1}{|G|}\sum_{s\in H\backslash G}\sum_{g\in s^{-1}Hs}\chi_1(g)\overline{\chi_2(sgs^{-1})}\\
&= \frac{1}{|G|}\sum_{s\in H\backslash G}\sum_{h\in H}\chi_1(s^{-1}hs)\overline{\chi_2(h)}\\
&= \frac{|H\backslash G|}{|G|}\sum_{h\in H}\chi_1(h)\overline{\chi_2(h)} = \langle \mathrm{Res}_H^G(\varrho_1), \varrho_2\rangle_H.
\end{aligned}$$

Note that, by symmetry, we also have

$$\langle \mathrm{Ind}_H^G(\varrho_2), \varrho_1\rangle_G = \langle \varrho_2, \mathrm{Res}_H^G\,\varrho_1\rangle_H$$

(something which is not universally true in the generality in which we defined induced representations).

For instance, if one thinks that ϱ_1 is an irreducible representation of G, and ϱ_2 is one of H, Frobenius reciprocity says, "the multiplicity of ϱ_1 in the representation induced from ϱ_2 is the same as the multiplicity of ϱ_2 in the restriction of ϱ_1 to H."

Here is an example of application. For a finite group G, we denote by $A(G)$ the maximal dimension of an irreducible complex representation of G. So, for instance, $A(G) = 1$ characterizes finite abelian groups (see Proposition 4.2.2). More generally, $A(G)$ can be seen to be some measure of the complexity of G.

Proposition 4.3.20. *For any finite group G and subgroup $H \subset G$, we have $A(H) \leqslant A(G)$.*

Proof. To see this, pick an irreducible representation π of H such that $\dim \pi = A(H)$. Now consider the induced representation $\varrho = \mathrm{Ind}_H^G(\pi)$. It may or may not be irreducible; in any case, let τ be any irreducible component of ϱ. Then we have

$$1 \leqslant \langle \tau, \varrho\rangle = \langle \tau, \mathrm{Ind}_H^G(\pi)\rangle = \langle \mathrm{Res}_H^G(\tau), \pi\rangle_H$$

by Frobenius reciprocity. This means that π occurs with multiplicity at least 1 in the restriction of τ, and therefore $\dim \tau \geqslant \dim \pi$. Thus τ is an irreducible representation of G of dimension at least $A(H)$; i.e., we have $A(G) \geqslant A(H)$. ∎

Exercise 4.3.21. For a finite group G and a real number $p \geqslant 0$, let

$$A_p(G) = \sum_{\pi} (\dim \pi)^p.$$

If $p \geqslant 1$, show that for any subgroup $H \subset G$, we have $A_p(H) \leqslant A_p(G)$.

Exercise 4.3.22. Let G be a finite group.

(1) If $H \lhd G$ is a normal subgroup of G, show that

$$\mathrm{Ind}_H^G(1) \simeq \bigoplus_{\varrho} \dim(\varrho)\varrho,$$

where the direct sum is over the irreducible complex representations of G such that $H \subset \ker(\varrho)$. (The point is that the multiplicity of such a representation ϱ is equal to its dimension.)

(2) Conversely, let H be a subgroup of G such that the multiplicity of any irreducible complex representation ϱ of G in $\mathrm{Ind}_H^G(1)$ is either 0 or $\dim(\varrho)$. Show that H is normal in G.

Here is a last, very cute, application of the multiplicity formula:

Proposition 4.3.23 (Where to find irreducible representations?). *Let G be a finite group, and let $\varrho : G \longrightarrow \mathrm{GL}(E)$ be any finite-dimensional* faithful *complex representation. Then any irreducible representation $\pi \in \widehat{G}$ can be found as a subrepresentation of a tensor power $\varrho \otimes \cdots \otimes \varrho$, with k factors, for some $k \geqslant 0$.*

Proof. Fix $\pi \in \widehat{G}$. For $k \geqslant 0$, we denote by $m_k \geqslant 0$ the multiplicity of π in $\varrho^{\otimes k}$ (with the convention that the 0-th tensor power is the trivial representation), in other words

$$m_k = \langle \varrho^{\otimes k}, \pi \rangle$$

for $k \geqslant 0$. The goal is to show that this multiplicity m_k is non-zero for some $k \geqslant 0$, and the clever idea is to consider the generating series

$$\sum_{k \geqslant 0} m_k X^k \in \mathbf{Z}[[X]]$$

and to show that this formal power series is non-zero. For this purpose, we write

$$\langle \varrho^{\otimes k}, \pi \rangle = \frac{1}{|G|} \sum_{g \in G} \chi_\varrho(g)^k \overline{\chi_\pi(g)},$$

and compute the power series by exchanging the two sums

$$\sum_{k \geqslant 0} m_k X^k = \frac{1}{|G|} \sum_{g \in G} \overline{\chi_\pi(g)} \sum_{k \geqslant 0} \chi_\varrho(g)^k X^k = \frac{1}{|G|} \sum_{g \in G} \frac{\overline{\chi_\pi(g)}}{1 - \chi_\varrho(g) X}.$$

This does not look like the zero power series, but there might be cancellations in the sum. However, we have not used the assumption that ϱ is faithful, and there is the cunning trick: the point $1/\chi_\varrho(1) = 1/\dim(\varrho)$ is a pole of the term corresponding to $g = 1$, and it *cannot* be cancelled because $\chi_\varrho(g) = \dim(\varrho)$ if and only if $g \in \operatorname{Ker} \varrho = 1$ (this is the easy Proposition 4.6.4 below; the point is that the character values are traces of unitary matrices, hence the sum of $\dim \varrho$ complex numbers of modulus 1). So it follows that the power series is non-zero, which certainly means that m_k is not always 0. ☐

Remark 4.3.24. (1) See Exercise 5.5.5 for another proof that generalizes more easily to compact groups.

(2) The choice of the generating series is not unique, and many other similar ideas lead to the result. For instance, one may consider the power series

$$g(z) = e^{-\dim(\varrho)z} \sum_{k \geqslant 0} m_k \frac{z^k}{k!},$$

which converges absolutely for all $z \in \mathbf{C}$, since $m_k \leqslant \dim(\varrho^{\otimes k}) = \dim(\varrho)^k$. One finds that

$$g(z) = \frac{1}{|G|} \sum_{g \in G} e^{(\chi_\varrho(g) - \dim \varrho)z}.$$

Consider z real. The term corresponding to $g = 1$ gives a summand equal to $1/|G|$, while the term corresponding to any $g \neq 1$ has modulus

$$\left| \frac{1}{|G|} e^{(\chi_\varrho(g) - \dim \varrho)z} \right| = \frac{1}{|G|} e^{-z(\operatorname{Re}(\chi_\varrho(g)) - \dim \varrho)} \leqslant \frac{e^{-cz}}{|G|},$$

where

$$c = \min |\operatorname{Re}(\chi_\varrho(g)) - \dim \varrho| > 0$$

because ϱ is faithful. Hence

$$\left| g(z) - \frac{1}{|G|} \right| \leqslant e^{-cz},$$

which shows that $g(z) \neq 0$ for z a large enough real number, and hence $m_k \neq 0$ for some k.

4.3.6. Isotypic projections. We now come back to the problem of determining the projections on all the isotypic components of a representation, not only the invariant subspace. In the language of the group algebra, Proposition 4.3.1 means that the *single* element

$$e = \frac{1}{|G|} \sum_{g \in G} g \in k(G)$$

of the group algebra has the property that its action on *any* representation of G gives *universally* the space of invariants. Since E^G is the same as the isotypic component of E with respect to the trivial representation, it is natural to ask for similar elements for the other irreducible representations of G. These exist indeed, and they are also remarkably simple: they are given by the characters.

Proposition 4.3.25 (Projections on isotypic components)**.** *Let G be a finite group, and let k be an algebraically closed field of characteristic $p \nmid |G|$. For any k-representation*

$$\varrho : G \longrightarrow \mathrm{GL}(E)$$

of G, and for any irreducible k-representation π of G, the element

$$e_\pi = \frac{\dim(\pi)}{|G|} \sum_{g \in G} \chi_\pi(g^{-1}) g \in k(G)$$

acts on E as a homomorphism in $\mathrm{Hom}_G(E, E)$ and is a projection onto the isotypic component $\mathrm{M}(\pi) \subset E$. In other words, the linear map

$$\begin{cases} E \longrightarrow E \\ v \mapsto \dfrac{\dim \pi}{|G|} \displaystyle\sum_{g \in G} \chi_\pi(g^{-1}) \varrho(g) v \end{cases} \tag{4.20}$$

is a G-homomorphism and is a projection onto $\mathrm{M}(\pi)$.

For $k = \mathbf{C}$, if we think of unitary representations, we get

Proposition 4.3.26 (Orthogonal projections on isotypic components)**.** *Let G be a finite group, and let $\varrho : G \longrightarrow \mathrm{U}(\mathrm{H})$ be a unitary representation of G. For any irreducible unitary representation π of G, the element*

$$e_\pi = \frac{\dim(\pi)}{|G|} \sum_{g \in G} \overline{\chi_\pi(g)} g \in \mathbf{C}(G)$$

acts on H as a homomorphism in $\mathrm{Hom}_G(\mathrm{H}, \mathrm{H})$ and is the orthogonal *projection onto the isotypic component $\mathrm{M}(\pi) \subset E$.*

We will explain how one can find this formula for e_π, instead of merely checking its properties; indeed, this leads to additional insights. The point is that, for a given irreducible representation π, the family of projections to the π-isotypic component, which maps all other isotypic components to 0, gives for every representation $\varrho : G \longrightarrow \mathrm{GL}(E)$ of G a linear map

$$\varepsilon_\varrho : E \longrightarrow E,$$

in a *functorial* manner, in the sense described in Exercise 3.1.5: for any representation τ on F and any G-homomorphism

$$E \xrightarrow{\Phi} F,$$

we have

$$\varepsilon_\tau \circ \Phi = \Phi \circ \varepsilon_\varrho.$$

Exercise 4.3.27. Check this fact.

The outcome of Exercise 3.1.5 is that the source of a *universal* linear map on all representations can only be the action of some *fixed* element a of the group algebra; even if you did not solve this exercise, it should be intuitively reasonable that this is the only obvious source of such maps. Thus, we know a priori that there *is* a formula for the projection in terms of $k(G)$. We only need to find it.

The projections are not just linear maps, but also intertwiners; according to the last part of Exercise 3.1.5, this corresponds to an element a of the group algebra $k(G)$ which is in its center $Z(k(G))$. (This is because a gives rise to G-homomorphism if and only if a satisfies

$$g \cdot a = a \cdot g \in k(G)$$

for all $g \in G$, which is equivalent to $a \in Z(k(G))$ because G generates $k(G)$ as a ring).

Remark 4.3.28. If we write

$$a = \sum_{x \in G} \alpha_x x, \qquad \alpha_x \in k,$$

the condition that a belongs to the center becomes

$$\alpha_{x^{-1}g} = \alpha_{gx^{-1}}, \qquad \text{for all } x \text{ and } g,$$

or, in other words, the function

$$x \mapsto \alpha_x$$

must be a *class function.*

Now we assume that $a \in Z(k(G))$, so that a acts as a G-morphism on every representation of G. In particular, the action of a on an irreducible representation π must be given by multiplication by some scalar $\omega_\pi(a) \in k$, according to Schur's Lemma. Because this is *universal*, we see that the element giving the projection on $\mathrm{M}(\pi)$ is the element $a \in k(G)$ such that $\omega_\pi(a) = 1$, and $\omega_\tau(a) = 0$ for all other (non-isomorphic) irreducible k-representations τ of G. Indeed, if a has this property, it follows that for a given representation of G on E, a acts as identity on all subrepresentations of E isomorphic to π, i.e., on $\mathrm{M}(\pi)$, and also a that annihilates all other isotypic components. This is exactly the desired behavior.

To determine a exactly, we observe that we can compute $\omega_\tau(a)$, as a function of the coefficients α_x of a and of the irreducible representation τ, by taking the trace: from

$$\omega_\tau(a)\mathrm{Id}_\tau = \sum_{x\in G} \alpha_x \tau(x)$$

we get

$$\omega_\tau(a)\dim\tau = \sum_{x\in G} \alpha_x \operatorname{Tr}(\tau(x)) = \sum_{x\in G} \alpha_x \chi_\tau(x).$$

Hence we are looking for coefficients α_x such that

$$\sum_{x\in G} \alpha_x \chi_\pi(x) = \dim\pi$$

(the case $\tau = \pi$) and

$$\sum_{x\in G} \alpha_x \chi_\tau(x) = 0,$$

if τ is an irreducible representation non-isomorphic to π. But the orthogonality of characters (4.13) precisely says that

$$\alpha_x = \dim(\pi)\chi_\pi(x^{-1})$$

satisfies these conditions, and when $k = \mathbf{C}$ and we have unitary representations, this becomes

$$\alpha_x = \dim(\pi)\overline{\chi_\pi(x)}$$

(see also Corollary 4.3.10). Thus Proposition 4.3.26 is proved.

Having obtained the formula for the projections on isotypic components of any representation, there is one important example that should come to mind where we can (and should) apply this: the group algebra itself, when G acts on $k(G)$ by multiplication on the left. The special feature of $k(G)$ is its algebra structure, which also gives some extra structure to the isotypic components.

Let $\mathrm{I}(\pi)$ be the π-isotypic component of $k(G)$. According to the above, the projection on $\mathrm{I}(\pi)$ is given by $a \mapsto e_\pi a$, where

$$e_\pi = \frac{\dim(\pi)}{|G|} \sum_{g\in G} \chi(g^{-1})g.$$

Taking $a = 1$, we deduce from this that $e_\pi \in \mathrm{I}(\pi)$. We deduce, for instance, that

$$e_\pi^2 = e_\pi e_\pi = e_\pi, \tag{4.21}$$

and also (since other projections map $\mathrm{I}(\pi)$ to 0) that

$$e_\varrho e_\pi = 0 \tag{4.22}$$

if ϱ is an irreducible representation not isomorphic to π. Note moreover that

$$1 = \sum_\pi e_\pi, \tag{4.23}$$

which is simply because of the isotypic decomposition

$$k(G) = \bigoplus_{\pi} \mathrm{I}(\pi).$$

In any ring A, a family (e_i) of elements satisfying the relations (4.21), (4.22), and (4.23) is known as a *complete system of orthogonal idempotents*. Their meaning is the following:

Corollary 4.3.29 (Product decomposition of the group algebra). *Let G be a finite group, and let k be an algebraically closed field of characteristic not dividing $|G|$. Then the subspaces $\mathrm{I}(\pi)$, where π runs over irreducible k-representations of G, are two-sided ideals in $k(G)$. Moreover, with $e_\pi \in \mathrm{I}(\pi)$ as unit, $\mathrm{I}(\pi)$ is a subalgebra of $k(G)$ isomorphic to the matrix algebra $\mathrm{End}_k(\pi)$, and we have a k-algebra isomorphism*

$$(4.24) \qquad \begin{cases} k(G) & \xrightarrow{\sim} & \prod_{\pi} \mathrm{End}(\pi) \\ a & \mapsto & (\pi(e_\pi a))_\pi. \end{cases}$$

Proof. The space $\mathrm{I}(\pi)$ is the image of the projection given by multiplication by e_π, i.e., we have

$$\mathrm{I}(\pi) = e_\pi k(G),$$

which is, a priori, a right ideal in $k(G)$. But if we remember that e_π is also in the center $Z(k(G))$ of the group algebra, we deduce that $\mathrm{I}(\pi) = k(G)e_\pi$, i.e., that $\mathrm{I}(\pi)$ is a two-sided ideal.

In particular, like any two-sided ideal, $\mathrm{I}(\pi)$ is stable under multiplication. Usually, $1 \notin \mathrm{I}(\pi)$, so that 1 does not provide a unit. But, for any $a \in \mathrm{I}(\pi)$, if we write $a = e_\pi a_1$, we find that

$$e_\pi a = e_\pi^2 a_1 = e_\pi a_1 = a$$

by (4.21), and similarly $ae_\pi = a$, so that e_π, which is in $\mathrm{I}(\pi)$, is a unit for this two-sided ideal!

The identity (4.23) means that, as algebras, we have

$$k(G) \simeq \prod_{\pi} \mathrm{I}(\pi),$$

where any $a \in k(G)$ corresponds to $(e_\pi a)_\pi$. Thus there only remains to prove that the map

$$\begin{cases} \mathrm{I}(\pi) & \longrightarrow & \mathrm{End}_k(\pi) \\ a & \mapsto & \pi(a) \end{cases}$$

is an algebra isomorphism.

This is certainly an algebra homomorphism (the unit e_π maps to the identity in $\mathrm{End}_k(\pi)$, since, by the above, the action of e_π is the projection on $\mathrm{M}(\pi)$, which is the identity for π itself). It is surjective, by Burnside's irreducibility criterion (the latter says, more precisely, that the image of all of $k(G)$ is $\mathrm{End}_k(\pi)$, but the other isotypic components map to 0). We can show that it is an isomorphism either by dimension count (since the representation of G on $k(G)$ is, for a finite group, isomorphic to that on $C_k(G)$, the isotypic component $\mathrm{I}(\pi)$ has the same dimension as the space of matrix coefficients of π, namely $(\dim \pi)^2$), or by proving injectivity directly: If $a \in \mathrm{I}(\pi)$ satisfies $\pi(a) = 0$, it follows that the action of a on every π-isotypic component of every representation is also zero. If we take the special case of $k(G)$ itself, this means in particular that $ae_\pi = 0$. However, $ae_\pi = e_\pi a = a$ if $a \in \mathrm{I}(\pi)$, and thus $a = 0$. □

Remark 4.3.30. This result also leads to Theorem 4.2.5. Indeed, the center $Z(k(G))$ of $k(G)$ has dimension equal to the number of conjugacy classes of G (since it is the space of all elements

$$a = \sum_{g \in G} \alpha_g g,$$

with $g \mapsto \alpha_g$ a class function, as we observed before), while from (4.24), we have

$$Z(k(G)) \simeq \prod_{\pi} Z(\mathrm{End}_k(\pi)) = \prod_{\pi} k\mathrm{Id}_\pi$$

(since any endomorphism ring has one-dimensional center spanned by the identity). Thus the dimension of $Z(k(G))$ is also equal to the number of irreducible k-representations of G.

Exercise 4.3.31 (How big can a cyclic representation be?). Let G be a finite group, and let k be a field of characteristic not dividing $|G|$. Let $\varrho : G \longrightarrow \mathrm{GL}(E)$ be a finite-dimensional k-representation of G, and let π be an irreducible k-representation.

(1) For $v \in \mathrm{M}_E(\pi) \subset E$, show that the subrepresentation F_v of E generated by v (which is a cyclic representation, see Remark 2.2.8) is the direct sum of at most $\dim(\pi)$ copies of π.

(2) Show that this cannot be improved (i.e., it is possible, for some ϱ and v, that F_v is the direct sum of exactly $\dim(\pi)$ copies of π).

(3) If you solved (1) using the group algebra $k(G)$, try to do it without (see [**53**, Ex. 2.10] if needed).

In the argument leading to the projection formula, we have also proved the following useful result.

Proposition 4.3.32 (Action of the center of the group algebra). *Let G be a finite group, and let k be an algebraically closed field of characteristic not dividing $|G|$. For any irreducible k-representation ϱ of G, there is an* *associated algebra homomorphism*

$$\omega_\varrho : \begin{cases} Z(k(G)) & \longrightarrow & k \\ a & \mapsto & \varrho(a), \end{cases}$$

i.e.,

(4.25) $$\varrho(a) = \omega_\varrho(a)\mathrm{Id}.$$

This is given by

(4.26) $$\omega_\varrho\Big(\sum_{g\in G} \alpha_g g\Big) = \frac{1}{\dim(\varrho)} \sum_{g\in G} \alpha_g \chi_\varrho(g).$$

The last formula is obtained, as usual, by taking the trace on both sides of (4.25). Note the following special case: if $c \subset G$ is a conjugacy class, the element

$$a_c = \sum_{g\in c} g$$

is in the center of the group algebra, and we get

(4.27) $$\omega_\varrho(a_c) = \frac{|c|\chi_\varrho(c)}{\dim \varrho}.$$

This can be used to show how to compute (in principle) all characters of irreducible representations of G; see Proposition 4.6.2 below.

4.4. Harmonic analysis on finite groups

The terminology *harmonic analysis* refers roughly to the use of specific orthonormal bases of a Hilbert space to analyze its elements, in particular in the setting of function spaces. In the case of finite groups, there are two main examples, which are related:

(1) either one considers the space $c(G)$ of complex-valued class functions, and the orthonormal basis of irreducible characters;

(2) or one considers the full space $C(G)$ of functions on the group, and a basis of matrix coefficients.

The second case is often more difficult to handle, because matrix coefficients are not entirely canonical objects. This explains also why the first case is worth considering separately, and not simply as a corollary of the theory of matrix coefficients.

Given a class function $f \in c(G)$, we have

$$f = \sum_{\varrho\in\hat{G}} \langle f, \chi_\varrho\rangle \chi_\varrho.$$

It is worth isolating the contribution of the trivial representation $\mathbf{1} \in \widehat{G}$, which is the constant function with value

$$\langle f, 1\rangle = \frac{1}{|G|} \sum_{g \in G} f(g),$$

i.e., the average value of f on G. It is characteristic of harmonic analysis to decompose f in such a way that its *average* behavior is clearly separated from the fluctuations around it, which are given by the sum of the contributions of non-trivial characters.

We now write down explicitly what is the outcome of this decomposition when f is especially simple: fix a conjugacy class $c \subset G$, and let f_c be its characteristic function, which is a class function. We then obtain

Proposition 4.4.1 (Decomposition of characteristic functions of conjugacy classes). *Let g and $h \in G$. We have*

$$\sum_{\varrho \in \widehat{G}} \chi_\varrho(h)\overline{\chi_\varrho(g)} = \begin{cases} \dfrac{|G|}{|g^\sharp|} & \text{if } g \text{ is conjugate to } h, \\ 0 & \text{otherwise,} \end{cases} \tag{4.28}$$

where $g^\sharp$ is the conjugacy class of g.

This corollary is often called the *second orthogonality formula*, and is usually proved by observing that the transpose of a unitary matrix (namely, the character table of G, see Section 4.6) is also unitary. Note that in the *diagonal* case, the value

$$\frac{|G|}{|g^\sharp|}$$

is also equal to $|C_G(g)|$, the size of the centralizer of g in G.

Proof. As indicated, we expand f_c in terms of characters

$$f_c = \sum_{\varrho \in \widehat{G}} \langle f_c, \chi_\varrho\rangle \chi_\varrho,$$

and we remark that, by definition, we have

$$\langle f_c, \chi_\varrho\rangle = \frac{1}{|G|} \sum_{g \in G} f_c(g)\overline{\chi_\varrho(g)} = \frac{|c|}{|G|}\overline{\chi_\varrho(h)},$$

since f_c is 1 on the conjugacy class c and 0 elsewhere. □

Remark 4.4.2 (The space of conjugacy classes). The space $c(G)$ of class functions can be identified with the space $C(G^\sharp)$ of complex-valued functions on the set $G^\sharp$ of *conjugacy classes* in G, since a class function is constant on each conjugacy class. It is often useful to think in these terms. However, one must be careful that the Hilbert space inner product on $C(G^\sharp)$ coming from this identification (i.e., the inner product such that the identification is an isometry) is *not* the inner product

$$\frac{1}{|G^\sharp|}\sum_{c\in G^\sharp} f_1(c)\overline{f_2(c)}$$

that might seem most natural on a finite set. Instead, we have

$$\langle f_1, f_2\rangle = \frac{1}{G}\sum_{c\in G^\sharp} |c| f_1(c)\overline{f_2(c)}$$

for any functions

$$f_1,\ f_2\ :\ G^\sharp \longrightarrow \mathbf{C}.$$

This means that each conjugacy class carries a weight which is proportional to its size as a subset of G, instead of being uniform over all classes.

Harmonic analysis often involves using the expansion of a characteristic function in order to replace a condition of the type "g is in such and such subset X of G" by its expansion in terms of some orthonormal basis, so that one can write

$$\sum_{x\in X} f(x) = \sum_{x\in G} f(x)1_X(x) = \sum_{\text{basis } (\varphi_i)} \langle 1_X, \varphi_i\rangle \sum_{x\in G} f(x)\varphi_i(x),$$

where 1_X is the characteristic function of X. Furthermore, it is usually the case that the constant function 1 is part of the orthonormal basis (this is the case for characters as well as for matrix coefficients), in which case the corresponding term is

$$\langle 1_X, 1\rangle \sum_{x\in G} f(x) = \frac{|X|}{|G|}\sum_{x\in X} f(x),$$

which may be interpreted as the term that would arise from a heuristic argument in which one sees $|X|/|G|$ as the rough *probability* that some element of G is in X.

We present a good illustration of this principle now, and another one will be found in Section 4.7.1 later on. The problem we consider is to detect whether an element $g \in G$ is a commutator. The following result in this direction is due to Frobenius.

Proposition 4.4.3 (Detecting commutators). *Let G be a finite group. The number $N(g)$ of pairs $(x, y) \in G \times G$ such that $g = [x, y] = xyx^{-1}y^{-1}$ is equal to*

$$|G|\sum_{\pi\in\hat{G}} \frac{\chi_\pi(g)}{\chi_\pi(1)}, \tag{4.29}$$

and in particular g is a commutator if and only if

$$\sum_{\pi \in \hat{G}} \frac{\chi_\pi(g)}{\chi_\pi(1)} \neq 0.$$

We begin with a lemma which has independent interest, and which is also an application of the formula for isotypic projections.

Lemma 4.4.4. *Let G be a finite group, and let k be an algebraically closed field with characteristic coprime to $|G|$. For any irreducible k-representation π of G and any x, $y \in G$, we have*

$$\chi_\pi(x)\chi_\pi(y) = \frac{\dim \pi}{|G|} \sum_{g \in G} \chi_\pi(xgyg^{-1}).$$

Proof. We can express the right-hand side as

$$\frac{\dim \pi}{|G|} \sum_{g \in G} \chi_\pi(xgyg^{-1}) = \frac{\dim \pi}{|y^\sharp|} \sum_{g \in y^\sharp} \chi_\pi(xg),$$

where $y^\sharp$ is the conjugacy class of y. In this second expression, we use harmonic analysis to detect the condition $g \in y^\sharp$, exactly as described above: We have

$$\sum_{g \in y^\sharp} \chi_\pi(xg) = \sum_{g \in G} \chi_\pi(xg) \mathbf{1}_{y^\sharp}(g) = \sum_{g \in G} \chi_\pi(xg) \frac{|y^\sharp|}{|G|} \sum_{\varrho \in \hat{G}} \chi_\varrho(y) \overline{\chi_\varrho(g)}$$

by the second orthogonality formula (Proposition 4.4.1). Exchanging the sums, we obtain

$$\sum_{g \in y^\sharp} \chi_\pi(xg) = |y^\sharp| \sum_{\varrho \in \hat{G}} \chi_\varrho(y) \frac{1}{|G|} \sum_{g \in G} \overline{\chi_\varrho(g)} \chi_\pi(xg).$$

But the function

$$x \mapsto \frac{\dim \varrho}{|G|} \sum_{g \in G} \overline{\chi_\varrho(g)} \chi_\pi(xg)$$

is equal to

$$\frac{\dim \varrho}{|G|} \sum_{g \in G} \overline{\chi_\varrho(g)} (\varrho_G(g) \chi_\pi),$$

i.e., to the ϱ-isotypic projection of χ_π for the regular representation. This is equal to 0 except when $\varrho = \pi$, in which case it is χ_π itself, and hence we obtain

$$\sum_{g \in y^\sharp} \chi_\pi(xg) = \frac{|y^\sharp|}{\dim \pi} \chi_\pi(x) \chi_\pi(y),$$

which gives the stated result. ∎

We can now prove the formula of Frobenius.

Proof of Proposition 4.4.3. This time, we won't try to motivate the arguments, so this might appear mysterious.... The first idea is to compute, instead of $N(g)$, the quantity $N^\sharp(g)$, which is defined to be the number of pairs $(x,y) \in G \times G$ such that $[x,y]$ is *conjugate* to g. The point is that

$$N^\sharp(g) = \sum_{h \in g^\sharp} N(h) = |g^\sharp| N(g) \tag{4.30}$$

(where $g^\sharp$ is the conjugacy class of g), simply because the expressions of conjugate elements as commutators are naturally in bijection:

$$[x,y] = g \text{ if and only if } [zxz^{-1}, zyz^{-1}] = zgz^{-1}.$$

This is so that one recovers easily $N(g)$ from $N(g^\sharp)$, while relaxing equality to conjugation allows us to detect the condition using characters instead of involving all matrix coefficients.

Now we start by fixing x, and we attempt to determine the number $n(x,g)$ of $y \in G$ such that g is conjugate to $[x,y]$, so that

$$N^\sharp(g) = \sum_{x \in G} n(x,g).$$

Now we can compute $n(x,g)$ using characters: we have

$$n(x,g) = \sum_{y \in G} \mathbf{1}_{g^\sharp}([x,y]).$$

Using again Proposition 4.4.1 and exchanging the order of the two sums, we get

$$n(x,g) = \frac{|g^\sharp|}{|G|} \sum_{\pi \in \hat{G}} \overline{\chi_\pi(g)} \sum_{y \in G} \chi_\pi(xyx^{-1}y^{-1}). \tag{4.31}$$

The sum over y is a special case of that considered in Lemma 4.4.4, and this lemma gives

$$\frac{\dim \pi}{G} \sum_{y \in G} \chi_\pi(xyx^{-1}y^{-1}) = \chi_\pi(x)\chi_\pi(x^{-1}) = |\chi_\pi(x)|^2,$$

so that

$$n(x,g) = |g^\sharp| \sum_{\pi \in \hat{G}} \frac{\overline{\chi_\pi(g)} |\chi_\pi(x)|^2}{\dim \pi}.$$

Summing over x, we get

$$N^{\sharp}(g) = \sum_{x \in G} n(x, g) = |g^{\sharp}| \sum_{\pi \in \widehat{G}} \frac{\overline{\chi_\pi(g)}}{\dim \pi} \sum_{x \in G} |\chi_\pi(x)|^2$$
$$= |G||g^{\sharp}| \sum_{\pi \in \widehat{G}} \frac{\overline{\chi_\pi(g)}}{\dim \pi} = |G||g^{\sharp}| \sum_{\pi \in \widehat{G}} \frac{\chi_\pi(g)}{\dim \pi},$$

and it follows that

$$N(g) = |G| \sum_{\pi \in \widehat{G}} \frac{\chi_\pi(g)}{\dim \pi},$$

using (4.30). This is what we wanted. □

Remark 4.4.5. (1) If we isolate the contribution of the trivial representation, we see that the number of (x, y) with $[x, y] = g$ is given by

$$|G|\Big(1 + \sum_{\pi \neq 1} \frac{\chi_\pi(g)}{\chi_\pi(1)}\Big).$$

Suppose the group G has no non-trivial one-dimensional representation (which means that the commutators *generate* G, or that G is a *perfect group*). If we apply the basic intuition of harmonic analysis, we can expect that in many circumstances the first term will dominate, and hence that many elements in G will be commutators. There are indeed many results in this direction. For instance, a recent theorem of Liebeck, O'Brien, Shalev, and Tiep [**43**], confirming a striking conjecture (or question) of Ore, shows that if G is a finite non-abelian simple group, *every* element of G is a commutator. One of the criteria used to detect commutators in this work is the one we just proved. As a simple example, the reader will be invited to determine the commutators in $\mathrm{GL}_2(\mathbf{F}_p)$ in Exercise 4.6.17.

On the other hand, the reader may check (!), using software packages such as Magma [**8**] or Gap [**21**], that there exists a perfect group of order 960, which fits into an exact sequence

$$1 \longrightarrow (\mathbf{Z}/2\mathbf{Z})^4 \longrightarrow G \longrightarrow A_5 \longrightarrow 1,$$

where not all elements of G are actual commutators. To be more precise, this group G is isomorphic to the commutator subgroup of the group W_5 discussed below in Exercise 4.7.13, with the homomorphism to A_5 defined as the restriction to $[W_5, W_5]$ of the natural surjection $W_5 \longrightarrow \mathfrak{S}_5$. It inherits from W_5 a faithful (irreducible) representation of dimension 5 by signed permutation matrices, and it turns out that the 120 elements in the conjugacy class of

$$g = \begin{pmatrix} 0 & -1 & 0 & 0 & 0 \\ 1 & 0 & 0 & 0 & 0 \\ 0 & 0 & 0 & 1 & 0 \\ 0 & 0 & 1 & 0 & 0 \\ 0 & 0 & 0 & 0 & -1 \end{pmatrix}$$

are not commutators, as one can check that all commutators have trace in $\{-3, -2, 0, 1, 2, 5\}$. On the other hand, one can see that g *is* a commutator in W_5. (Note that, because G contains at least 481 commutators, it also follows that any $g \in G$ is the product of at most two commutators, by the following well-known, but clever, argument. In any finite group G, if S_1 and S_2 are subsets of G with $|S_i| > |G|/2$, and if $g \in G$ is arbitrary, the fact that[5] $|S_1| + |gS_2^{-1}| > |G|$ implies that $S_1 \cap gS_2^{-1} \neq \emptyset$, so that g is always of the form $s_1 s_2$ with $s_i \in S_i$. The end of the proof of Theorem 4.7.1 in Section 4.7.1 will use an even more clever variant of this argument involving three subsets....)

(2) If we take $g = 1$ in (4.29), we see that the number of (x, y) in $G \times G$ which commutes, i.e., such that $xy = yx$ is equal to

$$|G||\widehat{G}| = |G||G^\sharp|.$$

The reader should attempt to prove this directly (without using characters).

The representations of a finite group G can also be used to understand other spaces of functions. We give two further examples, by showing how to construct fairly convenient orthonormal bases of functions on a quotient G/H, as well as on a given coset of a suitable subgroup.

Proposition 4.4.6 (Functions on G/H). *Let G be a finite group, and let H be a subgroup of G. Let V be the space of complex-valued functions on the quotient G/H with the inner product*

$$\langle \varphi_1, \varphi_2 \rangle_V = \frac{1}{|G/H|} \sum_{x \in G/H} \varphi_1(x)\overline{\varphi_2(x)}.$$

For any representation $\pi : G \longrightarrow \mathrm{GL}(E_\pi)$ and $v \in E_\pi^H$, $w \in E_\pi$, define

$$\varphi_{\pi,v,w} : gH \mapsto \sqrt{\dim \pi}\langle \pi(g)v, w\rangle_\pi.$$

Then the family of functions $(\varphi_{\pi,v,w})$, where π runs over $\widehat{G}$, w runs over an orthonormal basis of the space E_π of π, and v runs over an orthonormal basis of the invariant space E_π^H, forms an orthonormal basis of V.

[5] We use here the notation $S^{-1} = \{x^{-1} \mid x \in S\}$.

Proof. For a representation $\pi : G \longrightarrow \mathrm{GL}(E_\pi)$, $v \in E_\pi^H$ and $w \in E_\pi$, we first denote by

$$\tilde{\varphi}_{\pi,v,w}(g) = \langle \pi(g)v, w\rangle_\pi$$

the corresponding matrix coefficients in $C(G)$. We first check that these are H-invariants on the right, so that the corresponding function $\varphi_{\pi,v,w}$ is well defined and is an element of V. For $g \in G$ and $h \in H$, we have

$$\tilde{\varphi}_{\pi,v,w}(gh) = \langle \pi(gh)v, w\rangle_\pi = \langle \pi(g)\pi(h)v, w\rangle_\pi = \langle \pi(g)v, w\rangle_\pi,$$

since $v \in E^H$ by definition, which proves our claim.

We next observe that

$$\langle \varphi_{\pi_1,v_1,w_1}, \varphi_{\pi_2,v_2,w_2}\rangle_V = \langle \tilde{\varphi}_{\pi_1,v_1,w_1}, \tilde{\varphi}_{\pi_2,v_2,w_2}\rangle,$$

where the inner-product on the right is that in $C(G)$. This shows that the restricted family of functions $\varphi_{\pi,v,w}$ in the statement is an orthonormal family in V, by the orthonormality of matrix coefficients of G.

It only remains to show that these functions span V. But their total number is

$$\begin{aligned}\sum_{\pi\in\hat{G}} (\dim \pi)(\dim \pi^H) &= \sum_{\pi\in\hat{G}} (\dim \pi)\langle \mathrm{Res}_H^G \pi, \mathbf{1}_H\rangle_H \\ &= \sum_{\pi\in\hat{G}} (\dim \pi)\langle \pi, \mathrm{Ind}_H^G(\mathbf{1}_H)\rangle_G\end{aligned}$$

by Frobenius reciprocity. However, for any representation ϱ of G, we have

$$\sum_{\pi\in\hat{G}} (\dim \pi)\langle \pi, \varrho\rangle_G = \sum_{\pi\in\hat{G}} (\dim \pi) n_\pi(\varrho) = \dim \varrho,$$

so that the number of functions in our orthonormal system is equal to

$$\dim \mathrm{Ind}_H^G(\mathbf{1}_H) = [G : H] = \dim(V),$$

which means that this system is in fact an orthonormal basis. □

The second case is a bit more subtle. We consider a finite group G, and a normal subgroup $H \lhd G$ such that the quotient A in the exact sequence

$$1 \longrightarrow H \longrightarrow G \xrightarrow{\phi} A \longrightarrow 1$$

is *abelian* (see also Section 2.8). Fixing $a \in A$, we want to describe an orthonormal basis of the space W of class functions supported on the H-coset $\phi^{-1}(a) = Y \subset G$, with the inner product

$$\langle \varphi_1, \varphi_2\rangle_W = \frac{1}{|H|}\sum_{y\in Y} \varphi_1(x)\overline{\varphi_2(x)}.$$

This makes sense because H is normal in G, which implies that any coset of H is a union of conjugacy classes.

The basic starting point is that the restrictions of characters to Y still form a generating set of W (because one can extend by zero any function in W, obtaining a class function on G, which becomes a linear combination of characters). For dimension reasons, this cannot be a basis (except if $H = G$). In order to extract a basis and to attempt to make it orthonormal, we therefore need to compute the inner product in W of characters restricted to Y. To detect the condition $y \in Y$, we use the orthogonality of (one-dimensional) irreducible characters of the quotient group A. We have

$$\frac{1}{|A|}\sum_{\psi\in\hat{A}}\overline{\psi(\alpha)}\psi(\phi(g)) = \begin{cases} 0 & \text{if } \phi(g) \neq \alpha, \text{ i.e., if } g \notin Y, \\ 1 & \text{if } g \in Y, \end{cases}$$

and hence

$$\begin{aligned} \langle \chi_{\pi_1}, \chi_{\pi_2} \rangle_W &= \frac{1}{|H|}\sum_{y\in Y}\chi_{\pi_1}(y)\overline{\chi_{\pi_2}(y)} \\ &= \frac{1}{|H||A|}\sum_{\psi\in\hat{A}}\overline{\psi(\alpha)}\sum_{y\in Y}\psi(\phi(y))\chi_{\pi_1}(y)\overline{\chi_{\pi_2}(y)} \\ &= \sum_{\psi\in\hat{A}}\overline{\psi(\alpha)}\langle(\psi\circ\phi)\otimes\pi_1, \pi_2\rangle_G \\ &= \sum_{\substack{\psi\in\hat{A} \\ \pi_2\simeq(\psi\circ\phi)\otimes\pi_1}}\overline{\psi(\alpha)}. \end{aligned}$$

This is more complicated than the usual orthogonality relation for characters, but it remains manageable. It shows that the characters remain orthogonal on H unless we have $\pi_1 \simeq (\psi\circ\phi)\otimes\pi_2$ for some $\psi \in \hat{A}$. This is natural, because evaluating the characters, we obtain in that case

$$\chi_{\pi_1}(y) = \psi(\alpha)\chi_{\pi_2}(y)$$

for $y \in Y$, i.e., χ_{π_1} and χ_{π_2} are then *proportional*. The factor $\psi(\alpha)$ is of modulus 1, and hence

$$\langle \chi_{\pi_1}, \chi_{\pi_2} \rangle_W = \langle \chi_{\pi_1}, \chi_{\pi_1} \rangle_W = \sum_{\substack{\psi\in\hat{A} \\ (\psi\circ\phi)\otimes\pi_1\simeq\pi_1}}\overline{\psi(\alpha)}$$

in this situation. This can still be simplified a bit. If ψ occurs in the sum, we obtain

$$\psi(\alpha)\chi_{\pi_1}(y) = \chi_{\pi_1}(y)$$

for all $y \in Y$, and therefore *either* $\psi(\alpha) = 1$ or $\chi_{\pi_1}(y) = 0$ for all $y \in Y$. In the second case, the character actually vanishes on all of Y (and will not help in constructing an orthonormal basis, so we can discard it!), while in the first case, we get

$$\langle \chi_{\pi_1}, \chi_{\pi_1} \rangle_W = |\{\psi \in \hat{A} \mid (\psi \circ \phi) \otimes \pi_1 \simeq \pi_1\}|.$$

Let us denote by

$$\kappa(\pi) = |\{\psi \in \hat{A} \mid (\psi \circ \phi) \otimes \pi \simeq \pi\}| \tag{4.32}$$

the right-hand side of this formula. It is an interesting invariant attached to any irreducible representation of G (see Proposition 7.1.14 and Exercise 7.1.15 for some properties when we define a similar invariant for all groups). We can then summarize as follows the discussion:

Proposition 4.4.7 (Functions on cosets). *Let G be a finite group, and let $H \lhd G$ be a normal subgroup with abelian quotient G/H. For $\alpha \in A$, Y and W as defined above, an orthonormal basis of W is obtained by considering the functions*

$$\varphi_\pi(y) = \frac{1}{\sqrt{\kappa(\pi)}} \chi_\pi(y)$$

for $y \in Y$, where π runs over a subset $\hat{G}_H$ defined by

(1) *removing from $\hat{G}$ those π such that the character of π is identically 0 on Y;*

(2) *considering among other characters only a set of representatives for the equivalence relation*

$$\pi_1 \sim_H \pi_2 \textit{ if and only if } \operatorname{Res}_H^G \pi_1 \simeq \operatorname{Res}_H^G \pi_2.$$

To completely prove this, we must simply say a few additional words to explain why the relation $\pi_1 \sim_H \pi_2$ in the statement is equivalent with the existence of $\psi \in \hat{A}$ such that $\pi_2 \simeq (\psi \circ \phi) \otimes \pi_1$. This is the content of Proposition 2.8.2, but we can also argue directly with characters: in one direction this is clear (evaluating the character, which is 1 on $H \supset \operatorname{Ker} \psi$), and, otherwise, if $\operatorname{Res}_H^G \pi_1 \simeq \operatorname{Res}_H^G \pi_2$, we apply the inner product formula with $\alpha = 0$ (so that $Y = H$) to get

$$0 \neq \langle \operatorname{Res}_H^G \pi_1, \operatorname{Res}_H^G \pi_2 \rangle_H = \sum_{\substack{\psi \in \hat{A} \\ \pi_2 \simeq (\psi \circ \phi) \otimes \pi_1}} \overline{\psi(\alpha)},$$

so that the sum cannot be empty, and the existence of ψ follows. This remark means that the functions described in the statement are an orthonormal system in W. We observed at the beginning of the computation that they generate W, and hence we are done.

Exercise 4.4.8. In the situation of Proposition 4.4.7, show how to obtain an orthonormal basis of the space of all functions $Y \longrightarrow \mathbf{C}$, with respect to the inner product on $C(G)$, using restrictions of matrix coefficients. (*Hint*: Example 3.4.8 can be useful.)

Exercise 4.4.9. Let $\mathbf{F}_q$ be a finite field with q elements, and let $n \geqslant 2$ be an integer.

(1) Show that taking $G = \mathrm{GL}_n(\mathbf{F}_q)$ and $H = \mathrm{SL}_n(\mathbf{F}_q)$ gives an example of the situation considered above. What is A in that case?

(2) Show that, in this case, the invariant defined in (4.32) satisfies

$$\kappa(\pi) \leqslant n$$

for any irreducible representation $\pi \in \widehat{G}$.

In Exercise 4.6.6, we will give examples of groups having representations where $\kappa(\pi) \neq 1$, and also examples where the set $\widehat{G}_H$ differs from $\widehat{G}$ (for both possible reasons—characters vanishing on Y or two characters being proportional on Y).

4.5. Finite abelian groups

Finite abelian groups are the easiest groups to deal with when it comes to representation theory. Since they are also very important in applications, we summarize here again the results of the previous sections, specialized to abelian groups, before discussing some features which are specific to this situation.

Theorem 4.5.1 (Finite abelian groups). *Let G be a finite abelian group.*

(1) *There are exactly $|G|$ one-dimensional complex representations, often simply called* characters *of G, namely group homomorphisms*

$$\chi : G \longrightarrow \mathbf{C}^{\times}.$$

(2) *Let $\widehat{G}$ be the set of characters of G. We have the orthogonality relations*

$$\sum_{x \in G} \chi_1(x)\overline{\chi_2(x)} = \begin{cases} |G| & \text{if } \chi_1 = \chi_2, \\ 0 & \text{if } \chi_1 \neq \chi_2 \end{cases} \tag{4.33}$$

for $\chi_1, \chi_2 \in \widehat{G}$, and

$$\sum_{\chi \in \widehat{G}} \chi(x)\overline{\chi(y)} = \begin{cases} |G| & \text{if } x = y, \\ 0 & \text{if } x \neq y \end{cases} \tag{4.34}$$

for $x, y \in G$.

(3) *Let* $\varphi : G \longrightarrow \mathbf{C}$ *be any function on* G*. We have the Fourier decomposition*

$$\varphi = \sum_{\chi \in \widehat{G}} \hat{\varphi}(\chi)\chi,$$

where

$$\hat{\varphi}(\chi) = \langle \varphi, \chi \rangle = \frac{1}{|G|} \sum_{x \in G} \varphi(x)\overline{\chi(x)},$$

and the Plancherel formula

$$\sum_{\chi \in \widehat{G}} |\hat{\varphi}(\chi)|^2 = \frac{1}{|G|} \sum_{x \in G} |\varphi(x)|^2.$$

The crucial feature which is specific to abelian groups is that, since all irreducible representations are of dimension 1, they form a *group* under pointwise multiplication: if χ_1, χ_2 are in $\widehat{G}$, the product

$$\chi_1\chi_2 : x \mapsto \chi_1(x)\chi_2(x)$$

is again in $\widehat{G}$. Similarly, the inverse

$$\chi^{-1} : x \mapsto \chi(x)^{-1} = \overline{\chi(x)}$$

(where the last formula holds because $|\chi(x)| = 1$ for all characters) is a character. Hence, with the trivial character as neutral element, the set $\widehat{G}$ is also a group, in fact a finite abelian group of the same order as G. Its properties are summarized by

Theorem 4.5.2 (Duality of finite abelian groups). *Let* G *be a finite abelian group, and let* $\widehat{G}$ *be the group of characters of* G*, called the* dual group.

(1) *There is a* canonical *isomorphism*

$$e \begin{cases} G & \longrightarrow & \widehat{\widehat{G}} \\ x & \mapsto & e_x, \end{cases}$$

where e_x *is the homomorphism of evaluation at* x *defined on* $\widehat{G}$*, i.e.*

$$e_x(\chi) = \chi(x).$$

(2) *The group* $\widehat{G}$ *is* non-canonically *isomorphic to* G*.*

Proof. (1) A simple check shows that e is a group homomorphism. To show that it is injective, we must show that if $x \neq 1$, there is at least one character χ with $\chi(x) \neq 1$. This follows, for instance, from the orthogonality relation

$$\sum_{\chi} \chi(x) = 0.$$

(2) The simplest argument is to use the structure theory of finite abelian groups (see, e.g., [**40**, Th. 10.2]): there exist integers $r \geqslant 0$ and positive integers

$$d_1 \mid d_2 \mid \cdots \mid d_r$$

such that

$$G \simeq \mathbf{Z}/d_1\mathbf{Z} \times \cdots \times \mathbf{Z}/d_r\mathbf{Z}.$$

Now we observe that for a direct product $G_1 \times G_2$, there is a natural isomorphism

$$\begin{cases} \widehat{G_1} \times \widehat{G_2} \longrightarrow \widehat{G_1 \times G_2} \\ (\chi_1, \chi_2) \mapsto \chi_1 \boxtimes \chi_2, \end{cases}$$

with $(\chi_1 \boxtimes \chi_2)(x_1, x_2) = \chi_1(x_1)\chi_2(x_2)$. Indeed, this is a group homomorphism, which is quite easily seen to be injective, and the two groups have the same order.[6] Thus we find

$$\widehat{G} \simeq \widehat{\mathbf{Z}/d_1\mathbf{Z}} \times \cdots \times \widehat{\mathbf{Z}/d_r\mathbf{Z}},$$

and this means that it is enough to prove that $\widehat{G} \simeq G$ when G is a finite cyclic group $\mathbf{Z}/d\mathbf{Z}$. But a homomorphism

$$\chi : \mathbf{Z}/d\mathbf{Z} \longrightarrow \mathbf{C}^\times$$

is determined uniquely by $e_1(\chi) = \chi(1)$. This complex number must be a d-th root of unity, and this means that we have an isomorphism

$$e_1 : \begin{cases} \widehat{\mathbf{Z}/d\mathbf{Z}} & \longrightarrow & \boldsymbol{\mu}_d = \{z \in \mathbf{C}^\times \mid z^d = 1\} \\ \chi & \mapsto & \chi(1). \end{cases}$$

Since the group of d-th roots of unity in $\mathbf{C}^\times$ is isomorphic to $\mathbf{Z}/d\mathbf{Z}$ (though non-canonically, if $d \geqslant 3$), we are done. □

Remark 4.5.3. In practice, one uses very often the explicit description of characters of $\mathbf{Z}/m\mathbf{Z}$ that appeared in this proof. Denoting $e(z) = e^{2i\pi z}$ for $z \in \mathbf{C}$, they are the functions of the form

$$e_a : x \mapsto e\Big(\frac{ax}{m}\Big),$$

where $x \in \mathbf{Z}/m\mathbf{Z}$ and $a \in \mathbf{Z}/m\mathbf{Z}$. In this description, of course, the exponential is to be interpreted as being computed using representatives in $\mathbf{Z}$ of x and a, but the result is independent of these choices (simply because $e(k) = 1$ if $k \in \mathbf{Z}$).

Exercise 4.5.4. We have derived the basic facts about representations of finite abelian groups from the general results of this chapter. However, one can also prove them using more specific arguments. This exercise discusses one possible approach (see [**54**, VI.1]).

[6] It is also surjective by an application of Proposition 2.3.23.

(1) Prove the orthogonality relation (4.33) directly.

(2) Show—without using anything other than the definition of one-dimensional characters—that if $H \subset G$ is a subgroup of a finite abelian group and $\chi_0 \in \widehat{H}$ is a character of H, there exists a character $\chi \in \widehat{G}$ of G such that χ restricted to H is equal to χ_0. (*Hint*: Use induction[7] on $|G/H|$.) Afterward, reprove this using Frobenius reciprocity, and compare with Exercise 2.3.17.

(3) Deduce from this the orthogonality relation (4.34).

(4) Deduce that $\widehat{G}$ is an abelian group of the same order as G, and that the homomorphism e is an isomorphism.

Example 4.5.5 (Dirichlet's Theorem on primes in arithmetic progressions). We sketch how Dirichlet succeeded in proving Theorem 1.2.2. Thus we have a positive integer $q \geqslant 1$ and an integer $a \geqslant 1$ coprime with q, and we want to find prime numbers $p \equiv a \pmod q$.

Dirichlet's proof is motivated by an earlier argument that Euler used to give a proof that there are infinitely many prime numbers, as follows. We know that

$$\lim_{\sigma \to 1} \sum_{n \geqslant 1} \frac{1}{n^\sigma} = +\infty,$$

e.g., by comparison of the series with the integral

$$\int_1^{+\infty} x^{-\sigma} dx = \frac{1}{\sigma - 1}, \qquad \text{for } \sigma > 1.$$

On the other hand, exploiting the unique factorization of positive integers in products of primes, Euler showed that

$$\sum_{n \geqslant 1} \frac{1}{n^\sigma} = \prod_p (1 - p^{-\sigma})^{-1} \tag{4.35}$$

for $\sigma > 1$, where the infinite product (which is called an *Euler product*) is over all prime numbers, and it is defined as the limit, as $x \to +\infty$, of the partial products

$$\prod_{p \leqslant x} (1 - p^{-\sigma})^{-1} = \prod_{p \leqslant x} \sum_{k \geqslant 0} p^{-k\sigma} = \sum_{n \in P(x)} n^{-\sigma},$$

where $P(x)$ is the sum of all positive integers with no prime divisor $> x$, and the unique factorization of integers has been used in the last step. Thus the absolute convergence of the series on the left of (4.35) is enough to justify that equality.

Now obviously, if there were only finitely many primes, the right-hand side of the formula would converge to some fixed real number as $\sigma \to 1$,

[7] *Not* induction of representations, but proof by induction....

which contradicts what we said about the left-hand side. Hence there are infinitely many primes.

An equivalent way to conclude is to take the logarithm on both sides. Denoting

$$\zeta(\sigma) = \sum_{n\geqslant 1} n^{-\sigma}$$

for $\sigma > 1$, we have

$$\log\zeta(\sigma) \to +\infty \qquad \text{as } \sigma \to 1,$$

on the one-hand, and on the other hand

$$\log\zeta(\sigma) = -\sum_p \log(1-p^{-\sigma}) = \sum_p p^{-\sigma} + \sum_{p,k\geqslant 2} k^{-1}p^{-k\sigma} = \sum_p p^{-\sigma} + O(1)$$

as $\sigma \to 1$ (where we have used the power series expansion

$$\log\Bigl(\frac{1}{1-x}\Bigr) = \sum_{k\geqslant 1} \frac{x^k}{k},$$

which converges absolutely and uniformly on compact sets for $|x| < 1$; recall that the meaning of $O(1)$ here is that the difference

$$\log\zeta(\sigma) - \sum_p p^{-\sigma}$$

is a bounded function of σ for $\sigma \to 1$). Thus Euler's argument can be phrased as

$$\lim_{\sigma\to 1} \sum_p p^{-\sigma} = +\infty.$$

Using this, it is rather tempting (isn't it?) to try to analyze similarly either the product

$$\prod_{p\equiv a \,(\mathrm{mod}\, q)} (1-p^{-\sigma})^{-1}$$

or the sum

$$\sum_{p\equiv a \,(\mathrm{mod}\, q)} p^{-\sigma}$$

and to show that, as $\sigma \to 1$, these functions tend to $+\infty$. But if we expand the product, we do *not* get the obvious series

$$\sum_{\substack{n\geqslant 1\\ n\equiv a \,(\mathrm{mod}\, q)}} n^{-\sigma}$$

because there is no reason that the primes dividing an integer congruent to a modulo q should have the same property. Also, if a is not $\equiv 1 \,(\mathrm{mod}\, q)$, the product of primes congruent to a modulo q is not necessarily $\equiv a \,(\mathrm{mod}\, q)$), e.g., $35 = 7 \times 5$ is congruent to 3 modulo 4, but $5 \equiv 1 \,(\mathrm{mod}\, 4)$.

In other words, we are seeing the effect of the fact that the characteristic function of the single element $a \in \mathbf{Z}/q\mathbf{Z}$, which is used to select the primes in the product or the series, is *not* multiplicative. Dirichlet's idea is to use, instead, some functions on $\mathbf{Z}/q\mathbf{Z}$ which *are* multiplicative, and to use them to recover the desired characteristic function.

A *Dirichlet character* modulo q is defined to be a map

$$\chi : \mathbf{Z} \longrightarrow \mathbf{C}$$

such that $\chi(n) = 0$ if n is not coprime to q, and otherwise

$$\chi(n) = \chi_*(n \pmod q)$$

for some character of the multiplicative group of invertible residue classes modulo q:

$$\chi_* : (\mathbf{Z}/q\mathbf{Z})^\times \longrightarrow \mathbf{C}^\times.$$

We recall the notation $\varphi(q) = |(\mathbf{Z}/q\mathbf{Z})^\times|$ for the order of this finite group; this is the *Euler totient function*.

For χ as above, it follows that $\chi(nm) = \chi(n)\chi(m)$ for all n, $m \geqslant 1$ (either because both sides are 0 or because χ_* is a homomorphism). Since Dirichlet characters modulo q correspond exactly to the characters of the group $(\mathbf{Z}/q\mathbf{Z})^\times$, the orthogonality relation implies

$$\sum_{\chi \pmod q} \overline{\chi(a)}\chi(n) = \begin{cases} |(\mathbf{Z}/q\mathbf{Z})^\times| = \varphi(q), & \text{if } n \equiv a \pmod q \\ 0, & \text{otherwise,} \end{cases}$$

for $n \geqslant 1$, where the sum ranges over all Dirichlet characters modulo q (because a is assumed to be coprime to q).

This is the first crucial point: the use of *suitable harmonics* to analyze the characteristic function of a residue class. We apply this to all primes p, and multiply by $p^{-\sigma}$ before summing over primes. We obtain the formula

$$\sum_{p \equiv a \pmod q} p^{-\sigma} = \frac{1}{\varphi(q)} \sum_{\chi \pmod q} \overline{\chi(a)} \sum_p \chi(p) p^{-\sigma}.$$

On the other hand, for each χ, the multiplicativity of χ leads to an analogue of the Euler product

$$\sum_{n \geqslant 1} \chi(n) n^{-\sigma} = \prod_p (1 - \chi(p)p^{-\sigma})^{-1}.$$

As is now classical, we denote by $L(\sigma, \chi)$ the function in this last formula. By the same reasoning used for $\zeta(\sigma)$, it satisfies

$$\log L(\sigma, \chi) = \sum_p \chi(p) p^{-\sigma} + O(1)$$

as $\sigma \to 1$. We have therefore

$$\sum_{p \equiv a \,(\mathrm{mod}\, q)} p^{-\sigma} = \frac{1}{\varphi(q)} \sum_{\chi \,(\mathrm{mod}\, q)} \overline{\chi(a)} \log L(\sigma, \chi) + O(1)$$

for all $\sigma > 1$, and now the idea is to imitate Euler by letting $\sigma \to 1$ and seeing a divergence emerge on the right-hand side, which then implies that the series on the left cannot have only finitely many non-zero terms.

On the right-hand side, for the character χ_0 corresponding to $\chi_* = 1$, we have

$$L(\sigma, \chi_0) = \prod_{p \nmid q} (1 - p^{-\sigma})^{-1}$$

(the primes dividing q have $\chi_0(p) = 0$), which therefore satisfies

$$\log L(\sigma, \chi_0) = \log(1/(\sigma - 1)) + O(1)$$

as $\sigma \to 1$, since only the finitely many terms where $p \mid q$ make this different from Euler's case of $\zeta(\sigma)$. This contribution therefore diverges, and we see that Dirichlet's Theorem follows from the second crucial ingredient: the fact that for a Dirichlet character χ associated to a character $\chi_* \neq 1$, the function

$$L(\sigma, \chi)$$

converges to a *non-zero* value as $\sigma \to 1$, so that its logarithm also has a limit. Showing that the function converges to some complex number is not too difficult; however, proving that this complex number is non-zero is more subtle, and we refer to, e.g., to [**54**, Ch. 6] for a very careful presentation of the details.

This argument of Dirichlet has been extraordinarily fruitful: besides being probably the first use of representation theory to prove a deep theorem, it is also closely related to reciprocity laws in algebraic number theory. Starting from there, one reaches naturally the point of view of Tate for L-functions and the amazing ideas of Langlands and their ramifications that extend over a large part of modern arithmetic. The book [**3**] provides good, and gentle, introductions to these ideas.

Exercise 4.5.6 (Burnside's inequality for cyclic groups)**.** Although, much of the time, one deals with irreducible characters of finite abelian groups, higher-dimensional representations do sometimes occur. Here is one result of Burnside which is used in the proof of Proposition 4.7.11 below.

For a finite cyclic group $G = \mathbf{Z}/m\mathbf{Z}$ with $m \geqslant 1$, we let $G^* \subset G$ be the set of generators of G. The goal is to prove that if ϱ is any finite-dimensional representation of G, we have

$$\sum_{x \in G^*} |\chi_\varrho(x)|^2 \geqslant |G^*| \tag{4.36}$$

unless $\chi_\varrho(x) = 0$ for all $x \in G^*$.

(1) If you know Galois theory, prove this directly. (*Hint*: Use the arithmetic-geometric mean inequality.)

The next steps present an alternative argument which does not require Galois theory.

(2) Show that there exists a non-negative quadratic form $Q_m(X)$, where $X = (X_a)_{a\in \mathbf{Z}/m\mathbf{Z}}$ are variables parametrized by $\mathbf{Z}/m\mathbf{Z}$, with integral coefficients, such that

$$\sum_{x\in G^*} |\chi_\varrho(x)|^2 = Q_m(\boldsymbol{n})$$

for any representation ϱ, where $\boldsymbol{n} = (n(a))$ and $n(a) \geqslant 0$ is the multiplicity in ϱ of the irreducible character

$$x \mapsto e\Big(\frac{ax}{m}\Big)$$

of $\mathbf{Z}/m\mathbf{Z}$.

(3) Show that if

$$m = \prod_{p|m} p^{k_p}$$

is the prime factorization of m, we have

$$Q_m = \bigotimes_{p|m} Q_{p^{k_p}}$$

(for some obvious notion of tensor product of quadratic forms).

(4) Show that for p prime and for any quadratic form Q' of rank $d \geqslant 1$, we have

$$(Q_p \otimes Q')(\boldsymbol{n}) = \frac{1}{2} \sum_{a,b\in \mathbf{Z}/p\mathbf{Z}} Q'(n(a) - n(b))$$

for any $\boldsymbol{n} = (n(a)) \in (\mathbf{Z}^d)^p$. (*Hint*: It may be useful to start with $Q'(n) = n^2$ of rank 1.)

(5) For Q' as above, non-negative, let $s(Q')$ denote the smallest non-zero value of $Q'(\boldsymbol{n})$ for $\boldsymbol{n} \in \mathbf{Z}^d$. Show that for any quadratic form Q' of rank $d \geqslant 1$, we have

$$s(Q_p \otimes Q') = (p-1)s(Q').$$

(6) For $k \geqslant 2$ and p prime, show that there exists a quadratic form Q' of rank p^{k-1} such that $Q_{p^k} = Q_p \otimes Q'$ and $s(Q') = p^{k-1}$. Then prove (4.36).

4.6. The character table

The *character table* of a finite group G is the name given to the matrix $(\chi_\varrho(c))_{\varrho,c}$ which gives the values of all the (complex) irreducible characters χ_ϱ of G evaluated at all conjugacy classes $c \in G^\sharp$. In particular, it is a square matrix which determines the irreducible characters as class functions, and hence encapsulates (in theory at least!) all the information given by representation theory over $\mathbf{C}$ for the group G. It is typically represented as a square matrix with rows given by the irreducible characters (in some order) and columns indexed by the conjugacy classes.

Table 4.1. Character table of $\mathfrak{S}_3$

	1	(12)	(123)
$\mathbf{1}$	1	1	1
ε	1	-1	1
ϱ_2	2	0	-1

Example 4.6.1. A very simple example is the character table of $G = \mathfrak{S}_3$ is shown in Table 4.1. Here the top line, as well as the left-most row, simply recall the chosen ordering of the conjugacy classes and characters. For the former, this is usually fairly self-explanatory, but for the characters, one often wants, if possible, a description of an actual representation which has the character values given in the row (if only to check that it is correctly described!)

This might be a complicated matter, but for this example, this is simple (see Example 4.2.4): $\mathbf{1}$ is the trivial representation; $\varepsilon : \mathfrak{S}_3 \longrightarrow \mathbf{C}^\times$ is the signature; and ϱ_2 is the 2-dimensional representation acting on

$$E = \{(x, y, z) \in \mathbf{C}^3 \mid x + y + z = 0\}.$$

Indeed, it is not hard to check that the character values are correct. Note that, once we know its character, one can check that ϱ_2 is irreducible by computing the norm $\langle \chi_{\varrho_2}, \chi_{\varrho_2} \rangle = 1$ (see Corollary 4.3.14; we emphasize again that this criterion is only rigorously applicable after constructing a representation with the given character). We also recall that one needs to know the order of the conjugacy classes in order to weigh properly the character values (see Remark 4.4.2). This extra information is often indicated in parallel with the character table, but it can in fact be recovered[8] from it

[8] This, of course, assumes that the full character table is known....

using (4.28): if we fix a class $c \in G^\sharp$ and take $h \in c$ there, we see that

$$|c| = \frac{|G|}{\sum_{\varrho} |\chi_\varrho(h)|^2}. \tag{4.37}$$

For instance, taking $c = (12)$ and $c' = (123)$ for $\mathfrak{S}_3$, we get

$$|c| = \frac{6}{1^2 + 1^2} = 3, \qquad |c'| = \frac{6}{1^2 + 1^2 + 1^2} = 2,$$

as it should.

Before we discuss which information concerning a group can be extracted from the character table, it is interesting to ask, Can we always compute it? Here is a first answer:

Proposition 4.6.2 (The character table is computable). *Let G be a finite group, given in such a way that one can enumerate all elements of G and one can compute the group law and the inverse.*[9] *Then there is an algorithm for computing the character table of G that will terminate in finite time.*

This is a fairly poor version of computability: we make no claim, or guarantee, about the amount of time the algorithm will require (an estimate can be obtained from the argument, but it will be very bad).

Proof. First of all, by enumerating all pairs of elements and computing all products xyx^{-1}, one can make the list of all the conjugacy classes of G and find means to associate its conjugacy class to any element of G.

The idea is then to see that the regular representation can be decomposed into isotypic components. Indeed, first of all, the regular representation is computable: a basis of $C(G)$ is given by characteristic functions of single points, and the action of G on the basis vectors is computable from the inverse map. Moreover, decomposing an arbitrary $f \in C(G)$ in this basis is immediate (given that the values of f are computable).

It is then enough to find an algorithm to compute the decomposition

$$C(G) = \bigoplus_{\pi \in \hat{G}} \mathrm{M}(\pi)$$

of the regular representation into isotypic components, in the sense of giving a list of bases of the spaces $\mathrm{M}(\pi)$. Indeed, given a subspace M among these, one can then compute the corresponding character by

$$\chi(g) = \frac{1}{\dim(M)} \operatorname{Tr}(\varrho_G(g)|_M).$$

[9] As an example of the subtleties that may be involved, note that having a generating set is not enough: one must be able to say whether two arbitrary products of elements from such a set are equal or not.

Now the crucial step: the subspaces $\mathrm{M}(\pi)$ are characterized as the common eigenspaces of all operators $\varrho_G(a)$, where $a \in Z(k(G))$ is an element in the center of the group algebra, or equivalently as the common eigenspaces of the operators $\varrho_G(a_c)$, where c runs over the conjugacy classes of G and

$$a_c = \sum_{g \in c} g.$$

In other words, assume a non-zero subspace $M \subset C(G)$ has the property that there exist eigenvalues $\lambda_c \in \mathbf{C}$, defined for all c, such that

$$M = \{v \in C(G) \mid \varrho_G(a_c)v = \lambda_c v \text{ for all } c \in G^\sharp\}. \tag{4.38}$$

Then we claim that M is one of the $\mathrm{M}(\pi)$, and conversely.

If this is granted, we proceed as follows. List the conjugacy classes, and for each of them find the eigenvalues and eigenspaces of the operator $\varrho_G(a_c)$ (by finding bases for them, using linear algebra, which is eminently computable). Then, compute all the possible intersections of these eigenspaces, and list the resulting subspaces: they are the isotypic components $\mathrm{M}(\pi)$.

We now check the claim. Let M be a non-zero common eigenspace of the $\varrho_G(a_c)$ given by (4.38). For any irreducible representation π, we have

$$\begin{aligned}
\varrho_G(e_\pi)v &= \frac{\dim \pi}{|G|} \sum_{g \in G} \overline{\chi_\pi(g)} \varrho_G(g)v \\
&= \frac{\dim \pi}{|G|} \sum_{c \in G^\sharp} \overline{\chi_\pi(c)} \sum_{g \in c} \varrho_G(g)v \\
&= \frac{\dim \pi}{|G|} \sum_{c \in G^\sharp} \overline{\chi_\pi(c)} \varrho_G(a_c)v \\
&= \Bigl(\frac{\dim \pi}{|G|} \sum_{c \in G^\sharp} \overline{\chi_\pi(c)} \lambda_c\Bigr) v,
\end{aligned}$$

since χ_π is conjugacy invariant (and we write $\chi_\pi(c)$ for the value of the character at any element of the conjugacy class). This means that M is contained in an eigenspace of the π-isotypic projection e_π.

However, we know these eigenspaces already: the isotypic component $\mathrm{M}(\pi)$ is the 1-eigenspace of e_π, and the sum of the other isotypic components is the 0-eigenspace. Since the sum over π of the projections e_π acts as the identity on $C(G)$, hence on M, there exists necessarily some $\pi \in \widehat{G}$ such that the eigenvalue of e_π acting on M is 1. But this means that $M \subset \mathrm{M}(\pi)$. Since $\mathrm{M}(\pi)$ itself is contained in a common eigenspace, namely

$$\mathrm{M}(\pi) \subset \Bigl\{v \in C(G) \mid \varrho_G(a_c)v = \frac{|c|\chi_\pi(c)}{|G|} v \text{ for all } c \in G^\sharp\Bigr\}$$

(by Proposition 4.7.11), this means that λ_c must coincide with $\frac{|c|\chi_\pi(c)}{|G|}$. But then

$$M \subset \mathrm{M}(\pi) \subset M,$$

and these inclusions must be equalities! $\square$

This algorithm is not at all practical if G is large, but at least it shows that, by trying to get information *from* the character table, we are not building castles completely in the air!

Note that besides this fact, the proof has led to the characterization

$$\mathrm{M}(\pi) = \Big\{ v \in C(G) \mid \varrho_G(a)v = \omega_\pi(a)v \text{ for all } a \in Z(k(G)) \Big\}$$

(where $\omega_\pi(a)$, as in Section 4.3.6, is the scalar giving the action of a on the the irreducible representation π) of the isotypic components of the regular representation, which is of independent interest.

Exercise 4.6.3. Proposition 4.6.2 shows how to compute, in principle, the character table of a finite group G. Explain how one can also, in principle, write matrix representations $\pi^m : G \longrightarrow \mathrm{GL}_{\dim(\pi)}(\mathbf{C})$ for each irreducible representation $\pi \in \widehat{G}$.

4.6.1. Some features of the character table. We present here some of the information that can be derived from the character table of a group, if it is known (other examples are given for instance in [**28**, Ch. 2]). As in Proposition 4.6.2, we do not attempt to measure the actual computational efficiency of the procedures we describe, many of which are quite impractical when implemented directly. In the next sections, we will compute the character tables of some concrete groups in detail.

• As already noted, the sizes of the conjugacy classes, or equivalently the sizes of the centralizers of $C_G(g) = |G|/|g^\sharp|$ of elements in G, can be computed from the character table using formula (4.37).

• The kernel of an irreducible representation ϱ can be determined from its row in the character table, because of the following:

Proposition 4.6.4 (Size of the character values)**.** *Let G be an arbitrary group, and let ϱ be a finite-dimensional unitary, or unitarizable, representation of G. For $g \in G$, we have*

$$|\chi_\varrho(g)| \leqslant \dim \varrho, \tag{4.39}$$

and there is equality if and only if $\varrho(g)$ is a scalar. *In particular, $\varrho(g) = 1$ if and only if $\chi_\varrho(g) = \chi_\varrho(1) = \dim(\varrho)$.*

Proof. Since ϱ is unitary, the eigenvalues of $\varrho(g)$ are of modulus 1, and hence the modulus of the trace $\chi_\varrho(g)$ of $\varrho(g)$ is at most $\dim(\varrho)$. Moreover, by the equality case of the triangle inequality, there can be equality only if all eigenvalues are equal, which means (since $\varrho(g)$ is diagonalizable) that $\varrho(g)$ is the multiplication by this common eigenvalue.

Finally, when this happens, we can compute the eigenvalue, which is equal to $\chi_\varrho(g)/\dim(\varrho)$, and this eigenvalue is equal to 1 if and only if $\varrho(g)$ is the identity, which is true if and only if $\chi_\varrho(g) = \dim \varrho$. □

Note, however, that in general a character, even for a faithful representation, has no reason to be *injective* (on G or on conjugacy classes): the regular representation gives an example of this (it is faithful but, for $|G| \geqslant 3$, its character is not an injective function on G (Example 2.7.40)). Another type of *failure of injectivity* related to characters is described in Exercise 4.6.15.

• More generally, all normal subgroups of G can be computed using the character table, as well as their possible inclusion relations. To do this, one can find the kernels of the irreducible representations using the lemma and the character table.

As a next step, for any normal subgroup $N \lhd G$, we have

$$N = \operatorname{Ker} \varrho_N,$$

where ϱ_N is the permutation representation associated to the left-action of G on G/N. Indeed, if e_{xN} are the basis vectors for the space of ϱ_N, to have $g \in \operatorname{Ker} \varrho_N$ means that $ge_{xN} = e_{xN}$ for all $x \in G$, i.e., $gxN = xN$ for all x, so $g \in N$, and the converse follows because N is normal. If we denote by

$$X = \{\varrho \in \hat{G} \mid \langle \varrho, \operatorname{Ind}_N^G(\mathbf{1})\rangle \geqslant 1\}$$

the set of those irreducible representations which occur in the induced representation, it follows that

$$N = \bigcap_{\varrho \in X} \operatorname{Ker} \varrho.$$

Thus, to determine all normal subgroups of G, from the character table, one can first list all kernels of irreducible representations, and then compute all intersections of finitely many such subgroups. In particular, once the conjugacy classes which form a normal subgroup are known, its order can of course be computed by summing their size. On the other and, it is interesting to know that there is *no way* to determine *all* subgroups of G from the character table (see Exercise 4.6.7).

• The character table of a quotient G/N of G by a normal subgroup $N \lhd G$ can also be determined from the character table, since irreducible representations of G/N correspond bijectively to those irreducible representations of G where $N \subset \operatorname{Ker} \varrho$.

• One can check if a group is abelian by checking whether all irreducible representations have dimension 1.

• Whether G is solvable can also, in principle, be determined from the character table. Indeed, if G is abelian, it is solvable. Otherwise, one can determine whether G contains a proper normal subgroup N with abelian quotient G/N. If N does not exist, the group is not solvable; otherwise, N is a non-trivial proper subgroup, and we can iterate with G replaced by G/N, which is solvable if and only if G itself is solvable.

• It is also natural to ask what *cannot* be determined from the character table. At first, one might hope that the answer would be "nothing at all!", i.e., that it may be used to characterize the group up to isomorphism. This is not the case, however,[10] as we will explain in Exercise 4.6.7 below with a very classical example of two non-isomorphic groups with the same character tables. It will follow, in particular, that the character table cannot be used to determine *all* subgroups of a finite group.

4.6.2. A nilpotent group. In order of structural group-theoretic complexity, after abelian groups come (non-abelian) nilpotent groups. We recall (see, e.g., [**51**, Ch. 5]) that G is nilpotent if, for some integer $i \geqslant 1$, we have $G_i = 1$, where the sequence of subgroups $(G_i)_{i\geqslant 1}$ is defined inductively by

$$G_1 = G, \qquad G_{i+1} = [G_i, G],$$

where $[G_i, G]$ is the subgroup generated by commutators $[x, y]$ with $x \in G_i$ and $y \in G$.

A good example is given by the family of finite Heisenberg groups H_p defined by

$$H_p = \left\{ \begin{pmatrix} 1 & x & z \\ 0 & 1 & y \\ 0 & 0 & 1 \end{pmatrix} \mid x, y, z \in \mathbf{F}_p = \mathbf{Z}/p\mathbf{Z} \right\}$$

for p prime. This is a p-group since $|H_p| = p^3$, and any finite p-group is nilpotent (see, e.g., [**51**, Th. 5.33]).

We will construct the character table of the groups H_p. We first gain some insight into the structure of the group H_p by computing its conjugacy classes. To simplify notation, we will use the shorthand

$$\{x, y, z\}_H = \begin{pmatrix} 1 & x & z \\ 0 & 1 & y \\ 0 & 0 & 1 \end{pmatrix}$$

[10] The reader should *not* add "unfortunately": there are no unfortunate events in mathematics. . . .

to denote the elements of H_p. Then straightforward computations yield the product formula

$$\{x, y, z\}_H\{a, b, c\}_H = \{x + a, y + b, xb + z + c\}_H,$$

the conjugacy formula

$$\{x, y, z\}_H\{a, b, c\}_H\{x, y, z\}_H^{-1} = \{a, b, xb - ya + c\}_H,$$

as well as the commutator relation

$$[\{x, y, z\}_H, \{a, b, c\}_H] = \{0, 0, xb - ya\}_H.$$

The last formula shows that

$$[H_p, H_p] = Z = \{\{0, 0, z\}_H \mid z \in \mathbf{F}_p\},$$

is the *center* of H_p. Each element of Z is a one-element conjugacy class in H_p; on the other hand, if $(a, b) \neq (0, 0)$, the conjugacy formula shows that, for any fixed $c \in \mathbf{F}_p$, the conjugacy class of $\{a, b, c\}_H$ is

$$X_{a,b} = \{\{a, b, z\}_H \mid z \in \mathbf{F}_p\}$$

(because the image of $(x, y) \mapsto xb - ya + c$ is all of $\mathbf{F}_p$ in that case, as the image of a non-constant affine map). We have therefore found all conjugacy classes:

- There are p central conjugacy classes of size 1;
- There are $p^2 - 1$ conjugacy classes $X_{a,b}$ of size p.

In particular, the character table of H_p has $p^2 + p - 1$ rows and columns. To start completing it, a good first step is to determine all one-dimensional representations χ. Not only is it a beginning to the table, but one can also hope to produce new representations later by considering the "twists" $\varrho \otimes \chi$ of a "brand new" representation ϱ with the one-dimensional representations (see Exercise 2.2.14).

The one-dimensional representations are determined by computing the abelianization $H_p/[H_p, H_p] = H_p/Z$, and we see here that we have an isomorphism

$$\begin{cases} H_p/Z \longrightarrow \mathbf{F}_p^2 \\ \{a, b, c\}_H \mapsto (a, b). \end{cases}$$

Thus we have p^2 distinct one-dimensional representations of H_p given by

$$\chi_{\psi_1,\psi_2} : \{a, b, c\}_H \mapsto \psi_1(a)\psi_2(b),$$

where ψ_1, ψ_2 are two (one-dimensional) characters of $\mathbf{F}_p$.

This now leaves us to find $p^2 + p - 1 - p^2 = p - 1$ irreducible representations, about which we know that the sum of the squares of their dimensions must be

$$|G| - p^2 = p^2(p - 1).$$

By comparison, it is very tempting to think that each of those new representations should be of dimension p. (Indeed, if we also use the fact that their dimension divides $|H_p| = p^3$, by Theorem 4.3.8, and hence must be a power of p, and not 1, this is the only possibility, since a representation of dimension p^2 would already have $(\dim \varrho)^2 = p^4 > p^2(p-1)\ldots.$)

One of the most common ways of finding irreducible representations is to try to construct them as induced representations, or at least to construct such induced representations which contain "new" irreducibles. In particular, inducing one-dimensional representations of a subgroup can be quite efficient. In the case of H_p, if we want to find representations of dimension p, we can look for a subgroup of index p. For instance, we consider

$$K = \{\{0, y, z\}_H \mid y, z \in \mathbf{F}_p\} = \left\{\begin{pmatrix} 1 & 0 & z \\ 0 & 1 & y \\ 0 & 0 & 1 \end{pmatrix}\right\} \subset H_p.$$

We see that $K \simeq \mathbf{F}_p^2$; thus we fix a one-dimensional character ψ of K, given by characters ψ_1, ψ_2 of $\mathbf{F}_p$ such that

$$\psi(\{0, y, z\}_H) = \psi_1(y)\psi_2(z),$$

and we consider the p-dimensional representation

$$\varrho = \mathrm{Ind}_K^{H_p}(\psi).$$

To determine whether this representation is irreducible or not, we will compute its character, using formula (2.49). For this purpose we need a set T of representatives of H_p/K, and we can take

$$T = \{t(x) = \{x, 0, 0\}_H \mid x \in \mathbf{F}_p\}.$$

Then the character of ϱ is given by

$$\chi_\varrho(g) = \sum_{\substack{x \in \mathbf{F}_p \\ t(x)gt(x)^{-1} \in K}} \psi(t(x)gt(x)^{-1}).$$

But for $g = \{a, b, c\}_H$, we have

$$t(x)gt(x)^{-1} = \begin{pmatrix} 1 & a & xb + c \\ 0 & 1 & b \\ 0 & 0 & 1 \end{pmatrix},$$

so that we see already that $\chi_\varrho(g) = 0$ if $a \neq 0$, i.e., if $g \notin K$ (in fact, since $K \lhd G$, we know this must be true from Example 2.7.44(3)). If $a = 0$, on the other hand, the condition $t(x)gt(x)^{-1} \in K$ is always satisfied, and thus

$$\chi_\varrho(g) = \sum_{x \in \mathbf{F}_p} \psi_1(b)\psi_2(xb + c) = \psi_1(b) \sum_{x \in \mathbf{F}_p} \psi_2(xb + c).$$

If $b \neq 0$, sending x to $xb+c$ is a bijection of $\mathbf{F}_p$, and the result is therefore

$$\chi_\varrho(g) = \begin{cases} p\psi_1(b) & \text{if } \psi_2 = \mathbf{1} \\ 0 & \text{if } \psi_2 \neq \mathbf{1}, \end{cases}$$

while for $b = 0$, which means $g = \{0, 0, c\}_H \in Z$, we have

$$\chi_\varrho(g) = p\psi_2(c).$$

Hence there are two cases for χ_ϱ, depending on whether ψ_2 is trivial or not:

	$\{0, 0, c\}_H$	$\{0, b, \star\}_H$, $b \neq 0$	$\{a, b, \star\}_H$, $a \neq 0$
$\psi_2 = \mathbf{1}$	p	$p\psi_1(b)$	0
$\psi_2 \neq \mathbf{1}$	$p\psi_2(c)$	0	0

The middle column concerns $p - 1$ non-central conjugacy classes, and the last concerns the remaining $p^2 - p$ classes, each having p elements. Thus the respective squared norms in the two cases are

$$\frac{1}{p^3}\Big(p \times p^2 + p^2 \times (p-1) \times p\Big) = p,$$

when $\psi_2 = \mathbf{1}$ and

$$\frac{p \times p^2}{p^3} = 1$$

when $\psi_2 \neq \mathbf{1}$. Hence, we have an irreducible representation whenever $\psi_2 \neq \mathbf{1}$. Moreover, the character values in that case show that ϱ is then independent of the choice of ψ_1, up to isomorphism; on the other hand, if we look at the character values for central classes, we see that different choices of ψ_2 lead to different induced representations of H_p. In other words, the $p - 1$ representations

$$\varrho_{\psi_2} = \mathrm{Ind}_K^{H_p}(\psi), \qquad \psi(b, c) = \psi_2(c),$$

with ψ_2 non-trivial, give the remaining p-dimensional irreducible representations of H_p.

We can then present the full character table, as shown in Table 4.2, where the sole restriction is that ψ_2 in the last row should be non-trivial.

Table 4.2. Character table of H_p

	$\{0, 0, c\}_H$	$\{a, b, \star\}_H$, $(a, b) \neq (0, 0)$
χ_{ψ_1, ψ_2}	1	$\psi_1(a)\psi_2(b)$
ϱ_{ψ_2}	$p\psi_2(c)$	0

Remark 4.6.5. Although it might well seem that we had an easy time finding the representations of the Heisenberg groups H_p because they are group-theoretically rather easy to understand, this turns out to be misleading. Indeed, if we consider the groups $U_n(\mathbf{F}_p)$ of upper-triangular matrices of size $n \geqslant 3$ with diagonal coefficients 1, so that $\mathbf{F}_p \simeq U_2(\mathbf{F}_p)$ and $H_p = U_3(\mathbf{F}_p)$ for all primes p, then we obtain again nilpotent groups (of nilpotency class $n-1$), but there is no known complete classification of the complex representations of $U_n(\mathbf{F}_p)$ in general.

4.6.3. Some solvable groups. Moving toward greater group-theoretic complexity, it is natural to consider some non-nilpotent solvable groups. Here a good example to handle is the family

$$B_p = \left\{ \begin{pmatrix} x & t \\ 0 & y \end{pmatrix} \mid t \in \mathbf{F}_p,\ x, y \in \mathbf{F}_p^\times \right\},$$

where p is a prime number. Thus $|B_p| = p(p-1)^2$, and the group is solvable because we have a surjective homomorphism

$$\begin{cases} B_p \longrightarrow (\mathbf{F}_p^\times)^2 \\ \begin{pmatrix} x & t \\ 0 & y \end{pmatrix} \mapsto (x, y), \end{cases} \tag{4.40}$$

with abelian kernel

$$U = \left\{ \begin{pmatrix} 1 & t \\ 0 & 1 \end{pmatrix} \mid t \in \mathbf{F}_p \right\} \simeq \mathbf{F}_p,$$

i.e., B_p is an extension of abelian groups.

As before, we compute the conjugacy classes, using the formula

$$\begin{pmatrix} x & t \\ 0 & y \end{pmatrix} \begin{pmatrix} a & u \\ 0 & b \end{pmatrix} \begin{pmatrix} x & t \\ 0 & y \end{pmatrix}^{-1} = \begin{pmatrix} a & y^{-1}\{t(b-a) + xu\} \\ 0 & b \end{pmatrix}.$$

We consider the middle matrix to be fixed, and we look for its conjugates. If $b \neq a$, the top-left coefficient can take any value when varying x, y and t, in fact even with $x = y = 1$, and this gives us $(p-1)(p-2)$ conjugacy classes of size p. If $a = b$, there are two cases:

(1) if $u \neq 0$, we can get all *non-zero* coefficients, thus we have $p-1$ conjugacy classes of size $p-1$;

(2) if $u = 0$, then the matrix is scalar and its conjugacy class is a single element.

To summarize, there are:

- $p-1$ central conjugacy classes of size 1;
- $p-1$ conjugacy classes with representatives
$$\begin{pmatrix} a & 1 \\ 0 & a \end{pmatrix}$$
of size $p-1$;
- $(p-1)(p-2)$ conjugacy classes with representatives
$$\begin{pmatrix} a & 0 \\ 0 & b \end{pmatrix}$$
of size p.

The number of conjugacy classes, and hence of irreducible representations, is now

$$(p-1)(p-2)+2(p-1)=p(p-1).$$

We proceed to a thorough search for the representations. First, as in the previous section, we can easily find the one-dimensional characters. The commutator formula

$$\begin{pmatrix} x & t \\ 0 & y \end{pmatrix}\begin{pmatrix} a & u \\ 0 & b \end{pmatrix}\begin{pmatrix} x & t \\ 0 & y \end{pmatrix}^{-1}\begin{pmatrix} a & u \\ 0 & b \end{pmatrix}^{-1}=\begin{pmatrix} 1 & \text{(something)} \\ 0 & 1 \end{pmatrix}$$

shows that the morphism (4.40) factors in fact through an isomorphism

$$B_p/[B_p,B_p]\simeq(\mathbf{F}_p^{\times})^2.$$

Thus we have $(p-1)^2$ one-dimensional representations

$$\varrho(\chi_1,\chi_2):\begin{pmatrix} a & u \\ 0 & b \end{pmatrix}\mapsto\chi_1(a)\chi_2(b), \tag{4.41}$$

where χ_1 and χ_2 are one-dimensional characters of $\mathbf{F}_p^{\times}$. Subtracting, we see that we now require

$$p(p-1)-(p-1)^2=p-1$$

other irreducible representations, and that the sums of the squares of their dimensions must be

$$|B_p|-(p-1)^2=p(p-1)^2-(p-1)^2=(p-1)^3.$$

This time, the natural guess is that there should be $p-1$ irreducible representations, each of dimension $p-1$, as this would fit the data very well. (Note also that $p-1 \mid |B_p|$, as we know it should by Theorem 4.3.8.)

This time, we will find these representations using a slightly different technique than induction. Namely, we consider some natural permutation representations attached to B_p. Let X_p be the set of all lines (passing through the origin) in $\mathbf{F}_p^2$ on which B_p acts naturally (an element g acts on a line by mapping it to its image under the associated linear map $\mathbf{F}_p^2 \longrightarrow \mathbf{F}_p^2$). By definition, the line $\mathbf{F}_p e_1$ spanned by $\binom{1}{0}$ is fixed by all elements in B_p, and thus we can consider the permutation representation associated to the action on the complement $Y_p = X_p - \{\mathbf{F}_p e_1\}$. This set has order p (it contains the "vertical" line with equation $x = 0$ and the lines $y = \lambda x$, where $\lambda \in \mathbf{F}_P^\times$), and thus the associated permutation representation π has dimension p. This is not the right dimension, but we know that a permutation representation of this type always contains the trivial representation, represented by the invariant element which is the sum of the basis vectors. Thus we have a representation τ of dimension $p-1$ on the space

$$E = \Big\{ (x_\ell)_{\ell \in Y_p} \mid \sum_{\ell} x_\ell = 0 \Big\} \subset \mathbf{C}^{Y_p}.$$

We proceed to compute its character. This is easy because

$$chi_\tau = \chi_\pi - 1,$$

and we know that for a permutation representation, such as π, we have

$$\chi_\pi(g) = |\{\ell \in Y_p \mid g \cdot \ell = \ell\}|,$$

the number of fixed points of the permutation associated to a given element of the group. We can easily compute this number by looking at the conjugacy classes described above:

- If g is central, it fixes every line, so $\chi_\pi(g) = p$, and $\chi_\tau(g) = p-1$.
- If
$$g = \begin{pmatrix} a & 1 \\ 0 & a \end{pmatrix},$$
there is no fixed point, since this would correspond to an eigenvector of this matrix independent from e_1, whereas the matrix is not diagonalizable. Hence $\chi_\pi(g) = 0$, and $\chi_\tau(g) = -1$.
- If
$$g = \begin{pmatrix} a & 0 \\ 0 & b \end{pmatrix},$$
with $a \neq b$, there is a unique fixed point (here, the line spanned by the second basis vector e_2). Thus $\chi_\pi(g) = 1$ and $\chi_\tau(g) = 0$.

We summarize the character of τ:

	$\begin{pmatrix} a & 0 \\ 0 & a \end{pmatrix}$	$\begin{pmatrix} a & 1 \\ 0 & a \end{pmatrix}$	$\begin{pmatrix} a & 0 \\ 0 & b \end{pmatrix}$, $a \neq b$
τ	$p-1$	-1	0

What is the squared norm of this character? We find

$$\langle \chi_\tau, \chi_\tau \rangle = \frac{1}{p(p-1)^2}\Big((p-1)^2 \times (p-1) + (p-1) \times (p-1)\Big) = 1$$

so that it *is* indeed irreducible, of dimension $p-1$. This is just one representation, but we know that we can twist it using one-dimensional characters. For $\chi = \varrho(\chi_1, \chi_2)$ as in (4.41), we find that the character values of $\tau \otimes \chi$ are as follows.

	$\begin{pmatrix} a & 0 \\ 0 & a \end{pmatrix}$	$\begin{pmatrix} a & 1 \\ 0 & a \end{pmatrix}$	$\begin{pmatrix} a & 0 \\ 0 & b \end{pmatrix}$, $a \neq b$
$\tau \otimes \chi$	$(p-1)\chi_1(a)\chi_2(a)$	$-\chi_1(a)\chi_2(a)$	0

These are all irreducible representations, but they depend only on the product character $\chi_1\chi_2$ of $\mathbf{F}_p^\times$, and thus there are only $p-1$ different irreducible representations of dimension $p-1$ that arise in this manner.

We have now found the right number of representations. We summarize all this in the character table (Table 4.3), using the characters $\varrho(\chi, \mathbf{1})$ to obtain the $(p-1)$-dimensional representations.

Table 4.3. Character table of B_p

	$\begin{pmatrix} a & 0 \\ 0 & a \end{pmatrix}$	$\begin{pmatrix} a & 1 \\ 0 & a \end{pmatrix}$	$\begin{pmatrix} a & 0 \\ 0 & b \end{pmatrix}$, $a \neq b$
$\varrho(\chi_1, \chi_2)$	$\chi_1(a)\chi_2(a)$	$\chi_1(a)\chi_2(a)$	$\chi_1(a)\chi_2(b)$
$\tau \otimes \varrho(\chi, \mathbf{1})$	$(p-1)\chi(a)$	$-\chi(a)$	0

Exercise 4.6.6 (Dihedral groups)**.** The *dihedral* groups D_n, of order $2n$, form another well-known family of solvable groups. These can be defined either as the subgroup of isometries of $\mathbf{R}^2$ fixing (globally, not pointwise) a regular n-sided polygon centered at the origin or as the group generated by a normal cyclic subgroup $C_n \simeq \mathbf{Z}/n\mathbf{Z}$ of order 2 and an element $i \in D_n - C_n$ of order 2 such that

$$ixi^{-1} = ixi = x^{-1}$$

for $x \in C_n$ (geometrically, C_n corresponds to rotations, generated by the rotation of angle $2\pi/n$, and i to an orientation-reversing isometry).

(1) Find the character table for D_n. (*Hint*: They are slightly different when n is even or odd; see, e.g., [**53**, §5.2] for the details.)

(2) In the notation of Exercise 4.3.21, show that if $p \geqslant 0$ is such that

$$A_p(C_n) \leqslant A_p(D_n),$$

for n large enough, then necessarily $p \geqslant 1$.

(3) Assume that n is odd. Show the following: There exist distinct irreducible representations of D_n which are proportional on the non-trivial coset $Y = D_n - C_n$ of C_n in D_n. There exist irreducible representations with character identically zero on Y. There exist irreducible representations such that the invariant

$$\kappa(\pi) = |\{\psi \in \hat{A} \mid (\psi \circ \phi) \otimes \pi \simeq \pi\}|$$

is not equal to 1, where $\phi : D_n \longrightarrow A = D_n/C_n \simeq \mathbf{Z}/2\mathbf{Z}$ is the projection. (This provides the examples mentioned in Exercise 4.4.9.)

Exercise 4.6.7 (Two non-isomorphic groups with the same character table)**.** Consider the dihedral group $G_1 = D_4$ of order 8, and the group G_2 defined as the subgroup of the multiplicative group of the Hamilton quaternions generated by i, j and k, or in other words (for readers unfamiliar with quaternions) the group generated by symbols i, j, k, subject to the relations

$$i^2 = j^2 = k^2 = ijk = -1.$$

(1) Show that G_2 is of order 8 (by enumerating its elements for instance), and that it contains a single element of order 2. Deduce that G_2 is not isomorphic to G_1.

(2) Compute the character table of G_2. Show that, up to possible re-ordering of the rows and columns, *it is identical with that of* G_1.

(3) Deduce from this a few things about a finite group that the character table *cannot* determine (try to find as many things as possible that are different in G_1 and G_2; note that (1) already gives examples, and you should try to find others). For instance, can one determine all subgroups of a finite group from the character table and not just the normal ones?

4.6.4. A family of finite linear groups. The building blocks of all finite groups are, in some precise sense, the simple groups. We now consider the representations of the first infinite family of finite groups which are almost non-abelian simple groups: the linear groups $G_p = \mathrm{GL}_2(\mathbf{F}_p)$ for p prime (these are not simple, but for $q \notin \{2, 3\}$, the quotient $\mathrm{PSL}_2(\mathbf{F}_q) = \mathrm{SL}_2(\mathbf{F}_q)/\{\pm 1\}$ of the subgroup of elements with determinant 1 is simple). In contrast to the previous case, some of the irreducible representations that arise cannot easily be described at the level of actual actions of G_p on specific vector spaces. We will first identify them only as *characters*.

The whole computation is rather more involved than in the previous cases, as can be expected, and the reader should be active in checking the details. For other fairly detailed accounts along the same lines, see [**20**, §5.2] or [**18**, §4.24], and for treatments from slightly different perspectives, see [**11**, §5.1] or [**48**]. We note however that these character tables were already known to Frobenius.

We begin in the usual way by finding the size of G_p and its conjugacy classes. For the first point, the number of elements of G_p is the same as the number of bases of the plane $\mathbf{F}_p^2$, i.e.,

$$|G_p| = (p^2-1)(p^2-p) = p(p-1)^2(p+1)$$

(there are p^2-1 choices for the first basis vector, and then the second may be any vector except the p which are linearly dependent on the first one).

We will assume that $p \geqslant 3$ (and briefly mention the simpler case of $\mathrm{GL}_2(\mathbf{F}_2)$ in a final remark). Determining the conjugacy classes is a question of linear algebra over $\mathbf{F}_p$, and we can argue using the characteristic polynomial of a given element $g \in G_p$.

- *Scalar classes.* If g has a multiple eigenvalue in $\mathbf{F}_p$, but is diagonalizable, then g is a scalar matrix

$$g = \begin{pmatrix} a & 0 \\ 0 & a \end{pmatrix}$$

  for some $a \in \mathbf{F}_p^\times$. There are $p-1$ such matrices, and each is a conjugacy class of size 1; the set of these matrices is the center Z of G_p.

- *Non-semisimple classes.* If g has a multiple eigenvalue but is *not* diagonalizable, we can find a basis of $\mathbf{F}_p^2$ in which g has the form

$$g = \begin{pmatrix} a & 1 \\ 0 & a \end{pmatrix}$$

  (first we can conjugate g to triangular form, but then the argument of the previous section gives a conjugate as above). There are $p-1$ such conjugacy classes, and to compute their size we leave it to the reader to check that the centralizer of such a matrix g is the subgroup

$$K = \left\{ \begin{pmatrix} x & t \\ 0 & x \end{pmatrix} \mid x \in \mathbf{F}_p^\times, t \in \mathbf{F}_p \right\}$$

  so that the size of the conjugacy class of g is $|G_p/K| = p^2-1$.

- *Split semisimple classes.* If g has two distinct eigenvalues *in* $\mathbf{F}_p$, it is diagonalizable over $\mathbf{F}_p$, i.e., it is conjugate to a matrix

$$g = \begin{pmatrix} a & 0 \\ 0 & b \end{pmatrix}$$

with $a \neq b$. However, there are only $\frac{1}{2}(p-1)(p-2)$ such classes because one can permute a and b by conjugating with

(4.42) $$w = \begin{pmatrix} 0 & 1 \\ 1 & 0 \end{pmatrix},$$

and each of these classes has size $|G_p|/(p-1)^2 = p(p+1)$ because the centralizer of g as above is easily checked to be the group

(4.43) $$T_1 = \left\{ \begin{pmatrix} x & 0 \\ 0 & y \end{pmatrix} \mid x,\ y \neq 0 \right\}$$

of diagonal (not necessarily scalar) matrices, which is isomorphic obviously to $\mathbf{F}_p^\times \times \mathbf{F}_p^\times$, and is of order $(p-1)^2$.

- *Non-split semisimple classes*. Finally, if g has two distinct eigenvalues, but they do not belong to the base field $\mathbf{F}_p$, the matrix can be diagonalized, but only over the extension field $k/\mathbf{F}_p$ generated by the eigenvalues. This is necessarily the unique (up to isomorphism) extension field of degree 2. It is generated by some element α such that $\varepsilon = \alpha^2$ is a fixed *non-square* in $\mathbf{F}_p^\times$ (the existence of ε uses the assumption $p \geqslant 3$; for $p \equiv 3 \pmod 4$, for instance, one may take $\varepsilon = -1$). We will write, by convention $\alpha = \sqrt{\varepsilon}$. Once ε is fixed, one shows (see the remark below) that g is conjugate to a matrix

(4.44) $$g = \begin{pmatrix} a & b \\ \varepsilon b & a \end{pmatrix} \in G_p$$

for some $a \in \mathbf{F}_p$ and $b \in \mathbf{F}_p^\times$. However, as before, the number of classes of this type is only $\frac{1}{2}p(p-1)$ (changing b into $-b$ does not change the conjugacy class).

By direct computation, one checks that the centralizer of an element g of this type is equal to the subgroup

(4.45) $$T_2 = \left\{ \begin{pmatrix} x & y \\ \varepsilon y & x \end{pmatrix} \mid (x,y) \neq (0,0) \right\}$$

of order p^2-1, so that the conjugacy classes are of size $p(p-1)$. We observe that T_2 is not at all a complicated group: it is abelian, and in fact the map

(4.46) $$\begin{cases} T_2 & \longrightarrow \quad \mathbf{F}_p(\sqrt{\varepsilon})^\times \\ \begin{pmatrix} x & y \\ \varepsilon y & x \end{pmatrix} & \mapsto \quad x + y\sqrt{\varepsilon} \end{cases}$$

is an isomorphism. The determinant on T_2 corresponds, under this isomorphism, to the map sending $x + y\sqrt{\varepsilon}$ to $x^2 - \varepsilon y^2$, which is the norm homomorphism $\mathbf{F}(\sqrt{\varepsilon})^\times \to \mathbf{F}_p^\times$ (this is because $\alpha^p = -\alpha$, since $\alpha = \sqrt{\varepsilon}$ generates the extension of $\mathbf{F}_p$ of degree 2).

Remark 4.6.8. (1) Here is the argument to obtain the representative (4.44) of a non-split semisimple conjugacy class. Let $k = \mathbf{F}(\sqrt{\varepsilon})$ be the quadratic extension of $\mathbf{F}$. We need the automorphism

$$\begin{cases} k & \longrightarrow & k \\ a + b\sqrt{\varepsilon} & \mapsto & a - b\sqrt{\varepsilon}, \end{cases}$$

which we denote by $x \mapsto \bar{x}$ for $x \in k$. We extend this map to k^2 by applying it coordinatewise. Observe that $v \in k^2$ is in $\mathbf{F}_p^2$ if and only if $\bar{v} = v$.

Let g be an element of the conjugacy class. There exist $a, b \in \mathbf{F}$ such that the eigenvalues of g are λ and $\bar{\lambda}$; indeed, let v denote an eigenvector of g for an eigenvalue $\lambda = a + b\sqrt{\varepsilon} \in k$. We have $b \neq 0$, since $\lambda \notin \mathbf{F}_p$. Then, because the coordinates of g are in $\mathbf{F}_p$, we have $g\bar{v} = \overline{gv} = \bar{\lambda}\bar{v}$, so that $\bar{v}$ is an eigenvector for $\bar{\lambda} \neq \lambda$, and hence $(v, \bar{v})$ is a basis of k^2 with respect to which g acts diagonally.

Now let $e = v + \bar{v}$ and $f = \frac{1}{\sqrt{\varepsilon}}(v - \bar{v})$. Using the criterion above, we see that $e \in \mathbf{F}_p^2$ and $f \in \mathbf{F}_p^2$. Solving for v and $\bar{v}$ in terms of e and f, we next see that (e, f) is a basis of $\mathbf{F}_p^2$, and that the matrix of g with respect to this basis is

$$\begin{pmatrix} a & b \\ \varepsilon b & a \end{pmatrix},$$

which means that g is, as claimed, conjugate to this matrix. For instance, we leave to the reader to check that

$$ge = g(v + w) = \frac{\lambda + \bar{\lambda}}{2}e + \sqrt{\varepsilon}\frac{\lambda - \bar{\lambda}}{2}f = ae + b\varepsilon f,$$

which gives the first column.

(2) Note the close formal similarity between the group T_2 and the group

$$\left\{\begin{pmatrix} x & y \\ -y & x \end{pmatrix} \mid x,\ y \in \mathbf{R},\ (x, y) \neq (0, 0)\right\} \subset \mathrm{GL}_2(\mathbf{R}),$$

which is isomorphic to $\mathbf{C}^\times$ by mapping an element as above to $x + iy$. Here, of course, the real number -1, which is not a square in $\mathbf{R}$, plays the role of ε. If $y \neq 0$, the corresponding matrix in $\mathrm{GL}_2(\mathbf{R})$ is not diagonalizable over $\mathbf{R}$, but it is over $\mathbf{C}$, with conjugate eigenvalues $x \pm iy$.

Tallying all this, we see that the number of conjugacy classes is

$$|G_p^\sharp| = 2(p - 1) + \tfrac{1}{2}(p - 1)(p - 2) + \tfrac{1}{2}p(p - 1) = p^2 - 1.$$

Remark 4.6.9. The terminology we used for these four types of conjugacy classes is standard. Note that the fourth type (non-split semisimple classes) did not appear in the previous section, whereas the first three intersect the upper-triangular subgroup B_p. Of course, one must be careful to avoid confusion with the other meaning of semisimple in representation theory....

It will not be very difficult to find three *families* of irreducible representations, using the type of methods that were successful in the previous sections. Once this is done, we will see what is missing.

First, the commutator group of G_p is $\mathrm{SL}_2(\mathbf{F}_p)$ (see, e.g., [**40**, XIII, Th. 8.3], or Exercise 4.7.3 below), and thus the determinant gives an isomorphism

$$\det\,:\,G_p/[G_p,G_p]\longrightarrow\mathbf{F}_p^\times,$$

so that we have $p-1$ characters of dimension 1 given by

$$\chi(g)=\chi_1(\det(g))$$

for some character χ_1 of $\mathbf{F}_p^\times$.

The next construction is based on induction. We use induction of the one-dimensional characters of the subgroup B_p of upper-triangular matrices (that appeared in Section 4.6.3). These are given by

$$\varrho(\chi_1,\chi_2)\,:\,\begin{pmatrix}a&t\\0&b\end{pmatrix}\mapsto\chi_1(a)\chi_2(b),$$

where χ_1 and χ_2 are again characters of $\mathbf{F}_p^\times$. We denote

$$\pi(\chi_1,\chi_2)=\mathrm{Ind}_{B_p}^{G_p}(\varrho(\chi_1,\chi_2)),$$

which has dimension $p+1$. To compute the character of this representation, we use the set of representatives

$$R=\left\{\begin{pmatrix}1&0\\t&1\end{pmatrix}\mid t\in\mathbf{F}_p\right\}\cup\left\{\begin{pmatrix}0&1\\1&0\end{pmatrix}\right\}$$

of the cosets $B_p\backslash G_p$. Thus, for $\pi=\pi(\chi_1,\chi_2)$ we have

$$\chi_\pi(g)=\sum_{\substack{r\in R\\ rgr^{-1}\in B_p}}\chi_1(a_r)\chi_2(b_r),\tag{4.47}$$

where we write a_r and b_r for the diagonal coefficients of rgr^{-1}. Before considering the four conjugacy types in turn, we observe the following very useful fact. For any $x\in G_p$, we have

$$xB_px^{-1}\cap B_p=\begin{cases}B_p \text{ if } x\in B_p,\\ \{g\mid g \text{ diagonal in the basis } (e_1,xe_1)\} \text{ if } x\notin B_p.\end{cases}\tag{4.48}$$

Indeed, by definition, an element of B_p has the first basis vector $e_1\in\mathbf{F}_p^2$ as eigenvector, and an element of xB_px^{-1} has xe_1 as eigenvector. If xe_1 is not proportional to e_1 (i.e., if $x\notin B_p$) this means that $g\in B_p\cap xB_px^{-1}$ if (and only if) g is diagonalizable in the fixed basis (e_1,xe_1). (Note that in this case, the intersection is a specific conjugate of the group T_1 of diagonal matrices.)

Now we compute the following:

- If $g = \begin{pmatrix} a & 0 \\ 0 & a \end{pmatrix}$ is scalar, we obtain $\chi_\pi(g) = (p+1)\chi_1(a)\chi_2(a)$.
- If $g = \begin{pmatrix} a & 1 \\ 0 & a \end{pmatrix}$ is not semisimple, then since $g \in B_p$, only $1 \in R$ contributes to the sum (since for $r \neq 1$ to contribute, it would be necessary that $g \in r^{-1}B_p r \cap B_p$, which is not possible by (4.48)). Thus we get

$$\chi_\pi(g) = \chi_1(a)\chi_2(a)$$

  in that case.
- If $g = \begin{pmatrix} a & 0 \\ 0 & b \end{pmatrix}$ with $a \neq b$, besides $r = 1$, the other contributions must come from $r \in R$ such that g is diagonal in the basis (e_1, re_1), which is only possible if $re_1 = e_2$, i.e., the only other possibility is the matrix $r = w$ in (4.42). This gives

$$\chi_\pi(g) = \chi_1(a)\chi_2(b) + \chi_1(b)\chi_2(a)$$

  for the split semisimple elements.
- If g is non-split semisimple, it has no conjugate at all in B_p (as this would mean that g has an eigenvalue in $\mathbf{F}_p$), and hence $\chi_\pi(g) = 0$.

The character values (shown in Table 4.4) are therefore quite simple.

Table 4.4. Character of $\pi(\chi_1, \chi_2)$

	$\begin{pmatrix} a & 0 \\ 0 & a \end{pmatrix}$	$\begin{pmatrix} a & 1 \\ 0 & a \end{pmatrix}$	$\begin{pmatrix} a & 0 \\ 0 & b \end{pmatrix}$	$\begin{pmatrix} a & b \\ \varepsilon b & a \end{pmatrix}$
$\pi(\chi_1, \chi_2)$	$(p+1)\chi_1(a)\chi_2(a)$	$\chi_1(a)\chi_2(a)$	$\chi_1(a)\chi_2(b)+\chi_2(b)\chi_1(a)$	0

It is now a straightforward computation to determine the squared norm of the character of these induced representations. We find

$$\langle \chi_\pi, \chi_\pi \rangle = \frac{1}{|G_p|}\Big\{(p-1)(p+1)^2 + (p-1)(p^2-1) + A\Big\},$$

where A is the contribution of the split semisimple classes, namely

$$A = \frac{p(p+1)}{2} \sum\sum_{\substack{a,b \in \mathbf{F}_p^\times \\ a \neq b}} |\chi_1(a)\chi_2(b) + \chi_1(b)\chi_2(a)|^2.$$

To compute A, one can expand the square, obtaining

$$A = \frac{p(p+1)}{2}\Big(2(p-1)(p-2) + 2\,\mathrm{Re}(B)\Big)$$

with

$$B = \sum_{\substack{a,b \in \mathbf{F}_p^\times \\ a \neq b}} \sum \chi_1(a)\overline{\chi_1(b)\chi_2(a)}\chi_2(b)$$
$$= \sum_{a,b} \sum \chi_1(a)\overline{\chi_1(b)\chi_2(a)}\chi_2(b) - (p-1)$$
$$= \Big(\sum_{x \in \mathbf{F}_p^\times} \chi_1(x)\overline{\chi_2(x)}\Big)^2 - (p-1).$$

Thus there are two cases: if $\chi_1 = \chi_2$, we have $B = (p-1)^2 - (p-1)$, whereas if $\chi_1 \neq \chi_2$, we get $B = -(p-1)$. This leads, if no mistake is made in gathering all the terms, to

$$\langle \pi(\chi_1, \chi_2), \pi(\chi_1, \chi_2) \rangle = \begin{cases} 2 & \text{if } \chi_1 = \chi_2 \\ 1 & \text{if } \chi_1 \neq \chi_2. \end{cases} \tag{4.49}$$

Thus $\pi(\chi_1, \chi_2)$ is irreducible if and only $\chi_1 \neq \chi_2$ (see Exercise 4.8.3 for another argument toward this result, which is less computational). This means that we have found many irreducible representations of dimension $p+1$. The precise number is $\frac{1}{2}(p-1)(p-2)$, because in addition to requiring $\chi_1 \neq \chi_2$, we must remove the possible isomorphisms between those representations, and the character values show that if $\chi_1 \neq \chi_2$, we have

$$\pi(\chi_1, \chi_2) \simeq \pi(\chi_1', \chi_2')$$

if and only if

$$(\chi_1', \chi_2') = (\chi_1, \chi_2) \text{ or } (\chi_1', \chi_2') = (\chi_2, \chi_1).$$

These $\frac{1}{2}(p-1)(p-2)$ representations are called the *principal series* representations for $\mathrm{GL}_2(\mathbf{F}_p)$.

Remark 4.6.10. The existence of the isomorphism $\pi(\chi_1, \chi_2) \simeq \pi(\chi_2, \chi_1)$ is guaranteed by the equality of characters. It is not immediate to *write down* an explicit isomorphism. (Note that, by Schur's Lemma, we have $\dim \mathrm{Hom}_{G_p}(\pi(\chi_1, \chi_2), \pi(\chi_2, \chi_1)) = 1$, so at least the isomorphism is unique, up to scalar; for an actual description, see, e.g., [**11**, p. 404].)

Even when $\chi_1 = \chi_2$, we are not far from having an irreducible representation: since

$$\langle \pi(\chi_1, \chi_1), \pi(\chi_1, \chi_1) \rangle = 2,$$

the induced representation has two irreducible components. Could it be that one of them is one-dimensional? Using Frobenius reciprocity, we see that for a one-dimensional character of the type $\chi \circ \det$, we have

$$\langle \pi(\chi_1, \chi_1), \chi \circ \det \rangle_{G_p} = \langle \varrho(\chi_1, \chi_1), \varrho(\chi, \chi) \rangle_{B_p} = \begin{cases} 0 & \text{if } \chi_1 \neq \chi \\ 1 & \text{if } \chi_1 = \chi, \end{cases}$$

since the restriction of $\chi \circ \det$ to B_p is

$$\begin{pmatrix} a & t \\ 0 & b \end{pmatrix} \mapsto \chi(a)\chi(b).$$

Switching notation, we see that $\pi(\chi, \chi)$ contains a unique one-dimensional representation, which is $\chi \circ \det$. Its other component, denoted $\mathrm{St}(\chi)$, is irreducible of dimension p, with character values given by

$$\chi_{\mathrm{St}(\chi)} = \chi_{\pi(\chi,\chi)} - \chi \circ \det,$$

namely the following.

	$\begin{pmatrix} a & 0 \\ 0 & a \end{pmatrix}$	$\begin{pmatrix} a & 1 \\ 0 & a \end{pmatrix}$	$\begin{pmatrix} a & 0 \\ 0 & b \end{pmatrix}$, $b \neq a$	$\begin{pmatrix} a & b \\ \varepsilon b & a \end{pmatrix}$, $b \neq 0$
$\mathrm{St}(\chi)$	$p\chi(a)^2$	0	$\chi(a)\chi(b)$	$-\chi(a^2 - \varepsilon b^2)$

From this, we see also that these representations are pairwise non-isomorphic (use the values for split semisimple elements).

Another description of these representations, which are called the *Steinberg* representations, is the following. First G_p acts on the set X_p of lines in $\mathbf{F}_p^2$, as did B_p in the previous section. By linear algebra, this action is doubly transitive (choosing two non-zero vectors on a pair of distinct lines gives a basis of $\mathbf{F}_p^2$, and any two bases can be mapped to one another using G_p), and therefore by Proposition 4.3.17, the permutation representation associated to X_p splits as

$$\mathbf{1} \oplus \mathrm{St}$$

for some irreducible representation St of dimension p. Then we get

$$\mathrm{St}(\chi) \simeq \mathrm{St} \otimes \chi(\det)$$

(e.g., because the permutation representation on X_p is isomorphic to the induced representation $\mathrm{Ind}_{B_p}^{G_p}(\mathbf{1}) = \pi(\mathbf{1}, \mathbf{1})$, as in Example 2.6.4(2), so that St is the same as $\mathrm{St}(\mathbf{1})$, and then one can use character values to check the effect of multiplying with $\chi \circ \det$). The character of "the" Steinberg representation is particularly nice:

	$\begin{pmatrix} a & 0 \\ 0 & a \end{pmatrix}$	$\begin{pmatrix} a & 1 \\ 0 & a \end{pmatrix}$	$\begin{pmatrix} a & 0 \\ 0 & b \end{pmatrix}, b \neq a$	$\begin{pmatrix} a & b \\ \varepsilon b & a \end{pmatrix}, b \neq 0$
St	p	0	1	-1

To summarize, we have found

$$p-1, \qquad \tfrac{1}{2}(p-1)(p-2), \qquad p-1$$

irreducible representations of dimension

$$1, \qquad p+1, \qquad p,$$

respectively. There remains to find $\frac{1}{2}p(p-1)$ representations, with sum of squares of dimensions equal to

$$|G_p| - (p-1) - (p-1)p^2 - \tfrac{1}{2}(p-1)(p-2)(p+1)^2 = \tfrac{1}{2}p(p-1)(p-1)^2.$$

It seems therefore to be an excellent guess that the representations in question should be of dimension $p-1$. These will be the *cuspidal* representations of G_p (also sometimes called *discrete series* representations, in analogy with the discrete series representations of Lie groups, such as those we will see in Section 7.4).

Note already the striking parallel between the (known) rows and columns of the evolving character table: For the first three families of conjugacy classes, we have found families consisting of the same number of irreducible representations, all with a common dimension. We can therefore indeed expect to find a last family, which should correspond somehow to the non-split semisimple conjugacy classes of G_p. Another clear reason for the existence of a link with these conjugacy classes is that, for the moment, any combination of "known" irreducible characters is, when evaluated on a non-split semisimple class (4.44), a function of the determinant $\det(g) = a^2 - \varepsilon b^2$.

As it turns out, just as the induced representations $\pi(\chi_1, \chi_2)$ are parametrized by the pair (χ_1, χ_2), which can be interpreted as a character of the centralizer T_1 of a split semisimple conjugacy class (see (4.43)), the cuspidal representations are parametrized by certain characters ϕ of the common (abelian) centralizer T_2 of the representatives we use for the non-split semisimple classes, defined in (4.45). Here, we will just pull the formula out of a hat, as a class function, but in Exercise 4.6.21, we explain an actual construction of the corresponding representations (though it does remain mysterious). We identify

$$\phi : T_2 \longrightarrow \mathbf{C}^\times$$

with a character $\mathbf{F}_p(\sqrt{\varepsilon})^\times \longrightarrow \mathbf{C}^\times$ using the isomorphism (4.46), and define a function $R(\phi)$ by the following.

	$\begin{pmatrix} a & 0 \\ 0 & a \end{pmatrix}$	$\begin{pmatrix} a & 1 \\ 0 & a \end{pmatrix}$	$\begin{pmatrix} a & 0 \\ 0 & b \end{pmatrix}$, $b \neq a$	$\begin{pmatrix} a & b \\ \varepsilon b & a \end{pmatrix}$, $b \neq 0$
$R(\phi)$	$(p-1)\phi(a)$	$-\phi(a)$	0	$-(\phi(a+b\sqrt{\varepsilon}) + \phi(a-b\sqrt{\varepsilon}))$

We claim that these give us the missing characters, for suitable ϕ.

Proposition 4.6.11. *Let ϕ be a character of T_2, or equivalently of $\mathbf{F}_p(\sqrt{\varepsilon})^\times$, such that $\phi \neq \phi'$, where the character ϕ' is defined by*[11]

$$\phi'(x + y\sqrt{\varepsilon}) = \phi(x - y\sqrt{\varepsilon}). \tag{4.50}$$

Then $R(\phi)$ is an irreducible character of G_p. Moreover, we have

$$R(\phi_1) = R(\phi_2)$$

if and only if either $\phi_1 = \phi_2$ or $\phi_1 = \phi_2'$.

Once this is known, we have all the characters we need. Indeed, these characters are of dimension $p-1$. To count them, we note that condition (4.50) is equivalent with

$$\mathrm{Ker}(\phi) \supset \left\{ w \in \mathbf{F}_p(\sqrt{\varepsilon})^\times \mid w = \frac{x + y\sqrt{\varepsilon}}{x - y\sqrt{\varepsilon}} \right\}$$

(when seeing ϕ as a character of $\mathbf{F}_p(\sqrt{\varepsilon})^\times$) and the right-hand side is the same as the kernel of the norm map

$$\mathbf{F}_p(\sqrt{\varepsilon})^\times \longrightarrow \mathbf{F}_p^\times.$$

Thus those ϕ which *do* satisfy the condition are in bijection with the characters of the image of the norm map, which is $\mathbf{F}_p^\times$ since the norm is surjective. There are therefore $p-1$ characters to be excluded from the p^2-1 characters of T_2 (namely, those of the form

$$x + y\sqrt{\varepsilon} \mapsto \chi(x^2 - \varepsilon y^2)$$

where χ is a character of $\mathbf{F}_p^\times$). Finally, the identities $R(\phi) = R(\phi')$ show that the total number of irreducible characters given by the proposition is, as expected, $\frac{1}{2}p(p-1)$.

Proof of Proposition 4.6.11. We see first that the identity $R(\phi) = R(\phi')$ does hold, as an equality of class functions. Similarly, the restriction $\phi \neq \phi'$

[11] A better notation would be $\phi' = \phi^p$, since this is what the operation $a + b\sqrt{\varepsilon} \mapsto a - b\sqrt{\varepsilon} = (a + b\sqrt{\varepsilon})^p$ amounts to.

is a necessary condition for $R(\phi)$ to be an irreducible character, as we see by computing the square norm, which should be equal to 1. We have

$$\langle R(\phi), R(\phi)\rangle = \frac{1}{|G_p|}\Big\{(p-1)^3 + (p-1)(p^2-1) + \frac{1}{2}p(p-1)\sum_{\substack{a,b\in \mathbf{F}_p\\ b\neq 0}} |\phi(a+b\sqrt{\varepsilon}) + \phi(a-b\sqrt{\varepsilon})|^2\Big\},$$

and the last sum (rather like what happened for the induced representation $\pi(\chi_1, \chi_2)$) is equal to

$$\begin{aligned}\sum_{\substack{a,b\in \mathbf{F}_p\\ b\neq 0}} &|\phi(a+b\sqrt{\varepsilon}) + \phi(a-b\sqrt{\varepsilon})|^2\\ &= 2p(p-1) + 2\,\mathrm{Re}\Big(\sum_{\substack{a,b\in \mathbf{F}_p\\ b\neq 0}} \phi(a+b\sqrt{\varepsilon})\overline{\phi(a-b\sqrt{\varepsilon})}\Big)\\ &= 2p(p-1) + 2\,\mathrm{Re}\Big(\sum_{x\in \mathbf{F}_p(\sqrt{\varepsilon})^\times} \phi(x)\overline{\phi'(x)} - \sum_{a\in \mathbf{F}_p^\times} 1\Big)\\ &= 2(p-1)^2 + (p^2-1)\langle \phi, \phi'\rangle,\end{aligned}$$

where the last inner product refers to the group T_2. Thus we carefully find

$$\langle R(\phi), R(\phi)\rangle = 1 + \langle \phi, \phi'\rangle,$$

which is 1 if and only if $\phi \neq \phi'$.

This result, and similar checks (one may verify for instance that $R(\phi)$, as a class function, is orthogonal to all the irreducible characters previously known), show that $R(\phi)$ behaves like the character of an irreducible representation. But this strong evidence is not, by itself, conclusive: Although it shows that $R(\phi)$, when expanded into a combination of characters, must only involve the missing ones. This does not by itself guarantee that it is one itself.

This is something we noticed already in Remark 4.3.18; and as in that remark, since $R(\phi)$ has norm 1 and $R(\phi)$ takes positive value at 1, we see at least that in order to conclude, it is enough to exhibit a linear combination of characters with *integral* coefficients which is equal to $R(\phi)$. This we do as in [**20**, p. 70], although with even less motivation: we claim that

$$R(\phi) = \chi_1 - \chi_2 - \chi_3,$$

where χ_i is the character of the representation ϱ_i given by

$$\varrho_1 = \pi(\phi, \mathbf{1}) \otimes \mathrm{St}(\phi) \quad \text{(where } \phi \text{ is restricted to } \mathbf{F}_p^\times\text{)},$$
$$\varrho_2 = \pi(\phi, \mathbf{1}),$$
$$\varrho_3 = \mathrm{Ind}_{T_2}^{G_p}(\phi).$$

Checking this is a matter of computation. Note at least that the dimension

$$p(p+1) - (p+1) - [G_p : T_2] = p^2 - 1 - p(p-1) = p - 1$$

is correct, the reader should of course make sure of the other values. We only give the character of the induced representation $\mathrm{Ind}_{T_2}^{G_p}(\phi)$ to facilitate the check if needed:

	$\begin{pmatrix} a & 0 \\ 0 & a \end{pmatrix}$	$\begin{pmatrix} a & 1 \\ 0 & a \end{pmatrix}$	$\begin{pmatrix} a & 0 \\ 0 & b \end{pmatrix}, b \neq a$	$\begin{pmatrix} a & b \\ \varepsilon b & a \end{pmatrix}, b \neq 0$
$\mathrm{Ind}_{T_2}^{G_p}(\phi)$	$p(p-1)\phi(a)$	0	0	$\phi(a + b\sqrt{\varepsilon}) + \phi(a - b\sqrt{\varepsilon})$

This is especially easy to evaluate because T_2 only intersects conjugacy classes of central and of non-split semisimple elements. □

We are thus done computing this character table! To summarize, we present it in Table 4.5.

Table 4.5. Character table of $\mathrm{GL}_2(\mathbf{F}_p)$

	$\begin{pmatrix} a & 0 \\ 0 & a \end{pmatrix}$	$\begin{pmatrix} a & 1 \\ 0 & a \end{pmatrix}$	$\begin{pmatrix} a & 0 \\ 0 & b \end{pmatrix}$	$\begin{pmatrix} a & b \\ \varepsilon b & a \end{pmatrix}$
$\chi \circ \det$	$\chi(a^2)$	$\chi(a^2)$	$\chi(ab)$	$\chi(a^2 - \varepsilon b^2)$
$\pi(\chi_1, \chi_2)$	$(p+1)\chi_1(a)\chi_2(a)$	$\chi_1(a)\chi_2(a)$	$\chi_1(a)\chi_2(b)+$ $\chi_2(b)\chi_1(a)$	0
$\mathrm{St}(\chi)$	$p\chi(a^2)$	0	$\chi(ab)$	$-\chi(a^2 - \varepsilon b^2)$
$R(\phi)$	$(p-1)\phi(a)$	$-\phi(a)$	0	$-(\phi(a + b\sqrt{\varepsilon})+$ $\phi(a - b\sqrt{\varepsilon}))$

Remark 4.6.12. (1) For $p = 2$, the only difference is that there are no split semisimple conjugacy classes, and correspondingly no principal series (induced) representations. Indeed, $\mathrm{GL}_2(\mathbf{F}_2)$ is isomorphic to $\mathfrak{S}_3$ (an isomorphism is obtained by looking at the permutations of the three lines in $\mathbf{F}_2^2$ induced by an element of $\mathrm{GL}_2(\mathbf{F}_2)$), and the character table of the latter in Example 4.6.1 corresponds to the one above when we remove the third line and column: the two-dimensional representation of $\mathfrak{S}_3$ corresponds to

the (unique) Steinberg representation and the signature corresponds to the (unique) cuspidal representation of $\mathrm{GL}_2(\mathbf{F}_2)$.

(2) The restriction to $\mathrm{GL}_2(k)$ where k is a field of prime order was merely for convenience; all the above, and in particular the full character table, are valid for an arbitrary finite field k, with characters of $k^\times$, and of the group of invertible elements in its quadratic extension, instead of those of $\mathbf{F}_p^\times$ and $\mathbf{F}_{p^2}^\times$.

This computation of the character table of $\mathrm{GL}_2(\mathbf{F}_p)$ was somewhat involved. The following series of exercises shows some of the things that can be done once it is known.

Exercise 4.6.13 (Characters of $\mathrm{SL}_2(\mathbf{F}_p)$). The group $\mathrm{SL}_2(\mathbf{F}_p)$ is quite closely related to $\mathrm{GL}_2(\mathbf{F}_p)$, and one can compute the character table of one from that of the other.

(1) For $p \geqslant 3$, show that $\mathrm{SL}_2(\mathbf{F}_p)$ has $p+4$ conjugacy classes, and describe representatives of them.

(2) By decomposing the restriction to $\mathrm{SL}_2(\mathbf{F}_p)$ of the irreducible representations of $\mathrm{GL}_2(\mathbf{F}_p)$, describe the character table of $\mathrm{SL}_2(\mathbf{F}_p)$ for $p \geqslant 3$. (See, for instance, [**20**, §5.2] for the results; there are two irreducible representations of $\mathrm{GL}_2(\mathbf{F}_p)$ that decompose as a direct sum of two representations whose characters are quite tricky to compute, and you may try at first to just compute the dimensions of the irreducible components.)

(3) Show in particular that

$$\min_{\pi \neq \mathbf{1}} \dim \pi = \frac{p-1}{2}, \tag{4.51}$$

where π runs over all non-trivial irreducible (complex) representations of $\mathrm{SL}_2(\mathbf{F}_p)$.

(In Section 4.7.1, we will see some striking applications of the fact that this dimension is large, in particular that it tends to infinity as p does, and we will prove (4.51) more directly, independently of the computation of the full character table.)

(4) For $G = \mathrm{GL}_2(\mathbf{F}_p)$, $H = \mathrm{SL}_2(\mathbf{F}_p)$, $A = G/H \simeq \mathbf{F}_p^\times$, compute the invariant $\kappa(\pi)$ defined in (4.32) for all representations π of G.

Exercise 4.6.14 (The Gelfand–Graev representation). Let

$$U = \left\{ \begin{pmatrix} 1 & t \\ 0 & 1 \end{pmatrix} \mid t \in \mathbf{F}_p \right\} \subset \mathrm{GL}_2(\mathbf{F}_p).$$

This is a subgroup of $\mathrm{GL}_2(\mathbf{F}_p)$, isomorphic to the additive group of $\mathbf{F}_p$. Let $\psi \neq 1$ be a non-trivial irreducible character of U.

(1) Compute the character of $\varrho = \mathrm{Ind}_U^{\mathrm{GL}_2(\mathbf{F}_p)}(\psi)$, and show that it is independent of $\psi \neq 1$.

(2) Show that if π is an irreducible representation of $\mathrm{GL}_2(\mathbf{F}_p)$, the multiplicity $\langle \varrho, \pi \rangle$ of π in ϱ is either 0 or 1, and that it is equal to 1 if and only if $\dim \pi \geqslant 2$.

This representation ϱ is called the *Gelfand–Graev* representation of the group $\mathrm{GL}_2(\mathbf{F}_p)$. In concrete terms, the result means that for any irreducible representation

$$\pi : \mathrm{GL}_2(\mathbf{F}_p) \longrightarrow \mathrm{GL}(E),$$

of dimension at least 2, there exists a unique linear form

$$\ell_\pi : E \longrightarrow \mathbf{C}$$

(up to scalar) such that

$$\ell_\pi\Bigl(\pi\begin{pmatrix} 1 & x \\ 0 & 1 \end{pmatrix} v\Bigr) = \psi(x)\ell_\pi(v)$$

for $x \in \mathbf{F}_p$ and $v \in E$. (This is because such a linear form is exactly an element of $\mathrm{Hom}_U(\pi, \psi)$, which is isomorphic to $\mathrm{Hom}_G(\pi, \varrho)$ by Frobenius reciprocity; such a linear form is called a *Whittaker functional* for π.)

Using the specific isomorphism that implements Frobenius reciprocity, we find that given such a linear form $\ell_\pi \neq 0$, the homomorphism

$$\pi \longrightarrow \mathrm{Ind}_U^G(\psi)$$

is given by mapping a vector v to the function

$$W_v(g) = \ell_\pi(\varrho(g)v).$$

Exercise 4.6.15 (Distinct characters that coincide on a generating set). Show that the following can happen for some finite groups: there may exist a group G, a *generating set* S, and two irreducible (even faithful) representations ϱ_1 and ϱ_2 which are non-isomorphic but satisfy

$$\chi_{\varrho_1}(s) = \chi_{\varrho_2}(s)$$

for all $s \in S$.

Exercise 4.6.16 (Contragredient). Let ϱ be an irreducible representation of $\mathrm{GL}_2(\mathbf{F}_p)$. Let χ be the character of $\mathbf{F}_p^\times$ such that

$$\varrho\Bigl(\begin{pmatrix} x & 0 \\ 0 & x \end{pmatrix}\Bigr) = \chi(x)\mathrm{Id}$$

(the central character of ϱ). Show that the contragredient $\check{\varrho}$ of ϱ is isomorphic to the representation $\hat{\varrho}$ given by

$$\hat{\varrho}(g) = \chi(\det(g))^{-1}\varrho(g).$$

(*Hint*: This can be done without using the character table, by looking at what is the transpose of g^{-1}.)

Exercise 4.6.17 (Commutators in $\mathrm{GL}_2(\mathbf{F}_p)$)**.** Using Proposition 4.4.3, show that the *set* of commutators (not only the subgroup they generate!) in $\mathrm{GL}_2(\mathbf{F}_p)$ is equal to $\mathrm{SL}_2(\mathbf{F}_p)$ for $p \geqslant 3$.

Exercise 4.6.18. (1) For $p \geqslant 3$ and π an irreducible representation of $\mathrm{GL}_2(\mathbf{F}_p)$, show that there exists a constant $c_\pi \in \mathbf{C}$ and a character χ of $\mathbf{F}_p^\times$ such that

$$\chi_\pi\left(\begin{pmatrix} x & 1 \\ & x \end{pmatrix}\right) = c_\pi \chi(x)$$

for all $x \in \mathbf{F}_p^\times$ (this can be done without the character table).

(2) Let f denote the characteristic function of the set of all $g \in \mathrm{GL}_2(\mathbf{F}_p)$ that are diagonalizable over an algebraic closure of $\mathbf{F}_p$. Show that

$$\langle f, \chi_\pi \rangle = 0$$

for all except p of the irreducible representations of $\mathrm{GL}_2(\mathbf{F}_p)$. (*Hint*: Here you should probably use the character table.)

In the last exercise, we present one of the known constructions of the cuspidal representations as actual representations: they will arise as irreducible components of the *Weil representation*. This uses the following fact.

Proposition 4.6.19. *Let k be a field. Then $\mathrm{SL}_2(k)$ is isomorphic to the group with generators given by* symbols $\alpha(y)$ *for* $y \in k^\times$, $n(x)$ *for* $x \in k$ *and* w, *subject to the relations*

$$\begin{aligned} &\alpha(y_1)\alpha(y_2) = \alpha(y_1 y_2), \qquad n(x_1)n(x_2) = n(x_1 + x_2),\\ &\alpha(y)n(x)\alpha(y^{-1}) = n(y^2 x), \qquad w\alpha(y)w = \alpha(-y^{-1}),\\ &wn(x)w = \alpha(-x^{-1})n(-x)wn(-x^{-1}) \text{ for } x \neq 0, \end{aligned} \tag{4.52}$$

where y, y_1, $y_2 \in k^\times$, x, x_1, $x_2 \in k$.

Sketch of proof. Let G be the group defined by these generators and relations. One first checks that $\mathrm{SL}_2(k)$ is isomorphic to a quotient of G by the homomorphism given by

$$\alpha(y) \mapsto \begin{pmatrix} y & 0 \\ 0 & y^{-1} \end{pmatrix}, \quad n(x) \mapsto \begin{pmatrix} 1 & x \\ 0 & 1 \end{pmatrix}, \quad w \mapsto \begin{pmatrix} 0 & 1 \\ -1 & 0 \end{pmatrix}$$

(i.e., one must check that the matrices above satisfy the stated relations, and that they generate $\mathrm{SL}_2(k)$). Then one can prove that this homomorphism is an isomorphism by the simple expedient of constructing its inverse. The latter is given by

$$\begin{pmatrix} a & b \\ 0 & d \end{pmatrix} \mapsto n(a)\alpha(b/a)$$

and, for $c \neq 0$, by

$$\begin{pmatrix} a & b \\ c & d \end{pmatrix} \mapsto n(a/c)\alpha(-c^{-1})wn(d/c).$$

Set-theoretically, it is very easy to see that this is indeed an inverse: for instance, one has to verify that

$$\begin{pmatrix} 1 & a/c \\ 0 & 1 \end{pmatrix}\begin{pmatrix} -c^{-1} & 0 \\ 0 & -c \end{pmatrix}\begin{pmatrix} 0 & 1 \\ -1 & 0 \end{pmatrix}\begin{pmatrix} 1 & -d/c \\ 0 & 1 \end{pmatrix} = \begin{pmatrix} a & b \\ c & d \end{pmatrix},$$

and this is straightforward. One must however ensure that this set-theoretic inverse is in fact a homomorphism; this is again, in principle, a matter of computations, but it is more involved since one has to handle different cases. The generic one (when the bottom-left entries are non-zero) is done in detail in [**11**, p. 406], and the reader may wish to handle another one (at least) before taking the result for granted. □

Exercise 4.6.20. Let k be a field.

(1) Show that $\mathrm{SL}_2(k)$ is generated by the elements

$$\begin{pmatrix} 1 & x \\ 0 & 1 \end{pmatrix}, \quad \begin{pmatrix} 1 & 0 \\ x & 1 \end{pmatrix}$$

for $x \in k$. (*Hint*: Show first that w is in the group generated by these elements.)

(2) For $k = \mathbf{F}_p$, show that $\mathrm{SL}_2(\mathbf{F}_p)$ is generated by

$$\begin{pmatrix} 1 & 1 \\ 0 & 1 \end{pmatrix}, \quad \begin{pmatrix} 1 & 0 \\ 1 & 1 \end{pmatrix}.$$

Exercise 4.6.21 (The Weil representation)**.** Let p be an odd prime, and let $k = \mathbf{F}_p(\sqrt{\varepsilon})$ be (as before) the quadratic extension of $\mathbf{F}_p$. We denote by σ the automorphism

$$\sigma(x + y\sqrt{\varepsilon}) = x - y\sqrt{\varepsilon}$$

of k, and by

$$\mathrm{Tr}(x + y\sqrt{\varepsilon}) = 2x, \qquad N(x + y\sqrt{\varepsilon}) = x^2 - \varepsilon y^2,$$

the trace (resp. norm) map from k to $\mathbf{F}_p$ (resp. $k^\times$ to $\mathbf{F}_p^\times$).

Let V be the space of functions $k \longrightarrow \mathbf{C}$. Define the *Fourier transform* on V by

$$\hat{f}(x) = -\frac{1}{p}\sum_{y \in k} f(y)e\Big(\frac{\mathrm{Tr}(x\sigma(y))}{p}\Big),$$

where $e(z) = e^{2i\pi z}$, as in Remark 4.5.3. This is a linear map $V \longrightarrow V$.

(1) Show that there exists a representation of $\mathrm{SL}_2(\mathbf{F}_p)$ on V such that

$$\Bigl(\begin{pmatrix} y & 0 \\ 0 & y^{-1} \end{pmatrix} \cdot f\Bigr)(x) = f(yx),$$
$$\Bigl(\begin{pmatrix} 1 & t \\ 0 & 1 \end{pmatrix} \cdot f\Bigr)(x) = e\Bigl(\frac{tN(x)}{p}\Bigr) f(x),$$
$$\Bigl(\begin{pmatrix} 0 & 1 \\ -1 & 0 \end{pmatrix} \cdot f\Bigr)(x) = \hat{f}(x)$$

for $y \in \mathbf{F}_p^\times$ and $x \in \mathbf{F}_p$. (*Hint*: Use the presentation of $\mathrm{SL}_2(\mathbf{F}_p)$ by generators and relations above; this requires some non-trivial identities, which may be checked in [**11**, p. 408].)

(2) Let ϕ be a character of T_2 such that $\phi \neq \phi'$, as in Proposition 4.6.11. Define

$$V_\phi = \{f \in V \mid f(yx) = \phi(y)^{-1} f(x) \text{ for all } y \in k \text{ with } N(y) = 1\}.$$

Show that V_ϕ is a subrepresentation of V of dimension $p - 1$.

(3) Show that the representation of $\mathrm{SL}_2(\mathbf{F}_p)$ on V_ϕ extends to a representation of $\mathrm{GL}_2(\mathbf{F}_p)$ determined by the condition

$$\Bigl(\begin{pmatrix} y & 0 \\ 0 & 1 \end{pmatrix} \cdot f\Bigr)(x) = \phi(\tilde{y}) f(\tilde{y}x),$$

where $\tilde{y} \in k^\times$ satisfies $N(\tilde{y}) = y$.

(4) Show that this representation of $\mathrm{GL}_2(\mathbf{F}_p)$ on V_ϕ is isomorphic to the cuspidal representation $R(\phi)$. (*Hint*: Show that V_ϕ has the correct central character, and that $R(\phi)$ has no non-zero vector invariant under $\left\{\begin{pmatrix} 1 & t \\ 0 & 1 \end{pmatrix}\right\}$; then use the character table.)

4.6.5. The symmetric groups. The irreducible characters of the symmetric groups $\mathfrak{S}_n$, $n \geqslant 1$, were already essentially determined by Frobenius. Since then, there have been many different interpretations and variants of the construction and the subject remains a very lively topic of current research, both for its own sake and because of its many applications.

We will content ourselves here by stating a description of the irreducible representations in the language of *Specht modules*, but we will not give the proofs. There are many excellent detailed treatments in the literature, including those in [**20**, Ch. 4] or [**12**, §28] and the very concise version in [**16**, Ch. 7].

The conjugacy classes in the symmetric group $\mathfrak{S}_n$ are naturally classified by the *partitions* of n, corresponding to the cycle decomposition of a permutation $\sigma \in \mathfrak{S}_n$: if σ is written as the product of disjoint cycles of lengths $\lambda_1, \ldots, \lambda_k$ (with $k \leqslant n$ and $\lambda_i = 1$ corresponding to a fixed point of σ), then the conjugacy class of σ consists of all permutations which are products of disjoint cycles of the same lengths. These lengths satisfy the relation

$$\lambda_1 + \cdots + \lambda_k = n, \tag{4.53}$$

and conversely, for any $1 \leqslant k \leqslant n$ and for any positive integers $\lambda_1, \ldots, \lambda_k$ summing to n, we can construct a conjugacy class of permutations in $\mathfrak{S}_n$ as the product of cycles of these lengths taken successively. To normalize the set of lengths, one may order them in such a way that

$$\lambda_1 \geqslant \lambda_2 \geqslant \cdots \geqslant \lambda_k \geqslant 1, \tag{4.54}$$

and by "$\lambda = (\lambda_1, \ldots, \lambda_k)$ is a partition of n", we mean that the λ_i are integers satisfying (4.53) and (4.54).

Remark 4.6.22. This elementary fact associated to Theorem 4.2.5 gives us a way to compute directly the number of irreducible representations of $\mathfrak{S}_n$, at least for small n. For instance, there are 56 different irreducible representations of $\mathfrak{S}_{11}$. Note that we only described three of them up to now: the trivial representation, the signature $\varepsilon : \mathfrak{S}_n \longrightarrow \{\pm 1\}$, and the ten-dimensional irreducible subrepresentation of the permutation action on $\mathbf{C}^n$ (see Example 4.3.16).

For large n, denoting by $p(n)$ the number of partitions of n, it was proved by Hardy and Ramanujan that

$$p(n) \sim \frac{1}{4\sqrt{3}n} \exp\Big(\frac{2\pi}{\sqrt{6}}\sqrt{n}\Big), \qquad \text{as } n \to +\infty$$

(which means that the ratio of the two sides tends to 1 as $n \to +\infty$), and even stronger exact expansions are known (the first being due to Rademacher). These facts are however proved without much, if any, reference to the group-theoretic interpretation of $p(n)$ (see for instance [**30**, §20.1]).

We now describe a construction that associates a complex representation S^λ of $\mathfrak{S}_n$ to any partition λ of n. The main result is that these are irreducible representations, and that they are pairwise non-isomorphic. Since there are as many such representations as the number of conjugacy classes of $\mathfrak{S}_n$, it follows that the S^λ give all the irreducible representations of the symmetric group $\mathfrak{S}_n$. The construction proceeds in four steps.

(1) *The diagram of* λ. Given λ, the associated *Young diagram* is a graphical representation of λ. It contains k rows of boxes, where the i-th row contains λ_i boxes, and the rows are aligned on the left. For $n = 11$ and $\lambda = (4, 3, 3, 1)$, for instance, the diagram follows.

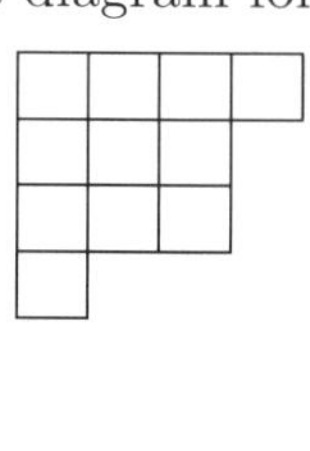

(2) *λ-tableaux.* A λ-tableau is obtained from the diagram of λ by filling each box with an integer from 1 to n, without repetitions. Hence there are $n!$ distinct λ-tableaux. For instance, the tableau

$$\begin{array}{|c|c|c|c|}\hline 3 & 7 & 1 & 11 \\ \hline 2 & 9 & 6 \\ \cline{1-3} 4 & 10 & 8 \\ \cline{1-3} 5 \\ \cline{1-1}\end{array}$$

is a $(4, 3, 3, 1)$-tableau.

(3) *λ-tabloids.* A *λ-tabloid* is an equivalence class of λ-tableaux for the equivalence relation $t_1 \sim t_2$ if and only if, for each i, the numbers in the i-th row of t_1 are the same as those in the i-th row of t_2. For instance, with t the tableau above, the first of the two tableaux below is equivalent to t, but the other is not.

$$\begin{array}{|c|c|c|c|}\hline 11 & 7 & 3 & 1 \\ \hline 2 & 6 & 9 \\ \cline{1-3} 4 & 8 & 10 \\ \cline{1-3} 5 \\ \cline{1-1}\end{array} \qquad \begin{array}{|c|c|c|c|}\hline 2 & 7 & 1 & 11 \\ \hline 3 & 9 & 6 \\ \cline{1-3} 4 & 10 & 8 \\ \cline{1-3} 5 \\ \cline{1-1}\end{array}$$

It is easy to see that the number of distinct λ-tabloids is equal to

$$\frac{n!}{\lambda_1! \cdots \lambda_k!}.$$

Indeed, $\mathfrak{S}_n$ acts simply transitively on the λ-tableaux (by permuting the integers in the boxes), and this action respects the equivalence relation, so $\mathfrak{S}_n$ acts transitively on the λ-tabloids. The stabilizer of any λ-tabloid is isomorphic to $\mathfrak{S}_{\lambda_1} \times \cdots \times \mathfrak{S}_{\lambda_k}$ (each factor permuting the integers in the corresponding row), and the formula follows.

(4) *The Specht module.* To define S^λ, first let M^λ be the permutation representation corresponding to the action of $\mathfrak{S}_n$ on the λ-tabloids. Thus we have

$$\dim M^\lambda = \frac{n!}{\lambda_1! \cdots \lambda_k!}.$$

Like any permutation representation, this is not irreducible. We construct S^λ as the subrepresentation of M^λ generated by the vectors defined by

$$e_t = \sum_{\sigma \in C_t} \varepsilon(\sigma)[\sigma \cdot t] \in M^\lambda,$$

where ε is the signature of a permutation, t runs over the λ-tableau and C_t is the subgroup of $\mathfrak{S}_n$ which permutes the integers in each *column* of the tableau. For instance, for the tableau t above, we see that C_t is the subgroup of order $4! \cdot 3! \cdot 3!$ of $\mathfrak{S}_{11}$ which permutes $(3, 2, 4, 5)$, $(7, 9, 10)$ and $(1, 6, 8)$, and fixes (11).

The subrepresentation S^λ generated by all e_t is in fact generated (as an $\mathfrak{S}_n$-module) by a single one of them, since one checks that

$$\sigma \cdot e_t = e_{\sigma \cdot t}.$$

Now we have

Theorem 4.6.23. *Each irreducible complex representation of $\mathfrak{S}_n$ is isomorphic to S^λ for a unique partition λ of n.*

For the proof, see for instance [**16**, §7.A]. It is remarkable that, in this important case, one has a *canonical* correspondence between the conjugacy classes of a group and its irreducible representations.

The following exercise illustrates some special examples of this construction.

Exercise 4.6.24. (1) Show that S^λ is the trivial representation for $\lambda = (n)$ with diagram given by a single row with n boxes:

(2) Show that S^λ is the signature homomorphism ε for $\lambda = (1, 1, \ldots, 1)$, with diagram given by n rows with one box per row:

(3) Show that for $\lambda = (n-1, 1)$, for $n \geqslant 2$, the Specht module S^λ is the irreducible subrepresentation (4.19) of dimension $n-1$ of the permutation representation associated to the standard action of $\mathfrak{S}_n$ on $\{1, \ldots, n\}$.

4.7. Applications

We present in this section some sample applications of the representation theory of finite groups, where the statements do not, by themselves, seem to depend on representations. The sections are independent of each other.

4.7.1. Quasirandom groups. Quite recently, Gowers [**23**] introduced a notion of *quasirandom* groups, motivated in part by similar ideas in the context of graph theory. Roughly speaking, a finite group G is said to be quasirandom when it has no non-trivial irreducible linear representation of *small* degree, where small must be interpreted in comparison with the order of G. Here is one of the simplest results that can be obtained in this area, showing one interesting consequence of this type of property:

Theorem 4.7.1 (Gowers; Nikolo and Pyber). *Let $G \neq 1$ be a non-trivial finite group, and let $k \geqslant 1$ be the smallest dimension of a non-trivial irreducible complex representation of G. For any subsets A, B, C in G such that*

$$\frac{|A||B||C|}{|G|^3} > \frac{1}{k}, \tag{4.55}$$

we have $ABC = G$ or, in other words, every *element $g \in G$ can be written as $g = abc$ with $a \in A$, $b \in B$ and $c \in C$.*

We use here the following product notation. For subsets A_1, $\ldots$, A_k of a group G (not necessarily distinct), the set $A_1 A_2 \cdots A_k$ is the set of all products

$$a = a_1 a_2 \cdots a_k$$

with $a_i \in A_i$ for all i. If some $A_i = \{a_i\}$ are singletons, we may just write the corresponding element a_i, e.g., in $a_1 A_2 a_3$. We also write A_1^{-1} for the set of all a^{-1} with $a \in A$.

It is also convenient to denote

$$\nu(A) = \frac{|A|}{|G|}$$

for $A \subset G$: this is the *density* of A in G. It can be interpreted intuitively as the probability that a *random element* in G belongs to A, and the hypothesis (4.55) of the theorem can be phrased as

$$\nu(A)\nu(B)\nu(C) > \frac{1}{k}. \tag{4.56}$$

Proof. The first step is due to Gowers [**23**, Lemma 5.1]: under the stated condition (4.56), we will show that $AB \cap C$ is not empty, i.e., that some $c \in C$ is of the form ab with $a \in A$ and $b \in B$.

To proceed with better motivation, fix only the two sets B and C. We try to find an upper bound on the size of the set D of those elements $g \in G$ such that the intersection

$$C \cap gB$$

is empty. Indeed, to say that a set A fails to satisfy $AB \cap C \neq \varnothing$ is to say that $A \subset D$, and if we know that D has a certain size, then it cannot contain any set of larger size.

The idea to control $|D|$ (or the density $\nu(D)$) is to look at the function

$$\varphi_{B,C} : g \mapsto \frac{|C \cap gB|}{|G|} = \nu(C \cap gB)$$

defined on G, and to show that it is non-zero on a relatively large set by finding an upper bound for its *variance*, i.e., the mean-square of $\varphi_{B,C}$ minus its average.

This average value is easy to determine. We have

$$\langle \varphi_{B,C}, 1\rangle = \frac{1}{|G|^2}\sum_{g\in G}|C\cap gB| = \frac{1}{|G|^2}\sum_{c\in C}\sum_{\substack{g\in G\\ c\in gB}} 1 = \frac{|B||C|}{|G|^2} = \nu(B)\nu(C),$$

since $c\in gB$ is equivalent with $g\in cB^{-1}$, which has order $|B|$. Hence, we wish to understand the quantity

$$\frac{1}{|G|}\sum_{a\in G}\Bigl(\varphi_{B,C}(a)-\nu(B)\nu(C)\Bigr)^2,$$

and if we know an upper-bound (say V) for it, we can argue by positivity[12] that the set X of those $g\in G$ with $\varphi_{B,C}=0$ (i.e., $C\cap gB=\varnothing$) satisfies

$$\frac{|X|}{|G|}(\nu(B)\nu(C))^2 = \frac{1}{|G|}\sum_{g\in X}\Bigl(\varphi_{B,C}(g)-\nu(B)\nu(C)\Bigr)^2 \leqslant V,$$

and in particular, if $A\subset G$ satisfies

$$\nu(A) > \frac{V}{(\nu(B)\nu(C))^2},$$

it must be the case that $\varphi_{B,C}$ is not identically zero on A, i.e., that $AB\cap C$ is not empty.

Now, in order to analyze $\varphi_{B,C}$, we observe that for any $g\in G$, we have

$$\varphi_{B,C}(g) = \frac{1}{|G|}\sum_{x\in G}\mathbf{1}_C(x)\mathbf{1}_{gB}(x) = \frac{1}{|G|}\sum_{x\in G}\mathbf{1}_C(x)\mathbf{1}_B(g^{-1}x),$$

where, for any subset $D\subset G$, we denote by $\mathbf{1}_D$ the characteristic function of D. In other words, defining $\psi(g)=\varphi_{B,C}(g^{-1})$, we have

$$\psi = \varrho_G(\Delta_C)\mathbf{1}_B,$$

where Δ_C is the element of the complex group algebra $\mathbf{C}(G)$ given by

$$\Delta_C = \frac{1}{|G|}\sum_{g\in C} g.$$

We now normalize ψ by subtracting the average, defining

$$\psi_0 = \psi - \langle\psi, 1\rangle = \varrho_G(\Delta_C)\mu_B, \qquad \mu_B = \mathbf{1}_B - \nu(B).$$

Our goal is then to bound from above the quantity

$$\langle\psi_0,\psi_0\rangle = \frac{1}{|G|}\sum_{g\in G}\Bigl(\psi(g)-\nu(B)\nu(C)\Bigr)^2 = \frac{1}{|G|}\sum_{a\in G}\Bigl(\varphi_{B,C}(a)-\nu(B)\nu(C)\Bigr)^2.$$

We do this by observing that $\varrho_G(\Delta_C)$ is a linear map acting on the subspace

$$C_0(G) = \{\varphi\in C(G)\mid \langle\varphi,1\rangle = 0\}\subset C(G),$$

[12] This is the trick known as *Chebychev's inequality* in probability theory.

and hence, by elementary Hilbert space theory, we have

$$\langle \psi_0, \psi_0 \rangle \leqslant \lambda^2 \langle \mu_B, \mu_B \rangle, \tag{4.57}$$

where $\lambda^2 \geqslant 0$ is the largest eigenvalue of the non-negative self-adjoint operator

$$\Delta_2 = \varrho_G(\Delta_C)^* \varrho_G(\Delta_C)$$

acting on $C_0(G)$, the adjoint $\varrho_G(\Delta_C)^*$ being computed for the inner product on $C(G)$.

We have not yet really used much representation theory. But here is the crux: Consider the λ-eigenspace of Δ_2, say $E \subset C_0(G)$. Then E is a subrepresentation of the left-regular representation, i.e., it is stable under the action of G such that

$$\lambda_G(g)\varphi(x) = \varphi(g^{-1}x),$$

for $\varphi \in C(G)$, simply because the two actions of G on itself by right and left multiplication commute. Since $\varrho_G(\Delta_C)$ is defined using right-multiplication, the operators $\lambda_G(g)$ commute with $\varrho_G(\Delta_C)$ and its adjoint, hence with Δ_2, and therefore stabilize its eigenspaces. Indeed, if $\Delta_2\varphi = \lambda\varphi$, we have

$$\Delta_2(\lambda_G(g)\varphi) = \lambda_G(g)\varrho_G(\Delta_2)\varphi = \lambda\lambda_G(g)\varphi.$$

Now our assumption shows that $\dim(E) \geqslant k$, because under λ_G, the invariant subspace of $C(G)$ is the space of constant functions, which is orthogonal to $C_0(G)$, so that λ_G cannot act trivially on any subspace of E. Thus the eigenvalues of Δ_2 have *large* multiplicity (if k is large).

How can this knowledge of the dimension of the eigenspace help bounding the eigenvalue? The point is that we can achieve some control of *all* the eigenvalues of Δ_2 using its trace, and because all eigenvalues are non-negative, we have

$$k\lambda^2 \leqslant (\dim E)\lambda^2 \leqslant \mathrm{Tr}(\Delta_2),$$

which we compute separately, using the relation

$$\varrho_G(g)^* = \varrho_G(g^{-1})$$

coming from unitarity, to obtain

$$\mathrm{Tr}(\Delta_2) = \frac{1}{|G|^2} \sum_{x,y \in G} 1_C(x) 1_C(y^{-1}) \,\mathrm{Tr}(\varrho_G(y^{-1}x)),$$

so that, by the character formula for the regular representation, we obtain

$$\mathrm{Tr}(\Delta_2) = \frac{1}{|G|} \sum_{\substack{x,y \in C \\ x=y}} 1 = \frac{|C|}{|G|} = \nu(C).$$

Thus we find an upper bound for λ^2, namely

$$\lambda^2 \leqslant \frac{\nu(C)}{k},$$

and hence by (4.57) we get

$$\frac{1}{|G|}\sum_{g\in G}\Big(\varphi_{B,C}(g)-\nu(B)\nu(C)\Big)^2 \leqslant \frac{\nu(C)}{k}\langle \mu_B,\mu_B\rangle.$$

But the last term is also easy to compute: we have

$$\begin{aligned}\langle \mu_B,\mu_B\rangle &= \langle 1_B,1_B\rangle - 2\nu(B)\langle 1_B,1\rangle + \nu(B)^2\\ &= \nu(B)(1-\nu(B)) \leqslant \nu(B),\end{aligned}$$

and therefore[13] the conclusion is

$$\frac{1}{|G|}\sum_{g\in G}\Big(\varphi_{B,C}(g)-\nu(B)\nu(C)\Big)^2 \leqslant \frac{\nu(B)\nu(C)}{k}.$$

Now the positivity argument shows that the number, say N, of those $g \in G$ with $C \cap gB = \varnothing$ satisfies

$$\nu(N) \leqslant \frac{1}{k\nu(B)\nu(C)}.$$

This gives the intermediate statement proved by Gowers: if A, B, C satisfy

$$\frac{|A||B||C|}{|G|^3} = \nu(A)\nu(B)\nu(C) > \frac{1}{k},$$

then the intersection $C \cap AB$ is not empty. Now we bootstrap this using a clever trick of Nikolov and Pyber [**47**, Prop. 1]: Consider again A, B, C as in the proposition, and *redefine*

$$C_1 = G - AB = \{g \in G \mid g \text{ is not of the form } ab \text{ with } a \in A,\ b \in B\}.$$

Then *by definition*, we have $C_1 \cap AB = \varnothing$. Using the negation of the result of Gowers, this means that we must have

$$\frac{|A||B||C_1|}{|G|^3} \leqslant \frac{1}{k},$$

and the assumption (4.56) now leads to

$$|C_1| < |C|.$$

This means that $|AB| + |C| > |G|$. Now for any $g \in G$, this means also that $|AB| + |gC^{-1}| = |AB| + |C| > |G|$. Therefore the sets AB and gC^{-1} must intersect; this precisely means that $g \in ABC$, and we are done. □

The next corollary does not mention representations at all.

---

[13] We could have kept the term $-\nu(B)^2$ to very slightly improve this estimate, but it does not seem to matter in any application.

Corollary 4.7.2 ($\mathrm{SL}_2(\mathbf{F}_p)$ is quasirandom). *If $p \geqslant 3$ is a prime number and $A \subset \mathrm{SL}_2(\mathbf{F}_p)$ is a subset such that*

$$|A| > 2^{1/3}p(p+1)(p-1)^{2/3} \tag{4.58}$$

or equivalently

$$\frac{|A|}{|\mathrm{SL}_2(\mathbf{F}_p)|} > \Big(\frac{2}{p-1}\Big)^{1/3},$$

then for any $g \in \mathrm{SL}_2(\mathbf{F}_p)$, there exist $a_1, a_2, a_3 \in A$ with $g = a_1a_2a_3$.

Proof. This follows from the proposition for $G = \mathrm{SL}_2(\mathbf{F}_p)$, where $|G| = p(p^2+p)$ and $B = C = A$, using (4.51), which shows that $k = \frac{1}{2}(p-1)$. □

This result shows that subsets A of $\mathrm{SL}_2(\mathbf{F}_p)$ satisfying (4.58) generate $\mathrm{SL}_2(\mathbf{F}_p)$, but in a very strong sense, since every element is the product of at most three elements of A. Results like this can be used to help with certain proofs of Theorem 1.2.5 in Chapter 1: to show that a very small subset such as $S = \{s_1, s_2\}$ (as in (1.1)) generates $\mathrm{SL}_2(\mathbf{F}_p)$ in at worst $C \log p$ steps, it suffices to find some C' such that the number of elements in $\mathrm{SL}_2(\mathbf{F}_p)$ obtained using products of $\leqslant C' \log p$ elements from S is $\geqslant 2(p+1)^{8/9}$, for instance. Indeed, if that is the case, the set A formed by these products satisfies the assumption of the corollary, and every element of G is the product of at most $3C' \log p$ elements from S.

Exercise 4.7.3. We indicate here how to prove that the minimal dimension of a (complex) non-trivial representation of $\mathrm{SL}_2(\mathbf{F}_p)$ is at least $(p-1)/2$ (see (4.51)), without invoking the full computation of the character table of $\mathrm{SL}_2(\mathbf{F}_p)$. Thus let $\varrho \neq \mathbf{1}$ be an irreducible representation of $\mathrm{SL}_2(\mathbf{F}_p)$, with $p \geqslant 3$ a prime.

(1) Show that one can assume that

$$T = \varrho\Big(\begin{pmatrix} 1 & 1 \\ 0 & 1 \end{pmatrix}\Big)$$

has an eigenvalue ξ which is a primitive p-th root of unity. (*Hint*: Use the fact that $\begin{pmatrix} 1 & 1 \\ 0 & 1 \end{pmatrix}$ and $\begin{pmatrix} 1 & 0 \\ 1 & 1 \end{pmatrix}$ generate $\mathrm{SL}_2(\mathbf{F}_p)$.)

(2) Show that, for all a coprime to p, ξ^{a^2} is also an eigenvalue of T. (*Hint*: Use a suitable conjugate of T.)

(3) Deduce that $\dim(\varrho) \geqslant (p-1)/2$.

(Note that it is only by constructing the cuspidal representations of $\mathrm{SL}_2(\mathbf{F})$ that it is possible to show that this bound is sharp, and also that if $\mathbf{F}_p$ is replaced with another finite field with q elements, of characteristic p, this argument does not give the correct lower bound $\frac{1}{2}(q-1)$.)

(4) Show that $\mathrm{SL}_2(\mathbf{F}_p)$ is perfect for $p \geqslant 5$ and deduce that the commutator subgroup of $\mathrm{GL}_2(\mathbf{F}_p)$ is equal to $\mathrm{SL}_2(\mathbf{F}_p)$ for $p \geqslant 3$.

The terminology "quasirandom" may seem mysterious at first, but it is well explained by the mechanism of the proof: the average of the function $\varphi_{B,C}$ corresponds precisely to the intuitive "probability" that an element $x \in G$ belongs to two subsets of density $|B|/|G|$ and $|C|/|G|$ if the corresponding conditions are genuinely random and independent. Hence, the fact that $\varphi_{B,C}$ is quite closely concentrated around its average value, when k is large, may be interpreted as saying that its elements and subsets behave as if they were random (in certain circumstances). The paper [**23**] of Gowers gives more examples of this philosophy.

To put the result in context, note that if $k = 1$ (for instance if G is abelian, or if $G = \mathrm{GL}_2(\mathbf{F}_p)$, which has many one-dimensional irreducible representations), the condition (4.56) cannot be satisfied unless $A = B = C = G$. And indeed a statement like Corollary 4.7.2 is completely false if, say, $G = \mathbf{Z}/p\mathbf{Z}$ with p large: for instance, if $A = B = C$ is the image modulo p of the set of integers $1 \leqslant n \leqslant \lfloor \frac{p}{3} \rfloor - 1$, we see that $A + B + C$ is not all of G, although the density of A is about $1/3$, for all p.

4.7.2. Burnside's two primes theorem. We prove here the theorem of Burnside mentioned in Chapter 1 (Theorem 1.2.7, known as "Burnside's $p^a q^b$ Theorem"): A finite group with order divisible by at most two distinct primes is necessarily solvable. The proof is remarkable, in that it does not depend on being able to write the character table of the group being investigated, but on subtler features about a finite group that may be found by looking at its irreducible characters. These are related to *integrality properties*, which have many other important applications.

The basic idea is to prove the following seemingly weaker statement:

Proposition 4.7.4 (Existence of normal subgroup). *Let G be a finite group of order $p^a q^b$ for some primes p and q and integers a, $b \geqslant 0$. If G is not abelian, it contains a normal subgroup $H \lhd G$ with $H \neq 1$ and $H \neq G$.*

To see that this implies Burnside's Theorem—that groups of order $p^a q^b$ are solvable—one can argue (for fixed primes $p \neq q$) by induction on $a+b$ for $a+b \geqslant 1$. The base case is when $a+b = 1$, in which case $|G|$ is prime, and in this case, G is cyclic, hence solvable. Now assume that Burnside's Theorem holds for non-trivial groups of order $|G| = p^c q^d$ with $1 \leqslant c+d < a+b$. The proposition shows that, either G is abelian (and therefore solvable), or there exists $H \lhd G$ such that $H \neq 1$ and $G/H \neq 1$. In that case we have an exact sequence

$$1 \longrightarrow H \longrightarrow G \longrightarrow G/H \longrightarrow 1,$$

and both H and G/H are non-trivial and have orders dividing $|G|$, and in particular divisible only (at most) by the primes p and q, with the sum of the exponents of p and q strictly smaller than $a+b$. By the induction hypothesis, they are therefore solvable, and this is well known to imply that G itself is solvable.

So we are reduced to a question of finding a non-trivial normal subgroup in a group G, one way or another, and Burnside's idea is to find it as the kernel of some suitable non-trivial irreducible representation

$$\varrho : G \longrightarrow \mathrm{GL}(E),$$

or of an associated homomorphism

$$\bar{\varrho} : G \xrightarrow{\varrho} \mathrm{GL}(E) \longrightarrow \mathrm{PGL}(E).$$

Indeed, it is a bit easier to ensure that $\operatorname{Ker}\bar{\varrho}$ is non-trivial (the kernel is enlarged modulo the scalars), and the possibility that $\operatorname{Ker}\bar{\varrho} = G$ is so special that its analysis is even simpler.

We will find the desired representation by means of the following result, which is itself of great interest:

Theorem 4.7.5 (Burnside). *Let G be a finite group,*[14] *and let*

$$\varrho : G \longrightarrow \mathrm{GL}(E)$$

be an irreducible complex representation of G. If $g \in G$ is such that its conjugacy class $g^\sharp \subset G$ has order coprime with $\dim \varrho$, then either $\chi_\varrho(g) = 0$ or $g \in \operatorname{Ker}\bar{\varrho}$, where $\bar{\varrho}$ is the composite homomorphism

$$G \xrightarrow{\varrho} \mathrm{GL}(E) \longrightarrow \mathrm{PGL}(E).$$

This may not be a result that is easy to guess or motivate, except that the statement may well come to mind after looking at many examples of character tables. For instance, in the case of the solvable groups B_p of order $p(p-1)^2$ (see Table 4.3 in Section 4.6.3), the characters of the irreducible representations of dimension $p-1$ vanish at all conjugacy classes of size $p-1$, and their values at conjugacy classes of size 1 are scalar matrices (hence in the kernel of $\bar{\varrho}$). Similarly, the character of the Steinberg representation of dimension p of $\mathrm{GL}_2(\mathbf{F}_p)$ is zero at all conjugacy classes of size p^2-1. (Note that if the reader did look, a striking fact would be remarkable: for any irreducible representation ϱ of dimension $\dim \varrho > 1$, there exists—or so it seems—some conjugacy class c with $\chi_\varrho(c) = 0$; this is indeed true, as we will explain in Remark 4.7.10 below....)

We can also check immediately that the statement of the theorem is true for conjugacy classes of size 1. This corresponds to elements of the center of

[14] Of any order.

G, for which $\varrho(g)$ is always a homothety for any irreducible representation ϱ (the central character, as in Corollary 2.7.17).

Proof of Proposition 4.7.4 using Theorem 4.7.5. Note first that we can certainly assume that $a \geqslant 1$ and $b \geqslant 1$, since a group of prime-power order has a non-trivial center (see, e.g., [**51**, Th. 4.4] for this basic feature of finite groups).

We attempt to find an element $g \in G$ and an irreducible representation ϱ so that Theorem 4.7.5 applies, while ensuring that the character value $\chi_\varrho(g)$ is non-zero. The difficulty is in ensuring the coprimality condition of $\dim(\varrho)$ with $|g^\sharp|$, and indeed this is where the assumption that $|G|$ is only divisible by two primes is important.

The following property holds for arbitrary finite groups and can be interpreted as one more attempt of conjugacy classes and irreducible representations to behave *dually* (see Remark 4.2.7):

Fact. Let $G \neq 1$ be a finite group, and let p, q be prime numbers. There exists a pair (g, ϱ), where $g \neq 1$ is an element of G, and $\varrho \neq \mathbf{1}$ is an irreducible complex representation of G, such that $\chi_\varrho(g) \neq 0$ and

$$p \nmid |g^\sharp|, \qquad q \nmid \dim(\varrho). \tag{4.59}$$

We first conclude the proof using this fact: the point is that if $|G| = p^a q^b$ with $a, b \geqslant 1$, then with g and ϱ as so conveniently given the conditions (4.59) mean that $|g^\sharp|$ and $\dim(\varrho)$ must be coprime (one does not need to know the—true—fact that $\dim(\varrho) \mid |G|$: the order of $g^\sharp$ does divide $|G|$, and hence must be a power of q, but q is coprime with $\dim(\varrho)$).[15] Hence we can apply Theorem 4.7.5 and conclude that $g \in \operatorname{Ker} \bar{\varrho}$, so that the latter is a *non-trivial* normal subgroup. The endgame is now straightforward. If $\operatorname{Ker} \bar{\varrho} \neq G$, this kernel is the required proper, non-trivial, normal subgroup, while otherwise the composition $\bar{\varrho}$ is trivial, and then ϱ takes scalar values, and must therefore be one-dimensional by irreducibility. We then get an isomorphism

$$G/\operatorname{Ker}(\varrho) \simeq \operatorname{Im}(\varrho) \subset \mathbf{C}^\times,$$

which shows in turn that $\operatorname{Ker} \varrho$ is a proper, non-trivial, normal subgroup... unless $G \simeq \operatorname{Im}(\varrho)$ is in fact abelian!

Now we prove the claim. We thus consider an arbitrary finite group G and primes p and q. We first show that there exists an element $g \neq 1$ in G such that $p \nmid |g^\sharp|$. For this, we note first that if the center of G is

[15] Of course, (4.59) does not exclude possibilities such as

$$|G| = pqr, \quad g^\sharp = qr, \quad \dim(\varrho) = pr,$$

where p, q, r are distinct primes; see Remark 4.7.6 for the case of the alternating group A_5.

non-trivial, then any $g \neq 1$ in the center satisfies $p \nmid |g^\sharp|$, since $g^\sharp = \{g\}$ in that case. Thus we may assume that the center is trivial. Then we use an averaging trick, which is familiar from arguments used in many proofs of Sylow's theorems: because G is partitioned into conjugacy classes, we have

$$\sum_{g^\sharp \in G^\sharp} |g^\sharp| = |G|.$$

Since $Z(G)$ is trivial by assumption, we obtain, by reduction modulo p and by isolating the contribution of the trivial conjugacy class of size 1, the identity

$$\sum_{\substack{g^\sharp \in G^\sharp \\ |g^\sharp| \neq 1}} |g^\sharp| \equiv -1 \,(\mathrm{mod}\, p).$$

In particular, one of the terms in the left-hand side must be non-zero modulo p, and this gives an element g with $p \nmid |g^\sharp|$.

The final step is to show that, given any $g \neq 1$, there exists some non-trivial irreducible representation ϱ with $q \nmid \dim \varrho$ and $\chi_\varrho(g) \neq 0$. Applying this to any of those g whose existence we just proved, this gives all of (4.59).

Given $g \neq 1$, the basic relation between the values of the irreducible characters at g and their dimensions is the orthogonality relation (4.28), which gives

$$\sum_{\varrho \in \hat{G}} \chi_\varrho(g)\overline{\chi_\varrho(1)} = \sum_{\varrho \in \hat{G}} (\dim \varrho)\chi_\varrho(g) = 0.$$

If we isolate, as usual, the contribution of the trivial representation, we find

$$\sum_{\varrho \neq \mathbf{1}} (\dim \varrho)\chi_\varrho(g) = -1, \tag{4.60}$$

which certainly tells us that there is some irreducible representation $\varrho \neq \mathbf{1}$ such that $\chi_\varrho(g) \neq 0$.

But even better, if we reduce this identity modulo the prime number q, it implies that there is some non-trivial irreducible representation with $\chi_\varrho(g) \neq 0$ and $q \nmid \dim \varrho$. This deduction relies on the fact that the values $\chi_\varrho(g)$ of irreducible characters, which are sums of roots of unity (the eigenvalues of $\varrho(g)$) are *algebraic integers*: modulo q, the right-hand side of (4.60) is non-zero, and some term in the sum is therefore not divisible by q.

Precisely, if it were the case that $q \mid \dim \varrho$ for all ϱ such that $\chi_\varrho(g) \neq 0$, we would get

$$-\frac{1}{q} = \sum_{\substack{\varrho \neq \mathbf{1} \\ q \mid \dim(\varrho)}} \Bigl(\frac{\dim(\varrho)}{q}\Bigr)\chi_\varrho(g), \tag{4.61}$$

where the right-hand side is an algebraic integer, and this is impossible since $1/q$ is not. In Section A.1 in the Appendix, we present a short discussion of the properties of algebraic integers that we use (here, Proposition A.1.1), and readers for whom this is not familiar may either read this now or continue while assuming that the character values involved are all actual integers in $\mathbf{Z}$, since in that case (4.61) is patently absurd. □

Remark 4.7.6 (Why A_5 is not solvable...). The simplest non-solvable group is the alternating group A_5 of order $60 = 2^2 \cdot 3 \cdot 5$ (one can show that all groups of order $30 = 2 \cdot 3 \cdot 5$—there are four up to isomorphism—are solvable). It is instructive to see *how* the argument fails in that case. The character table of A_5 is computed, e.g., in [**20**, §3.1, Ex. 3.5], and we just list it here in Table 4.6, subscripting the conjugacy classes with their sizes (the reader who has not seen it might think of finding natural linear actions corresponding to the representations displayed):

Table 4.6. Character table of A_5

	1_1	$(12)(34)_{15}$	$(123)_{20}$	$(12345)_{12}$	$(13452)_{12}$
$\mathbf{1}$	1	1	1	1	1
ϱ_3	3	-1	0	$\frac{1+\sqrt{5}}{2}$	$\frac{1-\sqrt{5}}{2}$
ϱ_3'	3	-1	0	$\frac{1-\sqrt{5}}{2}$	$\frac{1+\sqrt{5}}{2}$
ϱ_4	4	0	1	-1	-1
ϱ_5	5	1	-1	0	0

One can then list the pairs (g, ϱ) for which $\chi_\varrho(g) \neq 0$ and (4.59) holds, and see that in all cases, the greatest common divisor of $|g^\sharp|$ and $\dim(\varrho)$ is different from 1. (For instance, we have the pairs

$$(p, q) = (2, 3), \qquad (g^\sharp, \varrho) = ((12)(34), \varrho_4),$$

$$(p, q) = (3, 2), \qquad (g^\sharp, \varrho) = ((123), \varrho_5),$$

$$(p, q) = (2, 5), \qquad (g^\sharp, \varrho) = ((12)(34), \varrho_3 \text{ or } \varrho_3'),$$

and some others.)

We now come to the proof of Theorem 4.7.5. Here again, the basic ingredient is of independent interest, as it provides more subtle integrality properties of character values:

Proposition 4.7.7 (Divisibility). *Let G be a finite group, and let*

$$a = \sum_{g \in G} \alpha(g) g \in \mathbf{C}(G)$$

be an element of the group algebra with coefficients $\alpha(g)$ which are algebraic integers. *Moreover, assume $a \in Z(\mathbf{C}(G))$ is in the center of the group* *algebra or, equivalently, that α is a class function on G. Then a acts on irreducible representations by multiplication by scalars, and those are all algebraic integers. In particular, for any $g \in G$ and $\varrho \in \widehat{G}$, we have*

$$\dim(\varrho) \text{ divides } \chi_\varrho(g)|g^\sharp| \tag{4.62}$$

in the ring $\bar{\mathbf{Z}}$ of algebraic integers.

Proof. The action of a central element a on an irreducible representation ϱ is given by the scalar $\omega_\varrho(a)$ of Proposition 4.3.32, and so we must show that this is in $\bar{\mathbf{Z}}$ under the stated conditions.

Since, as a function of a, this scalar is a ring homomorphism and $\bar{\mathbf{Z}}$ is itself a ring (Proposition A.1.2), it is enough to prove the integrality of $\omega_\varrho(a)$ when a runs over a set of elements which span the subring $Z(\bar{\mathbf{Z}}(G))$ (defined as the center of the $\bar{\mathbf{Z}}$-module generated by the basis vectors g of $\mathbf{C}(G)$). For instance, one can take the elements

$$a_c = \sum_{g \in c} g,$$

where c runs over conjugacy classes in G. This means, in practice, that we may assume that the coefficients $\alpha(g)$ are in fact in $\mathbf{Z}$.

Under this condition, we consider the element $e_\varrho \in \mathbf{C}(G)$ giving the ϱ-isotypic projection. Using the left-multiplication action of G on the group ring, we have

$$ae_\varrho = \omega_\varrho(a)e_\varrho,$$

i.e., multiplication by a, as a map on $\mathbf{C}(G)$, has $\omega_\varrho(a)$ as an eigenvalue. We claim that this linear map

$$\Phi_a \begin{cases} \mathbf{C}(G) & \longrightarrow & \mathbf{C}(G) \\ x & \mapsto & ax \end{cases}$$

can be represented by an integral matrix in a suitable basis. In fact, the elements $x \in G$ form a basis in $\mathbf{C}(G)$ which does the job: we have

$$ax = \sum_{g \in G} \alpha(g)gx = \sum_{g \in G} \alpha(gx^{-1})g,$$

where the relevant coefficients, namely the $\alpha(gx^{-1})$, are indeed integers.

We conclude that $\omega_\varrho(a)$ is a root of the characteristic polynomial $\det(X - \Phi_a)$ of Φ_a, and if we use the basis above, we can see that this polynomial is monic with integral coefficients; hence, we deduce that $\omega_\varrho(a)$ is indeed an algebraic integer. (We are using here one part of the criterion in Proposition A.1.2.)

Now for the last part, we use the expression (4.26) for $\omega_\varrho(a)$ in the special case where $a = a_c$, which is

$$\omega_\varrho(a_c) = \frac{1}{\dim(\varrho)} \sum_{g \in c} \chi_\varrho(g) = \frac{|g^\sharp| \chi_\varrho(g)}{\dim(\varrho)},$$

and the fact that this is an algebraic integer is equivalent with the divisibility relation (4.62). □

Before using Proposition 4.7.7 to finish the proof of Theorem 4.7.5, the reader is probably tempted to apply the general fact that $\omega_\varrho(a)$ is an algebraic integer to other elements a. Doing this leads almost immediately to a proof of an observation we already mentioned (see Remark 4.3.7 for instance):

Proposition 4.7.8 (Dimensions of irreducible representations divide the order). *If G is a finite group and $\varrho \in \widehat{G}$ is an irreducible complex representation of G, then the dimension of ϱ divides $|G|$.*

Proof. We are looking for a suitable $a \in \mathbf{C}(G)$ to apply the proposition; since

$$\omega_\varrho(a) = \frac{1}{\dim(\varrho)} \sum_{g \in G} \alpha(g) \chi_\varrho(g),$$

the most convenient would be to have $\alpha(g)$ such that the sum is equal to $|G|$. But there does exist such a choice: by the orthogonality relation, we can take $\alpha(g) = \overline{\chi_\varrho(g)}$ and then

$$\omega_\varrho(a) = \frac{1}{\dim(\varrho)} \sum_{g \in G} \alpha(g) \chi_\varrho(g) = \frac{|G|}{\dim(\varrho)}.$$

Since $\alpha(g) \in \bar{\mathbf{Z}}$, this is indeed an algebraic integer, by the proposition. Hence

$$\frac{|G|}{\dim(\varrho)} \in \bar{\mathbf{Z}} \cap \mathbf{Q} = \mathbf{Z},$$

which is the desired result. □

We can finally finish:

Proof of Theorem 4.7.5. With (4.62) in hand, what to do is quite clear: The dimension $\dim(\varrho)$ divides the product

$$\chi_\varrho(g) |g^\sharp|,$$

and it is assumed that it is coprime with the second factor $|g^\sharp|$. So it must divide the character value $\chi_\varrho(g)$ (this happens in the ring $\bar{\mathbf{Z}}$ of algebraic integers, always; we are using Proposition A.1.6.)

Such a relation, we claim, is in fact equivalent with the conclusion of the theorem. This would again be clear if $\chi_\varrho(g)$ were in $\mathbf{Z}$, since the bound

$$|\chi_\varrho(g)| \leqslant \dim(\varrho)$$

and the divisibility $\dim(\varrho) \mid \chi_\varrho(g) \in \mathbf{Z}$ lead to

$$\chi_\varrho(g) \in \{-\dim(\varrho), 0, \dim(\varrho)\},$$

and we know that $|\chi_\varrho(g)| = \dim(\varrho)$ is equivalent with $\varrho(g)$ being a scalar matrix, i.e., $g \in \operatorname{Ker} \bar{\varrho}$ (Proposition 4.6.4).

To deal with the general case, we must be careful because if we have non-zero algebraic integers z_1, z_2 with

$$z_1 \mid z_2,$$

we cannot always conclude that $|z_1| \leqslant |z_2|$ (e.g., take $z_1 = 1$ and $z_2 = -1 + \sqrt{2}$). What we do is note that the divisibility relation implies that

$$\dim \varrho \mid z$$

for any conjugate z of $\chi_\varrho(g)$. Taking the product of these relations, we derive

$$\dim(\varrho)^r \mid N(\chi_\varrho(g)),$$

where r is the number of conjugates of $\chi_\varrho(g)$ (this is just Corollary A.1.9). We have now a divisibility relation among integers, and if $\chi_\varrho(g) \neq 0$, we deduce that inequality

$$\dim(\varrho)^r \leqslant |N(\chi_\varrho(g))|.$$

But since each conjugate z of $\chi_\varrho(g)$ is a sum of $\dim(\varrho)$ roots of unity,[16] it is of modulus $\leqslant \dim(\varrho)$. This implies

$$|N(\chi_\varrho(g))| \leqslant \dim(\varrho)^r,$$

and by comparison we must have equality in all the terms of the product, in particular

$$|\chi_\varrho(g)| = \dim(\varrho),$$

which, as before, gives $g \in \operatorname{Ker} \bar{\varrho}$. □

Remark 4.7.9. We used the divisibility relation (4.62) in the previous proof by assuming that $\dim(\varrho)$ is coprime with the factor $|g^\sharp|$ on the right-hand side. What happens if we assume instead that $\dim(\varrho)$ is coprime with the second factor, $\chi_\varrho(g)$? One gets the conclusion that $\dim(\varrho)$ divides the size of the conjugacy class of g. This is of some interest; in particular, if there exists some g with $\chi_\varrho(g) = \pm 1$ (or even $\chi_\varrho(g)$ a root of unity), we have

$$\dim(\varrho) \mid |g^\sharp|.$$

[16] In fact, any conjugate is also a character value $\chi_\varrho(x)$ for some $x \in G$, but checking this fact requires the Galois theoretic interpretation of the conjugates, which we do not wish to assume.

We can see this concretely in the Steinberg representations of $\mathrm{GL}_2(\mathbf{F}_p)$, of dimension p: the values at semisimple conjugacy classes are roots of unity, and indeed $\dim(\mathrm{St}) = p \mid p(p+1)$, $p(p-1)$, which are the sizes of the split (resp. non-split) semisimple classes. On the other hand, it is not clear if this dual statement has any interesting applications in group theory.

Remark 4.7.10 (Characters have zeros)**.** We come back to the following observation, which is certainly experimentally true for those groups for which we computed the character table:

Proposition 4.7.11 (Burnside)**.** *Let G be a finite group, and let $\varrho \in \widehat{G}$ be an irreducible representation of dimension at least 2. Then there exists some $g \in G$ such that $\chi_\varrho(g) = 0$.*

Proof. This is once again obvious if the character takes actual integer values in $\mathbf{Z}$: the orthonormality relation for ϱ gives

$$\frac{1}{|G|} \sum_{g \in G} |\chi_\varrho(g)|^2 = 1,$$

i.e., the mean-square average over G of the character of ϱ is 1. Hence either $|\chi_\varrho(g)|^2 = 1$ for all g, which can only happen when $\dim(\varrho) = 1$, or else some element g must have

$$|\chi_\varrho(g)| < 1,$$

which gives immediately $\chi_\varrho(g) = 0$ if $\chi_\varrho(g) \in \mathbf{Z}$.

In the general case, we must again be careful, since there are many non-zero algebraic integers z with $|z| < 1$ (e.g., $-1 + \sqrt{2}$). However, one can partition G into subsets for which the sum of the character values is an actual integer. To be precise, we write G as the union of the equivalence classes for the relation defined by $x \sim y$ if and only if x and y generate the same (finite cyclic) subgroup of G. Hence

$$\sum_{g \in G} |\chi_\varrho(g)|^2 = \sum_{S \in G/\sim} \sum_{x \in S} |\chi_\varrho(x)|^2.$$

Each class S is the set of generators of some finite cyclic subgroup H of G. Applying (to H and the restriction of ϱ to H) the inequality (4.36) from Exercise 4.5.6, we deduce that

$$\sum_{x \in S} |\chi_\varrho(x)|^2 \geqslant |S|$$

unless some (in fact, all) character values $\chi_\varrho(x)$ are zero for $x \in S$. Summing over S, and comparing with the orthonormality relation, it follows that when χ_ϱ has no zero, there must be equality in each of these inequalities. But $S = \{1\}$ is one of the classes, and therefore $|\chi_\varrho(1)|^2 = 1$, which gives the desired result by contraposition. □

This fact is about the rows of the character table. Is there another, dual, property of the columns? If there is, it is not the existence of at least one zero entry in each column, except for those of central elements (for which the modulus of the character value is the dimension). Although this property holds in a number of examples, for instance the groups $\mathrm{GL}_2(\mathbf{F}_p)$, we can see that it is false for the solvable groups B_p of Section 4.6.3. Further, we see in Table 4.3 that for the non-diagonalizable elements $\begin{pmatrix} a & 1 \\ 0 & a \end{pmatrix}$, with conjugacy classes of size $p-1$, every character value is a root of unity.

4.7.3. Relations between roots of polynomials. Our last application is to a purely algebraic problem about polynomials. Given a field k (arbitrary to begin with) and a non-zero irreducible polynomial $P \in k[X]$ of degree $d \geqslant 1$, the question is whether the roots

$$x_1, \ldots, x_d$$

of P (in some algebraic closure of k) satisfy any non-trivial k-linear relation, or any non-trivial multiplicative relation? By this, we mean, for linear relations, do there exist coefficients $\alpha_i \in k$, not all zero, such that

$$\alpha_1 x_1 + \cdots + \alpha_d x_d = 0\ ?$$

Or (for multiplicative relations), do there exist integers $n_i \in \mathbf{Z}$, not all zero, such that

$$x_1^{n_1} \cdots x_d^{n_d} = 1\ ?$$

For instance, since

$$x_1 + \cdots + x_d = a_{d-1}, \quad x_1 \cdots x_d = (-1)^d a_0,$$

for

$$P = X^d + a_{d-1} X^{d-1} + \cdots + a_1 X + a_0,$$

we have a non-trivial linear relation

$$x_1 + \cdots + x_d = 0,$$

whenever the coefficient of degree $d-1$ of P is zero, and a non-trivial multiplicative relation

$$x_1^2 \cdots x_d^2 = 1$$

whenever $a_0 = P(0) = \pm 1$.

A general method to investigate such questions was found by Girstmair (see [**22**] and the references therein, for instance), based on representation theory. We present here the basic idea and the simplest results.

As in the (first) proof of Burnside's irreducibility criterion (see Section 2.7.3), the basic idea is to define the set of all relations (linear or multiplicative) between the roots of P, and to show that it carries a natural representation of a certain finite group G. If we can decompose this representation in terms of irreducible representations, we will obtain a classification of all the possible relations that may occur. As was the case for the Burnside criterion, this is often feasible because the relation space is a subrepresentation of a well-understood representation of G.

With notation as before for the roots of P, we denote by

$$R_a = \Big\{(\alpha_i)_i \in k^d \mid \sum_{i=1}^{d} \alpha_i x_i = 0\Big\},$$

$$R_m = \Big\{(n_i)_i \in \mathbf{Z}^d \mid \prod_{i=1}^{d} x_i^{n_i} = 1\Big\}$$

the spaces of linear or multiplicative relations between the roots. We see immediately that R_a is a k-vector subspace of k^d, while R_m is a subgroup of the abelian group $\mathbf{Z}^d$, so that it is a free abelian group of rank at most d.

The group G that acts naturally on these spaces is the *Galois group* of the polynomial P, which means the Galois group of its splitting field

$$k_P = k(x_1, \ldots, x_d).$$

In the remainder of this section, we will assume known the basic statements of Galois theory.

To ensure that this Galois group is well defined, we must assume that P is separable, for instance that k has characteristic zero ($k = \mathbf{Q}$ will do). The elements of G are therefore field automorphisms

$$\sigma : k_P \longrightarrow k_P.$$

By acting on a relation (linear or multiplicative) using G, we see that the Galois group acts indeed on R_a and R_m. More precisely, recall that $\sigma \in G$ permutes the roots (x_i), so that there exists a group homomorphism

$$\begin{cases} G \longrightarrow \mathfrak{S}_d \\ \sigma \mapsto \hat{\sigma} \end{cases}$$

characterized by

$$\sigma(x_i) = x_{\hat{\sigma}(i)}$$

for all roots of P. This homomorphism is injective since the roots of P generate the splitting field k_P.

If $\boldsymbol{\alpha} = (\alpha_i)$ is in R_a, acting by σ on the relation

$$\alpha_1 x_1 + \cdots + \alpha_d x_d = 0,$$

we get

$$0 = \sigma(\alpha_1 x_1 + \cdots + \alpha_d x_d) = \alpha_1 x_{\hat{\sigma}(1)} + \cdots + \alpha_d x_{\hat{\sigma}(d)}$$

(since σ is the identity on k) or in other words the vector

$$\sigma \cdot \boldsymbol{\alpha} = (\alpha_{\hat{\sigma}^{-1}(i)})_{1 \leqslant i \leqslant d}$$

is also in R_a. But note that we can define

$$\sigma \cdot \boldsymbol{\alpha} = (\alpha_{\hat{\sigma}^{-1}(i)})_{1 \leqslant i \leqslant d}$$

for *arbitrary* $\alpha \in k^d$, and $\sigma \in G$. This is in fact simply the *permutation k-representation* π_k of G on k^d constructed from the action of G on the set $\{x_i\}$ of roots of G (see Section 2.6.2), and hence we see that R_a is a subrepresentation of the permutation representation π_k.

Similarly, we can act with G on multiplicative relations. However, since R_m is only an abelian group, it is only the $\mathbf{Q}$-vector space $R_m \otimes \mathbf{Q}$ that we can view as a subrepresentation of the (same!) permutation representation $\pi_{\mathbf{Q}}$ of G over $\mathbf{Q}$.

If we succeed in decomposing π_k, we can hope to see which subrepresentations can arise as relation spaces R_a, and similarly for $\pi_{\mathbf{Q}}$ and R_m. The simplest example of this idea is the following:

Proposition 4.7.12. *Let k be a field of characteristic zero, and let $P \in k[X]$ be an irreducible polynomial of degree $d \geqslant 2$ with Galois group isomorphic to the full symmetric group $\mathfrak{S}_d$.*

(1) *Either $R_a = 0$, i.e., there are no non-trivial linear relations between the roots of P, or R_a is one-dimensional and is spanned by the element $e_0 = (1, \ldots, 1)$ corresponding to the relation*

$$x_1 + \cdots + x_d = 0.$$

There exists a polynomial for which this second case occurs, for a given field k, if and only if there exists a polynomial in $k[X]$ with Galois group $\mathfrak{S}_d$.

(2) *Either $R_m = 0$, i.e., there are non non-trivial multiplicative relations between the roots of P, or R_m is a free $\mathbf{Z}$-module of rank 1 generated by ne_0 for some $n \geqslant 1$, or the splitting field of P is contained in the splitting field of a Kummer polynomial $X^n - b$. There exists a polynomial for which the first case occurs, for a given field k, if and only if there exists a polynomial with Galois group $\mathfrak{S}_d$. The second and third cases occur for $n = 2$ and $n = 3$ in the case $k = \mathbf{Q}$.*

Proof. As we have already observed, the space k^d of π_k decomposes as a direct sum of subrepresentations

$$k^d = ke_0 \oplus V,$$

where

$$V = \Big\{ v = (v_i) \in k^d \mid \sum_i v_i = 0 \Big\}.$$

When $G \simeq \mathfrak{S}_d$, although k is not algebraically closed (otherwise an irreducible polynomial P of degree $\geqslant 2$ would not exist!), these are irreducible subrepresentations. Indeed, we must only check this for V, and we immediately see that if V could be decomposed into two or more subrepresentations (recall that Maschke's Theorem does apply for any field of characteristic 0), then the same would be true for the representation of G on $V \otimes \bar{k}$, which contradicts the fact that it is irreducible (though we have only directly proved this for $k \subset \mathbf{C}$, see (4.19), it is in fact valid for any algebraically closed field of characteristic 0).

Because ke_0 and V are non-isomorphic as representations of $G \simeq \mathfrak{S}_d$ (even for $d = 2$, where V is also one dimensional), the only possibilities for R_a are therefore

$$R_a = 0, \text{ or } R_a = ke_0, \text{ or } R_a = V, \text{ or } R_a = V \oplus ke_0 = k^d.$$

This is the uniqueness of isotypic subspaces (see the second part of Proposition 2.7.9). The cases $R_a = 0$ and $R_a = ke_0$ are precisely the two possibilities of the statement we are trying to prove, and we now check that the others cannot occur. For this, it is enough to show that $R_a \supset V$ is impossible. But V is spanned by the vectors

$$(4.63) \qquad f_2 = (1, -1, 0, \ldots, 0), \quad \ldots, \quad f_d = (1, 0, \ldots, 0, -1).$$

Even if only the first were to be in R_a, this would translate into $x_1 = x_2$, which is impossible, and rules out both other cases.

There only remains to prove the existence part. Provided *some* polynomial in $k[X]$ with Galois group $\mathfrak{S}_d$ exists,[17] one can find instances where both cases $R_a = 0$ or $R_a = ke_0$ occur. This is easy: if we fix such a polynomial

$$P = X^d + a_{d-1}X^{d-1} + \cdots + a_1X + a_0 \in k[X]$$

with Galois group $\mathfrak{S}_d$, the polynomials $P(X + a)$, for any $a \in k$, have the same Galois group (because the splitting field has not changed!), and the sum of the roots $y_i = x_i - a$ is

$$\sum_{i=1}^{d} x_i - da,$$

which takes all values in k when a varies. When it is zero, the polynomial $P(X + a)$ satisfies $R_a = ke_0$, and otherwise $R_a = 0$.

We now deal with the multiplicative case; the argument is similar, but some of the cases excluded in the additive case become possible. First, since $R_m \otimes \mathbf{Q}$ is a subrepresentation of $\pi_{\mathbf{Q}}$, we see again that there are the same four possibilities for the subspace $R_m \otimes \mathbf{Q}$. Of course, if it is zero, we have $R_m = 0$ (because R_m is a free abelian group); if $R_m \otimes \mathbf{Q} = \mathbf{Q}e_0$, on the other hand, we can only conclude that $R_m = n\mathbf{Z}e_0$ for some integer

[17] This may not be the case or only for some d (in the case of $k = \mathbf{R}$, only $d = 2$ is possible).

$n \geqslant 1$ (examples below show that it is indeed possible that this happens with $n \neq 1$).

Continuing with the other possibilities, we have $R_m \otimes \mathbf{Q} \supset V$ if and only, for some $n \geqslant 1$, the vectors $nf_2, \ldots, nf_d$ are in R_m, where f_i is defined in (4.63). This means that we have

$$\left(\frac{x_1}{x_2}\right)^n = \cdots = \left(\frac{x_1}{x_d}\right)^n = 1,$$

and from this we deduce that

$$\sigma(x_1^n) = x_{\hat{\sigma}(1)}^n = x_1^n,$$

for all $\sigma \in G$. By Galois theory, this translates to $x_1^n \in k$. Therefore x_1 is a root of a *Kummer polynomial* $X^n - b \in k[X]$, which is the last possible conclusion we claimed. Note that b could be a root of unity (belonging to k): this is a special case, which corresponds to $R_m \otimes \mathbf{Q} = \mathbf{Q}^d$ (instead of V), since each x_i is then a root of unity.

In terms of existence of polynomials with these types of multiplicative relations, we first note that $R_m = 0$ is always possible if there exists at least *one* irreducible polynomial $P \in k[X]$ with Galois group $\mathfrak{S}_d$. Indeed, we have $P(0) \neq 0$, and as before, we may replace P with $Q = P(aX)$ for $a \in k$, without changing the Galois group; the roots of Q are $y_i = a^{-1}x_i$, and

$$y_1 \cdots y_d = \frac{x_1 \cdots x_d}{a^d}.$$

Then, if we pick $a \in k^\times$ so that this is expression is not a root of unity, we obtain a polynomial Q with $R_m = 0$ (such an $a \neq 0$ exists; otherwise, taking $a = 1$ would show that $x_1 \cdots x_d$ is itself a root of unity, and then it would follow that *any* $a \in k^\times$ is a root of unity, which is absurd since $\mathbf{Q} \subset k$).

For the case of $R_m \otimes \mathbf{Q} = \mathbf{Q}e_0$, we will just give examples for $k = \mathbf{Q}$. It is known (see, e.g., [**56**, p. 42]) that the polynomial

$$P = X^d - X - 1$$

has Galois group $\mathfrak{S}_d$ for $d \geqslant 2$. Since the product of its roots is $(-1)^d P(0) = (-1)^{d+1}$, it satisfies the relation

$$\prod_i x_i^2 = 1,$$

so $R_m \otimes \mathbf{Q} = \mathbf{Q}e_0$, and in fact $R_m = n\mathbf{Z}e_0$ with $n = 1$ if d is odd, and $n = 2$ if d is even.

For the Kummer cases, we take $k = \mathbf{Q}$ for simplicity (it will be clear that many fields will do). For $n = 2$, any quadratic polynomial $X^2 - b$ with b not a square of a rational number has Galois group $\mathfrak{S}_2$. If $b = -1$, noting

$x_1 = i$, $x_2 = -i$, we have

$$R_m = \{(n_1, n_2) \in \mathbf{Z}^2 \mid n_1 + 2n_2 \equiv 0 \,(\mathrm{mod}\, 4)\},$$

which has rank 2 (so $R_m \otimes \mathbf{Q} = \mathbf{Q}^2$), and if $b \neq -1$, we have

$$R_m = \{(n_1, n_2) \in \mathbf{Z}^2 \mid n_i \equiv 0 \,(\mathrm{mod}\, 2),\ n_1 + n_2 = 0\},$$

with $R_m \otimes \mathbf{Q} = \mathbf{Q}e_0$. For $n = 3$, any Kummer equation $X^3 - b = 0$, with b not a perfect cube, will have splitting field with Galois group $\mathfrak{S}_3$, and a quick computation with the roots $\sqrt[3]{b}$, $j\sqrt[3]{b}$, $j^2\sqrt[3]{b}$, where j is a primitive cube root of unity in $\mathbf{C}$, leads to

$$R_m = \{(n_1, n_2, n_3) \in \mathbf{Z}^3 \mid n_1 + n_2 + n_3 = 0, \quad n_i \equiv n_j \,(\mathrm{mod}\, 3) \text{ for all } i, j\},$$

so that again $R_m \otimes \mathbf{Q} = \mathbf{Q}e_0$. □

Exercise 4.7.13 (Palindromic polynomials)**.** We consider in this exercise the case where d is even and the Galois group of P is the group W_d defined as the subgroup of $\mathfrak{S}_d$ that respects a partition of $\{1, \ldots, d\}$ into $d/2$ pairs. More precisely, let X be a finite set of cardinality $d = 2n$, and let $i : X \to X$ be an involution on X (i.e., $i \circ i = \mathrm{Id}_X$) with no fixed points, for instance $X = \{1, \ldots, d\}$ and $i(x) = d + 1 - x$. The $n = d/2$ pairs $\{x, i(x)\}$ partition X, and one defines

$$W_d = \{\sigma \in \mathfrak{S}_d \mid \sigma(i(x)) = i(\sigma(x)) \text{ for all } x \in X\},$$

which means concretely that an element of W_d permutes the pairs $\{x, i(x)\}$, and may (or not) switch x and $i(x)$. This group is sometimes called the group of *signed permutations* of $\{1, \ldots, n\}$.

(1) Show that W_d is of order $2^n n!$ for $d = 2n \geqslant 2$ even, and that there is an exact sequence

$$1 \longrightarrow (\mathbf{Z}/2\mathbf{Z})^n \longrightarrow W_d \longrightarrow \mathfrak{S}_n \longrightarrow 1.$$

Find a faithful representation of W_d in $\mathrm{GL}_n(\mathbf{C})$ where the matrix representation has values in $\mathrm{GL}_n(\mathbf{Z})$.

(2) Let k be a field of characteristic 0, and let $P \in k[X]$ be an irreducible polynomial of degree $d = 2n$ even, $d \geqslant 2$, of the form

$$P = X^d + a_{d-1}X^{d-1} + \cdots + a_{n+1}X^n + a_{n+1}X^{n-1} + \cdots + a_{d-1}X + 1,$$

i.e., with the same coefficients for X^j and X^{d+1-j} for all j (such polynomials are called *palindromic* or *self-reciprocal*). Show that the Galois group of P can be identified with a subgroup of W_d. (*Hint*: If x is a root of P, then $1/x$ is also one.)

(3) Show that the permutation k-representation π_k of W_d associated to the action of W_d on X splits as a direct sum

$$E_0 \oplus E_1 \oplus E_2,$$

where $E_0 = ke_0$ and, for a suitable numbering of the roots, we have

$$E_1 = \Big\{(\alpha_i) \mid \alpha_{d+1-i} - \alpha_i = 0,\ 1 \leqslant i \leqslant d,\quad \sum \alpha_i = 0\Big\},$$

$$E_2 = \Big\{(\alpha_i) \mid \alpha_{d+1-i} + \alpha_i = 0,\ 1 \leqslant i \leqslant d\Big\},$$

and the three spaces are irreducible. (*Hint*: You may assume $k \subset \mathbf{C}$; compute the orbits of W_d on X, and deduce the value of the squared norm of the character of π_k.)

(4) In the situation of (2), assume that the Galois group of $P \in k[X]$ is equal to W_d. Show that the only possible spaces of linear relation between roots of a polynomial P as above are $R_a = 0$ and $R_a = ke_0$.

It is known that "many" palindromic polynomials with Galois groups W_d exist; for more information and some applications, the reader may refer to [**38**, §2].

Remark 4.7.14. Both Proposition 4.7.12 and this exercise are in the direction of showing that linear or multiplicative relations are rare when the Galois group of the polynomial is very large. However, for some other Galois groups, interesting things can happen. For instance, Girstmair showed that there exists a group G of order 72 which can arise as the Galois group (for $k = \mathbf{Q}$) of some polynomial P of degree 9 for which the roots, suitably numbered, satisfy

$$4x_1 + x_2 + x_3 + x_4 + x_5 - 2(x_6 + x_7 + x_8 + x_9) = 0.$$

Another example is the group G usually denoted $W(E_8)$, the "Weyl group of E_8", which can be defined as the group with eight generators

$$w_1, \ldots, w_8,$$

which are subject to the relations

$$w_i^2 = 1 \qquad (w_i w_j)^{m(i,j)} = 1, \qquad 1 \leqslant i < j \leqslant 8,$$

where $m(i,j) = 2$ *unless* the vertices numbered i and j in the diagram below

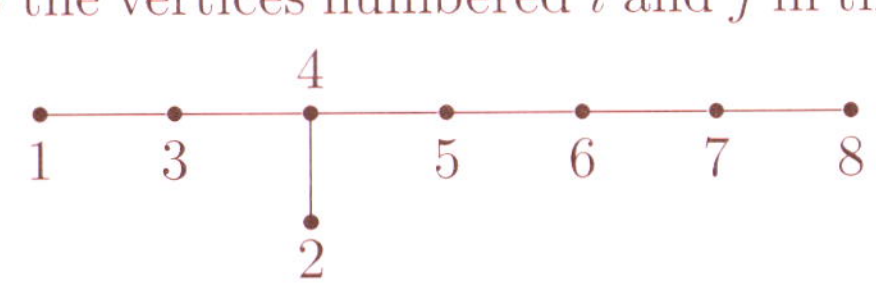

are linked with an edge, in which case $m(i,j) = 3$, e.g. $m(1,3) = m(2,4) = 3$. (This diagram is called the *Dynkin diagram* of E_8. This definition is given here only in order to be definite. Of course, this presentation is justified by the many other definitions and properties of this group, which is an example of a *Coxeter group*.)

The group $W(E_8)$ has order

$$W(E_8) = 696,729,600 = 2^{14} \cdot 3^5 \cdot 5^2 \cdot 7,$$

and one can construct irreducible polynomials $P \in \mathbf{Q}[X]$, of degree 240, with Galois group $W(E_8)$, such that

$$\dim(R_m \otimes \mathbf{Q}) = 232,$$

or in other words, there are eight roots of P, out of 240, such that all others are in the multiplicative group generated by those (see [**5**, §5] or [**31**, Rem. 2.4]).

4.8. Further topics

We finish this chapter with a short discussion of some further topics concerning representations of finite groups. These—and their developments—are of great interest and importance in some applications, and although we only consider basic facts, we will give references where more details can be found. One last important notion, the *Frobenius–Schur indicator* of an irreducible representation, will be considered in Section 6.2, because it makes sense, and is treated exactly the same way, for all compact groups.

4.8.1. Intertwiners between induced representations. We have used induced representations quite often, either in a general way (typically to exploit Frobenius reciprocity) or to construct specific representations of concrete groups. In the second role, in particular, we see that it is useful to understand intertwiners between two induced representations. In particular in Section 4.6.4, we computed the dimension of such spaces "by hand", as inner products of induced characters. The answers are rather clean, as (4.49) illustrates, and it should not be a surprise to see that there is a general approach to these computations.

Proposition 4.8.1 (Intertwiners between induced representations). *Let G be a finite group, and let H_1, H_2 be subgroups of G. Let ϱ_1, ϱ_2 be complex finite-dimensional representations of H_1 and H_2 acting on the vector spaces E_1 and E_2, respectively. There is an isomorphism*

$$\mathrm{Hom}_G(\mathrm{Ind}_{H_1}^G \varrho_1, \mathrm{Ind}_{H_2}^G(\varrho_2)) \simeq I(\varrho_1, \varrho_2),$$

where

$$(4.64) \quad \begin{aligned} I(\varrho_1, \varrho_2) = \{\alpha : G \to \mathrm{Hom}_{\mathbf{C}}(E_1, E_2) \mid \\ \alpha(h_1 x h_2) = \varrho_2(h_2)^{-1} \circ \alpha(x) \circ \varrho_1(h_1)^{-1}, \\ \textit{for all } h_1 \in H_1,\ x \in G,\ h_2 \in H_2\}. \end{aligned}$$

Proof. We start naturally by applying Frobenius reciprocity to "remove" one induced representation: we have an isomorphism

$$\mathrm{Hom}_G(\mathrm{Ind}_{H_1}^G \varrho_1, \mathrm{Ind}_{H_2}^G(\varrho_2)) \simeq \mathrm{Hom}_{H_2}(\mathrm{Res}_{H_2}^G \mathrm{Ind}_{H_1}^G \varrho_1, \varrho_2).$$

The next idea is to find first a convenient model for the space

$$\mathrm{Hom}(\mathrm{Res}^G_{H_2}\mathrm{Ind}^G_{H_1}\varrho_1, \varrho_2),$$

and then to isolate the H_2-intertwiners inside this space. Let F_1 denote the space on which $\mathrm{Ind}^G_{H_1}\varrho_1$ acts, as well as its restriction to H_2. By construction, F_1 is a subspace of the space V_1 of all functions from G to E_1.

Any $f \in V_1$ can be expressed uniquely as

$$f = \sum_{x\in G} f(x)\delta_x,$$

where $f(x) \in E_1$ and, for any vector $v \in E_1$, we denote by $v\delta_x$ the function $G \longrightarrow V_1$ equal to v at the point x and to 0 elsewhere.

It follows by linearity that any linear map T from V_1 to E_2 can be expressed in the form

$$Tf = T_\alpha f = \sum_{x\in G} \alpha(x)(f(x)),$$

where $\alpha(x) \in \mathrm{Hom}(E_1, E_2)$ is the linear map defined by

$$\alpha(x) : v \mapsto T(v\delta_x),$$

We claim that this gives an isomorphism

$$T : \begin{cases} I & \longrightarrow & \mathrm{Hom}(F_1, E_2) \\ \alpha & \mapsto & T_\alpha, \end{cases}$$

where

$$I = \{\alpha : G \longrightarrow \mathrm{Hom}(E_1, E_2) \mid \alpha(h_1 x) = \alpha(x) \circ \varrho_1(h_1)^{-1}\}.$$

In general, since F_1 is a subspace of V_1, $\mathrm{Hom}(F_1, E_2)$ would be a quotient of $\mathrm{Hom}(V_1, E_2)$, and what we are doing is finding a good representative subspace for it in $\mathrm{Hom}(V_1, E_2)$.

To check this, let $f \in F_1$ be given, and compute

$$\begin{aligned} T_\alpha(f) &= \sum_{y\in H_1\backslash G}\sum_{h_1\in H_1} \alpha(h_1 y)(f(h_1 y)) \\ &= \sum_{y\in H_1\backslash G}\sum_{h_1\in H_1} \alpha(y)\circ\varrho_1(h_1)^{-1}(\varrho_1(h_1)f(y)) \\ &= |H_1| \sum_{y\in H_1\backslash G} \alpha(y)(f(y)), \end{aligned}$$

which quickly shows that T is injective, since one can prescribe the values of an element $f \in F_1$ arbitrarily on each coset in $H_1\backslash G$. Since we also know that

$$\dim I = [G : H](\dim E_1)(\dim E_2)$$

(by an argument similar to the computation of the dimension of an induced representation) and this is also the dimension of $\mathrm{Hom}(F_1, E_2)$ (by Proposition 2.3.11), we conclude that T is indeed an isomorphism.

Now we can easily answer the question. Given $\alpha \in I$, when is $T_\alpha \in \mathrm{Hom}(F_1, E_2)$ an H_2-intertwiner? For $h_2 \in H_2$ and $f \in F_1$, we have

$$T_\alpha(h_2 \cdot f) = \sum_{x \in G} \alpha(x)((h_2 \cdot f)(x)) = \sum_{x \in G} \alpha(x)(f(xh_2)),$$

and we want this to be equal to

$$\varrho_2(h_2)T_\alpha(f) = \sum_{x \in G} (\varrho_2(h_2) \circ \alpha(x))(f(x))$$

for all h_2 and f. We fix h_2; by change of variable, the first expression is

$$T_\alpha(h_2 \cdot f) = \sum_{x \in G} \alpha(xh_2^{-1})(f(x)) = T_\beta(f)$$

for $\beta(x) = \alpha(xh_2^{-1})$. The second is $T_{\varrho_2(h_2)\alpha}$, and since β and $\varrho_2(h_2)\alpha$ are both still elements of I, the injectivity of T shows that the equality

$$T_\alpha(h_2 \cdot f) = \varrho_2(h_2)T_\alpha(f), \qquad f \in F_1,$$

is equivalent to

$$\alpha(xh_2^{-1}) = \varrho_2(h_2)\alpha(x)$$

for all $x \in G$. Replacing h_2 by h_2^{-1}, and combining these for all $h_2 \in H_2$, we find that $I(\varrho_1, \varrho_2)$ defined in (4.64) is the subspace of I isomorphic, under T, to $\mathrm{Hom}_{H_2}(F_1, E_2)$. □

If, as in the case of $\mathrm{GL}_2(\mathbf{F}_p)$ in Section 4.6.4, we induce one-dimensional characters, we get a very general irreducibility criterion for such representations:

Corollary 4.8.2 (Irreducibility of induced representations). *Let G be a finite group, let H be a subgroup of G, and let χ be a one-dimensional complex representation of H. The induced representation $\varrho = \mathrm{Ind}_H^G \chi$ is irreducible if and only if the one-dimensional representations χ_s of $H_s = H \cap sHs^{-1}$ defined by*

$$\chi_s(h) = \chi(s^{-1}hs)$$

are distinct from $\mathrm{Res}_{H_s}^G \chi$ as s runs over the complement of H in G.

Proof. By the irreducibility criterion of Corollary 4.3.14, which is the converse of Schur's Lemma, it suffices to determine when the space $I_{\chi,\chi}$ of intertwiners of $\varrho = \mathrm{Ind}_H^G(\chi)$ with itself is one dimensional. We apply Proposition 4.8.1 to compute this space; if we note that the space $\mathrm{Hom}(E_1, E_2)$

can be identified with $\mathbf{C}$ when $E_1 = E_2$ is one dimensional, we see that $I_{\chi,\chi}$ is isomorphic to the space I of functions $\alpha : G \longrightarrow \mathbf{C}$ such that

$$\alpha(h_1 x h_2) = \chi(h_1 h_2)^{-1}\alpha(x)$$

for all $x \in G$ and $h_1, h_2 \in H$. These conditions seem similar to those defining an induced representation, and this would suggest at first that the dimension of I is $|H\backslash G/H| = |S|$, but there is a subtlety. A representation $x = h_1 s h_2$ with $h_i \in H$ need *not* be unique, which creates additional relations to be satisfied. As a consequence, the dimension can be smaller than this guess.

The one-dimensional subspace of I corresponding to $\mathbf{C}\mathrm{Id}$ in $\mathrm{End}_G(\varrho)$ is spanned by the function α_0 such that

$$\alpha_0(x) = \begin{cases} \chi(x)^{-1} & \text{if } x \in H, \\ 0 & \text{otherwise,} \end{cases}$$

as one can easily check using the explicit form of the Frobenius reciprocity isomorphism.

We now determine the condition under which I is actually spanned by this special function α_0. If we denote by S a set of representatives for the double cosets $HsH \subset G$ (taking $s = 1$ for the double coset $H \cdot H = H$), we see first of all, from the relations defining I, that the map

$$\begin{cases} I & \longrightarrow & \mathbf{C}^S \\ \alpha & \mapsto & (\alpha(s))_{s \in S} \end{cases}$$

is *injective*. Similarly, we get

$$\alpha(hs) = \chi(h)^{-1}\alpha(s) = \alpha(sh) \tag{4.65}$$

for all $h \in H$ and $s \in S$.

Now fix $s \in S$. We claim that either $\alpha(s) = 0$ or $\chi_s = \chi$ on $H_s = H \cap sHs^{-1}$ (note in passing that χ_s is indeed a well-defined representation of this subgroup, since $s^{-1}H_s s \subset H$). Indeed, for $x \in H_s = H \cap sHs^{-1}$, we have

$$\alpha(xs) = \alpha(s(s^{-1}xs)) = \alpha((s^{-1}xs)s)$$

by (4.65), applied with $h = s^{-1}xs \in H$, and this gives

$$\chi(x)^{-1}\alpha = \chi(s^{-1}xs)^{-1}\alpha,$$

which is valid for all $x \in H_s$, hence the claim.

Consequently, if no χ_s coincides with χ on H_s when $s \notin H$ (corresponding to the double cosets which are distinct from H), any $\alpha \in I$ must vanish on the non-trivial double cosets and, hence, be a multiple of α_0. This proves the sufficiency part of the irreducibility criterion, and we leave the necessity to the reader. □

Exercise 4.8.3 (Principal series). Let $n \geqslant 2$ be an integer, and let p be a prime number. Let $G = \mathrm{GL}_n(\mathbf{F}_p)$ and define $B \subset G$ to be the subgroup of all upper-triangular matrices. Moreover, let $W \subset G$ be the subgroup of permutation matrices.

(1) *Bruhat decomposition.* Show that

$$G = \bigcup_{w \in W} BwB$$

and that this is a disjoint union. (*Hint*: Think of using Gaussian elimination to put a matrix in triangular form in some basis.)

(2) Let $\chi : B \longrightarrow \mathbf{C}^\times$ be the one-dimensional representation given by

$$\chi(g) = \chi_1(g_{1,1})\chi_2(g_{2,2}) \cdots \chi_n(g_{n,n}),$$

where χ_i, $1 \leqslant i \leqslant n$, are characters of $\mathbf{F}_p^\times$ and $g_{i,j}$ is the (i,j)-th entry of $g \in B$. Let $\varrho = \mathrm{Ind}_B^G \chi$. Show that ϱ is irreducible whenever all characters χ_i are distinct.

(3) For $n = 2$, show without using characters that the induced representations $\pi(\chi_1, \chi_2)$ and $\pi(\chi_2, \chi_1)$ (with $\chi_1 \neq \chi_2$) are isomorphic, and write down a formula for an isomorphism. (*Hint*: Follow the construction in Proposition 4.8.1 and the Frobenius reciprocity isomorphism.)

The irreducible representations of $\mathrm{GL}_n(\mathbf{F}_p)$ constructed in this exercise are called the *principal series*. For $n = 2$, they are exactly those whose irreducibility was proved in Section 4.6.4 using their characters.

Note that there is no principal series unless there are at least n distinct characters of $\mathbf{F}_p^\times$, i.e., unless $p-1 \geqslant n$. This suggests—and this is indeed the case!—that the character tables of $\mathrm{GL}_n(\mathbf{F}_p)$, when p is fixed and n varies, behave rather differently from those of $\mathrm{GL}_n(\mathbf{F}_p)$ when n is fixed and p varies.

Exercise 4.8.4. We explain here a different approach to the corollary. Let G be a finite group, let H_1 and H_2 be subgroups of G, and let χ be a one-dimensional complex representation of H_1.

(1) Show that

$$\mathrm{Res}_{H_2}^G \mathrm{Ind}_{H_1}^G \chi \simeq \bigoplus_{s \in S} \mathrm{Ind}_{H_{2,s}}^{H_2} \chi_s,$$

where S is a set of representatives of the double cosets $H_2 g H_1$, $H_{2,s} = H_2 \cap s^{-1} H_1 s$ and χ_s is the one-dimensional character of $H_{2,s}$ given by

$$\chi_s(x) = \chi(s x s^{-1}).$$

(*Hint*: This can be done with character theory.)

(2) Prove the corollary, and recover the irreducibility result (2) of the previous exercise, using (1).

4.8.2. Artin's theorem on induced representations. Our second topic concerning induction takes up the following question: We have seen, in concrete examples, that many irreducible representations of certain finite groups arise as induced representations from subgroups and, indeed, from one-dimensional characters of abelian subgroups. How general is this property? It turns out that, provided some leeway is allowed in the statement, one can in fact recover *all* irreducible representations using induced representations of cyclic subgroups. This is the content of the following theorem of Artin, which is an excellent illustration of the usefulness of the character ring $R(G) = R_{\mathbf{C}}(G)$ introduced in Definition 2.7.48, and of the virtual characters that it contains.

Theorem 4.8.5 (Artin). *Let G be a finite group, and let $\varrho \in \widehat{G}$ be an irreducible complex representation of G. There exists a decomposition*

$$\varrho = \sum_{i \in I} \alpha_i \operatorname{Ind}_{H_i}^{G} \chi_i \tag{4.66}$$

in $R(G) \otimes \mathbf{Q}$, where I is a finite index set, $\alpha_i \in \mathbf{Q}$, $H_i \subset G$ is a cyclic subgroup of G and χ_i is an irreducible character of H_i, and where we identify representations of G with their image in $R(G)$.

Concretely, recall that the meaning of (4.66) is the following: there exist an integer $m \geqslant 1$, finitely many non-negative integers $n_i \geqslant 0$ and $m_j \geqslant 0$, corresponding cyclic subgroups H_i and H_j, and one-dimensional characters χ_i and χ_j of H_i and H_j, such that we have an isomorphism of actual representations

$$m\varrho \oplus \bigoplus_i n_i \operatorname{Ind}_{H_i}^{G} \chi_i \simeq \bigoplus_j m_j \operatorname{Ind}_{H_j}^{G} \chi_j$$

(precisely, m is a common denominator of all α_i in (4.66), $n_i = -m\alpha_i$ if $\alpha_i < 0$, while $m_j = m\alpha_j$ if $\alpha_j \geqslant 0$).

Proof. There is a surprisingly easy proof: The $\mathbf{Q}$-vector space $R(G) \otimes \mathbf{Q}$ has a basis corresponding to the irreducible representations of G, and it inherits from the characters a non-degenerate symmetric bilinear form $\langle \cdot, \cdot \rangle_G$ for which those characters form an orthonormal basis. By duality of vector spaces, a subspace $V \subset R(G) \otimes \mathbf{Q}$ is equal to the whole space if and only if its orthogonal $V^{\perp}$ is zero. In particular, if V is generated by certain elements (χ_i) in $R(G) \otimes \mathbf{Q}$, we have $V = R(G) \otimes \mathbf{Q}$ if and only if no $\xi \in R(G) \otimes \mathbf{Q}$ is orthogonal to all χ_i.

We apply this now to the family (χ_i) of all representations induced from irreducible representations of cyclic subgroups of G. Suppose $\xi \in R(G) \otimes \mathbf{Q}$ is orthogonal to all such χ_i. Then, using the Frobenius reciprocity formula (which holds in $R(G) \otimes \mathbf{Q}$ by *linearity* of induction and restriction with respect to direct sums), we get

$$\langle \mathrm{Ind}_H^G \chi, \xi \rangle_G = \langle \chi, \mathrm{Res}_H^G \xi \rangle_H$$

for any cyclic subgroup H and $\chi \in \widehat{H}$. Varying χ for a fixed H, it follows that $\mathrm{Res}_H^G \xi$ is zero for all H. If we identify ξ with its virtual character in $C(G)$, this means that ξ vanishes on all cyclic subgroups of G. But since any element $x \in G$ belongs to at least one cyclic subgroup, this gives $\xi = 0$. ☐

It is natural to wonder whether a result like Artin's Theorem might not also be valid without involving virtual characters, i.e., with only $\alpha_i \geqslant 0$ in (4.66), but examples show that this is not the case. The simplest is the four-dimensional irreducible representation ϱ of the alternating group A_5 (the character table of which is found in Remark 4.7.6); see [**53**, Exercise 10.5] for the details.

Exercise 4.8.6. Find an Artin decomposition of all irreducible representations of A_5.

On the other hand, there is an important theorem of Brauer which shows that one may find decompositions with *integral* coefficients $\alpha_i \in \mathbf{Z}$, provided one allows representations induced by one-dimensional characters of a slightly larger class of subgroups than cyclic subgroups, the so-called *elementary subgroups*. We refer, here also, to the treatment in Serre's book [**53**, Ch. 10] for the details.

Chapter 5

Abstract representation theory of compact groups

In this chapter, we consider the representation theory of compact topological groups. Our goal is to present the basic facts from the general theory, which is due to Peter and Weyl, and to do so by highlighting the close parallel with the case of finite groups (which is a special case, in fact, if we consider a finite group as a compact group with discrete topology). This requires some care in the analytic set up, but the reader should appreciate how the work in getting the right definition of continuity and of the regular representation (for instance) are justified when, in the end, the character formalism and the decomposition of the regular representation look formally very much the same as they do for finite groups.

5.1. An example: the circle group

We begin with an example, where it turns out that the basic facts are already well known from the theory of Fourier series. This corresponds to what is probably the simplest infinite compact group, the unit circle

$$G = \{z \in \mathbf{C}^\times \mid |z| = 1\} \subset \mathbf{C}^\times$$

with its topology as a subset of $\mathbf{C}$. This group is often best understood in the equivalent representation as the quotient $\mathbf{R}/\mathbf{Z}$, or $\mathbf{R}/2\pi\mathbf{Z}$, with the quotient topology, where the isomorphism is given by

$$\begin{cases} \mathbf{R}/\mathbf{Z} & \longrightarrow & G \\ x & \mapsto & e^{2i\pi x} \end{cases}$$

(since it is important to view G as a topological group, one should note that this is a *homeomorphism*).[1]

We already know explicitly what are the irreducible unitary representation of G: by Example 3.4.10 and Schur's Lemma (Proposition 3.4.17), these are one dimensional (because G is abelian) and are precisely the characters

$$\chi_m : \begin{cases} \mathbf{R}/\mathbf{Z} & \longrightarrow & \mathbf{C}^\times \\ t & \mapsto & e^{2i\pi mt} \end{cases}$$

for $m \in \mathbf{Z}$. The finite linear combinations of these characters are simply trigonometric polynomials. This is a rather special subspace of functions on G, but it is dense in many important function spaces, including the space of continuous functions, by the Weierstrass approximation theorem.

The problem of expressing an *arbitrary* function in a series involving the χ_m is a basic problem of Fourier analysis, and it is one of the most classical (and beautiful) problems of analysis. The Fourier series of an integrable function φ is the series

$$\sum_{m\in\mathbf{Z}} \alpha(m)e^{2i\pi mt} = \sum_{m\in\mathbf{Z}} \alpha(m)\chi_m(t),$$

where

$$\alpha(m) = \int_{\mathbf{R}/\mathbf{Z}} \varphi(x)e^{-2i\pi mx}dt$$

are the Fourier coefficients. Many results are known about the convergence of Fourier series toward φ, and they reveal a wide variety of behavior. For instance, it was shown by Kolmogorov that there exist integrable functions $\varphi \in L^1(\mathbf{R}/\mathbf{Z})$ such that the Fourier series above diverges *for all* $x \in \mathbf{R}/\mathbf{Z}$. (For the classical theory of Fourier series, one can look at Zygmund's treatise [**67**]; for instance, Kolmogorov's Theorem is found in [**67**, Th. VIII.4.1].)

However, it is also classical that a very good theory emerges when considering the square-integrable functions: if $\varphi \in L^2(\mathbf{R}/\mathbf{Z})$, the Fourier series converges in L^2-norm, i.e.,

$$\Big\| \varphi - \sum_{|m|\leqslant M} \alpha(m)\chi_m \Big\|_{L^2} \longrightarrow 0,$$

as $M \longrightarrow +\infty$, and in particular the Parseval formula

$$\int_{\mathbf{R}/\mathbf{Z}} |\varphi(x)|^2dx = \sum_{m\in\mathbf{Z}} |\alpha(m)|^2 = \sum_{m\in\mathbf{Z}} |\langle\varphi, \chi_m\rangle|^2$$

holds.

[1] It is useful here to remember that a continuous bijection between compact topological spaces is necessarily a homeomorphism, i.e., the inverse is also automatically continuous.

This can be interpreted in terms of a unitary representation of G on $L^2(G)$: Under the regular action

$$\varrho(t)\varphi(x) = \varphi(x+t),$$

the condition of square-integrability is preserved, and in fact

$$\|\varrho(t)\varphi\|^2 = \int_{\mathbf{R}/\mathbf{Z}} |\varphi(x+t)|^2 dx = \int_{\mathbf{R}/\mathbf{Z}} |\varphi(x)|^2 dx = \|\varphi\|^2$$

because the Lebesgue measure is invariant under translations. Thus $L^2(G)$ is formally a unitary representation (the continuity requirement also holds; we will verify this later in a more general case). The L^2-theory of Fourier series says that the family of characters (χ_m) is an orthonormal basis of the space $L^2(G)$, and this is, quite recognizably, a suitable analogue of the decomposition of the regular representation: each of the characters χ_m appears once (a 1 that is the dimension of χ_m!) in $L^2(G)$.

It is rather natural to explore how facts about Fourier series can be interpreted in terms of the representation theory of $\mathbf{R}/\mathbf{Z}$. Although this is quite straightforward, this brings out a few facts which are useful to motivate some parts of the next section, where arbitrary compact groups enter the picture. For instance, let us consider the orthogonal projection map

$$p_m : L^2(G) \longrightarrow L^2(G)$$

onto the χ_m-isotypic component of G. Since this space is one dimensional, with χ_m itself as a unit vector, we have simply

$$p_m(\varphi) = \langle \varphi, \chi_m \rangle \chi_m,$$

i.e., for $t \in \mathbf{R}/\mathbf{Z}$, we have

$$p_m(\varphi)(t) = \Bigl(\int_{\mathbf{R}/\mathbf{Z}} \varphi(x) e^{-2i\pi m x} dx\Bigr) e^{2i\pi m t}. \tag{5.1}$$

This seems to be a complicated way of writing the m-th Fourier coefficient of φ, which is the integral that appears here. However, we can write this as

$$\begin{aligned} p_m(\varphi)(t) &= \int_{\mathbf{R}/\mathbf{Z}} e^{2i\pi m(t-x)} \varphi(x) dx \\ &= \int_{\mathbf{R}/\mathbf{Z}} e^{-2i\pi m y} \varphi(y+t) dy = \int_{\mathbf{R}/\mathbf{Z}} \overline{\chi_m(y)} (\varrho(y)\varphi)(t) dy, \end{aligned}$$

or in other words we have (formally) the formula

$$p_m = \int_{\mathbf{R}/\mathbf{Z}} \overline{\chi_m(y)} \varrho(y) dy,$$

which is clearly similar to (4.20).

There is yet another instructive way to express the projection, where we consider φ as fixed and m as varying. The formula

$$p_m(\varphi)(t) = \int_{\mathbf{R}/\mathbf{Z}} \varphi(x) e^{2i\pi m(t-x)} dx$$

can be written

$$p_m(\varphi) = \varphi \star \chi_m,$$

where $\cdot \star \cdot$ denotes the operation of convolution of functions on G

$$(\varphi_1 \star \varphi_2)(t) = \int_{\mathbf{R}/\mathbf{Z}} \varphi_1(x)\varphi_2(t-x) dx$$

(when this makes sense, of course). In particular, (5.1) means that the characters χ_m are *eigenfunctions* of the convolution operator

$$T_\varphi : \begin{cases} L^2(G) & \longrightarrow & L^2(G) \\ f & \mapsto & \varphi \star f, \end{cases}$$

with eigenvalues given precisely by the Fourier coefficients $\langle \varphi, \chi_m \rangle$. Note finally that this convolution operator is an intertwiner of the regular representation. This follows either from a direct computation or from the fact that T_φ acts by scalar multiplication on the characters.

5.2. The Haar measure and the regular representation of a locally compact group

In order to adapt the arguments which succeeded in the case of finite groups and which are suggested by the example of the circle group, we see that we need first to define the analogue of the regular representation. In order for this to be a unitary representation, it seems natural to look at the space of L^2-functions on G, with respect to some natural measure μ. Which measure to select is dictated by the requirement that the usual action

$$\varrho_G(g)\varphi(x) = \varphi(xg)$$

of the regular representation should be unitary. What is required is that

$$\int_G f(xg) d\mu(x) = \int_G f(x) d\mu(x)$$

for f integrable and $g \in G$. This property, for instance, holds for the Lebesgue measure on $\mathbf{R}/\mathbf{Z}$, or on $\mathbf{R}^d$.

It is by no means obvious that a measure μ exists with this property, and indeed, this imposes a restriction on the topological groups that one may consider in this manner. The relevant existence statement is given by the following theorem.

Theorem 5.2.1 (Existence and properties of Haar measure). *Let G be a locally compact topological group.*

(1) *There exists, up to multiplication by a scalar $\alpha > 0$, a unique non-zero Radon measure μ on G which is* right invariant, *i.e., which is such that, for any fixed $g \in G$, we have*

$$\int_G f(xg)d\mu(x) = \int_G f(x)d\mu(x) \tag{5.2}$$

for $f \geqslant 0$ measurable and for $f \in L^1(G, d\mu)$. Such a measure is called a Haar measure on G.

(2) *If G is compact, there exists a unique Haar measure such that $\mu(G) = 1$, which is called the* probability Haar measure *on G.*

(3) *Any Haar measure on a* compact *group G is also left invariant, i.e., we have*

$$\int_G f(gx)d\mu(x) = \int_G f(x)d\mu(x) \tag{5.3}$$

for fixed $g \in G$, and it is invariant under inversion, i.e.,

$$\int_G f(x^{-1})d\mu(x) = \int_G f(x)d\mu(x),$$

both for $f \geqslant 0$ measurable or $f \in L^1(G, d\mu)$.

(4) *The support of any Haar measure μ on G is equal to G, i.e., for any non-empty open set $U \subset G$, we have $\mu(U) > 0$.*

(5) *Let $C_c(G)$ be the space of continuous and compactly supported functions on G. Then, for any $p \geqslant 1$, the natural map $C_c(G) \to L^p(G, \mu)$ is injective, and if $p \neq +\infty$, the space $C_c(G)$ is dense in $L^p(G, \mu)$ for the L^p-norm.*

Remark 5.2.2. (1) We have written the definition of Haar measure in terms of integrals of functions. In terms of sets, μ satisfies (5.2) if, for any Borel subset of G and any $g \in G$, we have

$$\mu(B) = \mu(Bg^{-1}) = \mu(Bg)$$

(the last by applying the previous one to g^{-1}).

(2) We will often simply write $L^1(G)$, or more generally $L^p(G)$, for $p \in [1, +\infty]$, instead of $L^1(G, \mu)$ or $L^p(G, \mu)$, the integrability condition being understood implicitly to refer to a Haar measure. Note that the spaces $L^p(G, \mu)$ are independent of the choice of Haar measure.

We will not prove the existence (and uniqueness) parts of this theorem (see, e.g., [**19**, Th. 10.5, Th. 10.14]). For many important classes of groups, it is in fact possible to write down somehow a non-zero measure μ which turns out to satisfy (5.2), and the uniqueness shows that μ is then a Haar measure. For instance, if G is a connected Lie group of dimension d (such as $\mathrm{GL}_n(\mathbf{R})$ for $d = n^2$), then one can pick a positive right-invariant differential form ω of degree d on G, and then the measure given by integrating against ω is a Haar measure, so that

$$\int_G f d\mu = \int_G f(x)\omega(x). \tag{5.4}$$

Proof of (2), (3), (4). Given the statement (1) of existence and uniqueness, part (2) follows as soon as we check that, for a compact group G, the total measure $\mu(G)$ is finite if μ is a Haar measure. This is in fact part of the definition of a Radon measure, namely such a measure is finite on compact sets.

For (3), we fix a Haar measure μ and then observe that for any fixed g, one can define a measure μ_g on G by

$$\int_G f(x)d\mu_g(x) = \int_G f(gx)d\mu(x),$$

and that (because left and right multiplications on G commute!) this is always right invariant on G. Thus, by (1), there exists a non-negative real number $\Delta(g)$ such that

$$\mu_g = \Delta(g)\mu, \tag{5.5}$$

or equivalently, such that

$$\int_G f(gx)d\mu(x) = \Delta(g)\int_G f(x)d\mu(x)$$

for all f.

Taking $f = 1$, which we can do because G is compact and, hence, $\mu(G) < +\infty$, we see that $\Delta(g)\mu(G) = \mu(G)$, and therefore $\Delta(g) = 1$. This means that μ was indeed already left invariant.

A similar argument applies to invariance under inversion. Define ν by

$$\int_G f(x)d\nu(x) = \int_G f(x^{-1})d\mu(x),$$

then (because of the left-invariance just proved), the measure ν is also a Haar measure, and taking $f = 1$ leads to $\nu = \mu$, which is the desired statement.

Finally, the support property (4) is due intuitively to the fact that μ "treats the same" every point, so its support must be either empty (which is excluded because it would correspond to the zero measure) or all of G. Since, properly speaking, the support of a measure is not always well behaved on an arbitrary locally compact space, we will prove the fact that $\mu(U) > 0$ for any open set U, which always makes sense. Suppose to the contrary that $\mu(U) = 0$ for some non-empty open subset U of G; we may assume that

$1 \in U$ (replace U by Ux^{-1} for any $x \in U$, which has the same measure by the invariance of Haar measure). Then, we deduce first that $\mu(K) = 0$ for any compact subset $K \subset G$, because K is contained in a finite union of translates Ux of U, and then using the regularity of Radon measures, we deduce

$$\mu(G) = \sup\{\mu(K) \mid K \subset G \text{ compact}\} = 0,$$

which contradicts the fact that μ is non-zero.

Finally, that continuous functions with compact support inject in L^p-spaces is a consequence of the fact that the support of μ is G, and that bounded functions with compact support are integrable (since μ is finite on compact sets). The deeper fact that the continuous functions with compact support are dense in $L^p(G, \mu)$ for $1 \leqslant p < +\infty$ is a general property of Radon measures, see, e.g., [**19**, Prop. 7.9]. □

Remark 5.2.3. (1) It is important to remember to what extent the Haar measure is unique. If G is compact, it can be normalized uniquely by selecting as the probability Haar measure, but for a general locally compact group, there is *no* preferred Haar measure. In applications where two or more groups are involved simultaneously, it may be quite important—and sometimes delicate!—to assign suitable Haar measures to all of them.

(2) In some texts, the Haar measure is defined as a left invariant measure (satisfying (5.3)), instead of a right-invariant one, as we did. Of course, sometimes the Haar measure as we defined it is also left invariant (this is always the case for compact groups, as we saw), in which case this issue does not matter. However, this property is not always true, and one must be careful about the convention which is used (see Exercise 5.2.5 for an example of a Haar measure which is not left invariant).

In any case, an analogue of Theorem 5.2.1 holds with left-invariant measures instead of right-invariant ones. Moreover, there is a simple link between the two. Indeed, if μ is a Haar measure given by Theorem 5.2.1 (right invariant!), then for any $g \in G$ we obtain a real number $\Delta(g) \geqslant 0$ by the formula (5.5), and one can fairly easily see that $\Delta(g) > 0$ for all g and that the resulting map

$$\begin{cases} G \longrightarrow \mathbf{R}^{+,\times} \\ g \mapsto \Delta(g) \end{cases}$$

is a continuous homomorphism, such that the measure

$$d\nu(g) = \Delta(g) d\mu(g)$$

is a *left-invariant* Haar measure on G (see again Exercise 5.2.5). This function is called the *modulus* of G or *modular character* of G.

Clearly, a Haar measure on G is also left invariant if and only if $\Delta(g) = 1$ for all $g \in G$. A locally compact group G with this property is called *unimodular*. Thus, we have shown that compact groups are always unimodular. Similarly, abelian groups are unimodular since the left and right multiplication operators are then the same. One can also show that connected semisimple Lie groups (for instance $\mathrm{SL}_n(\mathbf{R})$) are unimodular (this follows, e.g., from [**2**, Prop. A.3.6] because the so-called Adjoint representation takes values in $\mathrm{SL}_{\dim(G)}(\mathbf{R})$).

Example 5.2.4 (Examples of Haar measure)**.** (1) If G is a finite group or more generally a discrete group, then a Haar measure is given by the counting measure

$$\mu(X) = |X|$$

for a subset $X \subset G$. For G finite, the probability Haar measure is then defined by

$$\mu(X) = \frac{|X|}{|G|}$$

for $X \subset G$, or in other words by the averaging formula

$$\int_G f(x)d\mu(x) = \frac{1}{|G|}\sum_{x \in G} f(x)$$

for $f \in C(G)$. We recognize a type of expression which was extensively used in Chapter 4!

(2) If $G = \mathbf{R}^d$, $d \geqslant 1$, or $G = (\mathbf{R}/\mathbf{Z})^d$, a Haar measure is obviously given by Lebesgue measure. (This shows that Theorem 5.2.1 is at least as deep as the existence of the Lebesgue measure!)

(3) Let $G = \mathbf{R}^\times$. Then a Haar measure on G is given by

$$d\mu(x) = \frac{dx}{|x|}$$

in terms of the Lebesgue measure dx on $\mathbf{R}$. (This is a special case of (5.4), with the differential form given by $\omega = |x|^{-1}dx$ on G.) This can be proved by a straightforward computation using the change of variable formula for the Lebesgue measure: for $a \in \mathbf{R}^\times$, putting $y = ax$ with $dy = |a|dx$, we have

$$\int_{\mathbf{R}^\times} f(ax)\frac{dx}{|x|} = \int_{\mathbf{R}^\times} f(y)\frac{dy}{|a||y/a|} = \int_{\mathbf{R}^\times} f(y)\frac{dy}{|y|}. \tag{5.6}$$

For the subgroup $\mathbf{R}^{+,\times}$, one may also argue conceptually using the exponential group isomorphism $\mathbf{R} \simeq \mathbf{R}^{+,\times}$ and the obvious fact that if

$$f : G_1 \to G_2$$

is an isomorphism of locally compact groups, i.e., a homeomorphism which is also a group isomorphism, then the direct image $f_*\mu_1$ of a Haar measure on G_1 is a Haar measure on G_2. One must then check that the direct image of the Lebesgue measure under the exponential is $x^{-1}dx$ on $\mathbf{R}^{+,\times}$.

If one restricts this measure to $\mathbf{R}^{+,\times}$, the multiplicative group of positive real numbers, we get the Haar measure $x^{-1}dx$ on the positive real numbers. Note that the invariance properties of the measure $x^{-1}dx$ explains the form of some classical formulas, such as the definition

$$\Gamma(s) = \int_0^{+\infty} x^{s-1}e^{-x}dx$$

of the Gamma function. It is really an integral of $x \mapsto x^s e^{-x}$ with respect to Haar measure on $\mathbf{R}^{+,\times}$.

(4) The example (3) can be generalized to the (genuinely non-abelian, non-compact) group $G = \mathrm{GL}_n(\mathbf{R})$ for $n \geqslant 1$. In terms of the Lebesgue measure $dx = \prod_{i,j} dx_{i,j}$ on the space $\mathrm{M}_n(\mathbf{R}) \simeq \mathbf{R}^{n^2}$ of matrices, a Haar measure on the subset G is given by

$$d\mu(x) = \frac{1}{|\det(x)|^n}dx. \tag{5.7}$$

This is again a simple application of the change of variable formula for multidimensional Lebesgue measure: the maps $x \mapsto xg$ on G extends to a diffeomorphism of $\mathrm{M}_n(\mathbf{R})$ with Jacobian $|\det(g)|^n$, and the result follows, formally, as in (5.6). Note that, although G is not compact (and not abelian), the left-invariance property (5.3) also holds for this Haar measure, so that G is unimodular.

(5) Let $G = \mathrm{SU}_2(\mathbf{C})$. Here, and in many similar circumstances, one needs first to find a convenient parametrization to describe the Haar measure on G. Typically, this means introducing finitely many continuous coordinates on G and using a measure defined in terms of these coordinates using Lebesgue measure or a related measure.

A convenient system of coordinates on $\mathrm{SU}_2(\mathbf{C})$ is obtained by remarking that any $g \in \mathrm{SU}_2(\mathbf{C})$ can be written uniquely

$$g = \begin{pmatrix} a & b \\ -\bar{b} & \bar{a} \end{pmatrix}, \tag{5.8}$$

where $a, b \in \mathbf{C}$ are arbitrary, subject to the unique condition $|a|^2 + |b|^2 = 1$ (we leave the verification of this fact as an exercise). Using the real and imaginary parts of a and b as coordinates, we see that G is homeomorphic to the unit sphere $\mathbf{S}^3$ in $\mathbf{R}^4$.

There is an obvious measure that comes to mind on this unit sphere, the *surface* Lebesgue measure μ, which can be defined as follows in terms of the Lebesgue measure μ_4 on $\mathbf{R}^4$

$$\mu(A) = \mu_4\Big(\Big\{x \in \mathbf{R}^4 - \{0\} \mid \|x\| \leqslant 1 \text{ and } \frac{x}{\|x\|} \in A\Big\}\Big)$$

for $A \subset \mathbf{S}^3$. This measure is natural because it is invariant under the obvious linear action of the orthogonal group $\mathrm{O}_4(\mathbf{R})$ on $\mathbf{S}^3$ (simply because so is the Lebesgue measure μ_4).

We claim that this also implies that μ is in fact a Haar measure on $\mathrm{SU}_2(\mathbf{C})$. Indeed, the map

$$\mathrm{SU}_2(\mathbf{C}) \longrightarrow \mathrm{SU}_2(\mathbf{C})$$

given by multiplication on the right by an element $g \in \mathrm{SU}_2(\mathbf{C})$, when seen as acting on $\mathbf{S}^3$ by the identification of $\mathrm{SU}_2(\mathbf{C})$ and $\mathbf{S}^3$ above, does so as the restriction of an element of $\mathrm{O}_4(\mathbf{R})$ (as an elementary computation reveals), which gives the result.

(6) In Section 7.4, we will discuss in some depth the unitary representations of $G = \mathrm{SL}_2(\mathbf{R})$. Thus it is useful to know that G is also unimodular. This is a special case of the last remark before the example (G is a connected and semisimple Lie group), but can also be seen as a consequence of another general useful fact: if G is a locally compact group such that $G/\overline{[G,G]}$ is compact (where $\overline{[G,G]}$ is the closure of the commutator group of G), then G is unimodular. Indeed, the modular character $g \mapsto \Delta(g)$ is a continuous homomorphism on G (as we have already mentioned, see the exercise below) with abelian image, so that it factors through this compact quotient. Then its image is a compact subgroup of $\mathbf{R}^{+,\times}$, so it must be trivial, which means that $\Delta = 1$.

Exercise 5.2.5. We present a few additional properties and examples of Haar measures in this exercise.

(1) *Compactness and Haar measure.* Show that if G is a locally compact topological group and μ is a Haar measure on G, we have $\mu(G) < +\infty$ if and only if G is compact.

Next, we will explain a slightly different proof of the left-invariance of Haar measure on a compact group, which applies to more general groups.

(2) Fix a Haar measure μ on G. Show that the function $\Delta : G \longrightarrow [0,+\infty[$ (defined by (5.5)) is nowhere zero and is a continuous homomorphism $G \longrightarrow \mathbf{R}^{+,\times}$.

(3) Show that if G is compact, such a homomorphism is necessarily trivial. Show directly (without using (5.7)) that Δ is also necessarily trivial for $G = \mathrm{SL}_n(\mathbf{R})$, $n \geqslant 2$.

(4) Let G be locally compact. Show that the measure

$$d\nu(y) = \Delta(y)d\mu(y)$$

is a non-zero left-invariant measure on G.

(5) Let

$$G = \left\{ \begin{pmatrix} a & b \\ 0 & 1 \end{pmatrix} \mid a \in \mathbf{R}^{\times},\ b \in \mathbf{R} \right\} \subset \mathrm{GL}_2(\mathbf{R})$$

be the affine group in one variable (this is a connected Lie group, but it is not semisimple).

Show that, in terms of the coordinates a, b, the measure

$$d\mu = \frac{da\, db}{|a|}$$

is a Haar measure on G. Check that it is *not* left invariant, i.e., (5.3) does not always hold. What is the function $\Delta(g)$ in this case? What is a non-zero left-invariant measure on G?

With the Haar measure in hand, we can define the regular representation of a locally compact group.

Proposition 5.2.6 (The regular representation). *Let G be a locally compact group, and let μ be a Haar measure on G. The regular action*

$$\varrho_G(g)\varphi(x) = \varphi(xg)$$

is a well-defined unitary representation of G on $L^2(G, \mu)$, which is strongly continuous.

Up to isometry, this representation is independent of the choice of Haar measure and is called "the" regular representation of G.

Similarly, the left-regular representation λ_G is defined on $L^2(G, \mu)$ by

$$\lambda_G(g)\varphi(x) = \varphi(g^{-1}x),$$

and the two representations commute: we have $\lambda_G(g)\varrho_G(h) = \varrho_G(h)\lambda_G(g)$ for all g, $h \in G$.

Proof. Although this sounds formal (and an important goal of the theory is to set it up in such a way that it becomes formally as easy and flexible as it is in the case of finite groups), it is important to see that there is actually some non-trivial subtleties involved.

First of all, the regular action is clearly well defined for functions, but an element of $L^2(G, \mu)$ is an equivalence class of functions on G, where functions which differ on a set of Haar measure zero are identified. Thus we must check that, if φ has this property, so does $\varrho_G(g)\varphi$. This follows directly from the invariance of Haar measure (but it is not a triviality). Once this is settled, we check that ϱ_G is unitary using once more the invariance of μ:

$$\|\varrho_G(\varphi)\|^2 = \int_G |\varphi(xg)|^2 d\mu(x) = \int_G |\varphi(x)|^2 d\mu(x) = \|\varphi\|^2.$$

What is by no means obvious is the continuity of the representation. We use the strong continuity criterion of Proposition 3.4.3 to check this.[2]

We first take $\varphi \in C(G)$, and we check the continuity at $g = 1$ of

$$g \mapsto \varrho_G(g)\varphi$$

as follows. We have

$$\|\varrho_G(g)\varphi - \varphi\|^2 = \int_G |\varphi(xg) - \varphi(x)|^2 d\mu(x),$$

and since $\varphi(xg) - \varphi(x) \to 0$ as $g \to 1$, for every $x \in G$, while

$$|\varphi(xg) - \varphi(x)|^2 \leqslant 4\|\varphi\|_\infty^2,$$

we see from the dominated convergence theorem that

$$\lim_{g \to 1} \|\varrho_G(g)\varphi - \varphi\|^2 = 0,$$

which is the desired statement in that case.

Now if φ is arbitrary, we use the density of $C(G)$ in $L^2(G, \mu)$. Let $\varepsilon > 0$ be arbitrarily small. We first find a continuous function $\varphi_\varepsilon \in C(G)$ such that $\|\varphi - \varphi_\varepsilon\| < \varepsilon$. Then for any $g \in G$, we have

$$\begin{aligned}\|\varrho_G(g)\varphi - \varphi\| &\leqslant \|\varrho_G(g)\varphi - \varrho_G(g)\varphi_\varepsilon\| + \|\varrho_G(g)\varphi_\varepsilon - \varphi_\varepsilon\| + \|\varphi_\varepsilon - \varphi\| \\ &= 2\|\varphi_\varepsilon - \varphi\| + \|\varrho_G(g)\varphi_\varepsilon - \varphi_\varepsilon\| \\ &\leqslant 2\varepsilon + \|\varrho_G(g)\varphi_\varepsilon - \varphi_\varepsilon\|.\end{aligned}$$

By the previous case, for all g in some open neighborhood of 1 in G, we have

$$\|\varrho_G(g)\varphi_\varepsilon - \varphi_\varepsilon\| < \varepsilon,$$

and hence for all such g we get

$$\|\varrho_G(g)\varphi - \varphi\| \leqslant 3\varepsilon,$$

and therefore the desired strong continuity. (Note that this argument spells out the result sketched in Exercise 3.4.13.)

Finally, we observe that if we replace the Haar measure μ with $\nu = \alpha\mu$, with $\alpha > 0$, the linear map

$$\begin{cases} L^2(G, \mu) & \longrightarrow & L^2(G, \nu) \\ f & \mapsto & \alpha^{-1/2} f \end{cases}$$

is an isometry that is immediately seen to intertwine the regular representations on the two spaces. Thus the regular representation is well defined up to isomorphism, independenly of the choice of Haar measure. □

[2] Except when G is finite, ϱ_G will not be continuous in the norm topology on the unitary group $\mathrm{U}(L^2(G, \mu))$.

Exercise 5.2.7 (Representations on L^p-spaces). Let G be a locally compact topological group, and let μ be a Haar measure on G. For any real number $p \geqslant 1$, show that there is a strongly continuous representation of G on $L^p(G,\mu)$, denoted ϱ, such that

$$\varrho(g)\varphi(x) = \varphi(xg)$$

for $\varphi \in L^p(G,\mu)$ and (almost all) $g \in G$. Check that if $\varphi \in L^2(G,\mu) \cap L^p(G,\mu)$, any L^p-translate $\varrho(g)\varphi$ coincides with the corresponding translate $\varrho_G(g)\varphi$ under the regular representation, in particular $\varrho_G(g)\varphi \in L^2(G,\mu) \cap L^p(G,\mu)$ also. (For this reason, in the very few cases where we use the L^p-regular representation, we will just use the same notation as for the L^2-representation.)

Exercise 5.2.8 (The regular representation is faithful).

(1) Show that the regular representation of a locally compact group G is faithful.

We now use this to give a simple application of representation theory to prove a general fact about *compact* topological groups. Let G be a compact group. The goal is to show the following: for any neighborhood U of 1 in G, there exists a neighborhood $V \subset U$ which is invariant under conjugacy, i.e., such that $xVx^{-1} \subset V$ for all $x \in G$. This looks deceptively simple but it is a bit tricky to prove directly (the reader may want to try!)

(2) Let $U \subset G$ be a neighborhood of 1. Show that there exist $\varepsilon > 0$ and finitely many functions $f_i \in L^2(G)$ with norm 1 such that

$$U \supset \{g \in G \mid \|\varrho_G(g)f_i - f_i\| < \varepsilon \text{ for all } i\}. \tag{5.9}$$

(*Hint*: Use (1) and the continuity of the regular representation of G, where the unitary group carries the strong operator topology, as in Remark 3.4.4.)

(3) For a fixed index i, let $A_i \subset L^2(G)$ be the set

$$A_i = \{\varrho_G(x)f_i \mid x \in G\}$$

of translates of f_i. Show that the set

$$V_i = \{g \in G \mid \|\varrho_G(g)f - f\| < \varepsilon \text{ for all } f \in A_i\}$$

is conjugacy-invariant and is equal to

$$V_i = \bigcap_{x \in G} xU_ix^{-1}, \qquad U_i = \{g \in G \mid \|\varrho_G(g)f_i - f_i\| < \varepsilon\}.$$

(4) To conclude, show that V_i is an open neighborhood of 1. (*Hint*: Use the fact that A_i is compact.)

(5) Show that in the non-compact group $\mathrm{SL}_2(\mathbf{R})$, it is not true that all neighborhoods of 1 contain a conjugacy-invariant neighborhood. Which parts of the argument above fail in that case?

What might be remembered of this exercise is the fact that the formula (5.9) shows how to use the regular representation of G to write down (a basis of) neighborhoods of 1 in G in such a way that they can be manipulated further.

Example 5.2.9 (Representations from measure-preserving actions). The regular representation of a finite group is a special case of a permutation representation. Similarly, the definition of the regular representation of a locally compact group can be generalized to analogues of more general permutation representations. Namely, consider a set X on which G acts (on the left) with the property that X carries a finite Radon measure ν and that G acts through measure-preserving transformations, i.e., with

$$\int_X f(g \cdot x) d\nu(x) = \int_X f(x) d\nu(x)$$

for any fixed $g \in G$ and f either $\geqslant 0$ measurable, or integrable. Then one can define a representation of G on $L^2(X, \nu)$ by

$$(g \cdot \varphi)(x) = \varphi(g^{-1} \cdot x).$$

Arguing as in the proof of Proposition 5.2.6, one sees that this is a unitary representation of G.

For a specific example, consider the group $G = \mathrm{SO}_3(\mathbf{R})$ of rotations of $\mathbf{R}^3$ acting (by matrix-vector multiplication) on the unit ball $B = \{x \in \mathbf{R}^3 \mid \|x\| \leqslant 1\}$. Taking for ν the restriction to B of Lebesgue measure on $\mathbf{R}^3$, we see that G preserves ν, and we obtain therefore a representation of $\mathrm{SO}_3(\mathbf{R})$ on $L^2(B, \nu)$.

Example 5.2.10 (Induced representations). Using the Haar measure, one can also define the proper analogue of induced representations in the setting of compact groups. Consider a compact group G, with Haar measure μ, and a compact (equivalently, closed) subgroup $K \subset G$. Given a unitary representation

$$\varrho : K \longrightarrow \mathrm{U}(\mathrm{H})$$

of K, one defines the induced representation $\pi = \mathrm{Ind}_K^G(\varrho)$ as follows. Define first the vector space

$$V_0 = \{f : G \longrightarrow \mathrm{H} \mid f \text{ is continuous, and } f(kg) = \varrho(k)f(g) \text{ for all } k \in K,\ g \in G\}, \tag{5.10}$$

on which G acts by the regular action

$$\pi(g)f(x) = f(xg).$$

Define an inner product on V_0 by

$$\langle f_1, f_2\rangle_0 = \int_G \langle f_1(x), f_2(x)\rangle_{\mathrm{H}} d\mu(x).$$

This is well defined (the integrand is a continuous function on G) and is positive definite on V_0. Moreover, it is G-invariant because of the invariance of Haar measure, namely

$$\langle \pi(g)f_1, \pi(g)f_2\rangle_0 = \int_G \langle f_1(xg), f_2(xg)\rangle_{\mathrm{H}} d\mu(x) = \langle f_1, f_2\rangle_0$$

for all $g \in G$.

So we almost have a unitary representation of G on V_0. But V_0 has no reason (in general) to be a Hilbert space, as completeness will usually fail. Still, one can check that π is strongly continuous on V_0 (so that it is a pre-unitary representation, as discussed in Exercise 3.4.13). Then, following the idea sketched in that exercise, we define V to be the completion of V_0 with respect to $\langle\cdot,\cdot\rangle_0$. This is a Hilbert space in which V_0 is a dense subspace, and since the $\pi(g)$ were unitary on V_0, they extend by continuity to unitary operators on V. Similarly, since the properties

$$\pi(gh) = \pi(g)\pi(h), \quad \pi(g^{-1}) = \pi(g)^{-1}$$

hold on the dense subspace $V_0 \subset V$, they are valid on all of V. The proof of the strong continuity of this representation is now obtained as in the case of the regular representation, using the fact that π is strongly continuous on V_0, unitarity, and the fact that V_0 is dense in V.

Just as was the case in Chapter 2 (see (2.23)), the regular representation can be identified with the induced representation $\mathrm{Ind}_1^G(\mathbf{1})$. Indeed, in that case the space V_0 is the space of continuous functions on G, with the inner product of $L^2(G,\mu)$ and the same action as the regular representation, so that the statement amounts to the fact that $C(G)$ is dense in $L^2(G,\mu)$ for the L^2-norm.[3]

Apart from its use in describing certain representations, the most important feature of induction is Frobenius reciprocity. Looking at the proof of Proposition 2.3.9, it may seem at first to be mostly formal, and likely to extend without much ado to compact groups. However, note that part of the construction in (2.27) involves evaluating functions in the space of the induced representation at a single point, which is not well defined in general in L^2-type spaces. Nevertheless, after proving the Peter–Weyl Theorem, we will be able to show that some important cases of Frobenius reciprocity hold (Proposition 5.4.9).

[3] More generally, one can give a description of the space V for an arbitrary induced representation in terms of square-integrable H-valued functions, defined in the spirit of what will be done in the next section.

We finish this section with a discussion of unitarizability. Using integration with respect to the Haar measure, one gets the following useful fact:

Theorem 5.2.11 (Unitarizability of representations of compact groups). *Let G be a compact topological group.*

(1) *Any finite-dimensional continuous representation of G is unitarizable. As a consequence, a finite-dimensional representation of G is semisimple.*

(2) *More generally, any continuous representation*

$$\varrho : G \longrightarrow \mathrm{BGL}(\mathrm{H})$$

with values in the group of invertible linear maps on a Hilbert space H *is unitarizable, i.e., there exists an inner product $\langle\cdot,\cdot\rangle_\varrho$ on* H *such that $\varrho(g) \in \mathrm{U}(\mathrm{H}, \langle\cdot,\cdot\rangle_\varrho)$ for all $g \in G$ and such that the topology defined by this inner product is the same as the original one on* H.

Proof. The first part is easy, as in the case of finite groups: we merely use integration with respect to a Haar measure μ, instead of averaging, to construct an invariant inner product. Precisely, let

$$\varrho : G \longrightarrow \mathrm{GL}(E)$$

be a finite-dimensional complex representation of G. Fix an arbitrary inner product $\langle\cdot,\cdot\rangle_0$ on E, and define

$$\langle v, w\rangle = \int_G \langle \varrho(x)v, \varrho(x)w\rangle_0 d\mu(x).$$

The continuity of ϱ shows that the integral is well defined (integral of a continuous bounded function on a compact space); it is obviously a non-negative hermitian form on E, and the invariance of the Haar measure shows that it is G-invariant. We have

$$\begin{aligned}\langle \varrho(g)v, \varrho(g)w\rangle &= \int_G \langle \varrho(x)\varrho(g)v, \varrho(x)\varrho(g)w\rangle_0 d\mu(x)\\ &= \int_G \langle \varrho(x)v, \varrho(x)w\rangle_0 d\mu(x).\end{aligned}$$

Moreover, note that

$$\|v\|^2 = \int_G \|\varrho(x)v\|_0^2 d\mu(x),$$

and if $v \neq 0$, this is the integral over G of a non-negative, continuous function which is non-zero since it takes the value $\|v\|_0^2 > 0$ at $x = 1$. Therefore, we have $\|v\|^2 > 0$ if $v \neq 0$, and the hermitian form is positive definite.

Thus ϱ can be seen as a homomorphism

$$\varrho : G \longrightarrow \mathrm{U}(\langle\cdot,\cdot\rangle).$$

As a final step, since all inner products on a finite-dimensional vector space define the same topology (which coincides with the strong topology), this representation is still strongly continuous.

For (2) everything in the above goes through identically, using the given norm $\|\cdot\|$ on H instead of $\|\cdot\|_0$, except this last step. It might conceivably be the case that

$$\langle v, w\rangle_\varrho = \int_G \langle \varrho(g)v, \varrho(g)w\rangle d\mu(g)$$

defines a different topology on an infinite-dimensional Hilbert space H. However, this is not the case. For every $v \in \mathrm{H}$, the map

$$g \mapsto \varrho(g)v$$

is continuous, and hence

$$\sup_{g\in G} \|\varrho(g)v\| < +\infty.$$

This means that the image of ϱ is *pointwise* bounded in L(H). The Banach–Steinhaus theorem (see, e.g., [**49**, Th. III.9]) implies that it is uniformly bounded on the unit ball, i.e., that

$$M = \sup_{g\in G} \|\varrho(g)\|_{\mathrm{L(H)}} < +\infty,$$

so that we get

$$\|\varrho(g)v\|^2 \leqslant M\|v\|^2, \qquad M^{-1}\|v\|^2 = M^{-1}\|\varrho(g^{-1})\varrho(g)v\|^2 \leqslant \|\varrho(g)v\|^2,$$

for every $g \in G$ and $v \in \mathrm{H}$. Integrating leads to

$$M^{-1}\|v\|^2 \leqslant \langle v, v\rangle_\varrho \leqslant M\|v\|^2,$$

so that the new norm is topologically equivalent to the old one. □

Example 5.2.12 (Inner product on finite-dimensional representations of $\mathrm{SU}_2(\mathbf{C})$)**.** We explain here how to compute an invariant inner product for the representation ϱ_m of $\mathrm{SU}_2(\mathbf{C})$ on the space V_m of homogeneous polynomials of degree m in $\mathbf{C}[X, Y]$.

One method is to follow the construction used in the proof of the theorem by picking an inner product on V_m and averaging it over G using a Haar measure on the group. In order to simplify the computation, it pays to make a careful choice. One observation that can help us choose wisely (here and in general) is the one in Lemma 3.4.21: unrelated unitary subrepresentations are orthogonal. This cannot be applied to V_m as a representation of $\mathrm{SU}_2(\mathbf{C})$ (of course), but we can apply it to the restriction to a suitable subgroup. Indeed, consider the diagonal subgroup

$$K = \left\{\begin{pmatrix} e^{i\theta} & 0 \\ 0 & e^{-i\theta}\end{pmatrix} \mid \theta \in \mathbf{R}\right\},$$

so that (as in (2.41)) the space V_m decomposes as the direct sum of the subspaces $\mathbf{C}e_j$ generated by the basis vectors $e_j = X^j Y^{m-j}$, $0 \leqslant j \leqslant m$, on which K acts by a one-dimensional representation, namely

$$\varrho_m\left(\begin{pmatrix} e^{i\theta} & 0 \\ 0 & e^{-i\theta} \end{pmatrix}\right) e_j = e^{i(2j-m)\theta} e_j.$$

Since the lines $\mathbf{C}e_j$ span distinct irreducible K-subrepresentations of V_m, it follows that for $j \neq k$, we have

$$\langle e_j, e_k \rangle = 0$$

for any invariant inner product on V_m.

Now we come back to the averaging procedure to compute the remaining inner products $\langle e_j, e_j \rangle$. However, in order to simplify again the computation, we adopt here the variant described at the end of Remark 4.1.3: it is enough to select a *non-negative* hermitian form $\langle \cdot, \cdot \rangle_0$ on V_m, not necessarily positive definite, provided the form defined by

$$\langle P_1, P_2 \rangle = \int_{\mathrm{SU}_2(\mathbf{C})} \langle \varrho_m(g) P_1, \varrho_m(g) P_2 \rangle_0 \, d\mu(g)$$

for polynomials P_1, $P_2 \in V_m$ is *itself* positive definite. We select

$$\langle P_1, P_2 \rangle_0 = P_1(1,0)\overline{P_2(1,0)}$$

and claim that this property does hold. Indeed, we obtain

$$\langle P, P \rangle = \int_{\mathrm{SU}_2(\mathbf{C})} |(\varrho_m(g)P)(1,0)|^2 d\mu(g) = \int_{\mathrm{SU}_2(\mathbf{C})} |P(a,b)|^2 d\mu(g),$$

where we write (as usual)

$$g = \begin{pmatrix} a & b \\ c & d \end{pmatrix},$$

so that $\|P\| = 0$ if and only if P vanishes on every $(a, b) \in \mathbf{C}^2$ which form the first row of a matrix in $\mathrm{SU}_2(\mathbf{C})$. As observed in (5.8), these are the pairs of complex numbers with $|a|^2 + |b|^2 = 1$, and by homogeneity of $P \in V_m$, it follows that in fact $P = 0$, proving that the inner product defined in this manner is positive definite.

It remains to compute the inner products $\langle e_j, e_j \rangle$. We get

$$\langle e_j, e_j \rangle = \int_{\mathrm{SU}_2(\mathbf{C})} |a|^{2j} |b|^{2(m-j)} d\mu(g),$$

and using the description of Haar measure from Example 5.2.4(5), one can show that this is equal to

$$\langle e_j, e_j \rangle = \int_0^1 t^j (1-t)^{m-j} dt = \frac{1}{(m+1)\binom{m}{j}} \tag{5.11}$$

(this is the value $B(j+1, m-j+1) = j!(m-j)!/(m+2)!$ of Euler's *beta function*).

Another argument for determining this invariant inner product is based on an a priori computation based on its (known) existence and its invariance property. Using the same basis vectors as above, we find, for instance, that for all $\theta \in \mathbf{R}$ and j, k, we must have

$$\begin{aligned}\langle e_j, e_k\rangle &= \left\langle \begin{pmatrix} e^{i\theta} & 0 \\ 0 & e^{-i\theta}\end{pmatrix} e_j, \begin{pmatrix} e^{i\theta} & 0 \\ 0 & e^{-i\theta}\end{pmatrix} e_k\right\rangle \\ &= \langle e^{i(2j-m)\theta} e_j, e^{i(2k-m)\theta} e_k\rangle = e^{2i(j-k)\theta}\langle e_j, e_k\rangle,\end{aligned}$$

which again immediately implies that e_j and e_k must be orthogonal when $j \neq k$. This means that (e_j) is an orthogonal basis, and (as before) that we need only determine the norms $\|e_j\|^2 = \langle e_j, e_j\rangle$ in order to find the inner product.

This must naturally rely on other elements of $\mathrm{SU}_2(\mathbf{C})$ than the diagonal ones (otherwise, we would be arguing with ϱ_m restricted to the diagonal subgroup K, where any choice of $\|e_j\|^2 > 0$ gives a K-invariant inner product). We explain the method used in [**63**, III.2.4].[4] We consider the elements

$$g_t = \begin{pmatrix} \cos t & -\sin t \\ \sin t & \cos t\end{pmatrix}, \qquad t \in \mathbf{R},$$

of $\mathrm{SO}_2(\mathbf{R}) \subset \mathrm{SU}_2(\mathbf{C})$, and we will differentiate with respect to t, and then evaluate at 0, the invariance formula

$$0 = \langle g_t \cdot e_j, g_t \cdot e_{j+1}\rangle. \tag{5.12}$$

Denote by A the linear operator defined on V_m by

$$AP = \frac{d}{dt}(g_t \cdot P)\Big|_{t=0}$$

for a polynomial $P \in V_m$. Because we differentiate an inner product (so that the Leibniz rule is valid), we obtain first the relation

$$\langle Ae_j, e_{j+1}\rangle + \langle e_j, Ae_{j+1}\rangle = 0 \tag{5.13}$$

by differentiating (5.12) at $t = 0$.

Now we compute that action of A on the basis vectors e_j. We have

$$g_t \cdot e_j = (X\cos t + Y \sin t)^j(-X\sin t + Y\cos t)^{m-j}$$

[4] Which can be seen a simple case of studying the group $\mathrm{SU}_2(\mathbf{C})$ through its *Lie algebra*.

and therefore, by differentiating, we get

$$\frac{d}{dt}g_t \cdot e_j = j(-X\sin t + Y\cos t)(X\cos t + Y\sin t)^{j-1}$$
$$\times (-X\sin t + Y\cos t)^{m-j} + (m-j)(X\cos t + Y\sin t)^j$$
$$\times (-X\cos t - Y\sin t)(-X\sin t + Y\cos t)^{m-j-1},$$

and therefore, evaluating at $t = 0$, we find that

$$Ae_j = je_{j-1} - (m-j)e_{j+1}.$$

Inserting this formula in (5.13) and using the orthogonality of the "non-diagonal" inner products, we derive the recurrence relation

$$(j+1)\langle e_j, e_j\rangle = (m-j)\langle e_{j+1}, e_{j+1}\rangle.$$

This determines, up to a constant scalar factor $c > 0$, the norms $\|e_j\|^2$, and indeed a quick induction leads to the formula

$$\langle e_j, e_j\rangle = cj!(m-j)! \qquad 0 \leqslant j \leqslant m,$$

which coincides—as it should!—with (5.11) when taking $c^{-1} = (m+1)!$.

5.3. The analogue of the group algebra

It is now natural to discuss the analogue of the action of the group algebra of a finite group. However, some readers may prefer to skip to the next section, and to come back once the proof of the Peter–Weyl Theorem has shown that this extension is a natural step.

The group algebra for a finite group G is the source of endomorphisms of a representation space ϱ which are linear combinations of the $\varrho(g)$. When G is compact, but possibly infinite, the group algebra (over $\mathbf{C}$) can still be defined (as finite linear combinations of symbols $[g]$, $g \in G$), but the endomorphisms it defines are not sufficient anymore. For instance, in a group like $\mathrm{SU}_2(\mathbf{C})$, the center of the group algebra is too small to create interesting intertwiners (e.g., on the regular representation) because all conjugacy classes in $\mathrm{SU}_2(\mathbf{C})$, except those of ± 1, are infinite.

It seems intuitively clear that one can solve this problem of paucity by replacing the sums defining elements of the group algebra with integrals (with respect to Haar measure). This means, that we want to consider suitable functions ψ on G and define

$$\int_G \psi(g)[g]d\mu(g),$$

in some sense. More concretely, we will consider a unitary representation ϱ and define $\varrho(\psi)$ as the linear map

$$\varrho(\psi) = \int_G \psi(g)\varrho(g)d\mu(g). \tag{5.14}$$

This is clearly well defined when ϱ is finite dimensional (the integration can be computed coordinate-wise after selecting a basis)[5] but needs some care when ϱ is not (e.g., when G is infinite and ϱ is the regular representation), since we are then integrating a function with values in the space L(H) of continuous linear maps on H, for some infinite-dimensional Hilbert space H.

Such an integral can indeed be defined. A natural approach would be to use the extension of Lebesgue's integration theory to Banach-space valued functions, but this is not so commonly considered in first integration courses. We use an alternate definition using *weak integrals* which is enough for our purposes and is much quicker to set up. The basic outcome is, for $\psi \in L^1(G)$ (with respect to Haar measure, as usual), one can define continuous linear operators $\varrho(\psi)$, for any unitary representation ϱ of G, which behaves exactly as one would expect from the formal expression (5.14) and the standard properties of integrals.

Proposition 5.3.1 (L^1-action on unitary representations). *Let G be a locally compact group with Haar measure μ. For any integrable function $\psi \in L^1(G)$ and any unitary representation $\varrho : G \longrightarrow \mathrm{U(H)}$, there exists a unique continuous linear operator*

$$\varrho(\psi) : \mathrm{H} \longrightarrow \mathrm{H}$$

such that

$$\langle \varrho(\psi)v, w\rangle = \int_G \psi(g)\langle \varrho(g)v, w\rangle d\mu(g) \tag{5.15}$$

for any vectors v, $w \in \mathrm{H}$. This has the following properties.

(1) *For a fixed ϱ, the map*

$$\begin{cases} L^1(G) & \longrightarrow & \mathrm{L(H)} \\ \psi & \mapsto & \varrho(\psi) \end{cases}$$

is linear and continuous, with norm at most 1, i.e., $\|\varrho(\psi)v\| \leqslant \|\psi\|_{L^1}\|v\|$ for all $\psi \in L^1(G)$. If $\Phi : \varrho \longrightarrow \pi$ is a homomorphism of unitary representations, we have

$$\Phi \circ \varrho(\psi) = \pi(\psi) \circ \Phi.$$

[5] Formally speaking, one should check that the result is independent of the basis, but that is of course easy, e.g., by computing the coordinates with respect to a fixed basis.

(2) *For a fixed $\psi \in L^1(G)$, the adjoint of $\varrho(\psi)$ is given by*

$$\varrho(\psi)^* = \varrho(\check{\psi}), \tag{5.16}$$

where the adjoint function $\check{\psi}$ *is defined by* $\check{\psi}(g) = \overline{\psi(g^{-1})}$.

(3) *For any ψ and ϱ, and for any subrepresentation π of ϱ acting on $\mathrm{H}_1 \subset \mathrm{H}$, the restriction of $\varrho(\psi)$ to H_1 is given by $\pi(\psi)$. In other words, H_1 is also invariant under the action of all operators $\varrho(\psi)$ for $\psi \in L^1(G)$.*

(4) *For any $g \in G$ and $\psi \in L^1(G)$, we have*

$$\varrho(g)\varrho(\psi) = \varrho(\lambda_G(g)\psi), \qquad \varrho(\psi)\varrho(g) = \varrho(\varrho_G(g^{-1})\psi), \tag{5.17}$$

where ϱ_G and λ_G represent here the right- and left-regular representations acting on $L^1(G)$, as in Exercise 5.2.7. *For $\psi \in L^2(G)$, this coincides with the usual regular representations.*

We will denote

$$\varrho(\psi) = \int_G \psi(x)\varrho(x)d\mu(x),$$

and for any vector $v \in \mathrm{H}$, we also write

$$\varrho(\psi)v = \int_G \psi(x)\varrho(x)v d\mu(x),$$

which is a vector in H. The properties of the proposition can then all be seen to be formally consequences of these expressions, and can be remembered (or recovered when needed) in this way. For instance, the inequality $\|\varrho(\psi)\|_{\mathrm{L(H)}} \leqslant \|\psi\|_{L^1(G)}$ can formally be recovered by writing

$$\Big\|\int_G \psi(x)\varrho(x)d\mu(x)\Big\| \leqslant \int_G |\psi(x)|\|\varrho(x)\|d\mu(x) = \int_G |\psi(x)|d\mu(x),$$

which is of course perfectly natural (this uses the fact that $\varrho(x)$ has norm 1 for all x.)

Proof. (1) The procedure is a familiar one in the theory of Hilbert spaces: a vector can be pinpointed (and shown to exist) by describing what its inner products with other vectors are, and these can be prescribed arbitrarily, provided only that they are linear and continuous functions of their argument. Precisely, given $\psi \in L^1(G)$, a unitary representation ϱ on H and a vector $v \in \mathrm{H}$, define

$$\ell_{\psi,v} \begin{cases} \mathrm{H} \longrightarrow \mathbf{C} \\ w \mapsto \displaystyle\int_G \psi(g)\langle w, \varrho(g)v\rangle d\mu(g). \end{cases}$$

This is well defined because ψ is integrable and the factor

$$|\langle w, \varrho(g)v\rangle| \leqslant \|w\|\|v\|$$

is bounded. It is also continuous, by strong continuity of ϱ, hence measurable. Moreover, the map $\ell_{\psi,v}$ is clearly a linear form on H, and it is continuous, since the same inequality leads to

$$|\ell_{\psi,v}(w)| \leqslant C\|w\|, \qquad C = \|v\|\|\psi\|_{L^1(G)},$$

for all $w \in$ H. According to the Riesz representation theorem for Hilbert spaces, there exists therefore a unique vector, which we denote $\varrho(\psi)v$, such that

$$\ell_{\psi,v}(w) = \langle w, \varrho(\psi)v\rangle$$

for all $w \in$ H. Taking the conjugate, we obtain (5.15), and the uniqueness is then also achieved.

From this, the remaining properties are quite easily checked. For (1), the linearity (both the linearity of $\varrho(\psi)$ as a map on H and then the linearity as a function of ψ) is deduced in a very standard way from the uniqueness and the linearity of $\ell_{\psi,v}$ as function of v and ψ.

Riesz's theorem also implies that $\|\varrho(\psi)v\| = \|\ell_{\psi,v}\|$, which is bounded by the constant $C = \|v\|\|\psi\|_{L^1}$ above. This inequality

$$\|\varrho(\psi)v\| \leqslant \|\psi\|_{L^1}\|v\|$$

shows that $\varrho(\psi)$ is continuous for a fixed ψ, and also that it is continuous as a map $L^1(G) \longrightarrow \mathrm{L(H)}$, with norm at most 1, as claimed.

(2) We leave this as an exercise.

(3) This plays an important role later on, so we give the following argument. Let $v \in \mathrm{H}_1$ be a vector in the stable subspace. We first check that $\varrho(\psi)v$ is also in H_1, using orthogonality. If $w \in \mathrm{H}_1^{\perp}$ is orthogonal to H_1, we have

$$\langle \varrho(\psi)v, w\rangle = \int_G \psi(g)\langle \varrho(g)v, w\rangle d\mu(g) = 0,$$

i.e., $\varrho(\psi)v \in (\mathrm{H}_1^{\perp})^{\perp} = \mathrm{H}_1$, as desired. But then, $\varrho(\psi)v$ is determined by its inner products with vectors $w \in \mathrm{H}_1$, and then we have

$$\langle \pi(\psi)v, w\rangle = \int_G \psi(g)\langle \pi(g)v, w\rangle d\mu(g)$$

by definition, which—since π *is* simply ϱ restricted to H_1— is equal to

$$\int_G \psi(g)\langle \varrho(g)v, w\rangle d\mu(g) = \langle \varrho(\psi)v, w\rangle$$

as expected.

(4) These formulas replace formal computations based on the invariance of the Haar measure under translation. We only write down one of these as

such a formal computation: For $g \in G$ and $\psi \in L^1(G)$, we have

$$\begin{aligned}\varrho(g)\varrho(\psi) &= \varrho(g)\int_G \psi(x)\varrho(x)d\mu(x)\\ &= \int_G \psi(x)\varrho(gx)d\mu(x) = \int_G \psi(g^{-1}y)\varrho(y)d\mu(y),\end{aligned}$$

which is is the first formula. We leave the other to the reader, as well as the task of translating it into a formal argument. ☐

Exercise 5.3.2 (More general integrals). Many variants of this construction are possible. For instance, let H be a separable Hilbert space (so there is a countable subset of H which is dense in H). For any function

$$f : G \longrightarrow \mathrm{H},$$

which is *weakly measurable* (in the sense that for every $w \in \mathrm{H}$, the function

$$g \mapsto \langle f(g), w\rangle$$

is measurable on G, and which has bounded values (in the norm of H)), show how to define the integrals

$$\int_G f(g)d\mu(g) \in \mathrm{H}$$

and show that this construction is linear with respect to f, and satisfies

$$\Big\|\int_G f(g)d\mu(g)\Big\| \leqslant \int_G \|f(g)\|d\mu(g) \tag{5.18}$$

(you will first have to show that $g \mapsto \|f(g)\|$ is integrable). Moreover, for a unitary representation ϱ of a locally compact group G on H and for any $\psi \in L^1(G)$ and $v \in \mathrm{H}$, show that

$$\varrho(\psi)v = \int_G f(g)d\mu(g)$$

for $f(g) = \psi(g)\varrho(g)v$.

Exercise 5.3.3 (Intertwiners from L^1-action). Let G be a locally compact group, and let μ be a Haar measure on G.

(1) For an integrable function φ on G, explain how to define the property that φ is a class function (i.e., formally, $\varphi(xyx^{-1}) = \varphi(y)$ for all x and y in G).

(2) Let φ be an integrable class function on G. For any unitary representation ϱ of G, show that the operator

$$\int_G \varphi(x)\varrho(x)d\mu(x)$$

is in $\mathrm{Hom}_G(\varrho, \varrho)$, i.e., is an intertwiner.

Exercise 5.3.4 (Convolution). Let G be a compact group, and let μ be a Haar measure on G. For any functions φ, $\psi \in L^2(G)$, show that

$$\lambda_G(\varphi)\psi = \varphi \star \psi,$$

where the convolution $\varphi \star \psi$ is defined by

$$(\varphi \star \psi)(g) = \int_G \varphi(x)\psi(x^{-1}g)d\mu(x).$$

Prove also that

$$(\varphi \star \psi)(g) = \langle \varphi, \lambda_G(g)\check{\psi} \rangle,$$

where the adjoint function $\check{\psi}$ is given by (5.16), and deduce that $\varphi \star \psi$ is a continuous function on G.

The formulas in (5.17) express the link between the action of L^1-functions on H (given by the implicit definition (5.15) using inner products) and the original representation ϱ. The following is another important relation. Crucially, it shows how to recover the original representation operators $\varrho(g)$ starting from the collection of linear maps $\varrho(\psi)$. This is trickier than for finite groups, because there is (in the infinite case) no integrable function supported only at a single point $g \in G$.

Proposition 5.3.5 (L^1-approximation of representation operators). *Let G be a compact topological group, and let $\varrho : G \longrightarrow$ U(H) be a unitary representation of G. Fix $g \in G$, and for any open neighborhood U of g in G, write ψ_U for the characteristic function of U, normalized so that $\|\psi_U\| = 1$, i.e., define*

$$\psi_U(x) = \begin{cases} \dfrac{1}{\mu(U)} & \text{if } x \in U \\ 0 & \text{otherwise.} \end{cases}$$

Then we have

$$\lim_{U \downarrow g} \varrho(\psi_U) = \varrho(g),$$

where the limit is a limit as U tends to g, taken in the strong topology in L(H), *i.e., it should be interpreted as follows: for any $v \in$ H, and for any $\varepsilon > 0$, there exists an open neighborhood V of g such that*

$$\|\varrho(\psi_U)v - \varrho(g)v\| < \varepsilon \tag{5.19}$$

for any $U \subset V$.

Proof. This is a simple analogue of the classical statements of *approximation by convolution* in integration theory. We use (5.15) and the fact that the integral of ψ_U is one, to write

$$\langle \varrho(\psi_U)v - \varrho(g)v, w \rangle = \int_G \psi_U(x)\langle \varrho(x)v - \varrho(g)v, w \rangle d\mu(x)$$

for any U and any $w \in \mathrm{H}$. Therefore

$$\begin{aligned}|\langle \varrho(\psi_U)v - \varrho(g)v, w\rangle| &\leqslant \|w\| \int_G \psi_U(x)\|\varrho(x)v - \varrho(g)v\| d\mu(x)\\ &\leqslant \|w\| \sup_{x\in U} \|\varrho(x)v - \varrho(g)v\|\end{aligned}$$

for all $w \in \mathrm{H}$. This implies that

$$\|\varrho(\psi_U)v - \varrho(g)v\| \leqslant \sup_{x\in U} \|\varrho(x)v - \varrho(g)v\|$$

(alternatively, this should be thought of as an application of the inequality (5.18) for the integral

$$\varrho(\psi_U)v - \varrho(g)v = \int_G \psi_U(x)(\varrho(x)v - \varrho(g)v)d\mu(x)$$

as defined in Exercise 5.3.2).

But now, the strong continuity of the representation ϱ means that $x \mapsto \|\varrho(x)v - \varrho(g)v\|$ is a continuous function on G taking the value 0 at $x = g$. Hence, for any $\varepsilon > 0$, we can find some neighborhood V of g such that

$$\|\varrho(x)v - \varrho(g)v\| < \varepsilon$$

for all $x \in V$. This choice of V gives the desired inequality (5.19). □

5.4. The Peter–Weyl Theorem

Once we have the regular representation ϱ_G of a compact group G at our disposal, we attack the study of unitary representations of the group by attempting to decompose it into irreducibles. This is done by the Peter–Weyl Theorem, from which all fundamental facts about the representations of general compact groups follow.

Theorem 5.4.1 (Peter and Weyl). *Let G be a compact topological group with probability Haar measure μ. Then the regular representation of G on the space $L^2(G,\mu)$ decomposes as a Hilbert space direct sum*[6]

$$L^2(G,\mu) = \widehat{\bigoplus_{\varrho}} \mathrm{M}(\varrho) \tag{5.20}$$

of isotypic components of the finite-dimensional *irreducible unitary representations of G, each $\mathrm{M}(\varrho)$ being isomorphic to a direct sum of* $\dim(\varrho)$ *copies of ϱ.*

Although this result addresses only the properties of the regular representation, it should not be surprising, in view of the importance of the latter in the case of finite groups, that it leads to results for arbitrary unitary representations:

[6] Recall that we defined a orthogonal direct sum of unitary representations in Section 3.4.

Corollary 5.4.2 (Decomposition of unitary representations). *Let G be a compact topological group.*

(1) *Any irreducible unitary representation of G is finite dimensional.*

(2) *Any unitary representation of G is isomorphic to a Hilbert direct sum of irreducible subrepresentations.*

We start with the proof of the Peter–Weyl Theorem. As usual, we attempt to motivate the arguments instead of trying to present the shortest proof possible.

The first observation we can make is that, essentially, we already know how to obtain the inclusion

$$\bigoplus_{\varrho} \mathrm{M}(\varrho) \subset L^2(G,\mu) \tag{5.21}$$

(where the direct sum is orthogonal) as well as the fact that $\mathrm{M}(\varrho)$ is, for any finite-dimensional irreducible unitary representation, isomorphic to a direct sum of $\dim(\varrho)$ copies of ϱ.

Indeed, we can follow the method of Section 2.7.3, and in particular Theorem 2.7.28, to embed irreducible finite-dimensional representations in $L^2(G,\mu)$. Given an irreducible unitary representation

$$\varrho : G \longrightarrow \mathrm{U}(\mathrm{H})$$

of G, we see that the unitary matrix coefficients

$$f_{v,w} : g \mapsto \langle \varrho(g)v, w\rangle$$

are bounded functions on G, by the Cauchy–Schwarz inequality, and are continuous, so that $f_{v,w} \in L^2(G)$ for all v, $w \in \mathrm{H}$ (this is (5) in Theorem 5.2.1, which also shows that distinct matrix coefficients are distinct in $L^2(G)$). A formal argument (as in Theorem 2.7.28) implies that, for a fixed $w \in \mathrm{H}$, the map

$$v \mapsto f_{v,w}$$

is an intertwiner of ϱ and ϱ_G. If $w \neq 0$, Schur's Lemma implies that this map is injective because it is then non-zero: we have $f_{w,w}(1) = \|w\|^2 \neq 0$, and the continuity of $f_{w,w}$ shows that it is a non-zero element of $L^2(G)$.

Now we assume in addition that ϱ is finite dimensional. In that case, the image of $v \mapsto f_{v,w}$ is a closed subspace of $L^2(G)$ (since any finite-dimensional subspace of a Banach space is closed) and hence it is a subrepresentation of ϱ_G which is isomorphic to ϱ. Varying w, again as in Theorem 2.7.28, we see furthermore that ϱ_G contains a subrepresentation isomorphic to a direct sum of $\dim(\varrho)$ copies of ϱ, and in fact isomorphic to $\varrho \otimes \check{\varrho}$.

If we let ϱ vary among non-isomorphic finite-dimensional unitary representations, we also see that, by Lemma 3.4.21, the corresponding spaces of matrix coefficients are orthogonal.

So to finish the proof of the first inclusion (5.21), we should only check that the ϱ-isotypic component coincides with the space of matrix coefficients of ϱ. We can expect this to hold, of course, from the case of finite groups, and it is indeed true. However, the argument in the proof of Theorem 2.7.28 needs some care before it can be applied, as it relies on the linear form $\delta : \varphi \mapsto \varphi(1)$ which is not necessarily continuous on a subspace of $L^2(G)$ (and it is certainly not continuous on all of $L^2(G)$, if G is infinite).

Still, δ is well defined on any space consisting of continuous functions. The following lemma will then prove to be ad hoc:

Lemma 5.4.3. *Let G be a compact group, and let $E \subset L^2(G)$ be a* finite-dimensional *subrepresentation of the regular representation. Then E is (the image of) a space of continuous functions.*

Assuming this, let $E \subset L^2(G)$ be any subrepresentation isomorphic to ϱ. We can view E, by the lemma, as a subspace of $C(G)$. Then the linear form

$$\delta : \varphi \mapsto \varphi(1)$$

is well defined on E. Using the inner product induced on E by the L^2-inner product, it follows that there exists a unique $\psi_0 \in E$ such that

$$\delta(\varphi) = \langle \varphi, \psi_0 \rangle$$

for all $\varphi \in E$. We conclude as in the proof of Theorem 2.7.28: for all $\varphi \in E$ and $x \in G$, we have

$$\varphi(x) = \delta(\varrho_G(x)\varphi) = \langle \varrho_G(x)\varphi, \psi_0 \rangle = f_{\varphi,\psi_0}(x),$$

so that $\varphi = f_{\varphi,\psi_0}$. This shows that E is contained in the space of matrix coefficients of ϱ.

Before going to the converse inclusion, we must prove Lemma 5.4.3:

Proof of Lemma 5.4.3. The basic idea is that continuous functions can be obtained by averaging integrable functions and that averaging translates of a given $\varphi \in E$ (under the regular representation) leads to another function in E. This way we will show that $E \cap C(G)$ is dense in E (for the L^2-norm), and since $\dim E < +\infty$, this implies that $E \cap C(G) = E$, which is what we want.

Thus, given $\varphi \in E$, consider functions of the type

$$\vartheta(g) = \int_G \psi(x)\varphi(gx)d\mu(x),$$

where $\psi \in L^2(G)$. If we write this as

$$\vartheta(g) = \langle \lambda_G(g^{-1})\varphi, \overline{\psi} \rangle,$$

the strong continuity of the left-regular representation shows that ϑ is continuous. But we can also write this expression as

$$\vartheta = \int_G \psi(x)\varrho_G(x)\varphi \, d\mu(x) = \varrho_G(\psi)\varphi,$$

using the action of L^1-functions defined in Proposition 5.3.1 (since G is compact, any square-integrable function is also integrable). This shows (using part (3) of the proposition) that $\vartheta \in E \cap C(G)$. Finally, by Proposition 5.3.5 applied to $g = 1$, for any $\varepsilon > 0$, we can find $\psi \in L^1(G)$ such that

$$\|\varphi - \vartheta\| = \|\varrho_G(1)\varphi - \varrho_G(\psi)\varphi\| \leqslant \varepsilon,$$

and we are done. □

We have now proved (5.21), and must consider the converse. The problem is that, for all we know, the set of finite-dimensional irreducible unitary representations of G might be reduced to the trivial one! In other words, to prove the reverse inclusion, we need a way to construct or show the existence of finite-dimensional representations of G.

The following exercise shows an "easy" way, which applies to certain important groups (compact subgroups of $\mathrm{U}_n(\mathbf{C})$).

Exercise 5.4.4 (Groups with a finite-dimensional faithful representation). Let G be a compact topological group which is a closed subgroup of $\mathrm{GL}_n(\mathbf{C})$ for some $n \geqslant 1$.

Show that the linear span of matrix coefficients of finite-dimensional irreducible representations of G is a dense subspace of $C(G)$, using the Stone–Weierstrass Theorem (we recall the statement of the latter in Section A.3). Deduce the Peter–Weyl Theorem from this.

The class of groups covered by this exercise is restricted. In the next chapter, we describe another characterization of these groups and give examples of compact groups which are not of this type. We now deal with the general case, by proving that the direct sum

$$\bigoplus_{\varrho} \mathrm{M}(\varrho)$$

is dense in $L^2(G)$. Equivalently, using Hilbert space theory, we must show that if $\varphi \in L^2(G)$ is non-zero, it is *not* orthogonal to all isotypic components $\mathrm{M}(\varrho)$.

The basic motivation for what follows comes from looking at the case of the circle group: When ϱ varies, we expect the projection onto $\mathrm{M}(\varrho)$ to be given by a convolution operator that commutes with the regular representation. If we can find a non-zero eigenvalue with finite multiplicity of this convolution operator, this will correspond to a non-trivial projection.

The details are a bit more involved because the group G is not necessarily abelian. We exploit the fact that $\mathrm{M}(\varrho)$ is stable under the left-regular representation λ_G: If $\varphi \perp \mathrm{M}(\varrho)$, we have

$$\langle \varphi, \lambda_G(g) f \rangle = 0$$

for all $g \in G$ and $f \in \mathrm{M}(\varrho)$. As a function of g, this is the convolution $\varphi \star \check{f}$ (see Exercise 5.3.4). If we further note that, for a basic matrix coefficient $f(g) = \langle \varrho(g)v, w \rangle$, we have

$$\check{f}(g) = \overline{\langle \varrho(g^{-1})v, w \rangle} = \langle \varrho(g)w, v \rangle,$$

which is also a matrix coefficient, we can deduce that $\varphi \perp \mathrm{M}(\varrho)$ implies that $\varphi \star f = 0$ for all $f \in \mathrm{M}(\varrho)$.

It is therefore sufficient to prove that, for some finite-dimensional subrepresentation $E \subset L^2(G)$, the convolution operator

$$T_\varphi : f \mapsto \varphi \star f$$

is non-zero on E. If this is true, then since E is semisimple (Proposition 3.4.14), it contains an irreducible subrepresentation ϱ such that φ is not orthogonal to it.

For an arbitrary operator, this property might be very tricky, but by Exercise 5.3.4 again, we have also

$$T_\varphi(f) = \lambda_G(\varphi) f,$$

so that T_φ is an intertwiner of the regular representation (Exercise 5.3.3). As such, by Schur's Lemma, it acts by a scalar on any finite-dimensional subrepresentation. The question is whether any of these scalars is non-zero, i.e., whether T_φ has a non-zero eigenvalue when acting on these finite-dimensional subspaces. Now here comes the trick: we basically want to claim that the convolution form of T_φ, *abstractly*, implies that (provided $\varphi \neq 0$) it has a non-zero eigenspace $\mathrm{Ker}(T_\varphi - \lambda)$ with non-zero eigenvalue λ, of finite dimension. If that is the case, then $\mathrm{Ker}(T_\varphi - \lambda)$ is a finite-dimensional subrepresentation on which T_φ is a non-zero scalar, and we deduce that φ is not orthogonal to it.

To actually implement this, we must change the operator to obtain a better-behaved one, more precisely a self-adjoint operator. We form the function $\psi = \check{\varphi} \star \varphi$ and consider the convolution operator

$$T_\psi = \lambda_G(\psi) = \lambda_G(\varphi)^* \lambda_G(\varphi).$$

Since the convolution product is associative, we see that $\mathrm{Ker}(T_\varphi) \subset \mathrm{Ker}(T_\psi)$, and hence the previous reasoning applies to T_ψ also: T_ψ intertwines the regular representation with itself, and if there exists a finite-dimensional subrepresentation E such that T_ψ acts as a non-zero scalar on E, the function φ is not orthogonal to E.

But now T_ψ is self-adjoint and even non-negative since

$$\langle T_\psi f, f\rangle = \|T_\varphi f\|^2 \geqslant 0$$

for $f \in L^2(G)$. Moreover, writing

$$T_\psi f(x) = \lambda_G(\psi)f(x) = \int_G \psi(y)f(y^{-1}x)d\mu(y) = \int_G \psi(xy^{-1})f(y)d\mu(y)$$

(by invariance of Haar measure), we see that T_ψ is an integral Hilbert–Schmidt operator on G with kernel

$$k(x,y) = \psi(xy^{-1})$$

(as defined in Proposition A.2.5); this kernel is in $L^2(G \times G, \mu \times \mu)$ (again by invariance of Haar measure, its L^2-norm is $\|\psi\|$). By Proposition A.2.5, it is therefore a compact operator.

The operator T_ψ is also non-zero, because ψ is continuous (Exercise 5.3.4 again) and $\psi(1) = \|\varphi\|^2 \neq 0$ (by assumption), and because by taking f to be a normalized characteristic function of a small enough neighborhood of 1, we have $\psi \star f$ close to ψ, hence non-zero (this is Proposition 5.3.5, applied to the left-regular representation). So we can apply the spectral theorem (Theorem A.2.3, or the minimal version stated in Proposition A.2.6) to deduce that T_ψ has a non-zero eigenvalue λ with finite-dimensional eigenspace $\mathrm{Ker}(T_\psi - \lambda)$, as desired.

Remark 5.4.5. At the end of Section A.2, we prove—essentially from scratch—the minimal part of the spectral theorem for compact self-adjoint operators which suffices for this argument, in the case when the space $L^2(G, \mu)$ is separable (which is true, for instance, whenever the topology of G is defined by a metric and thus for most groups of practical interest; see [**49**, Pb. IV.43] for this.) Alternatively, Exercise A.2.4 asks us to prove the spectral theorem for compact self-adjoint operators starting from the general version of Theorem 3.4.18; this illustrates the philosophy that this version encapsulates all that is really needed to apply the spectral theorem in simple situations.

Exercise 5.4.6 (Another argument). Let G be a compact topological group with probability Haar measure μ.

(1) Show directly that, for any $g \neq 1$, there exists a finite-dimensional unitary representation ϱ of G such that $\varrho(g) \neq 1$. (We also state this fact

formally in Corollary 5.4.8.) (*Hint*: One can also use compact operators for this purpose.)

(2) Use this and the Stone–Weierstrass Theorem (Theorem A.3.1) to give a proof of the Peter–Weyl Theorem in the general case. (See Exercise 5.4.4.)

We now come to the proof of Corollary 5.4.2. This also requires the construction of some finite-dimensional subrepresentations. The following lemma is therefore clearly useful.

Lemma 5.4.7. *Let G be a compact topological group, and let*

$$\varrho : G \longrightarrow \mathrm{U}(\mathrm{H})$$

be any non-zero unitary representation of G. Then ϱ contains an irreducible finite-dimensional subrepresentation.

Proof. It suffices to find a *non-zero* finite-dimensional subrepresentation of H, since it will in turn contain an irreducible one for dimension reasons. But we cannot hope to construct the desired subrepresentation using kernels or eigenspaces of intertwiners this time. For instance, this is certainly not possible if ϱ is an infinite orthogonal direct sum of copies of the same irreducible representation. Instead we bootstrap the knowledge we just acquired of the regular representation and use the following remark. If $E \subset L^2(G)$ is a finite-dimensional subrepresentation of the left-regular representation and $v \in \mathrm{H}$, then the image

$$F = \{\lambda_G(\varphi)v \mid \varphi \in E\}$$

of the action of E on v is a subrepresentation of H. Indeed, by (5.17), we have

$$\varrho(g)\varrho(\varphi)v = \varrho(\lambda_G(g)\varphi)v \in F$$

for all $\varphi \in E$ and $g \in G$. The subspace F is a quotient of E (by the obvious surjection $\varphi \mapsto \varrho(\varphi)v$), and hence is also finite-dimensional, and we will be done once we ensure that we can find E such that the quotient space F is non-zero.

To do this, we fix any $v \neq 0$, and basically use the fact that $\varrho(1)v = v \neq 0$ is "almost" in F. So we approximate $\varrho(1)$ using the L^2-action. First of all, by Proposition 5.3.5, we can find $\psi \in L^2(G)$ such that $\varrho(\psi)v$ is arbitrarily close to $\varrho(1)v$, in particular, we can ensure that $\varrho(\psi)v \neq 0$. Further, using the Peter–Weyl Theorem, we can find a $\psi_1 \in L^2(G)$, which is a (finite) linear combination of matrix coefficients of finite-dimensional representations and which approximates ψ arbitrarily closely in $L^2(G)$. Since (Proposition 5.3.1(1)) we have

$$\|\varrho(\psi) - \varrho(\psi_1)\|_{\mathrm{L}(\mathrm{H})} \leqslant \|\psi - \psi_1\|_{L^1} \leqslant \|\psi - \psi_1\|_{L^2},$$

we get

$$\|\varrho(\psi_1)v\|_{\mathrm{H}} \geqslant \|\varrho(\psi)v\|_{\mathrm{H}} - \|\psi - \psi_1\|_{L^2}\|v\|_{\mathrm{H}},$$

which will be > 0 if the approximation ψ_1 is suitably chosen. Now take for E the finite direct sum of the spaces $\mathrm{M}(\varrho)$ for the ϱ which occur in the expression of ψ_1 as a combination of matrix coefficients: we have $\psi_1 \in E$, and E is a subrepresentation of the left-regular representation with $F \neq 0$ since it contains $\varrho(\psi_1)v \neq 0$. □

Proof of Corollary 5.4.2. First of all, Lemma 5.4.7 shows by contraposition that all irreducible representations must be finite dimensional (if ϱ is infinite dimensional, the statement shows it is not irreducible). Then we also see that the *complete reducibility* must be true: if reducibility failed, the (non-zero) orthogonal complement of a *maximal* completely reducible subrepresentation could not satisfy the conclusion of the lemma. To be rigorous, this reasoning can be expressed using Zorn's Lemma. We give the details, though the reader may certainly think that this is quite obvious (or may rather write a proof individually). Let $\varrho : G \longrightarrow \mathrm{U}(\mathrm{H})$ be a unitary representation. Consider the set

$$\mathcal{O} = \{(I, (\mathrm{H}_i)_{i\in I})\}$$

where I is an arbitrary index set and $\mathrm{H}_i \subset \mathrm{H}$ are pairwise orthogonal finite-dimensional irreducible subrepresentations. This set is not empty, by Lemma 5.4.7. We order it by inclusion: $(I, (\mathrm{H}_i)) \leqslant (J, (\mathrm{H}'_j))$ if and only if $I \subset J$ and $\mathrm{H}'_i = \mathrm{H}_i$ for $i \in I \subset J$. This complicated-looking ordered set is set up so that it is easy to verify that every totally ordered subset $\mathcal{P} \subset \mathcal{O}$ has an upper bound,[7] as we invite the reader to check. Then by Zorn's Lemma, we can find a maximal element $(I, (\mathrm{H}_i))$ in $\mathcal{O}$. Then we claim that the subspace

$$\mathrm{H}' = \widehat{\bigoplus}^{\perp}_{i\in I} \mathrm{H}_i$$

is in fact equal to H, which is then exhibited as a Hilbert orthogonal sum of finite-dimensional subspaces. Indeed, consider $\mathrm{H}'' = (\mathrm{H}')^{\perp} \subset H$. If this is non-zero, H'' contains a finite-dimensional subrepresentation, say H_0, again by Lemma 5.4.7. But then (assuming the index 0 is not in $I\ldots$) we have

$$(I \cup \{0\}, (\mathrm{H}_i, \mathrm{H}_0)) \in \mathcal{O},$$

contradicting the maximality of $(I, (\mathrm{H}_i))$. Thus $\mathrm{H}'' = 0$, which means that $\mathrm{H}' = \mathrm{H}$ as claimed. □

We finally deduce some further corollaries of the Peter–Weyl Theorem.

[7] For the more natural set $\mathcal{O}'$ of all "completely reducible subrepresentations" of H, ordered by inclusion, checking this is more painful, because one is not keeping track of consistent decompositions of the subspaces to use to construct an upper bound.

Corollary 5.4.8 (Separating points and completeness criteria). *Let G be a compact topological group with probability Haar measure μ.*

(1) *If $g \neq 1$ is a non-trivial element of G, there exists a finite-dimensional unitary, indeed irreducible, representation ϱ of G such that $\varrho(g) \neq 1$.*

(2) *Suppose $\mathcal{C}$ is a given set of finite-dimensional irreducible unitary representations of G. Then $\mathcal{C}$ is complete, i.e., contains all irreducible representations of G, up to isomorphism, if and only if the linear span M of the matrix coefficients of representations $\varrho \in \mathcal{C}$ is dense in $L^2(G, \mu)$, or equivalently, if M is dense in $C(G)$ for the L^∞-norm.*

(3) *Suppose $\mathcal{C}$ is a given set of finite-dimensional irreducible unitary representations of G. Then $\mathcal{C}$ is complete if and only if we have the Plancherel formula*

$$\|\varphi\|^2 = \sum_{\varrho \in \mathcal{C}} \|p_\varrho(\varphi)\|^2$$

for all $\varphi \in L^2(G)$, where p_ϱ is the orthogonal projection on the isotypic component $\mathrm{M}(\varrho)$.

As a matter of fact, the statement of the Peter–Weyl Theorem in the original paper is that (3) holds (and the statement in Pontryagin's version, a few years later, was the density of M in $C(G)$!)

Proof. (1) The regular representation is faithful (Exercise 5.2.8), so for $g \neq 1$, the operator $\varrho_G(g)$ is not the identity. However, by the Peter–Weyl Theorem, $\varrho_G(g)$ is the direct sum of the operators obtained by restriction to each $\mathrm{M}(\varrho)$, which are just direct sums of $\dim(\varrho)$ copies of $\varrho(g)$. Hence, at least one of the operators $\varrho(g)$ must be distinct from the identity.

The statement of (2) concerning the L^2-norm is an immediate consequence of the Peter–Weyl Theorem and Hilbert space theory: if M is not dense in $L^2(G, \mu)$, the orthogonal complement of its closure is a non-zero subrepresentation of the regular representation, which must therefore contain some irreducible subrepresentation π. Because the definition of M means that it is spanned by the $\mathrm{M}(\varrho)$ where ϱ ranges over $\mathcal{C}$, we must have $\pi \notin \mathcal{C}$.

Similarly, (3) is a direct translation of the Peter–Weyl Theorem using the theory of Hilbert spaces.

For the part of (2) involving continuous functions, we recall that M is indeed a subspace of $C(G)$. We must show that M is dense in $C(G)$ for the L^∞-norm. This can be thought of as an analogue of the classical Weierstrass approximation theorem for trigonometric polynomials (which, indeed, corresponds to $G = \mathbf{S}^1$), and one can indeed use (1) and the Stone–Weierstrass Theorem (this is the content of Exercise 5.4.6) to establish this. ∎

Our last result in this section is a special case of Frobenius reciprocity for induced representations.

Proposition 5.4.9 (Frobenius reciprocity for compact groups). *Let G be a compact topological group with probability Haar measure μ, and let $K \subset G$ be a compact subgroup. For any* finite-dimensional *unitary representations*

$$\varrho_1 : G \longrightarrow \mathrm{U}(\mathrm{H}_1), \qquad \varrho_2 : K \longrightarrow \mathrm{U}(\mathrm{H}_2),$$

we have a natural isomorphism

$$\mathrm{Hom}_G(\varrho_1, \mathrm{Ind}_K^G(\varrho_2)) \simeq \mathrm{Hom}_K(\mathrm{Res}_K^G(\varrho_1), \varrho_2).$$

In particular, if ϱ_1 and ϱ_2 are irreducible representations of G and H respectively, then the multiplicity of ϱ_1 in the induced representation $\mathrm{Ind}_K^G(\varrho_2)$ is equal to the multiplicity of ϱ_2 in the restriction of ϱ_1 to H.

Proof. We leave it to the reader to check that the proof of Proposition 2.3.9 can carry through formally unchanged, provided one proves first that the image of any intertwiner

$$\Phi : \varrho_1 \longrightarrow \mathrm{Ind}_K^G(\varrho_2)$$

lies in the (image of) the subspace V_0 of continuous functions used in the definition of induced representations (see (5.10); note that all intertwiners constructed from right to left by (2.28) have this property, because of the strong continuity of unitary representations, so that Frobenius reciprocity can only hold in this form when this property is true).

Since ϱ_1 is finite dimensional, so is the image of Φ, and thus the statement is an analogue of Lemma 5.4.3. We can reduce to this case: fixing an orthonormal basis (e_i) of H_2, and expressing any square-integrable function $f : G \longrightarrow \mathrm{H}_2$ in the form

$$f = \sum_i f_i e_i$$

with $f_i \in L^2(G)$, the definition of the induced representation shows that we obtain an embedding

$$\begin{cases} \mathrm{Ind}_K^G(\varrho_2) & \longrightarrow & L^2(G)^{\dim \varrho_2} \\ f & \mapsto & (f_i), \end{cases}$$

where the right-hand side is the orthogonal direct sum of $\dim \varrho_2$ copies of the regular representation.

If $\Phi : \varrho_1 \longrightarrow \mathrm{Ind}_K^G(\varrho_2)$ is an intertwiner, then the projection onto each component of the image of Φ is a finite-dimensional subrepresentation of $L^2(G)$. By Lemma 5.4.3, each component f_i of any $f \in \mathrm{Im}(\Phi)$ is therefore continuous, and we deduce that $\mathrm{Im}(\Phi) \subset V_0$. ☐

Exercise 5.4.10. Which of the other properties of induction can you establish for compact groups?

5.5. Characters and matrix coefficients for compact groups

The reward for carefully selecting the conditions defining unitary representations of compact groups and proving the analytic side of the Peter–Weyl Theorem is that, with this in hand, character theory becomes available. Indeed, since any irreducible unitary representation π is finite dimensional, it has a well-defined character $\chi_\pi(g) = \operatorname{Tr}\pi(g)$.

For finite-dimensional representations, character theory is just as remarkably efficient as in the case of finite groups. We summarize the most important properties, using as before the notation $\widehat{G}$ for the set of isomorphism classes of irreducible unitary representations of G.

Theorem 5.5.1 (Character theory). *Let G be a compact topological group, and let μ be the probability Haar measure on G. Let $G^\sharp$ be the set of conjugacy classes in G.*

(1) *The characters of irreducible unitary representations of G are continuous functions on G, which form an orthonormal basis of the space $L^2(G^\sharp)$ of square-integrable conjugacy-invariant functions on G. In particular, a set $\mathcal{C}$ of finite-dimensional irreducible representations of G is complete, in the sense of Corollary* 5.4.8, *if and only if the linear combinations of characters of representations in $\mathcal{C}$ are dense in $L^2(G^\sharp)$.*

(2) *For any irreducible unitary representation $\pi \in \widehat{G}$ of G and any unitary representation*

$$\varrho : G \longrightarrow \mathrm{U}(\mathrm{H})$$

of G, the orthogonal projection onto the π-isotypic component of H *is given by*

$$\Phi_\pi = (\dim \pi) \int_G \overline{\chi_\pi(g)} \varrho(g) d\mu(g),$$

where

$$\chi_\pi(g) = \operatorname{Tr}\pi(g)$$

is the character of π, using the L^1-action of Proposition 5.3.1. *In particular, if ϱ is* finite dimensional, *the dimension of the space ϱ^G of G-invariant vectors in* H *is given by the average over G of the values of the character of ϱ, i.e.,*

$$\dim \varrho^G = \int_G \chi_\varrho(g) d\mu(g). \tag{5.22}$$

(3) *A unitary finite-dimensional representation ϱ of G is irreducible if and only if*

$$\int_G |\chi_\varrho(g)|^2 d\mu(g) = 1.$$

More generally, if ϱ_1 and ϱ_2 are finite-dimensional *unitary representation of G, we have*

$$\langle \chi_{\varrho_1}, \chi_{\varrho_2} \rangle = \sum_{\pi \in \hat{G}} n_\pi(\varrho_1) n_\pi(\varrho_2), \tag{5.23}$$

where $n_\pi(\varrho) = \langle \chi_\varrho, \chi_\pi \rangle$ is the multiplicity of π in ϱ.

The reader should definitely try checking all these facts, without looking at the proof, since they are analogues of things we know for finite groups. For the sake of variety, we use slightly different arguments (which can also be applied to finite groups.) First we compute the inner products of matrix coefficients.

Lemma 5.5.2. *Let G be a compact group with probability Haar measure μ.*

(1) *If π_1 and π_2 are non-isomorphic irreducible unitary representations of G, any two matrix coefficients of π_1 and π_2 are orthogonal in $L^2(G)$.*

(2) *If $\pi : G \longrightarrow \mathrm{U}(\mathrm{H})$ is an irreducible unitary representation of G and v_1, w_1, v_2, w_2 are vectors in H, we have*

$$\int_G \langle \pi(g)v_1, w_1 \rangle \overline{\langle \pi(g)v_2, w_2 \rangle} d\mu(g) = \frac{\langle v_1, v_2 \rangle_{\mathrm{H}} \overline{\langle w_2, w_1 \rangle_{\mathrm{H}}}}{\dim(\mathrm{H})}. \tag{5.24}$$

Proof. (1) A matrix coefficient of π_1 (resp. π_2) is a vector in the π_1-isotypic component (resp. π_2-isotypic component) of the regular representation of G. These isotypic components are orthogonal if π_1 and π_2 are not isomorphic (Lemma 3.4.21).

(2) Instead of arguing as we did in Chapter 4, we sketch a different argument: we use the fact that the isotypic component $\mathrm{M}(\pi) \subset L^2(G)$, under the action of $\varrho_G \boxtimes \lambda_G$, is isomorphic to the external tensor product $\pi \boxtimes \bar{\pi}$, as a representation of $G \times G$ (this is the unitary version of the computation done in the proof of Theorem 4.2.5 for finite groups, and the argument extends easily to compact groups.)

By Proposition 2.3.23, $\pi \boxtimes \bar{\pi}$ is an irreducible, and unitary, representation of $G \times G$. In particular, there exists a *unique* $(G \times G)$-invariant inner product on $\mathrm{M}(\pi)$, up to multiplication by a positive scalar (see Exercise 3.4.20). The L^2-inner product, restricted to $\mathrm{M}(\pi)$, is such an inner product, but so is the inner product induced by

$$\langle v_1 \otimes w_1, v_2 \otimes w_2 \rangle = \langle v_1, v_2 \rangle_{\mathrm{H}} \langle w_1, w_2 \rangle_{\bar{\mathrm{H}}} = \langle v_1, v_2 \rangle_{\mathrm{H}} \overline{\langle w_2, w_1 \rangle_{\mathrm{H}}}$$

on $H \otimes \bar{H}$. Hence there exists some $\alpha > 0$ such that

$$\int_G \langle \pi(g)v_1, w_1\rangle \overline{\langle \pi(g)v_2, w_2\rangle} d\mu(g) = \alpha \langle v_1, v_2\rangle_H \overline{\langle w_2, w_1\rangle_H} \tag{5.25}$$

for all vectors v_1, w_1, v_2, w_2. In order to compute α, we use the following trick. We fix an arbitrary non-zero vector $v \in H$, and take $v_1 = v_2 = v$ and, successively, $w_1 = w_2 = e_i$, the elements of an orthonormal basis of H. We then sum the identity (5.25) over i, and obtain

$$\sum_i \int_G |\langle \varrho(g)v, e_i\rangle|^2 d\mu(g) = \alpha \|v\|^2 \dim(H).$$

But the left-hand side is equal to

$$\int_G \sum_i |\langle \varrho(g)v, e_i\rangle|^2 d\mu(g) = \int_G \|\pi(g)v\|^2 d\mu(g) = \|v\|^2,$$

using the orthonormality of the basis and the unitarity of $\pi(g)$. Hence, by comparison, we get $\alpha = 1/\dim(H)$, which gives the statement (5.24). □

Proof of Theorem 5.5.1. (1) The space $L^2(G^\sharp)$ is a closed subspace of $L^2(G)$. The characters of finite-dimensional representations are (non-zero) continuous functions, invariant under conjugation, and therefore belong to $L^2(G^\sharp)$. Since the character of $\pi \in \widehat{G}$ lies in $M(\pi)$ (as a sum of matrix coefficients), the distinct characters are orthogonal, and Lemma 5.5.2 actually shows that they form an orthonormal system in $L^2(G^\sharp)$.

In order to show its completeness, we need only check that if $\varphi \in L^2(G^\sharp)$ is conjugacy-invariant, its isotypic components, say φ_π, are multiples of the character of π for all irreducible representations π. This can be done by direct computation, just as in the case of finite groups (Section 4.3.3), but we sketch again a different argument.

It is enough to prove that the space $M(\pi) \cap L^2(G^\sharp)$, in which φ_π lies, is one dimensional, since we already know that χ_π is a non-zero element of it. But we can see this space $M(\pi) \cap L^2(G^\sharp)$ as the invariant subspace of $M(\pi)$ when G acts on $L^2(G)$ by the diagonal or conjugation combination of the two regular representations, i.e.,

$$\varrho(g)\varphi(x) = \varphi(x^{-1}gx)$$

for $\varphi \in L^2(G)$. Let H be the space on which π acts. The isotypic component $M(\pi)$ is isomorphic to $H \otimes \bar{H} \simeq \mathrm{End}(H)$ as a vector space, and the corresponding action on End(H) is the usual representation of G on an endomorphism space. Thus the G-invariants of $M(\pi)$ under the action ϱ correspond to the space of G-invariants in End(H), which we know is $\mathrm{End}_G(H) = \mathbf{C}\,\mathrm{Id}$, by Schur's Lemma. This shows that it has dimension 1, as claimed.

(2) Because characters are conjugacy-invariant, the action of Φ_π on a unitary representation is an intertwiner (Exercise 5.3.3). In particular, Φ_π acts by multiplication by a scalar on every irreducible unitary representation $\varrho \in \widehat{G}$. This scalar is equal to the trace of Φ_π acting on ϱ, divided by $\dim \pi$. Since the trace is given by

$$(\dim \pi) \int_G \overline{\chi_\pi(g)} \chi_\varrho(g) d\mu(g),$$

which is equal to 0 if π is not isomorphic to ϱ and to $\dim \pi$ otherwise (by orthonormality of characters), we see that Φ_π is the identity on the π-isotypic component of any unitary representation, while it is zero on all other isotypic components. This means that it is the desired projection.

(3) If ϱ is a finite-dimensional unitary representation of G, we have

$$\chi_\varrho = \sum_{\pi \in \widehat{G}} n_\pi(\varrho) \chi_\pi,$$

and by orthonormality of the characters we find

$$\langle \chi_\varrho, \chi_\pi \rangle = n_\pi(\varrho)$$

for $\pi \in \widehat{G}$. Again orthonormality implies that the formula (5.23) is valid. Applied to $\varrho_1 = \varrho_2 = \varrho$, this gives

$$\|\chi_\varrho\|^2 = \sum_{\pi \in \widehat{G}} n_\pi(\varrho)^2.$$

Each term in this sum is a non-negative integer, hence the L^2-norm is equal to 1 if and only if a single term, say $n_\pi(\varrho)$, is non-zero, and in fact equal to 1, which means that ϱ is isomorphic to π. □

Exercise 5.5.3 (Paradox?). Explain why part (2) of Theorem 5.5.1 does not conflict with the result of Exercise 3.1.5 when G is infinite. (Note that the action of the projection Φ_π is not given by an element of the group algebra $\mathbf{C}(G)$ as defined in Section 3.1.)

Exercise 5.5.4. Let $\varrho : G \longrightarrow \mathrm{U}(\mathrm{H})$ be a unitary representation of a compact group G such that, for any $f \in L^2(G)$, the operator $\varrho(f)$ on H is *compact*. Show that the multiplicity of any irreducible representation $\pi \in \widehat{G}$ is finite in H.

Exercise 5.5.5. Let $\varrho : G \longrightarrow \mathrm{U}_n(\mathbf{C})$ be a faithful unitary representation of a compact group G. Let π be an irreducible unitary representation of G.

(1) Show that there exist integers $a \geqslant 0$ and $b \geqslant 0$ such that π occurs as a subrepresentation of $\varrho^{\otimes a} \otimes (\check{\varrho})^{\otimes b}$. (*Hint*: Use the completeness criterion of Corollary 5.4.8 and apply the Stone–Weierstrass Theorem to the span of matrix coefficients of all representations $\varrho^{\otimes a} \otimes (\check{\varrho})^{\otimes b}$.)

(2) Assume that there exists an integer $m \geqslant 0$ such that $\det(\varrho(g))^m = 1$ for all $g \in G$. Show that π occurs as a subrepresentation of $\varrho^{\otimes a}$ for some $a \geqslant 0$. (This applies, in particular, to any finite group G, and hence recovers Proposition 4.3.23.)

Remark 5.5.6 (Less duality)**.** All the statements, except for the care needed with L^2-theory, are identical with those which are valid for finite groups. There is, however, at least one sharp difference: there is no good analogue of the second orthogonality formula of Proposition 4.4.1, which expresses the orthogonality of the columns of the character table of a finite group. The expression

$$\sum_{\varrho \in \widehat{G}} \chi_\varrho(h)\overline{\chi_\varrho(g)}$$

for g, $h \in G$, does not make sense—in general—in any usual way (i.e., when G is infinite, this is usually a divergent series). In other words, the duality between conjugacy classes and irreducible representations is even fuzzier than was the case for finite groups.

Other features of the representations of finite groups that are missing when G is infinite are those having to do with integrality properties (though it is tempting to think that maybe there should be some analogue?)

In addition to character theory, matrix coefficients remain available to describe orthonormal bases of $L^2(G, \mu)$. We present this in two forms, one of which is more intrinsic since it does not require a choice of basis. However, it gives an expansion of a different nature than an orthonormal basis in Hilbert space, which is not so well known in general.

Theorem 5.5.7 (Decomposition of the regular representation)**.** *Let G be a compact topological group, and let μ be the probability Haar measure on G.*

(1) *For $\pi \in \widehat{G}$, fix an orthonormal basis $(e_{\pi,i})_{1 \leqslant i \leqslant \dim(\pi)}$ of the space of π. Then the family of matrix coefficients*

$$\varphi_{\pi,i,j}(g) = \sqrt{\dim(\pi)}\langle \pi(g)e_{\pi,i}, e_{\pi,j}\rangle$$

forms an orthonormal basis of $L^2(G, \mu)$, where π runs over $\widehat{G}$ and $1 \leqslant i, j \leqslant \dim \pi$.

(2) *For $\pi \in \widehat{G}$, acting on the space H_π, consider the map*

$$A_\pi \begin{cases} L^2(G) \longrightarrow \mathrm{End}(\mathrm{H}_\pi) \\ \varphi \mapsto \displaystyle\int_G \varphi(g)\pi(g^{-1})d\mu(g), \end{cases}$$

Then the A_π give matrix-valued *Fourier coefficients for φ, in the sense that*

$$\varphi(x) = \sum_{\pi \in \widehat{G}} (\dim \pi) \operatorname{Tr}(A_\pi(\varphi) \circ \pi(x))$$

for all $\varphi \in L^2(G)$, where this series converges in $L^2(G, \mu)$.

Proof. The first statement (1) is a consequence of the Peter–Weyl Theorem and the orthogonality of matrix coefficients. The second statement follows easily by the following computation: given $\varphi \in L^2(G)$, we have an L^2-convergent series

$$\varphi = \sum_{\pi \in \widehat{G}} \varphi_\pi,$$

where φ_π is the orthogonal projection of φ on the π-isotypic component. We compute it using the projection formula of Theorem 5.5.1 applied to the regular representation and to the irreducible representation π. Using the unitarity, this gives

$$\begin{aligned}
\varphi_\pi(x) &= (\dim \pi) \int_G \overline{\chi_\pi(g)} \varrho_G(g)\varphi(x) d\mu(g) \\
&= (\dim \pi) \int_G \chi_\pi(g^{-1})\varphi(xg) d\mu(g) \\
&= (\dim \pi) \operatorname{Tr}\Bigl(\int_G \varphi(xg)\pi(g^{-1}) d\mu(g)\Bigr) \\
&= (\dim \pi) \operatorname{Tr}\Bigl(\int_G \varphi(y)\pi(y^{-1}x) d\mu(y)\Bigr) \\
&= (\dim \pi) \operatorname{Tr}\Bigl(\Bigl(\int_G \varphi(y)\pi(y^{-1}) d\mu(y)\Bigr)\pi(x)\Bigr) \\
&= (\dim \pi) \operatorname{Tr}(A_\pi(\varphi)\pi(x)),
\end{aligned}$$

as claimed. □

We can see the second statement as another formulation of the Peter–Weyl decomposition, since the summands $g \mapsto \operatorname{Tr}(A_\pi(\varphi)\chi_\pi(g))$ are elements of $\mathrm{M}(\pi)$. The advantage of this version is that we obtain an intrinsic decomposition without having to select a basis of the spaces of irreducible representations. This is very helpful in some contexts (including for finite groups).

Exercise 5.5.8 (G-finite vectors). Let G be a compact group with probability Haar measure μ, and let

$$\varrho : G \longrightarrow \mathrm{U}(\mathrm{H})$$

be a unitary representation of G. A vector $v \in \mathrm{H}$ is called *G-finite* if the subrepresentation generated by v is finite dimensional.

(1) Show that the space H_1 of G-finite vectors is stable under G and that it is dense in H. When is it closed?

(2) Show that a function $f \in L^2(G, \mu)$ is a G-finite vector of the regular representation if and only if f is a finite linear combination of matrix coefficients of irreducible unitary representations of G. (These functions are the analogues, for G, of the trigonometric polynomials in the case of the circle group $\mathbf{S}^1$.)

(3) Prove the analogue for a unitary representation ϱ of a compact group G of the property described in Exercise 4.3.31 for finite groups. For any irreducible representation π and any vector v in the π-isotypic component of ϱ, the subrepresentation generated by v is the direct sum of at most $\dim(\pi)$ copies of π.

5.6. Some first examples

We present in this section some simple examples of characters of compact groups.

Example 5.6.1 (Representations of $\mathrm{SU}_2(\mathbf{C})$). The most basic example of a non-abelian compact group is the group $\mathrm{SU}_2(\mathbf{C}) \subset \mathrm{SL}_2(\mathbf{C})$. We have already seen that it has irreducible representations ϱ_m of degree $m+1$ for all integers $m \geqslant 0$, obtained by restricting the representation of $\mathrm{SL}_2(\mathbf{C})$ on homogeneous polynomials in two variables (see Section 2.6.1 and Exercise 2.7.13).

Using character theory, we can first check the Clebsch–Gordan formula for $\mathrm{SU}_2(\mathbf{C})$ in almost a single stroke of the pen (compare with the proof of Theorem 2.6.3 that we sketched earlier):

Corollary 5.6.2 (Clebsch–Gordan formula for $\mathrm{SU}_2(\mathbf{C})$). *For all $m \geqslant n \geqslant 0$, the representations ϱ_m of $\mathrm{SU}_2(\mathbf{C})$ satisfy*

$$\varrho_m \otimes \varrho_n \simeq \bigoplus_{0 \leqslant i \leqslant n} \varrho_{m+n-2i}.$$

Proof. One might say it is a matter of checking the identity at the level of characters, i.e., of proving the elementary formula

$$\frac{\sin((m+1)\theta)}{\sin\theta}\frac{\sin((n+1)\theta)}{\sin\theta} = \sum_{0 \leqslant i \leqslant n} \frac{\sin((m+n-2i)\theta)}{\sin\theta}$$

for all $\theta \in [0, \pi]$. But more to the point, character theory explains *how to guess* (or find) such a formula: by complete reducibility and orthonormality of characters, we know that

$$\varrho_m \otimes \varrho_n \simeq \bigoplus_{0 \leqslant k \leqslant mn-1} \langle \chi_m \chi_n, \chi_k \rangle \varrho_k$$

(the restriction of the range comes from dimension considerations), and we are therefore reduced to computing the multiplicities

$$\langle \chi_m \chi_n, \chi_k \rangle = \frac{2}{\pi} \int_0^{\pi} \frac{\sin((m+1)\theta)}{\sin \theta} \frac{\sin((n+1)\theta)}{\sin \theta} \frac{\sin((k+1)\theta)}{\sin \theta} \sin^2(\theta) d\theta.$$

This is, of course, an elementary—if possibly boring–exercise. □

Now we come to the concrete incarnation of the Peter–Weyl Theorem for $\mathrm{SU}_2(\mathbf{C})$: it is the statement that these represent *all* irreducible representations of $\mathrm{SU}_2(\mathbf{C})$.

Theorem 5.6.3 (Irreducible representations of $\mathrm{SU}_2(\mathbf{C})$). *The only irreducible unitary representations of* $\mathrm{SU}_2(\mathbf{C})$*, up to isomorphism, are given by the representations* ϱ_m *described above. We have* $\dim \varrho_m = m + 1$*, and the character* χ_m *of* ϱ_m *is given by*

$$\chi_m\left(\begin{pmatrix} e^{i\theta} & 0 \\ 0 & e^{-i\theta} \end{pmatrix}\right) = \frac{\sin((m+1)\theta)}{\sin \theta} \tag{5.26}$$

for $\theta \in [0, \pi]$.

Proof. The definition of ϱ_m makes it clear that it is a continuous representation. Thus we must check that there are no other irreducible unitary representations of $\mathrm{SU}_2(\mathbf{C})$ than those.

We give two proofs of this fact based on our previous results. There is another elegant proof, which is important because it generalizes to other compact subgroups of $\mathrm{U}_n(\mathbf{C})$, and which is based on the relation between representations of $\mathrm{SU}_2(\mathbf{C})$ and those of its Lie algebra, which are computed in Theorem 3.2.5. We refer, for instance, to [**36**, Prop. II.2.1, Th. II.2.4] for this.

(1) We first use the completeness criterion from character theory (Theorem 5.5.1). The set $\mathrm{SU}_2(\mathbf{C})^\sharp$ of conjugacy classes in $\mathrm{SU}_2(\mathbf{C})$ can be identified with the interval $[0, \pi]$ using the map

$$c \begin{cases} [0, \pi] & \longrightarrow & \mathrm{SU}_2(\mathbf{C})^\sharp \\ \theta & \mapsto & \begin{pmatrix} e^{i\theta} & 0 \\ 0 & e^{-i\theta} \end{pmatrix}. \end{cases}$$

Indeed, this follows from the diagonalizability of unitary matrices in an orthonormal basis. Precisely, any $g \in \mathrm{SU}_2(\mathbf{C})$ is conjugate in $\mathrm{U}_2(\mathbf{C})$ to such a matrix for some $\theta \in \mathbf{R}$, say $g = xc(\theta)x^{-1}$. By replacing x with αx for some $\alpha \in \mathbf{C}$, we can ensure that $\det(x) = 1$, i.e., that $x \in \mathrm{SU}_2(\mathbf{C})$. Finally, by periodicity, we can assume $\theta \in [-\pi, \pi]$, and by conjugating by $\left(\begin{smallmatrix} 0 & 1 \\ -1 & 0 \end{smallmatrix}\right)$, which replaces θ by $-\theta$, we see that we may assume that $\theta \in [0, \pi]$. Finally, the element θ is then unique, because the trace of $c(\theta)$ is a conjugacy invariant, and $\operatorname{Tr} c(\theta) = 2\cos\theta$, which is an injective function on $[0, \pi]$.

In terms of this identification of $\mathrm{SU}_2(\mathbf{C})^\sharp$, it is known that the inner product with respect to the probability Haar measure for two (square-integrable) class functions φ_1, φ_2 is given by

$$\int_G \varphi_1(g)\overline{\varphi_2(g)}d\mu(g) = \frac{2}{\pi}\int_0^\pi \varphi_1(c(\theta))\overline{\varphi_2(c(\theta))}\sin^2\theta d\theta.$$

This formula is a special case of the *Weyl integration formula*; see e.g. [**36**, Th. 4.45].

On the other hand, we already checked (in (2.51)) that

$$\chi_m(c(\theta)) = \frac{\sin((m+1)\theta)}{\sin\theta}.$$

This means that we are reduced to showing that the functions

$$\varphi_m(\theta) = \frac{\sin((m+1)\theta)}{\sin\theta}, \qquad m \geqslant 0,$$

form an orthonormal basis of the space $\mathrm{H} = L^2([0,\pi], \frac{2}{\pi}\sin^2\theta d\theta)$. Because this is close enough to classical Fourier series, we can do this by hand, by reduction to a Fourier expansion, and therefore finish the proof.

Let $\varphi \in \mathrm{H}$ be given. We define a function ψ on $[-\pi,\pi]$ by

$$\psi(\theta) = \begin{cases} \varphi(\theta)\sin\theta & \text{if } \theta \geqslant 0, \\ \varphi(-\theta)\sin\theta & \text{if } \theta \leqslant 0 \end{cases}$$

(i.e., we extend the function $\varphi(\theta)\sin\theta$ so that it is an odd function). By definition of H, we see that $\psi \in L^2([-\pi,\pi], d\theta)$. We can therefore expand ψ in Fourier series in this space: we have

$$\psi(\theta) = \sum_{h\in\mathbf{Z}} \alpha_h e^{ih\theta}$$

in $L^2([-\pi,\pi])$, with

$$\alpha_h = \frac{1}{2\pi}\int_{-\pi}^{\pi} \psi(\theta)e^{-ih\theta}d\theta.$$

Since ψ is odd, we have $\alpha_h = -\alpha_{-h}$, and in particular $\alpha_0 = 0$, hence the Fourier expansion on $[0,\pi]$ takes the form

$$\varphi(\theta)\sin\theta = \sum_{h\geqslant 1} \alpha_h 2i\sin(h\theta) = 2i\sum_{m\geqslant 0}\alpha_{m+1}\sin((m+1)\theta),$$

i.e.,

$$\varphi = 2i\sum_{m\geqslant 0}\alpha_{m+1}\varphi_m$$

in $L^2([0,\pi], \frac{2}{\pi}\sin^2\theta d\theta)$. Since we already know that the φ_m are an orthonormal system in this space, it follows that they form an orthonormal basis, as we wanted to prove.

(2) For the second proof, we exploit the result of Exercise 5.5.5 and the Clebsch–Gordan formula. The representation ϱ_1, of dimension 2, is faithful (it is isomorphic to the inclusion of $\mathrm{SU}_2(\mathbf{C})$ in $\mathrm{U}_2(\mathbf{C})$) and isomorphic to its contragredient (because $\check{\varrho}_1$ has the same character). Hence any irreducible unitary representation π of $\mathrm{SU}_2(\mathbf{C})$ occurs as a subrepresentation of a tensor power $\varrho_1^{\otimes a}$ for some integer $a \geqslant 0$. But from the Clebsch–Gordan formula, it follows easily by induction that $\varrho_1^{\otimes}$ decomposes as a direct sum of irreducible representations which are all among the ϱ_m, $m \geqslant 0$. Thus π must be isomorphic to one of these representations. □

Using the coordinates in (5.8) on $\mathrm{SU}_2(\mathbf{C})$, it is also possible to see how the representations ϱ_m, defined as acting on polynomials, can be embedded in the regular representation. Indeed, if we see the coordinates $(a, b) \in \mathbf{C}^2$ of some element $g \in \mathrm{SU}_2(\mathbf{C})$ as a row vector (with coefficients subject to the condition $|a|^2 + |b|^2 = 1$), a direct matrix multiplication shows that the row vector corresponding to a product gh is the same as the vector-matrix product $(a, b)h$. If we restrict a polynomial $P \in \mathbf{C}[X, Y]$ to $(X, Y) = (a, b)$, this gives an intertwiner from V_m to a space of continuous functions on $\mathrm{SU}_2(\mathbf{C})$. Since this map is non-zero (any basis vector $X^i Y^{m-i}$ restricts to a non-zero function), it is an injection

$$V_m \hookrightarrow L^2(\mathrm{SU}_2(\mathbf{C})).$$

According to Vilenkin [**63**], it was first observed by É. Cartan that matrix coefficients or characters of irreducible representations of certain important groups lead to most of the "classical" special functions[8] (of course, the fact that the exponential function is a representation of the additive group of $\mathbf{C}$, or of $\mathbf{R}$ by restriction, is an even older phenomenon that has the same flavor). We present here some simple instances, related to the group $\mathrm{SU}_2(\mathbf{C})$, and another case of this phenomenon will be described in Exercise 7.4.15 in Section 7.4; more information about these, as well as many additional examples, are found in [**63**].

We begin with the characters of $\mathrm{SU}_2(\mathbf{C})$. As functions of the parameter θ describing the conjugacy class, they are of course completely elementary, but a change of variable adds more subtlety:

Definition 5.6.4 (Chebychev polynomials). For $m \geqslant 0$, there exists a polynomial $P_m \in \mathbf{R}[X]$, of degree m, such that

$$\chi_m(\theta) = P_m(2\cos\theta), \qquad 0 \leqslant \theta \leqslant \pi,$$

i.e., such that

$$\frac{\sin((m+1)\theta)}{\sin(\theta)} = P_m(2\cos\theta). \tag{5.27}$$

[8] Especially the functions that arise in mathematical physics, e.g., Bessel functions.

The polynomial $U_m = P_m(X/2)$ is called the *m-th Chebychev polynomial of the second kind.*

The fact that the characters of $\mathrm{SU}_2(\mathbf{C})$ form an orthonormal basis of the space of class functions translates into the following fact:

Proposition 5.6.5 (Chebychev polynomials as orthogonal polynomials). *The restrictions to $[-1,1]$ of the polynomials U_m, $m \geqslant 0$, form an orthonormal basis of the space $L^2([-1,1], d\nu)$, where ν is the measure supported on $[-1,1]$ given by*

$$d\nu(t) = \frac{2}{\pi}\sqrt{1-t^2}dt.$$

The justification for this substitution is that the Chebychev polynomials arise in many applications completely independently of any (apparent) consideration of the group $\mathrm{SU}_2(\mathbf{C})$. On the other hand, algebraic properties of the representations ϱ_m can lead to very simple (or very natural) proofs of identities among Chebychev polynomials which might otherwise look quite forbidding if one starts from the definition (5.27), and even more if one begins with an explicit (!) polynomial expansion.

Example 5.6.6. The first few Chebychev polynomials U_m are

$$U_0 = 1, \qquad U_1 = 2X, \qquad U_2 = 4X^2 - 1,$$
$$U_3 = 8X^3 - 4X, \qquad U_4 = 16X^4 - 12X^2 + 1, \qquad \ldots$$

(as one can prove, e.g., by straightforward trigonometric manipulations...).

Exercise 5.6.7 (Playing with Chebychev polynomials). In this exercise, we express the Clebsch–Gordan formula for $\varrho_m \otimes \varrho_n$ in terms of expansions of Chebychev polynomials, and we deduce some combinatorial identities.

(1) Show that we have

$$U_m = \sum_{0 \leqslant j \leqslant m/2} (-1)^j \binom{m-j}{m-2j} (2X)^{m-2j}$$

for all $n \geqslant 0$. (*Hint*: It is helpful here to interpret the Chebychev polynomials in terms of characters of the larger group $\mathrm{SL}_2(\mathbf{C})$.)

(2) Using the Clebsch–Gordan decomposition, show that for $m \geqslant n \geqslant 0$, we have

$$U_m U_n = \sum_{k=0}^{n} U_{m+n-2k}.$$

(3) Deduce that for $m \geqslant n \geqslant 0$ and $k \leqslant (m+n)/2$, we have

$$\sum_{\substack{i+j=k \\ i \leqslant m/2, j \leqslant n/2}} (-1)^{i+j} \binom{m-i}{m-2i} \binom{n-j}{n-2j} = \sum_{\substack{\ell+t=k \\ \ell \leqslant n}} (-1)^t \binom{m+n-2\ell-t}{m+n-2k}.$$

Example 5.6.8 (Representations of $SO_3(\mathbf{R})$). The group $SO_3(\mathbf{R})$ of rotations in a three-dimensional euclidean space is also very important in applications. We will see it appear prominently in the analysis of the hydrogen atom in Section 6.4, for instance. As it turns out, it is very closely related to the group $SU_2(\mathbf{C})$, and this allows us to find easily the representations of $SO_3(\mathbf{R})$ from those of $SU_2(\mathbf{C})$.

Proposition 5.6.9. *There exists a continuous surjective group homomorphism*

$$p : SU_2(\mathbf{C}) \longrightarrow SO_3(\mathbf{R})$$

such that $\operatorname{Ker} p = \{\pm 1\} \subset SU_2(\mathbf{C})$ *has order* 2. *As a consequence, the irreducible unitary representations of the group* $SO_3(\mathbf{R})$ *are representations* π_ℓ, $\ell \geqslant 0$, *of dimension* $2\ell + 1$, *determined by the condition* $\pi_\ell \circ p = \varrho_{2l}$.

Partial proof. One can write explicitly p in terms of matrices; crossing fingers to avoid typing mistakes, and using the usual expression

$$\begin{pmatrix} a & b \\ -\bar{b} & \bar{a} \end{pmatrix}$$

of a matrix $g \in SU_2(\mathbf{C})$, for some $a, b \in \mathbf{C}$ with $|a|^2 + |b|^2 = 1$, this takes the form

(5.28)

$$p\Big(\begin{pmatrix} x+iy & u+iv \\ -u+iv & x-iy \end{pmatrix}\Big) = \begin{pmatrix} (x^2+y^2)-(u^2+v^2) & 2(yv+xu) & -2(xv-yu) \\ -2(xu-yv) & x^2-y^2-u^2+v^2 & 2(xy+uv) \\ 2(xv+yu) & -2(xy-uv) & x^2-y^2+u^2-v^2 \end{pmatrix},$$

but that is about as enlightening as checking by hand the associativity of the product of matrices of fixed size four or more. A good explanation for the existence of p is explained in Remark 6.3.10 in the next chapter, so we defer a reasonable discussion of this part of the result (see also [**58**, §4.3] for a down-to-earth approach). Note at least that the formula makes it easy to check that $p(g) = 1$ if and only if $g = 1$ or -1.

On the other hand, given the existence of p, we notice that if ϱ is any irreducible unitary representation of $SO_3(\mathbf{R})$, then $\varrho \circ p$ is an irreducible unitary representation of $SU_2(\mathbf{C})$. By the previous classification, it is therefore of the form ϱ_m for some $m \geqslant 0$. The question is therefore, For which integers $m \geqslant 0$ is ϱ_m of the form $\varrho \circ p$? The answer is elementary: this happens if and only if $\operatorname{Ker}(p) \subset \operatorname{Ker}(\varrho_m)$, and since $\operatorname{Ker}(p)$ has only two elements, this amounts to asking that $\varrho_m(-1) = 1$. The answer can then be obtained from the explicit description of ϱ_m, or by character theory using (5.26): -1

corresponds to $\theta = \pi$ and we have

$$\chi_m(-1) = \lim_{\theta \to \pi} \frac{\sin((m+1)\theta)}{\sin(\theta)} = (-1)^m,$$

so that ϱ_m is obtained from an irreducible representation of $\mathrm{SO}_3(\mathbf{R})$ if and only if $m = 2\ell$ is even. Since different values of $\ell \geqslant 0$ lead to representations of different dimension, this gives the desired correspondence. □

Example 5.6.10 (Infinite direct products)**.** The following class of examples does not occur very often in basic applications, but it is enlightening. By Proposition 2.3.23 (and unitarizability, and the fact that irreducible representations of compact groups are finite dimensional), we see that the irreducible unitary representations of a direct product $G_1 \times G_2$ of compact groups are of the form

$$\varrho_1 \boxtimes \varrho_2,$$

where ϱ_1 (resp. ϱ_2) ranges over the representations in $\widehat{G_1}$ (resp. $\widehat{G_2}$). This extends naturally to a similar statement for an arbitrary finite product $G_1 \times \cdots \times G_k$, with irreducible representations

$$\varrho_1 \boxtimes \cdots \boxtimes \varrho_k.$$

We will now extend this idea to infinite products of compact groups. Let I be an arbitrary index set (the interesting case being when I is infinite, say the positive integers, but it can be any set, including possibly an uncountable one), and let $(G_i)_{i \in I}$ be any family of compact topological groups indexed by I (for instance, the family $(\mathrm{GL}_2(\mathbf{F}_p))_p$, where p runs over primes). The product group

$$G = \prod_{i \in I} G_i$$

is given the product topology. By Tychonov's Theorem (see, e.g., [**49**, Th. IV.5]), this is a compact topological space. One can check that the product on G is continuous, and hence G is a compact topological group. We now determine its irreducible unitary representations.

There is an abundance of irreducible representations arising from the finite products of the groups: for any finite subset $J \subset I$, we have the projection homomorphism

$$(5.29) \qquad p_J : \begin{cases} G & \longrightarrow \quad G_J = \prod_{i \in J} G_i \\ (g_i)_{i \in I} & \mapsto \quad (g_j)_{i \in J}, \end{cases}$$

which is continuous (by definition of the product topology), so that any irreducible unitary representation $\boxtimes_{i \in J} \varrho_i$ of the finite product G_J gives by composition an irreducible representation of G, with character

$$\chi((g_i)) = \prod_{i \in J} \chi_{\varrho_i}(g_i).$$

Some of these representations are isomorphic, but this only happens when a component ϱ_i is trivial, in which case we might as well have constructed the character using $G_{J-\{i\}}$ (we leave a formal proof to the reader!) In other words, we have a family of irreducible representations parametrized by a finite—possibly empty—subset J of I, and a family $(\varrho_i)_{i\in J}$ of *non-trivial* irreducible unitary representations of the groups G_i, $i \in J$. In particular, the trivial representation of G arises from $J = \emptyset$, in which case $G_J = 1$ (an empty product of groups being the trivial group).

We now claim that these are the only irreducible unitary representations of G. This statement is easy to prove, but it depends crucially on the topological structure of G. For the proof, we use the completeness criterion from Peter–Weyl theory (Corollary 5.4.8) by proving that the linear span (say V) of the matrix coefficients of those known representations is dense in $L^2(G)$.

For this purpose, it is enough to show that the closure of V contains the continuous functions, since $C(G)$ is itself dense in $L^2(G)$. Let therefore φ be continuous on G. The main point is that the product topology imposes that φ depends "essentially" only on finitely many coordinates. Precisely, let $\varepsilon > 0$ be arbitrary. Since G is compact, the function φ is in fact *uniformly* continuous, in the sense that there exists an open neighborhood U of 1 such that

$$|\varphi(g) - \varphi(h)| \leqslant \varepsilon$$

if $gh^{-1} \in U$.[9] The definition of the product topology shows that, for some finite subset $J \subset I$, the open set U contains a product set

$$V = \{(g_i)_{i\in I} \mid g_i \in V_i \text{ for } i \in J\}$$

for suitable open neighborhoods V_i of 1 in G_i. Intuitively, up to a precision ε, it follows that φ "only depends on the coordinates in J". Let then

$$\varphi_J(g) = \varphi(\tilde{g}),$$

where $\tilde{g}_i = g_i$ for $i \in J$ and $\tilde{g}_i = 1$ otherwise. Since $g\tilde{g}^{-1} \in V$, this function on G satisfies

$$|\varphi(g) - \varphi_J(g)| \leqslant \varepsilon$$

for all $g \in G$.

But now, φ_J can be identified with a function on G_J. Then, by the Peter–Weyl Theorem, we can find a linear combination ψ of matrix coefficients of representations of G_J such that

$$\|\psi - \varphi_J\|_{L^2(G_J)} \leqslant \varepsilon.$$

[9] We leave the proof as an exercise, adapting the classical case of functions on compact subsets of **R**.

But it is quite easy to see that the probability Haar measure μ on G is such that its image under the projection p_J is the probability Haar measure μ_J on G_J. This means that when we see ψ (and again φ_J) as functions on G, we still have

$$\|\psi - \varphi_J\|_{L^2(G)} \leqslant \varepsilon.$$

Putting these inequalities together, we obtain

$$\|\varphi - \psi\|_{L^2(G)} \leqslant 2\varepsilon,$$

and as ε was arbitrary, we are done.

In Exercise 6.1.4 in the next chapter, we present a slightly different proof, which directly shows that an irreducible unitary representation of G must be of the "known" type.

Chapter 6

Applications of representations of compact groups

This chapter presents some applications of the representation theory of compact groups.

6.1. Compact Lie groups are matrix groups

The first application we present is a rather striking fact of differential geometry—the identification of compact Lie groups with compact subgroups of the linear matrix groups $\mathrm{GL}_n(\mathbf{C})$. We recall the definition of Lie groups first.

Definition 6.1.1 (Lie group). A *Lie group* G is a topological group which is also a topological manifold, i.e., for every $g \in G$, there exists an open neighborhood of g which is homeomorphic to an open ball in some euclidean space $\mathbf{R}^n$, $n \geqslant 0$.

To be more precise, this is the definition of a *topological* Lie group, which we use because it is self-contained. One can also define smooth (or real-analytic) Lie groups by asking that a topological group also be a smooth (resp. real-analytic) manifold such that the product and inverse maps are smooth (resp. analytic). This differential aspect will be used in Section 6.2 below.

The main result of this section is the fact that a much stronger-looking definition leads to the same class of groups in the compact case.

Theorem 6.1.2 (Compact Lie groups are matrix groups). *A topological group G is a compact Lie group if and only if there exists some $n \geqslant 1$ such that G is homeomorphic to a closed subgroup of* $\mathrm{U}_n(\mathbf{C})$ *or, equivalently, to a compact subgroup of* $\mathrm{GL}_n(\mathbf{C})$.

The proof of this result is a very nice combination of basic facts of the theory of Lie groups and of the Peter–Weyl theory. The statement is very powerful: in particular, note that for a closed subgroup of $\mathrm{GL}_n(\mathbf{C})$, the multiplication map is not only continuous (as required by the condition that G be a topological group) but smooth in an obvious sense, or indeed even polynomial in terms of the coefficients of the matrix arguments.

Example 6.1.3. (1) Because of the theorem, it is not surprising (!) that all examples of compact Lie groups that one can write directly are, in fact, obviously compact matrix groups. Examples are $\mathrm{U}_n(\mathbf{C})$, its subgroup $\mathrm{SU}_n(\mathbf{C})$, or the subgroup $\mathbf{T}^n \subset \mathrm{U}_n(\mathbf{C})$ of diagonal matrices, which can be identified also with the torus $(\mathbf{R}/\mathbf{Z})^n$. One can also consider the group of real orthogonal matrices $\mathrm{O}_n(\mathbf{R})$, which can be seen as $\mathrm{U}_n(\mathbf{C}) \cap \mathrm{GL}_n(\mathbf{R})$, or the unitary symplectic group $\mathrm{USp}_{2g}(\mathbf{C}) \subset \mathrm{GL}_{2g}(\mathbf{C})$, for $g \geqslant 1$, which consists of unitary matrices of size $2g$ preserving a fixed non-degenerate alternating bilinear form.

(2) One can also give many examples of compact topological groups which are not Lie groups, though it is maybe not immediately obvious that they cannot be embedded in a matrix group in any way—this becomes another consequence of the theorem. The infinite products considered in Example 5.6.10, namely

$$G = \prod_{i \in I} G_i, \qquad G_i \neq 1,$$

are of this type, provided I is infinite and infinitely many among the G_i are non-trivial (for the simplest example, take I to be countable and $G_i = \mathbf{Z}/2\mathbf{Z}$ for all i). Indeed, from the description of the irreducible unitary representations of G in Example 5.6.10, we see that for every irreducible representation ϱ of G, there exists a finite subset $J \subset I$ such that

$$\ker \varrho \supset G^{(J)} = \prod_{i \notin J} G_i.$$

A finite-dimensional representation of G, which is a direct sum of finitely many such representations, therefore has kernel containing a finite intersection

$$G^{(J_1)} \cap \cdots \cap G^{(J_k)}$$

of such subgroups, and the latter contains $G^{(J)}$ for $J = J_1 \cup \cdots \cup J_k$, which is a non-trivial group since J is finite, and hence $I - J$ is not empty. Thus G has no finite-dimensional faithful representation.

(3) Another example is the following group, which is called the group of p-adic integers. This group and its relatives are of considerable importance in number theory. Let p be a fixed prime number, and consider first the infinite product

$$G_p = \prod_{k \geqslant 1} \mathbf{Z}/p^k\mathbf{Z}.$$

Then the subgroup

$$\mathbf{Z}_p = \{(x_k) \in G_p \mid x_{k+1} \equiv x_k \,(\mathrm{mod}\, p^k) \text{ for all } k \geqslant 0\} \subset G_p$$

is called the (additive) group of p-adic integers.

To see examples of elements of $\mathbf{Z}_p$, take any $x \in \mathbf{Z}$, and consider $x_k = x\,(\mathrm{mod}\, p^k)$ for all $k \geqslant 0$; then $(x_k) \in \mathbf{Z}_p$. Also, the reader might find it enjoyable to check that if p is odd and congruent to 1 modulo 4, then there exists a p-adic integer $x \in \mathbf{Z}_p$ such that $x^2 = -1$. More information and discussion can be found, for instance, in [**54**, Chapter 2].

It is again an exercise to check that $\mathbf{Z}_p$ is a closed subgroup of G_p, and hence a compact topological group. It is an abelian group, and one can see as follows that it does not have a faithful finite-dimensional representation. First, since $\mathbf{Z}_p$ is abelian, its irreducible unitary representations are one dimensional. Then we note that if χ is a character of $\mathbf{Z}_p$, there exists an integer $j \geqslant 1$ such that

$$\operatorname{Ker} \chi \supset \{(x_k) \in \mathbf{Z}_p \mid x_j = 0\},$$

where it is useful to remark that if $x_j = 0$, then $x_1 = 0, \ldots, x_{j-1} = 0$, each in its respective group $\mathbf{Z}/p\mathbf{Z}, \ldots, \mathbf{Z}/p^{j-1}\mathbf{Z}$.

To see this, note that the sets

$$U_j = \{(x_k) \in \mathbf{Z}_p \mid x_j = 0\}$$

form a fundamental system of neighborhoods of 0 in $\mathbf{Z}_p$, and hence, by the continuity of χ, there exists $j \geqslant 0$ such that $\chi(U_j)$ is contained in the open neighborhood

$$V = \mathbf{S}^1 \cap \{z \in \mathbf{C} \mid |z-1| < 1/2\}$$

of 1 in $\mathbf{S}^1$. But then, since U_j is a subgroup of $\mathbf{Z}_p$, the set $\chi(U_j)$ is a subgroup of $\mathbf{S}^1$ contained in V. But it is elementary that $\{1\}$ is the only such subgroup, which means that $U_j \subset \operatorname{Ker} \chi$, as claimed.

We conclude with an argument quite similar to that in Example 5.6.10: the kernel of any finite direct sum of characters of $\mathbf{Z}_p$ will also contain a subgroup of the type U_j, and hence such a representation is not faithful (since, for instance, the image in $\mathbf{Z}_p$ of a non-zero integer $x \equiv 0\,(\mathrm{mod}\, p^j)$ is a non-trivial element of U_j).

Exercise 6.1.4 (No small subgroups in unitary groups). (1) Show that in any unitary group $U_n(\mathbf{C})$, there is a neighborhood V of 1 which contains no non-trivial subgroup.

(2) Use this to reprove the result of Example 5.6.10 by directly showing that any irreducible representation of an infinite product of compact groups is of the "known" form $\varrho \circ p_J$, with notation as in (5.29).

The group $\mathbf{Z}_p$ is abelian, and also carries an extra structure: $\mathbf{Z}_p$ is also a topological ring, with coordinate-wise multiplication, and one can therefore define other groups, such as

$$\mathrm{SL}_2(\mathbf{Z}_p) = \left\{ \begin{pmatrix} a & b \\ c & d \end{pmatrix} \mid a,\ b,\ c,\ d \in \mathbf{Z}_p, \quad ad - bc = 1 \right\}.$$

With the group structure coming from matrix multiplication and the induced topology from the product topology on $\mathbf{Z}_p^4$, this is again a compact topological group (it is a closed subset of the compact space $\mathbf{Z}_p^4$). Of course, it is non-abelian.

We conclude by mentioning that the similarity between the two counter-examples is not accidental. In fact, deep results in three works by Gleason, Montgomery and Zippin, and Yamabe imply that a locally compact topological group G is a Lie group if and only if contains no small subgroups, i.e., if there is a neighborhood of 1 in G containing no non-trivial subgroup.

We come now to the proof of the theorem. For this, we will need to use without proof the following facts concerning Lie groups (which are proved in all textbooks on Lie theory, for instance [**35**]):

- A Lie group G has a well-defined *dimension* $\dim(G)$, a non-negative integer, which is the dimension of G as a manifold; for instance

$$\dim(\mathbf{R}) = 1, \qquad \dim(\mathrm{GL}_n(\mathbf{R})) = n^2, \qquad \dim(\mathrm{GL}_n(\mathbf{C})) = 2n^2,$$

  where the last case illustrates that the dimension involved is that of G as a *real* manifold.
- A compact Lie group G has only finitely many connected components; in particular, if G is compact, we have $\dim(G) = 0$ if and only if G is a finite group (which is a compact Lie group with the discrete topology).
- If $H \subset G$ is a closed subgroup of a Lie group, then in fact H is a submanifold, and H is itself (with the induced topology) a Lie group.
- In the same situation where $H \subset G$ is a closed subgroup, we have in particular $\dim(H) \leqslant \dim(G)$, and if there is equality, the subgroup

H is a union of some of the connected components of G; especially, if H and G are both connected, we have $H = G$.

Now we embark on the proof....

Proof of Theorem 6.1.2. We will prove the following statement, which introduces representation theory. If G is a compact Lie group, there exists a finite-dimensional *faithful* representation

$$\varrho : G \longrightarrow \mathrm{GL}(E)$$

of G. This is enough to imply Theorem 6.1.2, since by fixing a basis of E, we can view ϱ as an injective homomorphism of G into $\mathrm{GL}_n(\mathbf{C})$, where $n = \dim(E)$. Since G is compact, ϱ is a homeomorphism onto its image, which is therefore a compact subgroup of $\mathrm{GL}_n(\mathbf{C})$, and with respect to an inner product for which ϱ is unitary (which exists by Theorem 5.2.11), this image is in fact a subgroup of $\mathrm{U}_n(\mathbf{C})$.

The basic idea is now to use the fact that the Peter–Weyl theory provides us with many finite-dimensional representations of G—indeed enough to separate points—which means that for every $g \in G$, $g \neq 1$, there is at least one finite-dimensional representation ϱ_g such that $\varrho_g(g) \neq 1$. If g is also in the connected component of 1 in G, then $\operatorname{Ker} \varrho_g$ will be a closed subgroup of G with strictly smaller dimension, and we can argue (essentially) by induction on the dimension.

We present this idea slightly differently, merely for the sake of diversity. Let $d \geqslant 0$ be the non-negative integer defined as the minimal d such that there exists a finite-dimensional representation ϱ of G with $\dim \ker \varrho = d$. This is well defined, since the trivial representation has kernel of dimension $\dim G$ (in particular, $d \leqslant \dim G$).

We claim that $d = 0$. If that is the case and ϱ is such that $\dim \operatorname{Ker} \varrho = 0$, we see by the facts above that the kernel is a finite subgroup of G (possibly trivial, of course). But then we can also consider the representation

$$\varrho \oplus \bigoplus_{\substack{g \in \operatorname{Ker} \varrho \\ g \neq 1}} \varrho_g,$$

which is still finite dimensional and is now faithful.

Let ϱ be such that the kernel H of ϱ has dimension d. Now assume, for contradiction, that $\dim H = \dim \operatorname{Ker} \varrho \geqslant 1$. Then we can find some $h \in H$ which is not trivial but is in the connected component of H containing 1. Then

$$\operatorname{Ker}(\varrho \oplus \varrho_h) = H \cap \operatorname{Ker} \varrho_h$$

is a proper subgroup of H. Its dimension is $\leqslant \dim(H)$, but it cannot be equal: by the facts recalled before the proof, this would only be possible if the connected component of 1 in H coincided with that in $H \cap \operatorname{Ker} \varrho_h$, which is not the case, as h is in the first one but not in the other! Thus

$$\dim \operatorname{Ker}(\varrho \oplus \varrho_h) < \dim H = d,$$

and this is a contradiction with the definition of d as the smallest possible dimension of the kernel of a finite-dimensional unitary representation of G, which means the supposition that $d \geqslant 1$ is untenable. □

6.2. The Frobenius–Schur indicator

The results in this section apply equally well (and are of interest!) for finite groups. The basic issue they address is the following. Given a finite-dimensional (complex) representation

$$\varrho : G \longrightarrow \mathrm{GL}(E)$$

of a compact group G, does there exist on E a symmetric, or alternating, non-degenerate bilinear form b which is invariant under G, i.e., such that

$$b(\varrho(g)v, \varrho(g)w) = b(v, w)$$

for all $g \in G$ and v, $w \in E$? This question should be contrasted with the unitarizability of ϱ, which can be interpreted partly as saying that there is always on E an invariant non-degenerate (positive-definite) hermitian form. As a first illustration of the techniques that will be used, we spell out the following algebraic version of this fact.

Proposition 6.2.1. *Let G be a compact group, and let $\varrho : G \longrightarrow \mathrm{GL}(E)$ be an* irreducible *finite-dimensional complex representation of G. Then there exists, up to multiplication by a positive scalar, a unique G-invariant hermitian form b on E.*

Proof. Because we know that ϱ is unitarizable, the only thing we need to explain is the uniqueness. We can deduce this from Schur's Lemma (in a way similar to Exercise 3.4.20) or use an argument from character theory which is also instructive. The group G acts on the vector space S of hermitian forms on E by

$$(g \cdot b)(v, w) = b(\varrho(g^{-1})v, \varrho(g^{-1})w),$$

and we wish to show that S^G is one dimensional.

The point is that the character of the action of G on S is $g \mapsto |\chi_\varrho(g)|^2$. (We leave this as an exercise; the arguments in the proof of Theorem 6.2.3 are very similar.) Then the dimension formula for invariants of representations of G (i.e., (5.22)) gives

$$\dim S^G = \int_G |\chi_\varrho(g)|^2 d\mu(g) = 1$$

(where μ is the probability Haar measure on G) by orthonormality of characters. $\square$

It is tempting to consider the real and imaginary parts of an invariant hermitian form to construct symmetric and alternating forms; however, these are only **R**-bilinear!

Definition 6.2.2. Let G be a group, and let $\varrho : G \longrightarrow \mathrm{GL}(E)$ be a finite-dimensional complex representation of G. One says that ϱ is *of orthogonal type* (or *of real type*) if there exists a G-invariant, non-degenerate, symmetric bilinear form on E, and that ϱ is *of symplectic type* (or *of quaternionic type*) if there exists a G-invariant, non-degenerate, alternating bilinear form on E. If neither of these possibilities holds, we say that ϱ is of *complex type*.

Note that we do not exclude a priori the possibility that a representation could be of more than one type. However, for irreducible representations of compact groups, this does not occur:

Theorem 6.2.3 (Frobenius–Schur indicator). *Let G be a compact group with probability Haar measure μ, and let*

$$\varrho : G \longrightarrow \mathrm{GL}(E)$$

be a finite-dimensional representation of G. Define the Frobenius–Schur indicator *of ϱ by*

$$\mathrm{FS}(\varrho) = \int_G \chi_\varrho(g^2)d\mu(g). \tag{6.1}$$

Then if ϱ is irreducible, we have $\mathrm{FS}(\varrho) \in \{-1, 0, 1\}$, and

(1) *the representation ϱ is of orthogonal type if and only if $\mathrm{FS}(\varrho) = 1$;*

(2) *the representation ϱ is of symplectic type if and only if $\mathrm{FS}(\varrho) = -1$;*

(3) *the representation ϱ is of complex type if and only if $\mathrm{FS}(\varrho) = 0$.*

In particular, one, and exactly one, of the three possibilities arises.

Proof. The first part of the argument is relatively similar to that used in Proposition 6.2.1. Only towards the end do we work out, using character theory, a numerical criterion that distinguishes the three possible types of representations. This naturally introduces the Frobenius–Schur indicator as the "right" tool to do this.

We let B denote the vector space of bilinear forms on the space E. The group G acts on B by the formula

$$(g \cdot b)(v, w) = b(\varrho(g^{-1})v, \varrho(g^{-1})w)$$

for $b \in B$ and v, $w \in E$, and an invariant bilinear form on E is therefore simply an element of the subspace B^G. We must attempt to compute this space and, in particular, determine its dimension.

For this, we compute first the character of the action of G on B. This is quite simple: the linear isomorphism

$$\begin{cases} E' \otimes E' \longrightarrow B \\ \lambda_1 \otimes \lambda_2 \mapsto b_{\lambda_1,\lambda_2}, \end{cases}$$

where

$$b_{\lambda_1,\lambda_2}(v,w) = \lambda_1(v)\lambda_2(w) \tag{6.2}$$

is an isomorphism of representations, where E' carries the contragredient of ϱ. Hence by the character formalism, we have

$$\chi_B(g) = \chi_{\check{\varrho}}(g)^2 = \overline{\chi_\varrho(g)}^2. \tag{6.3}$$

The next step looks innocuous: by the projection formula on invariants, we have

$$\dim B^G = \int_G \chi_B(g) d\mu(g),$$

and we can bound this from above by

$$\dim B^G \leqslant \int_G |\chi_B(g)| d\mu(g) = \int_G |\chi_\varrho(g)|^2 d\mu(g) = 1, \tag{6.4}$$

so that the space B^G is either zero or one dimensional, i.e., if there exists a non-zero invariant bilinear form on E, it is unique up to scalar.[1]

What remains to be done is to understand when the dimension is 0 and when it is 1, and this will lead to the refined statement of the theorem. The key is that B has an a priori decomposition

$$B = B_{\text{sym}} \oplus B_{\text{alt}} \tag{6.5}$$

into two subrepresentations, where B_{sym} is the space of symmetric bilinear forms and B_{alt} is the space of alternating bilinear forms. It is indeed clear that B_{sym} and B_{alt} are G-invariant in B, and the decomposition of B as a vector space is well known: $B_{\text{sym}} \cap B_{\text{alt}} = 0$, and one can write any $b \in B$ in the form $b = b_s + b_a$, where

$$b_s(v,w) = \frac{1}{2}(b(v,w) + b(w,v)), \quad b_a(v,w) = \frac{1}{2}(b(v,w) - b(w,v)), \tag{6.6}$$

with $b_s \in B_{\text{sym}}$ and $b_a \in B_{\text{alt}}$. (Note in passing that (6.5) can be interpreted as the decomposition of B under the representation of the group $\mathfrak{S}_2 = \mathbf{Z}/2\mathbf{Z}$ on B by permutation of the arguments, i.e., the action where the generator $1 \in \mathbf{Z}/2\mathbf{Z}$ acts by $1 \cdot b(v,w) = b(w,v)$.)

---

[1] As mentioned, this might pass unnoticed, but we have obtained here an upper bound for the invariants in a representation (the bilinear forms) using information concerning those of a space which seems, a priori, unrelated (the hermitian form S). This is all done through the remarkable effect of character theory.

It follows from (6.5) that

$$B^G = B^G_{\text{sym}} \oplus B^G_{\text{alt}},$$

with the summands being the spaces of invariant symmetric or alternating bilinear forms. Since $\dim B^G \leqslant 1$, we get the basic trichotomy in terms of bilinear forms: either $\dim B^G_{\text{sym}} = 1$, $\dim B^G_{\text{alt}} = 0$ (orthogonal type); or $\dim B^G_{\text{sym}} = 0$, $\dim B^G_{\text{alt}} = 1$ (symplectic type); or $B^G_{\text{sym}} = B^G_{\text{alt}} = B^G = 0$ (complex type). One might object that (for instance) it is possible that $\dim B^G_{\text{alt}} = 1$ but that a non-zero $b \in B^G_{\text{alt}}$ is degenerate, whereas the definition of symplectic type asks for a non-degenerate bilinear form. But for any non-zero $b \in B^G$, the kernel of b, i.e., the subspace

$$\operatorname{Ker} b = \{v \in E \mid b(v, w) = 0 \text{ for all } w \in E\},$$

is a subrepresentation of E, since for all $v \in \operatorname{Ker} b$ and $w \in E$, we have

$$b(\varrho(g)v, w) = b(v, \varrho(g)^{-1}w) = 0$$

so that $\varrho(g)v \in \operatorname{Ker} b$. Thus, since E is irreducible and $b \neq 0$ (so that $\operatorname{Ker} b \neq E$), we have $\operatorname{Ker} b = 0$, and b is non-degenerate.

The numerical criterion for the trichotomy, involving the Frobenius–Schur indicator, arises by noting the following clever way of encapsulating it. The three possibilities are characterized by the value of

$$\dim B^G_{\text{sym}} - \dim B^G_{\text{alt}} = \begin{cases} 1 - 0 = 1 & \text{for orthogonal type,} \\ 0 - 0 = 0 & \text{for unitary type,} \\ 0 - 1 = -1 & \text{for symplectic type,} \end{cases}$$

which is therefore the "explanation" for the Frobenius–Schur indicator.[2] Again from character theory, we get that the desired invariant is

$$\dim B^G_{\text{sym}} - \dim B^G_{\text{alt}} = \int_G (\chi_{\text{sym}}(g) - \chi_{\text{alt}}(g))\, d\mu(g),$$

where, for simplicity, we denote by χ_{sym} and χ_{alt} the characters of B_{sym} and B_{alt}. Therefore we proceed to compute the difference of the two characters.

This is quite easy. For a fixed element $g \in G$, we can diagonalize the unitary operator $\varrho(g)$ in some basis $(e_i)_{1 \leqslant i \leqslant n}$ of E, with dual basis (λ_i) of E', so that

$$\varrho(g)e_i = \theta_i e_i, \qquad \check{\varrho}(g)\lambda_i = \bar{\theta}_i \lambda_i,$$

for some eigenvalues θ_i, whose sum is $\chi_\varrho(g)$ or $\chi_{\check{\varrho}}(g)$, respectively. Then the bilinear forms

$$b_{i,j} = b_{\lambda_i, \lambda_j}$$

[2] Note that we could have exchanged the sign of the two terms, which would just have changed the meaning of the indicators ± 1; the choice we made is the standard one, but it is, to some extent, just a convention.

given by (6.2) form a basis of B, with

$$g \cdot b_{i,j} = \overline{\theta_i \theta_j} b_{i,j},$$

by definition of the action on B. Applying the decomposition (6.6), a basis of B_{sym} is given by the symmetric bilinear forms

$$\frac{1}{2}(b_{i,j} + b_{j,i}),$$

where i and j are arbitrary (but of course (i,j) and (j,i) give the same basic bilinear form, so we may assume $i \leqslant j$), and a basis of B_{alt} is given by the alternating forms

$$\frac{1}{2}(b_{i,j} - b_{j,i}),$$

where this time $i < j$. Notice that in both case, these are eigenvectors for the action of g with the same eigenvalue $\overline{\theta_i \theta_j}$ (because $b_{i,j}$ and $b_{j,i}$ are in the same eigenspace). Thus we find that

$$\chi_{\mathrm{sym}}(g) = \sum_{1 \leqslant i \leqslant j \leqslant n} \overline{\theta_i \theta_j}, \qquad \chi_{\mathrm{alt}}(g) = \sum_{1 \leqslant i < j \leqslant n} \overline{\theta_i \theta_j}.$$

Only the diagonal terms are missing from the second sum compared to the first; hence we get

$$\chi_{\mathrm{sym}}(g) - \chi_{\mathrm{alt}}(g) = \sum_{1 \leqslant i \leqslant n} \overline{\theta_i}^2 = \overline{\chi_\varrho(g^2)}$$

(since the matrix $\varrho(g^2)$ has eigenvalues θ_i^2 in the basis (e_i)), and to conclude, and recover the formula (6.1), we may simply observe that

$$\int_G \overline{\chi_\varrho(g^2)} d\mu(g) = \int_G \chi_\varrho(g^2) d\mu(g),$$

since we know already that the integral is a real number. □

We will now give a few examples. Before doing this, the following result illustrates another interesting meaning of the Frobenius–Schur indicator.

Proposition 6.2.4 (Self-dual representations). *Let G be a compact topological group, and let ϱ be an irreducible representation of G. Then* $\mathrm{FS}(\varrho) \neq 0$ *if and only if the character of ϱ is real-valued, and this is the case if and only if ϱ is isomorphic to its contragredient representation $\check{\varrho}$. Such a representation is called* self-dual.

Proof. According to the proof above, ϱ is symplectic or orthogonal if and only if the space B^G of invariant bilinear forms on the space of ϱ is one dimensional, and (in view of the character formula (6.3)) this is the case if and only if

$$\int_G \overline{\chi_\varrho(g)}^2 d\mu(g) = 1.$$

As we did earlier, we argue that

$$\Big|\int_G \overline{\chi_\varrho(g)}^2 d\mu(g)\Big| \leqslant \int_G |\chi_\varrho(g)|^2 d\mu(g) = 1,$$

but now we continue by noticing that *if* there is equality, it must be the case that $\overline{\chi_\varrho(g)}^2$ is proportional to $|\chi_\varrho(g)|^2$, with a scalar multiple of modulus 1. Taking $g = 1$ shows that the scalar must be equal to 1, i.e. (taking conjugate), we have

$$\chi_\varrho(g)^2 = |\chi_\varrho(g)|^2 \geqslant 0$$

for all $g \in G$. Since, among complex numbers, only real numbers have a non-negative square, we obtain the first result.

Now the last (and possibly most interesting!) conclusion is easy: since the character of $\check{\varrho}$ is $\overline{\chi_\varrho}$, it follows from character theory that ϱ has a real-valued character if and only if it is isomorphic to its contragredient. □

Example 6.2.5. (1) Let $G = \mathrm{SL}_2(\mathbf{C})$ or $\mathrm{SU}_2(\mathbf{C})$. Among the representations ϱ_m of G, $m \geqslant 0$, those with m even are of orthogonal type while those with m odd are of symplectic type. This can be checked in different ways. For $\mathrm{SU}_2(\mathbf{C})$, one may use the integration and character formulas (see (5.26)) to check that this amounts to proving the identity

$$\frac{2}{\pi}\int_0^\pi \frac{\sin(2(m+1)\theta)}{\sin 2\theta}\sin^2\theta d\theta = (-1)^m.$$

For either group, one may also define explicitly an invariant non-degenerate bilinear form b on the representation space V_m of homogeneous polynomials of degree m in two variables, by putting

$$b(e_i, e_j) = \frac{(-1)^i\delta(i, m-j)}{\binom{m}{i}}$$

for the basis vectors $e_i = X^iY^{m-i}$.

Such a definition certainly defines a bilinear form on V_m, and it is symmetric for m even and alternating for m odd (where $\delta(i, m-i)$ is always zero). To see that it is non-degenerate, observe that the non-zero coefficients of the matrix $(b(e_i, e_j))_{i,j}$ are exactly the antidiagonal ones, so that the determinant is their product, up to sign, which is non-zero.

What is by no means immediately obvious, on the other hand, is that b is $\mathrm{SL}_2(\mathbf{C})$-invariant! A fairly quick algebraic proof of this is explained in [**59**, 3.1.4, 3.1.5]: we know that the group $\mathrm{SL}_2(\mathbf{C})$ is generated by the elements

$$u(t) = \begin{pmatrix} 1 & t \\ 0 & 1\end{pmatrix},\quad t \in \mathbf{C},\quad a(x) = \begin{pmatrix} x & 0 \\ 0 & x^{-1}\end{pmatrix},\quad x \in \mathbf{C}^\times,\quad w = \begin{pmatrix} 0 & 1 \\ -1 & 0\end{pmatrix}$$

(see Proposition 4.6.19), so that it is enough to check that

$$b(\varrho_m(g)e_i, \varrho_m(g)e_j) = b(e_i, e_j)$$

for g in one of these three classes (and all basis vectors e_i, e_j). We leave the easy cases of $a(x)$ and w to the reader, and just present the (maybe somewhat mysterious) case of $g = u(t)$. In that case, we have

$$\varrho_m(g)e_i = X^i(tX+Y)^{m-i} = \sum_{k=0}^{m-i} \binom{m-i}{k} t^k e_{i+k} \tag{6.7}$$

by the binomial theorem, and hence

$$b(\varrho_m(g)e_i, \varrho_m(g)e_j) = \sum_{k=0}^{m-i}\sum_{\ell=0}^{m-j} t^{k+\ell}\binom{m-i}{k}\binom{m-j}{\ell} b(e_{i+k}, e_{j+\ell}).$$

Using the definition of b, only terms with $i+k+j+\ell = m$ remain, and this can be rearranged as

$$b(\varrho_m(g)e_i, \varrho_m(g)e_j) = (-1)^i t^{m-i-j} \sum_{k=0}^{m-i-j} (-1)^k \frac{\binom{m-i}{k}\binom{m-j}{m-i-j-k}}{\binom{m}{i+k}},$$

where it is possible that the sum be empty. If one rearranges the ratio of binomial coefficients in terms of factorials, this becomes

$$\begin{aligned} b(\varrho_m(g)e_i, \varrho_m(g)e_j) &= \frac{(-1)^i t^{m-i-j}}{(m-i-j)!} \\ &\quad \times \frac{(m-i)!(m-j)!}{m!} \sum_{k=0}^{m-i-j} (-1)^k \binom{m-i-j}{k}. \end{aligned}$$

The inner sum over k is zero (hence equal to $b(e_i, e_j)$) except when $m = i+j$, and in that last case we obtain

$$b(\varrho_m(g)e_i, \varrho_m(g)e_j) = \frac{(-1)^i(m-i)!i!}{m!} = b(e_i, e_j).$$

Overall, this concrete example can be considered as a rather striking illustration of the power of character theory: the existence of a symmetric or alternating invariant bilinear form on the space of ϱ_m is obtained by a simple integral of trigonometric functions, but the actual bilinear form is quite intricate and it is not straightforward at all to guess its expression. In particular, note that it is quite different from the $\mathrm{SU}_2(\mathbf{C})$-invariant inner product, for which the vectors e_i are orthogonal (see Example 5.2.12). Of course, this inner product $\langle\cdot,\cdot\rangle$ on V_m is *not* invariant under the larger group $\mathrm{SL}_2(\mathbf{C})$, since ϱ_m is not unitary as a representation of $\mathrm{SL}_2(\mathbf{C})$.

(2) This last remark illustrates again the strength of character theory: Let us write the crucial inequality (6.4) in the form

$$\dim B^G \leqslant \dim S^G$$

(with B the space of bilinear forms and S the space of hermitian forms). Now although both sides are purely algebraic invariants that may be defined for any finite-dimensional complex representations of any group and although the inequality is valid for all compact groups, *it is not* universally valid for finite-dimensional representations of topological groups! Indeed, already for $G = \mathrm{SL}_2(\mathbf{C})$ and $\varrho = \varrho_m$, the right-hand side is 0, while the left-hand side is always 1 (since we checked that the bilinear form b above was $\mathrm{SL}_2(\mathbf{C})$-invariant, and not merely in $B^{\mathrm{SU}_2(\mathbf{C})}$).

(3) But there is even more to this story. If we write (6.4) in the form

$$\dim B^G \leqslant 1,$$

then this inequality *is* valid for any finite-dimensional irreducible representation of any group G (even without imposing continuity conditions)! This is another variant of Schur's Lemma. Indeed, we can assume the existence of one non-zero invariant bilinear form b on the space E of the irreducible representation ϱ of G under consideration, since otherwise $B^G = 0$. As in the beginning of the proof of Theorem 6.2.3, any $b \neq 0$ in B^G is non-degenerate, because its kernel is a proper subrepresentation of ϱ. Now, if b_1 is any element in B^G, linear algebra shows that there exists a linear map $\Phi : E \longrightarrow E$ such that

$$b_1(v, w) = b(v, \Phi(w))$$

for all $v, w \in E$. Algebraically, this linear map Φ is simply the *transpose* of the identity map $E \longrightarrow E$, when the latter is seen as mapping the space E given with the bilinear form b to E given with b_1. It is natural from this point of view that Φ is an *intertwiner* $E \longrightarrow E$, and this is easily checked using the fact that b and b_1 are non-degenerate. By Schur's Lemma, there exists $\lambda \in \mathbf{C}$ such that $\Phi = \lambda \mathrm{Id}$, and this gives $b_1 = \lambda b$, so that b spans B^G.

(4) For some important classes of finite groups, all representations are of orthogonal type. This applies, for instance, to the symmetric groups because one can show that they can be constructed as matrix representations with values in $\mathrm{GL}_n(\mathbf{Q})$, or even $\mathrm{GL}_n(\mathbf{Z})$, so that their characters are real-valued. An interesting consequence of this arises as the application of the following simple lemma.

Lemma 6.2.6 (Groups with all representations orthogonal). *Let G be a finite group such that $\mathrm{FS}(\varrho) = 1$ for all irreducible complex representations ϱ of G. Then the sum*

$$\sum_{\varrho \in \hat{G}} \dim(\varrho)$$

of the dimensions of irreducible representations of G is equal to the number of elements of order 2 in G.

Proof. This is quite a cute argument: By assumption, we have

$$\frac{1}{|G|}\sum_{g\in G}\chi_\varrho(g^2) = 1$$

for all $\varrho \in \widehat{G}$. Multiplying by $\dim(\varrho)$ and then summing over all ϱ, we obtain

$$\begin{aligned}\sum_{\varrho\in\widehat{G}}\dim(\varrho) &= \frac{1}{|G|}\sum_{\varrho\in\widehat{G}}\sum_{g\in G}\chi_\varrho(g^2)\dim(\varrho)\\ &= \frac{1}{|G|}\sum_{g\in G}\sum_{\varrho\in\widehat{G}}\chi_\varrho(g^2)\chi_\varrho(1).\end{aligned}$$

By the second orthogonality relation (4.28), the inner sum vanishes unless g^2 is conjugate to 1, i.e., unless $g^2 = 1$, and in that case it is equal to $|G|$. Thus we get

$$\sum_{\varrho\in\widehat{G}}\dim(\varrho) = |\{g \in G \mid g^2 = 1\}|,$$

as claimed. □

The problem of evaluating this sum was mentioned in Remark 4.2.6.

Exercise 6.2.7. Consider the examples of finite groups for which we computed the full character table (in Section 4.6.2, 4.6.3 and 4.6.4), and for each of them determine the Frobenius–Schur indicators (in particular determine which are self-dual). (*Hint*: For $\mathrm{GL}_2(\mathbf{F}_p)$, one can use Exercise 4.6.16 to find rather easily the self-dual representations.)

6.3. The Larsen alternative

Our next application has some common features with the Frobenius–Schur theory, but it is a much more recent development which is really a fact about compact, infinite, Lie groups. The results are due to M. Larsen [**42**, §3] and have been extensively developed by N. Katz (for instance in [**32**]).

Their basic motivation can be described as follows. A compact group $G \subset \mathrm{U}_n(\mathbf{C})$ is given, by some means or other, and the question that arises is to identify it, in particular, to prove that it is big in some sense. Here, "big" has roughly the following meaning: either one would like to prove that $G \supset \mathrm{SU}_n(\mathbf{C})$ or one knows—again, one way or another—that G preserves either a symmetric or alternating non-degenerate bilinear form, and the goal is to prove that G contains either the corresponding (real) special orthogonal group or the unitary symplectic group. For this, Larsen found a beautiful numerical criterion. We present it here as an interesting and relatively elementary fact about representations of compact groups. It might not be clear whether this is actually applicable in practice, but we will describe quickly in a later remark how the problem appears in concrete applications.

The invariant introduced by Larsen is the following:

Definition 6.3.1 (Fourth moment of a representation). Let G be a compact subgroup of $\mathrm{U}_n(\mathbf{C})$ for some $n \geqslant 1$ with probability Haar measure μ. The *fourth moment* of G is defined by

$$\mathrm{M}_4(G) = \int_G |\operatorname{Tr}(g)|^4 d\mu(g). \tag{6.8}$$

More generally, given a finite-dimensional representation ϱ of G, the *fourth moment of* ϱ is defined by

$$\mathrm{M}_4(\varrho) = \int_G |\chi_\varrho(g)|^4 d\mu(g).$$

Thus $\mathrm{M}_4(G)$ is the fourth moment of the *tautological* (faithful) representation $\varrho : G \hookrightarrow \mathrm{U}_n(\mathbf{C})$.

A priori, this might be an arbitrary non-negative real number. However, as in the case of the Frobenius–Schur indicator (6.1), it is in fact an integer, and certain of its values carry important meaning. More precisely, we have the following rather remarkable result of Larsen.

Theorem 6.3.2 (Larsen alternative for unitary groups). *Let $n \geqslant 2$, and let G be a compact subgroup of $\mathrm{SU}_n(\mathbf{C})$. If the fourth moment $\mathrm{M}_4(G)$ is equal to 2, then either G is* finite *or $G = \mathrm{SU}_n(\mathbf{C})$. In particular, if G is connected, we have $G = \mathrm{SU}_n(\mathbf{C})$.*

The proof is a very nice application of basic character theory and representation theory, together with some facts of Lie theory. The first step, which we take backwards in comparison with Section 6.2, is to interpret the fourth moment in purely algebraic terms.

Lemma 6.3.3. *Let G be a compact group, and let*

$$\varrho : G \longrightarrow \mathrm{GL}(E)$$

be a finite-dimensional representation of G.

(1) *We have*

$$\mathrm{M}_4(\varrho) = \dim(\operatorname{End}(\varrho) \otimes \operatorname{End}(\varrho))^G = \dim(\operatorname{End}(\varrho \otimes \check{\varrho}))^G. \tag{6.9}$$

(2) *Let π be any of the representations of G on $\varrho \otimes \varrho$, $\varrho \otimes \check{\varrho}$ or* $\operatorname{End}(\varrho)$. *If we have a decomposition*

$$\pi \simeq \bigoplus_i n_i \varrho_i, \qquad n_i \geqslant 0,$$

into G-stable subspaces where the subrepresentations ϱ_i are not necessarily irreducible, then we have

$$\mathrm{M}_4(\varrho) \geqslant \sum_i n_i^2,$$

with equality if and only if the ϱ_i are pairwise distinct irreducible representations.

(3) *If $G \subset H$ are compact subgroups of $\mathrm{U}_n(\mathbf{C})$, then we have*

$$\mathrm{M}_4(H) \leqslant \mathrm{M}_4(G). \tag{6.10}$$

Proof. Note that the fourth moment is an inner product

$$\mathrm{M}_4(\varrho) = \langle |\chi_\varrho|^4, 1\rangle.$$

By the formalism of characters, the function $|\chi_\varrho|^4$ is the character of the representation

$$\tau = \varrho \otimes \varrho \otimes \check{\varrho} \otimes \check{\varrho},$$

so that $\mathrm{M}_4(\varrho)$ is the dimension of the invariant space τ^G. But using the associativity of the tensor product and the relations

$$(\varrho_1 \otimes \varrho_2)\check{} = \check{\varrho}_1 \otimes \check{\varrho}_2, \qquad \check{\check{\varrho}} = \varrho,$$

we can arrange the tensor product τ in two ways: either

$$\tau = (\varrho \otimes \varrho) \otimes (\varrho \otimes \varrho)\check{} \simeq \mathrm{End}(\varrho \otimes \varrho),$$

which gives

$$\mathrm{M}_4(\varrho) = \dim(\mathrm{End}(\varrho \otimes \varrho))^G,$$

or

$$\tau = (\varrho \otimes \check{\varrho}) \otimes (\varrho \otimes \check{\varrho})\check{} \simeq \mathrm{End}(\varrho \otimes \check{\varrho}) = \mathrm{End}(\mathrm{End}(\varrho)),$$

and therefore

$$\mathrm{M}_4(\varrho) = \dim(\mathrm{End}(\varrho \otimes \check{\varrho}))^G = \dim \mathrm{End}(\mathrm{End}(\varrho))^G.$$

This proves (1), and (2) is a general fact about $\dim \mathrm{End}(\pi)^G$ for any representation π. We have

$$\langle \mathrm{End}(\pi), 1\rangle = \sum_{i,j} n_i n_j \langle \varrho_i, \varrho_j\rangle$$

by linearity. Each term is a non-negative integer, and hence

$$\langle \mathrm{End}(\pi), 1\rangle \geqslant \sum_i n_i^2 \langle \varrho_i, \varrho_i\rangle \geqslant \sum_i n_i^2,$$

by keeping only the diagonal terms $i = j$. If there is equality, we see that we must have $\langle \varrho_i, \varrho_j\rangle = \delta(i,j)$, which means that the ϱ_i are irreducible (taking $i = j$) and distinct (for $i \neq j$).

Finally the inequality (6.10), though not at all obvious from the definition (6.8), is clear from (1): if $G \subset H$, then, for any representation of H, the space of G-invariants contains the space of H-invariants. □

Remark 6.3.4. The reader may have noted that the two quantities

$$\dim(\mathrm{End}(\varrho) \otimes \mathrm{End}(\varrho))^G, \qquad \dim \mathrm{End}(\varrho \otimes \check{\varrho})^G$$

in (6.9) make sense for any (finite-dimensional) representation of any group, and the algebraic argument with associativity of the tensor product used in the proof of the lemma shows that they are equal in this generality. One may therefore *define* an abstract "fourth moment" of a representation using either of them. It is natural to ask, if G is not compact, what is the meaning of this abstract $\mathrm{M}_4(\varrho)$, and in particular (in view of the Larsen alternative) to ask what the equality $\mathrm{M}_4(\varrho) = 2$ means in general. We will discuss this in Section 7.1.

The proof will use a little bit of differential geometry. Readers who are unfamiliar with the notion of manifolds and with the basic definition of Lie groups can simply skim (or skip) the proof.

Proof of the Larsen alternative. To study $\mathrm{M}_4(G)$, we use part (2) of Lemma 6.2.3 for the representation of G on the linear space $\mathrm{End}(\mathbf{C}^n)$, i.e., on $\mathrm{End}(\varrho)$, where

$$\varrho : G \hookrightarrow \mathrm{U}_n(\mathbf{C})$$

is the representation defining G as a subgroup of $\mathrm{U}_n(\mathbf{C})$.

We recall that this representation is the *conjugation* action, i.e., that

$$g \cdot A = gAg^{-1}$$

for $g \in G$ and $A \in E = \mathrm{End}(\mathbf{C}^n)$ (it is the restriction of the corresponding action for $\mathrm{SU}_n(\mathbf{C})$, or indeed for $\mathrm{GL}_n(\mathbf{C})$). There is, as usual, a canonical invariant subspace of dimension one, namely $\mathbf{C}\mathrm{Id} \subset E$. Moreover, a stable (orthogonal) complement is

$$E_0 = \{A \in E \mid \mathrm{Tr}(A) = 0\},$$

the space of endomorphisms of trace 0. Hence we have a first decomposition into subrepresentations

$$E = \mathbf{C}\mathrm{Id} \oplus E_0. \tag{6.11}$$

If only for dimension reasons, the two components are non-isomorphic; therefore, by the previous lemma, we get automatically

$$\mathrm{M}_4(G) \geqslant 1^2 + 1^2 = 2.$$

We deduce first from this that $\mathrm{M}_4(\mathrm{SU}_n(\mathbf{C})) = 2$; indeed, the lemma shows that this simply means that the decomposition (6.11) is a decomposition into irreducible representations of $\mathrm{SU}_n(\mathbf{C})$, which we know is true for the first component because it is one dimensional, and for the second by Exercise 2.7.14.

By the same token, we see that if $G \subset \mathrm{SU}_n(\mathbf{C})$, we can only have $\mathrm{M}_4(G) = 2$ if E_0 is also irreducible as a representation of G. We assume that this is the case and will deduce that G is either finite or equal to $\mathrm{SU}_n(\mathbf{C})$.

To do this, we must appeal to the fact that G is a Lie group, and in fact that it is a smooth manifold.[3] Thus we may consider the tangent space of G at the identity element, which is its Lie algebra,[4] denoted $\mathrm{Lie}(G)$. This is a *real* vector space, of dimension equal to the dimension of G as a manifold. The point is that G acts linearly on $\mathrm{Lie}(G)$, by means of the so-called *adjoint* representation,[5] which is obtained by differentiating at the identity the conjugation action of G on itself: denoting by $I(g)$ the inner automorphism that maps x to gxg^{-1}, the adjoint representation is given by

$$\mathrm{Ad}\begin{cases} G & \longrightarrow & \mathrm{GL}(\mathrm{Lie}(G)) \\ g & \mapsto & dI(g)_e, \end{cases}$$

where $dI(g)_e$ denotes the tangent map of the diffeomorphism $I(g)$ at the identity element e of G. This is a well-defined linear map on $\mathrm{Lie}(G)$ (since $I(g)(e) = e$ for each g), and it is a representation, by the chain rule, because $I(gh) = I(g)I(h)$ and $I(g^{-1}) = I(g)^{-1}$.

This representation is a *real* representation[6] of G, since $\mathrm{Lie}(G)$ is a real vector space. Most crucial for us, it has the following property, which is almost immediate: if $G \subset H$ with H also a compact Lie group, then $\mathrm{Lie}(G) \subset \mathrm{Lie}(H)$, and the adjoint representation of G is the restriction of the adjoint representation of H. Applied to $G \subset \mathrm{SU}_n(\mathbf{C})$, it follows that $\mathrm{Lie}(G)$ is a subrepresentation of $\mathrm{Lie}(\mathrm{SU}_n(\mathbf{C}))$.

This is the source of the desired subrepresentation of E_0. We will check below the following facts:

- The Lie algebra L_n of $\mathrm{SU}_n(\mathbf{C})$ is a real subspace of E_0, such that $L_n \oplus iL_n = L_n \otimes \mathbf{C} = E_0$;

[3] It does not suffice here to know that G is a topological manifold.

[4] Although we will in fact not need the structure of Lie algebra (see Section 3.2) that exists on this space.

[5] It may be confusing at first that there there exist an adjoint representation for a Lie group and one for a Lie algebra (see Example 3.2.3) and that neither has much to do with the adjoint of a linear map.

[6] Not to be confused with representations of *real type*, which were mentioned briefly in Definition 6.2.2.

- The adjoint representation of $\mathrm{SU}_n(\mathbf{C})$ on L_n is a real subrepresentation of E_0, i.e., on $L_n \subset E_0$, the adjoint representation is given by $g \cdot A = gAg^{-1}$ for $A \in L_n \subset E_0$.

If we assume these, we can conclude as follows. For our compact subgroup $G \subset \mathrm{SU}_n(\mathbf{C})$, we have the subrepresentation

$$\mathrm{Lie}(G) \otimes \mathbf{C} \subset L_n \otimes \mathbf{C} = E_0$$

of the G-action on E_0. Since we are assuming that the latter is irreducible, this means that either $\mathrm{Lie}(G)$ is 0, in which case G is finite, or that $\mathrm{Lie}(G)$ is equal to L_n, in which case, by Lie theory, we have $G = \mathrm{SU}_n(\mathbf{C})$. Hence the Larsen alternative is proved.

Now we explain the facts mentioned above—these are quite standard and the reader may well have already encountered them. To begin with, the special unitary group is defined by the conditions

$$\det(g) = 1, \qquad gg^* = 1,$$

in $\mathrm{GL}_n(\mathbf{C})$. The tangent space at 1 is obtained by considering the linearized forms of these equations, viewed as applying to matrices A in $\mathrm{M}_n(\mathbf{C})$, which form the tangent space at 1 of $\mathrm{GL}_n(\mathbf{C})$. The first equation becomes $\mathrm{Tr}(A) = 0$, which means $A \in E_0$, and the second becomes

$$A + A^* = 0,$$

i.e., A is skew-hermitian, so

$$L_n = \{A \in \mathrm{M}_n(\mathbf{C}) \mid A = -A^*, \text{ and } \mathrm{Tr}(A) = 0\} \subset E_0. \tag{6.12}$$

(Note that since the adjoint operation $A \mapsto A^*$ is not complex linear, this is indeed only a real vector space.)

We can easily check explicitly that $E_0 = L_n \otimes \mathbf{C}$: for $A \in E_0$, we write

$$A = \frac{A + A^*}{2} + \frac{A - A^*}{2} = iB + C \quad \text{(say)}.$$

Then $C^* = -C$, so C is skew-hermitian, and $B = (A + A^*)/(2i)$ has also $B^* = -(A^* + A)/(2i) = -B$, so that B is skew-hermitian. Since $\mathrm{Tr}(B) = \mathrm{Re}(\mathrm{Tr}(A))$ and $\mathrm{Tr}(C) = i\,\mathrm{Im}(\mathrm{Tr}(A))$, we deduce $\mathrm{Tr}(B) = \mathrm{Tr}(C) = 0$, so that we have found a decomposition of A as $C + iB$ with C, B both in L_n. This decomposition is unique, because $L_n \cap iL_n = 0$ (in E_0): any matrix in the intersection is both hermitian and skew-hermitian. So this proves the first claim.

The second one is not too surprising since the adjoint representation is defined using conjugation. To be precise, let $A \in L_n$ be a tangent vector. Then elementary differential geometry tells us that $\mathrm{Ad}(g)(A)$ can be computed as

$$\frac{d}{dt}I(g)(x_t)\Big|_{t=0},$$

where $x_t \in \mathrm{SU}_n(\mathbf{C})$ defines any smooth curve with tangent vector A at $t=0$. As usual, one takes $x_t = \exp(tA)$, where the exponential is that of matrices; then we have

$$I(g)x_t = g\exp(tA)g^{-1} = \exp(tgAg^{-1})$$

(e.g., using the Taylor series expansion), and the derivative at $t=0$ gives $\mathrm{Ad}(g)A = gAg^{-1}$, as desired. □

Example 6.3.5 (Finite groups with $\mathrm{M}_4 = 2$)**.** As observed by Katz [**32**, 1.6.1], there do exist finite groups $G \subset \mathrm{SU}_n(\mathbf{C})$, for some $n \geqslant 2$, for which $\mathrm{M}_4(G) = 2$. For instance, let $G = \mathrm{PSL}_2(\mathbf{F}_7)$. It follows from the character table of $\mathrm{SL}_2(\mathbf{F}_7)$ that G has two distinct irreducible representations π_1 and π_2 of dimension $3 = (7-1)/2$. Unitarized, either of these gives a homomorphism

$$G \longrightarrow \mathrm{U}_3(\mathbf{C}).$$

Since G is a simple group, this is necessarily a faithful representation, and (for the same reason) the composite $G \hookrightarrow \mathrm{U}_3(\mathbf{C}) \xrightarrow{\det} \mathbf{C}^\times$, which cannot be injective, is trivial. Thus the image of either of these representations is a finite subgroup of $\mathrm{U}_3(\mathbf{C})$, and one can check that these have a fourth moment equal to 2.

In addition to the case of the unitary group considered above, there are criteria for orthogonal and symplectic groups. Let $n \geqslant 4$ be an even integer, and let $G \subset \mathrm{U}_n(\mathbf{C})$ be a connected compact group which is contained in the subgroup $\mathrm{USp}_n(\mathbf{C}) = \mathrm{Sp}_n(\mathbf{C}) \cap \mathrm{U}_n(\mathbf{C})$ of unitary matrices that leave invariant a non-degenerate alternating bilinear form. Then Larsen showed that $G = \mathrm{USp}_n(\mathbf{C})$ if and only if $\mathrm{M}_4(G) = 3$. Similarly, if $n \geqslant 2$ and $G \subset \mathrm{U}_n(\mathbf{C})$ is a compact connected group contained in the compact group $\mathrm{O}_n(\mathbf{R}) = \mathrm{U}_n(\mathbf{C}) \cap \mathrm{O}_n(\mathbf{C})$ of unitary matrices leaving invariant a non-degenerate symmetric bilinear form, we have $G \supset \mathrm{SO}_n(\mathbf{R})$ if and only if $\mathrm{M}_4(G) = 3$. One cannot, however, distinguish between $\mathrm{SO}_n(\mathbf{R})$ and $\mathrm{O}_n(\mathbf{R})$ using the fourth moment.

The point in these results is that $\mathrm{USp}_n(\mathbf{C})$ and $\mathrm{O}_n(\mathbf{R})$ are maximal compact subgroups, respectively, of $\mathrm{Sp}_n(\mathbf{C})$ and $\mathrm{O}_n(\mathbf{C})$.

We sketch the argument for the symplectic case: denoting $V = \mathbf{C}^n$, one has a decomposition

$$V^{\otimes 2} = \mathrm{Sym}^2(V) \oplus \mathbf{C} \oplus V_1$$

as the $\mathrm{USp}_n(\mathbf{C})$-representation, where the trivial one-dimensional component $\mathbf{C}$ corresponds to the (dual of the) invariant alternating form on V, and $V_1 \neq 0$ because $\dim V \geqslant 4$. As representations of $\mathrm{USp}_n(\mathbf{C})$, the three pieces are known to be irreducible, so that $\mathrm{M}_4(\mathrm{USp}_n(\mathbf{C})) = 3$, and by Lemma 6.3.3(1), we have $\mathrm{M}_4(G) = 3$ if and only if all three representations are G-irreducible. It also turns out that $\mathrm{Sym}^2(V)$ is isomorphic to the adjoint representation of $\mathrm{USp}_n(\mathbf{C})$ on its Lie algebra, and hence it contains as a G-invariant subspace the Lie algebra of G itself. Therefore, irreducibility of $\mathrm{Sym}^2(V)$ implies that $G = \mathrm{USp}_n(\mathbf{C})$ since both are connected with the same Lie algebra.

Finally, we address the problem of applications of the Larsen alternative. We explain here, with a specific example, some of the situations where results like this are very valuable tools. As already hinted, sometimes theory gives the existence of some group which carries information concerning objects of interest. A very good example, though it is not directly relevant to the Larsen alternative, is the Galois group of the splitting field of a polynomial. This is a finite group, which is (usually) defined rather abstractly, so that if one knows the coefficients of the polynomial, it is not easy at all to determine the Galois group. In fact, often the only obvious information is that it is isomorphic to a subgroup of $\mathfrak{S}_n$, where n is the degree of the polynomial. For instance, can you guess the Galois group of the splitting field of

$$X^8 - 4X^7 + 8X^6 - 11X^5 + 12X^4 - 10X^3 + 6X^2 - 3X + 2$$

over $\mathbf{Q}$?

Now for the example, which is based on deep work and ideas of P. Deligne [**15**] and of N. Katz [**33**], and for which we assume some familiarity with the basic theory of finite fields (again, [**54**, Chapter 1] is an excellent reference).

Let $q = p^d$ be a prime power, and let $\mathbf{F}_q$ be a finite field with q elements. The map

$$\mathrm{Tr}_{\mathbf{F}_q/\mathbf{F}_d} \begin{cases} \mathbf{F}_q & \longrightarrow & \mathbf{F}_p \\ x & \mapsto & x + x^p + \cdots + x^{p^{d-1}} \end{cases}$$

is well defined and is a homomorphism of additive groups (the element $y = \mathrm{Tr}_{\mathbf{F}_q/\mathbf{F}_p}$ is in $\mathbf{F}_p$ because it satisfies $y^p = y$). We denote by ψ the non-trivial additive character of the additive group of $\mathbf{F}_p$ defined by

$$\psi(x) = e\Big(\frac{x}{p}\Big)$$

for $x \in \mathbf{F}_p$ (see Remark 4.5.3). The composition

$$\psi_{\mathbf{F}_q/\mathbf{F}_d} = \psi \circ \mathrm{Tr}_{\mathbf{F}_q/\mathbf{F}_p}$$

is then a character of the additive group of $\mathbf{F}_q$, and since one can show that the trace map $\mathrm{Tr}_{\mathbf{F}_q/\mathbf{F}_p}$ is surjective, it follows that $\psi_{\mathbf{F}_q/\mathbf{F}_p}$ is also non-trivial.

Note that if d is not divisible by p, the surjectivity follows easily from the fact that $\mathrm{Tr}_{\mathbf{F}_q/\mathbf{F}_p}(1) = d$ is then non-zero in $\mathbf{F}_p$.

Now, for any element $a \in \mathbf{F}_q$, one defines

$$S(a;q) = \sum_{x \in \mathbf{F}_q^\times} \psi_{\mathbf{F}_q/\mathbf{F}_p}(ax + x^{-1}), \tag{6.13}$$

a sum which is called a *Kloosterman sum.*

These sums are apparently just complex numbers, but they turn out to be related to some compact Lie groups. Indeed, it follows from the work of A. Weil that for every $a \in \mathbf{F}_q$ there exists a well-defined *conjugacy class* $\theta(a;q)$ in the unitary group $\mathrm{U}_2(\mathbf{C})$ such that

$$\mathrm{Tr}\,\theta(a;q) = -\frac{S(a;q)}{\sqrt{q}}, \tag{6.14}$$

where the trace on the left-hand side is just the ordinary trace of matrices, which can be applied to conjugacy classes, since the trace is invariant under conjugation. In particular, note that this formula immediately implies the bound

$$|S(a;q)| \leqslant 2\sqrt{q},$$

for $a \in \mathbf{F}_q^\times$, which is a deep theorem of A. Weil. In fact, it is a special case of the *Riemann hypothesis for curves over finite fields*; we refer, e.g., to [**30**, Th. 11.11] for one of the simplest proofs.

The connection with the Larsen alternative arises from the following fact, which is a special case of a famous theorem of Deligne (Deligne's equidistribution theorem) applied to the so-called Kloosterman sheaves. There exists a compact subgroup $K \subset \mathrm{U}_2(\mathbf{C})$, depending a priori on p, such that, first, all $\theta(a;q)$ are in fact naturally conjugacy classes of K, and second, they become *equidistributed* among conjugacy classes of K, in the sense that for any continuous class function $f : K \longrightarrow \mathbf{C}$, we have

$$\int_K f(x)d\mu(x) = \lim_{q \to +\infty} \frac{1}{q-1} \sum_{a \in \mathbf{F}_q^\times} f(\theta(a;q)), \tag{6.15}$$

where μ is the probability Haar measure on K, and q tends to infinity through powers of p. See the discussion in [**33**, Ch. 3].

Thus, if one succeeds in determining what the group K is—something which, just as was the case for Galois group, is by no means clear by just looking at (6.13)!—one can answer many questions about the asymptotic distribution of Kloosterman sums, something which is of great interest in number theory.

Now it is clear why the Larsen alternative is useful: applying first (6.15) with $f(x) = |\operatorname{Tr}(x)|^4$ and then (6.14), we get the alternative formula

$$\begin{aligned}\mathrm{M}_4(K) &= \lim_{q\to+\infty} \frac{1}{q-1} \sum_{a\in\mathbf{F}_q^\times} |\operatorname{Tr}\theta(a;q)|^4 \\ &= \lim_{q\to+\infty} \frac{1}{q^2(q-1)} \sum_{a\in\mathbf{F}_q^\times} \Big|\sum_{x\in\mathbf{F}_q^\times} \psi_{\mathbf{F}_q/\mathbf{F}_p}(ax+x^{-1})\Big|^4\end{aligned}$$

for the fourth moment of K, which involves the given concrete data defining the problem. We may have a chance to evaluate this....

As it turns out, one can evaluate directly this limit in this case (this is a relatively elementary computation, see Exercise 6.3.6 below) and see that it exists and is equal to 2. In other words, the compact group K satisfies $\mathrm{M}_4(K) = 2$. Hence the Larsen alternative shows that either K is finite, or $K \supset \mathrm{SU}_2(\mathbf{C})$. In fact, one can analyze the situation further, and show that K is equal to the special unitary group $\mathrm{SU}_2(\mathbf{C})$. It follows then from (6.15), for instance, that

$$\lim_{q\to+\infty} \frac{1}{q^{k+1/2}(q-1)} \sum_{a\in\mathbf{F}_q^\times} \Big|\sum_{x\in\mathbf{F}_q^\times} \psi_{\mathbf{F}_q/\mathbf{F}_p}(ax+x^{-1})\Big|^{2k+1} = 0$$

for any integer $k \geqslant 0$ and

$$\lim_{q\to+\infty} \frac{1}{q^{k}(q-1)} \sum_{a\in\mathbf{F}_q^\times} \Big|\sum_{x\in\mathbf{F}_q^\times} \psi_{\mathbf{F}_q/\mathbf{F}_p}(ax+x^{-1})\Big|^{2k} = \frac{1}{k+1}\binom{2k}{k}$$

for any integer $k \geqslant 1$ (see Exercise 6.3.7(1) below).

This may seem too good to be true, but we remind the reader however that the *existence* of this group K lies extremely deep: the equidistribution formula (6.15) cannot be proven without first knowing its existence, and only later can one attempt to determine the group K.

In the works of Katz, many other (more general) situations are considered, leading to extremely general and beautiful equidistribution theorems. But even though the statements can be extremely concrete, there is no known elementary proof of the deep connection between Kloosterman sums (or other similar sums) and a compact Lie group!

Exercise 6.3.6 (Fourth moment of Kloosterman sums)**.** For a finite field $\mathbf{F}_q$ with q elements and ψ as above, define

$$T(a,b;q) = \sum_{x\in\mathbf{F}_q^\times} \psi_{\mathbf{F}_q/\mathbf{F}_p}(ax+bx^{-1}).$$

(1) Show that $T(a,b;q) = S(ab;q)$ if a, $b \in \mathbf{F}_q^\times$.

(2) Let

$$N_4(q) = \sum_{a \in \mathbf{F}_q^\times} |S(a;q)|^4.$$

Deduce from (1) that

$$(q-1)N_4(q) = q^2 \mathcal{N}(q) - (q-1)^4 - 2(q-1),$$

where

$$\mathcal{N}(q) = \Big|\Big\{(x_1, x_2, y_1, y_2) \in (\mathbf{F}_q^\times)^4 \mid x_1+x_2 = y_1+y_2 \text{ and } \frac{1}{x_1}+\frac{1}{x_2} = \frac{1}{y_1}+\frac{1}{y_2}\Big\}\Big|.$$

(3) Prove that

$$\mathcal{N}(q) = 3(q-2)(q-1),$$

and deduce that

$$\lim_{q \to +\infty} \frac{N_4(q)}{q-1} = 2.$$

(*Hint*: Use the fact that a pair $(x+y, xy)$ determines x and y up to order.)

Exercise 6.3.7 (Other moments)**.** One can define other types of moments. For instance, given a compact group G (with probability Haar measure μ) and a finite-dimensional unitary representation ϱ of G, the k-th moment of ϱ is defined to be

$$M_k(\varrho) = \int_G \chi_\varrho(g)^k d\mu(g)$$

for an integer $k \geqslant 0$. If $G \subset \mathrm{GL}_n(\mathbf{C})$ is a compact subgroup, we denote by $M_k(G)$ the k-th moment of this inclusion.

It is an elementary consequence of character theory, which is not necessarily clear at first when expressed for a concrete group, that $M_k(\varrho)$ is a non-negative integer, since it is the multiplicity of the trivial representation in the finite-dimensional representation $\varrho^{\otimes k}$ (see (5.12)).

The sequence of moments $(M_k(\varrho))_{a \geqslant 0}$, as $k \geqslant 0$ varies, can be quite interesting:

(1) Take $G = \mathrm{SU}_2(\mathbf{C}) \subset \mathrm{GL}_2(\mathbf{C})$. Show that

$$M_k(G) = 0$$

if k is odd and

$$M_{2k}(G) = \frac{1}{k+1}\binom{2k}{k}$$

for $k \geqslant 0$. Can you prove directly that the right-hand side is an integer?

(2) Compute the first few terms and identify this sequence in the *Online Encyclopedia of Integer Sequences* (`http://oeis.org`). (These are called the *Catalan numbers* and have extremely varied interpretations.)

(3) Let $G = \mathrm{SU}_n(\mathbf{C}) \subset \mathrm{GL}_n(\mathbf{C})$. Show that $M_k(G) \neq 0$ if and only if k is divisible by n. What happens when $G = \mathrm{U}_n(\mathbf{C})$?

Exercise 6.3.8 (Another application of the Larsen alternative). For a prime number p and an element $a \in \mathbf{F}_p^{\times}$, let

$$S_3(a;p) = \frac{1}{p} \sum\sum_{x,y \in \mathbf{F}_p^{\times}} e\Big(\frac{x + y + a(xy)^{-1}}{p}\Big),$$

which is called a a hyper-Kloosterman sum in two variables (the inverse $(xy)^{-1}$ is computed in $\mathbf{F}_p^{\times}$).

(1) For reasonably large values of p (say $p \leqslant 100000$) and the first few $k \geqslant 0$, compute (using a computer) the *empirical* moments

$$m_{k,p} = \frac{1}{p-1} \sum_{\alpha \in \mathbf{F}_p^{\times}} S_3(a,p)^k.$$

Discuss the behavior of the result as p grows.

(2) Can you make a guess concerning some analogue of the equidistribution result for Kloosterman sums discussed above? Check in [**33**] whether this guess is correct.

Exercise 6.3.9 (Maximal fourth moment). (1) Let $G \subset \mathrm{U}_n(\mathbf{C})$ be a compact subgroup of $\mathrm{U}_n(\mathbf{C})$, and let $\varrho : G \longrightarrow \mathrm{U}_m(\mathbf{C})$ be an irreducible unitary representation of G. Show that $\mathrm{M}_4(\varrho) \leqslant m^2$. If G is connected, show that equality holds if and only if $m = 1$.

(2) Show that the dihedral group D_4 of order 8 has a two-dimensional irreducible representation ϱ with $\mathrm{M}_4(\varrho) = 4$.

Remark 6.3.10 (From $\mathrm{SU}_2(\mathbf{C})$ to $\mathrm{SO}_3(\mathbf{R})$). The adjoint representation turns out to provide a conceptual explanation of the projection homomorphism

$$p : \mathrm{SU}_2(\mathbf{C}) \longrightarrow \mathrm{SO}_3(\mathbf{R})$$

of Proposition 5.6.9. Indeed, for the compact Lie group $G = \mathrm{SU}_2(\mathbf{C})$, the Lie algebra is a three-dimensional real vector space (by (6.12) $\mathrm{M}_2(\mathbf{C})$ has dimension 8, the skew-hermitian condition implies that the bottom-left coefficient is minus the conjugate of the top-right one, and the diagonal ones are purely imaginary, leaving $8 - 2 - 2 = 4$ dimensions, and the matrices of trace zero form a three-dimensional subspace). In fact,

$$L_2 = \Big\{ \begin{pmatrix} ia & c + id \\ -c + id & -ia \end{pmatrix} \mid a,\ c,\ d \in \mathbf{R} \Big\},$$

so that a matrix representation for the adjoint representation of $\mathrm{SU}_2(\mathbf{C})$ on L_2 is a homomorphism

$$\mathrm{Ad}^m : \mathrm{SU}_2(\mathbf{C}) \longrightarrow \mathrm{GL}_3(\mathbf{R}).$$

This *is* the desired projection, in the sense that it has kernel $\{\pm 1\}$, and image conjugate to $\mathrm{SO}_3(\mathbf{R})$ in $\mathrm{GL}_3(\mathbf{R})$ (depending on which basis of the Lie algebra L_2 is used to compute the matrix form of the representation).

In topological terms, the projection p is a non-trivial covering map of $\mathrm{SO}_3(\mathbf{R})$ (since $\mathrm{SU}_2(\mathbf{C})$ is connected). Thus $\mathrm{SO}_3(\mathbf{R})$ is *not simply connected* (in fact, one can show that $\mathrm{SU}_2(\mathbf{C})$ is simply connected, so it is the universal covering of $\mathrm{SO}_3(\mathbf{R})$). There are well-known *physical* demonstrations of this property of the rotation group (due in particular to Dirac); see, e.g., [**6**] for an accessible mathematical account, though seeing movies on the web might be even more enlightening....

6.4. The hydrogen atom

We now come to the discussion of Example 1.2.3, i.e., of the basic invariants of simple quantum-mechanical systems, and in particular of the hydrogen atom.

In order to do this, we summarize briefly the fundamental formalism of (non-relativistic) quantum mechanics, contrasting it with classical Newtonian mechanics, in the simplest situation of a single (point-like) particle evolving in $\mathbf{R}^3$, under the influence of some force (or forces):

- The *state* of the system at a given time t is represented by a unit vector ψ (i.e., with $\|\psi\| = 1$) in some fixed complex Hilbert space H. In contrast, in Newtonian mechanics, the state of the particle is represented by an element $(x, p) \in \mathbf{R}^6$, where x represents the position of the particle and p its momentum $p = mv$, where $v \in \mathbf{R}^3$ is the the velocity at t and m is the mass of the particle.
- Two vectors ψ_1, ψ_2 in H correspond to the same state if and only if there exists $\theta \in \mathbf{R}$ such that $\psi_1 = e^{i\theta}\psi_2$, i.e., if the vectors are proportional.
- An *observable quantity* (or just an *observable*), such as position or momentum, is represented by a linear operator A defined on a dense subspace D_A of H. If A is continuous, it can be defined on all of H, but many interesting observables are not continuous on D_A. Moreover A must be *self-adjoint*, which has the usual meaning when A is continuous on H, and has a more technical definition otherwise (see Exercise 6.4.1). On the other hand, in Newtonian mechanics, an observable quantity is simply a real-valued function $f : P \longrightarrow \mathbf{R}$, where $P \subset \mathbf{R}^6$ is the set of possible states of the system.
- The physical interaction of the system described by the state ψ with an observable A must result, through experiments, in some actual

numerical (approximate) value. The crucial prediction of quantum mechanics is that this value λ will be an element of the *spectrum* $\sigma(A) \subset \mathbf{R}$ of A, but that its value *cannot* be predicted beforehand. Instead, one defines (purely mathematically) a probability measure $\mu_{\psi,A}$ on $\mathbf{R}$ such that $\mu_{\psi,A}(B)$ is the *probability* that the measurement will give a value in $B \subset \mathbf{R}$. The measure $\mu_{\psi,A}$ is called the *spectral measure of A with respect to ψ*. In the important case that A has (at most) countably many distinct eigenvalues $\lambda_i \in \mathbf{R}$, $i \geqslant 0$, whose eigenspaces span H, so that

$$\mathrm{H} = \widehat{\bigoplus_{i \geqslant 0}} \mathrm{Ker}(A - \lambda_i), \tag{6.16}$$

the spectral measure is defined by

$$\mu_{\psi,A}(B) = \sum_{\lambda_i \in B} \|p_i(\psi)\|^2, \tag{6.17}$$

where $p_i : \mathrm{H} \longrightarrow \mathrm{Ker}(A - \lambda_i)$ is the orthogonal projection on the i-th eigenspace. This is a probability measure since ψ is a unit vector and it satisfies the condition that any physically observed values would be among the eigenvalues λ_i of A, since the measure $\mu_{\psi,A}$ has support given by these eigenvalues.

In particular, suppose A is an orthogonal projection and that it is non-trivial, i.e., that $A \neq 0$, $A \neq \mathrm{Id}$. Its spectrum is $\{0, 1\}$, and the corresponding projections are just $p_1 = A$ itself and $p_0 = \mathrm{Id} - A$. Thus *measuring* the observable A will result in either of these values, with probability $\|A\psi\|^2$ of getting 1 and probability $1 - \|A\psi\|^2$ of getting 0. In probabilistic terms, this corresponds to a Bernoulli random variable with *probability of success* $\|A\psi\|^2$.

- The probability can be understood experimentally, and the prediction checked, as follows. If the measurement is repeated a large number of times (say N times), each time after preparing the system to be in state ψ, then the proportion N_B/N of the number N_B of measurements for which the experimental value λ is in B will be close to $\mu_{\psi,A}(B)$. This is in striking contrast with Newtonian mechanics: given that the particle is in the state $(x, p) \in P$, the value of the observation f is simply the exact value $f(x, p) \in \mathbf{R}$.

  In the example where the spectral measure is given by (6.17), one will therefore *observe* the eigenvalue λ_i with relative frequency given by

$$\mu_{\psi,A}(\{\lambda_i\}) = \|p_i(\psi)\|^2.$$

This property makes the link between the mathematical model and the natural world; it can, in principle, be experimentally falsified, but the probabilistic interpretation has turned out to be confirmed by test after test. Not only is it the case that the relative frequencies of various results (especially zero/one tests corresponding to projections) are found to be close to the theoretical values, but no method (either practical or even theoretical) has been found to predict exactly the values of the measurements as they are performed one after the other.

- Finally, the basic dynamical equation is Schrödinger's equation: there exists a particular observable E, the Hamiltonian, such that the state of the system evolves in time as a solution (ψ_t) of the equation

$$i\frac{h}{2\pi}\frac{d}{dt}\psi_t = E\psi_t, \tag{6.18}$$

where h is Planck's constant. The Hamiltonian encapsulates the forces acting on the particle. In Newtonian mechanics, the particle evolves according to the differential equation

$$m\frac{d^2x}{dt^2} = \text{sum of the forces}$$

instead. We will not discuss dynamics of quantum systems here, but it turns out that there is a connection between this equation and unitary representations of the (additive, non-compact) group $\mathbf{R}$; see Section 7.3.

For more information, from a mathematical point of view, the reader may refer to [**58, 61, 62, 64**]. There are also, of course, many physics books on quantum mechanics that are worth reading.

Exercise 6.4.1 (Unbounded self-adjoint operator). Let H be a Hilbert space, and let $A : D_A \longrightarrow \mathrm{H}$ be a linear operator defined on $D_A \subset \mathrm{H}$, a dense subspace of H. The pair (D_A, A) is called a *densely defined* operator on H. If there exists $C \geqslant 0$ such that

$$\|A(\psi)\| \leqslant C\|\psi\|$$

for all $\psi \in \mathrm{H}$, then A extends by continuity to a bounded linear operator on H, and otherwise (D_A, A) is called an *unbounded* linear operator with domain D_A. Furthermore, (D_A, A) is called *self-adjoint* if the following two conditions hold: (1) we have

$$D_A = \{\psi \in \mathrm{H} \mid \phi \mapsto \langle A\phi, \psi\rangle \text{ extends to a } \textit{continuous} \text{ linear form on } \mathrm{H}\},$$

and (2) for all ψ_1, $\psi_2 \in D_A$, we have

$$\langle A\psi_1, \psi_2\rangle = \langle \psi_1, A\psi_2\rangle.$$

Show that the following defines an unbounded, self-adjoint, operator:

$$\begin{aligned} \mathrm{H} &= L^2(\mathbf{R}, dx),\\ D_A &= \Big\{\psi \in \mathrm{H} \mid \int_{\mathbf{R}} x^2 |\psi(x)|^2 dx < +\infty\Big\},\\ (A\psi)(x) &= x\psi(x) \text{ for } \psi \in D_A. \end{aligned}$$

This observable is interpreted as the *position* of a particle constrained to move on the line $\mathbf{R}$. Given $\psi \in L^2(\mathbf{R})$ with $\|\psi\|^2 = 1$, the spectral measure $\mu_{\psi,A}$ is the probability measure $|\psi(x)|^2 dx$ on $\mathbf{R}$, such that the probability that a particle in the state described by ψ be located inside a set B is given by

$$\int_B |\psi(x)|^2 dx.$$

Much more about the general theory of unbounded linear operators can be found, for instance, in the books of Reed and Simon [**49, 50**].

How does representation theory enter the picture? The answer has to do with possible *symmetries* of the system, which must be compatible with the *linear structure* of the underlying Hilbert space. If an observable A of interest is also compatible with the symmetries of the system, and it can be described using only eigenvalues (as in (6.16)), it follows that the eigenspaces must be invariant under these symmetries. In other words, if there is a symmetry group G of the system, the eigenspaces of observables are (unitary) *representations* of G.

Now consider such an eigenspace, say $V = \mathrm{Ker}(A-\lambda)$. For states $\psi \in V$, the observable A has the specific, exact, value λ. If the representation V is *not* irreducible, we can find another observable B such that B commutes with A, and some eigenspace of B is a proper subspace W of V. For the states in W, both A and B have determined value. In the opposite direction, if V is an irreducible representation of G, nothing more may be said (without more information) concerning the states in V. This is the physical interpretation of Schur's Lemma.

What this shows is that, given a quantum system with symmetry group G, we should attempt to decompose the corresponding representation of G on H into irreducible representations. The states in each subrepresentation will be fundamental building blocks for all states (by linearity), which cannot be further analyzed unless more information is available.

We now illustrate these general principles with concrete examples. We consider a particle evolving in $\mathbf{R}^3$, which is constrained to lie on the unit sphere $\mathbf{S}^2 \subset \mathbf{R}^3$. (This restriction is not very physical, but it helps at first with the mathematical analysis.) The underlying Hilbert space is taken to be $\mathrm{H} = L^2(\mathbf{S}^2, \nu)$, where ν is the surface Lebesgue measure on the sphere, defined similarly as in Example 5.2.4(5). The operators of multiplication of a function by the coordinates can play the role of position observables, as in Exercise 6.4.1, but since the coordinates are bounded, the corresponding operators are continuous on H. Suppose now that the system evolves according to a homogeneous force, compatible with the rotational symmetry of the sphere $\mathbf{S}^2$. We are then dealing with the representation of the rotation group $\mathrm{SO}_3(\mathbf{R})$ on H, given by

$$(g \cdot \psi)(x) = \psi(g^{-1}x).$$

(This is indeed a unitary representation since the measure μ is $\mathrm{SO}_3(\mathbf{R})$-invariant; see Example 5.2.9.)

Even without knowing anything about certain observables such as energy, it is intuitively clear that if the system is rotation-invariant, these observables must be compatible with the symmetries. In some sense, applying a rotation to a state ψ amounts to observing this state from a different direction in space, and rotation-invariance implies the absence of privileged directions!

Thus we attempt to decompose the representation of $\mathrm{SO}_3(\mathbf{R})$ that we just introduced. Recall from Example 5.6.8 that the irreducible unitary representations of $\mathrm{SO}_3(\mathbf{R})$ are obtained, by composing with the projection

$$\mathrm{SU}_2(\mathbf{C}) \longrightarrow \mathrm{SO}_3(\mathbf{R}),$$

from the odd-dimensional irreducible representations of $\mathrm{SU}_2(\mathbf{C})$. Hence, for each integer $\ell \geqslant 0$, there exists a unique irreducible representation of $\mathrm{SO}_3(\mathbf{R})$ of dimension $2\ell + 1$, which we will denote V_ℓ.

Proposition 6.4.2 (Decomposition of $L^2(\mathbf{S}^2)$)**.** *The space $L^2(\mathbf{S}^2)$ is isomorphic, as a representation of* $\mathrm{SO}_3(\mathbf{R})$ *to the Hilbert direct sum*

$$\widehat{\bigoplus_{\ell\geqslant 0}} V_\ell$$

of all irreducible representations of $\mathrm{SO}_3(\mathbf{R})$*, each occurring with multiplicity* 1.

Proof. There is a quick proof coming from Frobenius reciprocity, which starts from the observation that the group $G = \mathrm{SO}_3(\mathbf{R})$ acts transitively on $\mathbf{S}^2$, and the stabilizer of the point $\boldsymbol{n} = (1, 0, 0)$ is the subgroup $K \simeq \mathrm{SO}_2(\mathbf{R})$ of rotations around the x-coordinate axis.[7] Hence we have a bijection

$$\phi \begin{cases} K\backslash G & \longrightarrow & \mathbf{S}^2 \\ g & \mapsto & g \cdot \boldsymbol{n}, \end{cases}$$

[7] One could use any other point.

which is in fact a homeomorphism (it is continuous, and both spaces are compact). It follows that functions on $\mathbf{S}^2$ are "the same" (by composition with this homeomorphism) as functions on G such that

$$f(kg) = f(g)$$

for $k \in K$, $g \in G$. Moreover, since the Lebesgue measure on $\mathbf{S}^2$ is known to be the unique rotation-invariant measure on $\mathbf{S}^2$, up to scalar, there exists a constant $c > 0$ such that the probability Haar measure μ satisfies

$$\phi_* \mu = c\nu.$$

We can therefore identify the representation on $L^2(\mathbf{S}^2, \nu)$ with the representation of $\mathrm{SO}_3(\mathbf{R})$ on

$$\begin{aligned} \mathrm{H}_1 = \{f \,:\, \mathrm{SO}_3(\mathbf{R}) \longrightarrow \mathbf{C} \mid f \in L^2(\mathrm{SO}_3(\mathbf{R})),\\ f(kg) = f(g) \text{ if } k \in K, g \in \mathrm{SO}_3(\mathbf{R})\} \end{aligned}$$

with action

$$(g \cdot f)(x) = f(xg),$$

and the restriction of the usual inner product. This means that

$$L^2(\mathbf{S}^2, \nu) \simeq \mathrm{Ind}_K^{\mathrm{SO}_3(\mathbf{R})} \mathbf{1}$$

as the unitary representation of $\mathrm{SO}_3(\mathbf{R})$ (see Example 5.2.10). Now, given an irreducible representation ϱ of $\mathrm{SO}_3(\mathbf{R})$, we can use the Frobenius reciprocity formula for compact groups (Proposition 5.4.9) to derive

$$\dim \mathrm{Hom}_{\mathrm{SO}_3(\mathbf{R})}(\varrho, L^2(\mathbf{S}^2)) = \dim \mathrm{Hom}_K(\mathrm{Res}_K^{\mathrm{SO}_3(\mathbf{R})} \varrho, \mathbf{1}),$$

which is the multiplicity of the trivial representation in the restriction of ϱ to K. We analyze this multiplicity by noting that the diagonal subgroup in $\mathrm{SU}_2(\mathbf{C})$ maps onto K via the projection. For instance (although there are more intrinsic ways to see this), this may be checked using the "ugly" formula (5.28): for any $\theta \in \mathbf{R}$, we have

$$\begin{aligned} \begin{pmatrix} e^{i\theta} & 0 \\ 0 & e^{-i\theta} \end{pmatrix} \mapsto & \begin{pmatrix} 1 & 0 & 0 \\ 0 & \cos^2(\theta) - \sin^2(\theta) & -2\cos(\theta)\sin(\theta) \\ 0 & 2\cos(\theta)\sin(\theta) & \cos^2(\theta) - \sin^2(\theta) \end{pmatrix} \\ & = \begin{pmatrix} 1 & 0 & 0 \\ 0 & \cos(2\theta) & -\sin(2\theta) \\ 0 & \sin(2\theta) & \cos(2\theta) \end{pmatrix}, \end{aligned}$$

which represents a rotation around the x-axis[8] with angle 2θ. Therefore, we get

$$\dim \mathrm{Hom}_{\mathrm{SO}_3(\mathbf{R})}(\varrho, L^2(\mathbf{S}^2)) = \dim \mathrm{Hom}_T(\mathrm{Res}_T^{\mathrm{SU}_2(\mathbf{C})} \varrho, \mathbf{1}),$$

[8] This calculation depends on a compatible normalization of K and the projection, but changing either would just require us to conjugate one or the other.

viewing ϱ as a representation of $\mathrm{SU}_2(\mathbf{C})$. But we computed the restriction of the representations of $\mathrm{SU}_2(\mathbf{C})$ to the diagonal subgroup a long time ago: for $\varrho = \varrho_\ell$, we have

$$\mathrm{Res}_T^{\mathrm{SU}_2(\mathbf{C})} \varrho = \chi_{-2\ell} \oplus \chi_{-2\ell+2} \oplus \cdots \oplus \chi_{2\ell-2} \oplus \chi_{2\ell}$$

by (2.41) (remember that the m there is 2ℓ here) where

$$\chi_j\left(\begin{pmatrix} e^{i\theta} & 0 \\ 0 & e^{-i\theta} \end{pmatrix}\right) = e^{ij\theta}.$$

By inspection, the multiplicity of the trivial representation $\mathbf{1} = \chi_0$ of T is indeed equal to 1. □

The physical meaning of this result, for our hypothetical quantum particle on the sphere, is that if it is in a "pure" state ψ with well-defined angular momentum, it has a natural invariant attached to it, namely the index ℓ such that ψ is in the subspace (say W_ℓ) of $L^2(\mathbf{S}^2)$ isomorphic to V_ℓ. This invariant is called the *angular momentum quantum number* of the state (or sometimes *orbital*, or *azimutal, quantum number*).

Continuing with this particle on the sphere, suppose it is (i.e., the state ψ is) in W_ℓ. If we want to pinpoint the state more precisely, or at least describe specific states which can combine to construct all the states in W_ℓ, we must *break* the rotational symmetry.

Suppose we consider observables B which are only symmetric with respect to rotations around a fixed axis (say, the x-axis). This means that the underlying symmetry group becomes the subgroup $K \simeq \mathrm{SO}_2(\mathbf{R})$ of $\mathrm{SO}_3(\mathbf{R})$ of rotations around this axis. If we start from states known to be in W_ℓ, we must then decompose this space *as a representation of* K, and the corresponding K-subrepresentations represent states for which the angular momentum and all K-invariant observables are fully known. Here K is abelian, so we know these subspaces are one dimensional and hence correspond to a unique state.

Precisely, as in the proof of Proposition 6.4.2, the space W_ℓ decomposes, as a representation of K, as the direct sum of the $2\ell+1$ characters of $\mathrm{SO}_2(\mathbf{R})$ given by

$$\chi_j : \begin{pmatrix} \cos\theta & -\sin\theta \\ \sin\theta & \cos\theta \end{pmatrix} \mapsto e^{ij\theta}$$

with $-\ell \leqslant j \leqslant \ell$. Each one of the (one-dimensional) subspaces $W_{\ell,j}$ on which K acts like χ_j therefore describes a unique quantum state, parametrized by the two quantum numbers $\ell \geqslant 0$ and $j \in \{-\ell, \ldots, \ell\}$. This second parameter is called the *magnetic quantum number*. Historically, this is because it can be experimentally detected by putting systems in magnetic fields which are symmetric with respect to the given axis, for instance in the so-called "Zeeman effect".

All this is still relevant for more realistic physical systems, where the particle evolves in $\mathbf{R}^3$, with state space given by $\mathrm{H} = L^2(\mathbf{R}^3)$, under conditions of spherical symmetry, so that the relevant unitary representation of $\mathrm{SO}_3(\mathbf{R})$ is given by

$$\varrho(g)\varphi(x) = \varphi(g^{-1} \cdot x)$$

for $g \in \mathrm{SO}_3(\mathbf{R})$ and $\varphi \in L^2(\mathbf{R}^3)$. The point is that, as a representation of $\mathrm{SO}_3(\mathbf{R})$, we can separate the radius-dependency of functions in $L^2(\mathbf{R}^3)$ (on which $\mathrm{SO}_3(\mathbf{R})$ acts trivially) and the spherical components. To be precise:

Proposition 6.4.3. *There is a linear map*

$$C(\mathbf{S}^2) \otimes C_0([0, +\infty[) \xrightarrow{\Phi} L^2(\mathbf{R}^3)$$

mapping $\varphi \otimes \psi$ *to the function*

$$f(\boldsymbol{x}) = \psi(\|\boldsymbol{x}\|)\varphi\Big(\frac{\boldsymbol{x}}{\|\boldsymbol{x}\|}\Big).$$

This map is an isometry for the inner product induced by

$$\langle \varphi_1 \otimes \psi_1, \varphi_2 \otimes \psi_2 \rangle_0 = \Big(\int_0^{+\infty} \psi_1(r)\overline{\psi_2(r)}rdr\Big)\Big(\int_{\mathbf{S}^2} \varphi_1(x)\overline{\varphi_2(x)}d\nu(x)\Big)$$

and has dense image. Moreover

$$\varrho(g)f(x) = \Phi((g \cdot \varphi) \otimes \psi). \tag{6.19}$$

Sketch of proof. Again, we leave some details to the reader, but the main point is the formula for integration of functions in spherical coordinates, which states that

$$\int_{\mathbf{R}^3} f(x, y, z)dxdydz = \int_{\mathbf{S}^2} \int_0^{+\infty} f(rx)rdrd\nu(x)$$

for integrable $f : \mathbf{R}^3 \longrightarrow \mathbf{C}$. This implies in particular the isometry condition

$$\langle \Phi(\varphi_1 \otimes \Psi_1), \Phi(\varphi_2 \otimes \psi_2) \rangle_{L^2(\mathbf{R}^3)} = \langle \varphi_1 \otimes \Psi_1, \varphi_2 \otimes \psi_2 \rangle_0$$

for all $\varphi_i \in C(\mathbf{S}^2)$ and $\psi_i \in C_0([0, +\infty[)$. The formula (6.19) can be checked easily, and it remains to prove that the image of Φ is dense in $L^2(\mathbf{R}^3)$. But suppose f is orthogonal to this image, so that

$$\langle f, \Phi(\varphi \otimes \psi) \rangle = 0$$

for all φ and ψ. This translates, by integrating in spherical coordinates, to

$$0 = \int_{\mathbf{S}^2} \overline{\varphi(x)}\Big(\int_0^{+\infty} \overline{\psi(r)}f(rx)rdr\Big)d\nu(x)$$

for all $\varphi \in C(\mathbf{S}^2)$ and $\psi \in C_0([0, +\infty[)$. For all x, this gives

$$\int_0^{+\infty} \overline{\psi(r)} f(rx) r dr = 0$$

for all ψ, and then we get $f = 0$ (to be precise, one must invoke Fubini's Theorem since one really gets this for almost all $x \in \mathbf{S}^2$, and then $f(rx)$ for almost all $r \geqslant 0$, depending possibly on $x \ldots$). □

In other words, in the notation of Exercise 3.4.7 we have an isometry

$$L^2(\mathbf{R}^3) \simeq L^2(\mathbf{S}^2, \nu) \,\hat{\otimes}\, L^2([0, +\infty[, rdr),$$

where the tensor product of Hilbert spaces on the right-hand side is the completion of $C(\mathbf{S}^2) \otimes C_0([0, +\infty[)$ with respect to the inner product $\langle \cdot, \cdot \rangle_0$. This isometry is an isomorphism of unitary representations of $\mathrm{SO}_3(\mathbf{R})$, where the action on $L^2([0, +\infty, rdr)$ is trivial. Consequently, we obtain

Corollary 6.4.4. *As a representation of* $\mathrm{SO}_3(\mathbf{R})$*, the space* $L^2(\mathbf{R}^3)$ *decomposes as a direct sum of infinitely many copies of every irreducible representation of* $\mathrm{SO}_3(\mathbf{R})$.

Thus, a state $f \in L^2(\mathbf{R}^3)$ lying in one irreducible subrepresentation still determines an angular momentum quantum number ℓ, and if the state is further compatible with breaking the spherical symmetry as described above, it has a magnetic quantum number m, $-\ell \leqslant m \leqslant \ell$.

Going further in understanding the hydrogen atom requires more physical information concerning the system, and this is not purely a question of symmetry (i.e., of representation theory!) anymore (though some non-obvious $\mathrm{SO}_4(\mathbf{R})$-symmetry of the system can be used to explain the results). Roughly speaking, the energy levels of the atom are themselves quantized because of the form that the Schrödinger equation takes in the radial direction, forming a decreasing sequence $E_n \to 0$, with $n \geqslant 1$ (the *principal quantum number*). The states with a given energy level E_n form a subrepresentation for the action of $\mathrm{SO}_3(\mathbf{R})$ on $L^2(\mathbf{R}^3)$, and one shows that it has dimension n^2, and is isomorphic to a direct sum of V_ℓ for $0 \leqslant \ell \leqslant n-1$ (dually, the Schrödinger equation can be solved on the V_ℓ-isotypic component of $L^2(\mathbf{R}^3)$, which is isomorphic to $V_\ell \otimes L^2([0, +\infty[)$, and is found to involve the energy levels E_n for $n > \ell$).

Comparing with Example 1.2.3, the only missing ingredient is the spin s. This is a purely quantum-mechanical invariant, which is best explained using the Dirac theory that combines the special theory of relativity with basic quantum mechanics.

We refer the reader to [**58**, Ch. 8, 9] and [**61**, §4.5, §4.8] for the details on the remainder of the story of the Hydrogen atom.

There are many other applications of representation theory in physics. For instance, as explained in detail in the very accessible survey [**1**] of Baez and Huerta, the Standard Model of particles interprets the basic types of particles of the universe (quarks, photons, bosons of various types, etc.) as basis vectors of certain irreducible unitary representations of an underlying compact Lie group that acts as symmetry group.

Chapter 7

Other groups: a few examples

The picture of representation theory beyond the case of compact groups changes quite dramatically. Even if one restricts to locally compact groups, it happens that non-compact groups have typically infinite-dimensional irreducible unitary representations, and from these one cannot usually produce all representations using Hilbert direct sums. Their study becomes in many respects more analytic and certainly requires quite different tools and techniques. These are mostly beyond the scope of this book (as people say!), and the very modest goal of this chapter is simply to present some concrete examples of groups and, even if without full proofs, a survey of some phenomena involved in the representation theory of such groups, emphasizing mostly those that were not visible in the previous chapters.

We begin however with some words about the more algebraic topic of *algebraic groups*.

7.1. Algebraic groups

In the discussion of the Larsen alternative in the previous chapter, we observed in Remark 6.3.4 that the fourth moment of a finite-dimensional complex representation $\varrho : G \longrightarrow \mathrm{GL}(E)$ can be defined algebraically and abstractly by

$$\mathrm{M}_4(\varrho) = \dim(\mathrm{End}(\varrho) \otimes \mathrm{End}(\varrho))^G = \dim \mathrm{End}(\varrho \otimes \check{\varrho})^G,$$

and in particular $\mathrm{M}_4(G)$ can be defined for any subgroup $G \subset \mathrm{GL}_n(\mathbf{C})$ using the tautological representation $\varrho : G \hookrightarrow \mathrm{GL}(\mathbf{C}^n)$.

In view of Theorem 6.3.2, we are naturally led to ask, What can we say about a group G such that $\mathrm{M}_4(G) = 2$, when the fourth moment is defined in this manner? In particular, What can we say if G is infinite, in which case Larsen's alternative states that, if G is compact and contained in $\mathrm{SU}_n(\mathbf{C})$, then it is equal to a conjugate of $\mathrm{U}_n(\mathbf{C})$?

The answer to that question leads naturally to the important topic of *algebraic groups* and to the Zariski closure of a linear group. We use this as a motivating example to introduce briefly these concepts.

The basic statement is that if $G \subset \mathrm{GL}_n(\mathbf{C})$ is infinite and satisfies $\mathrm{M}_4(G) = 2$, then it is "very large" in $\mathrm{GL}_n(\mathbf{C})$. But one cannot take "very large" to have the same meaning as in the case of compact groups; indeed, we are thinking of groups such as $\mathrm{SL}_n(\mathbf{Z}) \subset \mathrm{GL}_n(\mathbf{C})$ or $\mathrm{GL}_n(\mathbf{Q})$, which may have only a finite intersection with $\mathrm{SU}_n(\mathbf{C})$, or of more or less random finitely generated subgroups of $\mathrm{GL}_n(\mathbf{C})$, which are essentially impossible to classify. However, it turns out that one can associate to any subgroup $G \subset \mathrm{GL}_n(\mathbf{C})$ a larger group $\mathbf{G} \supset G$ in a natural way, and the conclusion will be that an infinite group $G \subset \mathrm{SL}_n(\mathbf{C})$ with $\mathrm{M}_4(G) = 2$ satisfies $\mathbf{G} = \mathrm{SL}_n(\mathbf{C})$.

The group $\mathbf{G}$ is called the *Zariski closure* of G in $\mathrm{GL}_n(\mathbf{C})$, and its informal description is that it is the smallest subgroup of $\mathrm{GL}_n(\mathbf{C})$ containing G which cannot be distinguished from G using only *polynomial* equations satisfied by the coordinates of the underlying matrices. It then follows that all purely algebraic invariants of G and $\mathbf{G}$ coincide (for instance, $\mathrm{M}_4(G) = \mathrm{M}_4(\mathbf{G})$), but $\mathbf{G}$ is a much more rigid object, and this leads to a form of the Larsen alternative for $\mathbf{G}$ which is as precise and clean as Theorem 6.3.2 for compact groups.

We first define $\mathbf{G}$. Given $G \subset \mathrm{GL}_n(\mathbf{C})$, we introduce two objects.

- The set $\mathcal{I}_G$ of all polynomials vanishing on G. To be more precise, let
$$A = \mathbf{C}[(X_{i,j})_{1 \leqslant i,j \leqslant n}, Y],$$
which is a polynomial algebra in $n^2 + 1$ variables.[1] Then let
$$\mathcal{I}_G = \{f \in A \mid f(g_{i,j}, \det(g)^{-1}) = 0 \text{ for all } g = (g_{i,j}) \in G\}.$$
- The set $\mathcal{V}_G$ of all elements in $\mathrm{GL}_n(\mathbf{C})$ which are common zeros of all these polynomials. That is,

(7.1) $$\mathcal{V}_G = \{x \in \mathrm{GL}_n(\mathbf{C}) \mid f(x, \det(x)^{-1}) = 0 \text{ for all } f \in \mathcal{I}_G\}.$$

Obviously, the definitions show that $G \subset \mathcal{V}_G$. But crucially, $\mathcal{V}_G$ is also a subgroup of $\mathrm{GL}_n(\mathbf{C})$:

[1] The usefulness of the presence of the extra variable Y, which is used to represent polynomially the inverse of an element $g \in \mathrm{GL}_n(\mathbf{C})$, will be clear very soon.

Lemma 7.1.1 (Zariski closure). (1) *The set* $\mathcal{V}_G \subset \mathrm{GL}(V)$ *is a* subgroup *of* $\mathrm{GL}_n(\mathbf{C})$ *containing* G. *It is denoted* $\mathbf{G}$ *and is called the* Zariski closure *of* G *in* $\mathrm{GL}_n(\mathbf{C})$.

(2) *Suppose* $H \supset G$ *is another subgroup of* $\mathrm{GL}_n(\mathbf{C})$. *Then* $\mathbf{H} \supset \mathbf{G}$, *where* $\mathbf{H}$ *is the Zariski closure of* H.

(3) *We have*

$$\mathrm{M}_4(G) = \mathrm{M}_4(\mathbf{G}).$$

Proof. (1) This is quite elementary, but a little bit long if one wishes to give all details. We start by showing that $\mathcal{V}_G$ is stable under inversion, since this is where one sees how the extra variable Y in the polynomial algebra A is useful.

Given any $f \in \mathcal{I}_G$, define a new polynomial $\tilde{f}$ by

$$\tilde{f}((X_{i,j}), Y) = f(Y \cdot (\tilde{X}_{i,j}), \det(X_{i,j})),$$

where Y is seen as a scalar in the first argument, and the matrix $\tilde{X} = (\tilde{X}_{i,j})$ that it multiplies is the transpose of the cofactor matrix of $X = (X_{i,j})$, i.e., the coefficients $\tilde{X}_{i,j}$ are the polynomials in $\mathbf{C}[(X_{i,j})]$ such that

$$\det(X) \times X^{-1} = \tilde{X}$$

(this encapsulates the Cramer formulas for solving linear equations). Thus $\tilde{f} \in A$. Now, for $g \in \mathrm{GL}_n(\mathbf{C})$, we have

$$\tilde{f}(g, \det(g)^{-1}) = f(\det(g)^{-1}(\tilde{g}_{i,j}), \det(g)) = f(g^{-1}, \det(g)),$$

and this vanishes for all $g \in G$, since $g^{-1} \in G$ and $f \in \mathcal{I}_G$. Thus we find that $\tilde{f}$ also belongs to $\mathcal{I}_G$, and this implies by definition that $\tilde{f}$ vanishes on $\mathcal{V}_G$, which means that

$$f(g^{-1}, \det(g)) = \tilde{f}(g, \det(g)^{-1}) = 0$$

for all $g \in \mathcal{V}_G$.

Finally, consider $g \in \mathcal{V}_G$ to be fixed, and vary $f \in \mathcal{I}_G$. We find that

$$f(g^{-1}, \det(g^{-1})^{-1}) = f(g^{-1}, \det(g)) = 0$$

for all $f \in \mathcal{I}_G$, which by definition means that $g^{-1} \in \mathcal{V}_G$, as desired.

We then proceed to show that $\mathcal{V}_G$ is stable under products, arguing along similar lines. First, we show that $\mathcal{V}_G$ is stable by multiplication by elements of G on both sides: if $g_1 \in G$ is given, then for all $g \in \mathcal{V}_G$, both $g_1 g$ and $g g_1$ are in $\mathcal{V}_G$.

Indeed, given $f \in \mathcal{I}_G$, we define

$$\tilde{f}(X_{i,j}, Y) = f(g_1 \cdot (X_{i,j}), \det(g_1)^{-1} Y),$$

where $g_1 \cdot (X_{i,j})$ denotes the matrix product. Since matrix multiplication is polynomial in the coordinates $(X_{i,j})$, this is an element of A. And as such, it belongs to $\mathcal{I}_G$, because for $g \in G$ we have $g_1 g \in G$—since G is itself a group—and hence

$$\tilde{f}(g, \det(g)^{-1}) = f(g_1 g, \det(g_1 g)^{-1}) = 0$$

by definition of $\mathcal{I}_G$. Hence $\tilde{f}$ vanishes in fact identically on all of $\mathcal{V}_G$, which means that $f(g_1 g) = 0$ for all $g \in \mathcal{V}_g$. Since $f \in \mathcal{I}_G$ is arbitrary, this property, applied to a fixed $g \in \mathcal{V}_G$, means that $g_1 g \in \mathcal{V}_G$. Reversing the order of a few products, we also obtain in this way that $g g_1 \in \mathcal{V}_G$ for all $g \in \mathcal{V}_G$.

Finally, we deduce the stability of $\mathcal{V}_G$ under all products: Let $g_1 \in \mathcal{V}_G$ be given. For $f \in \mathcal{I}_G$, define again

$$\tilde{f}((X_{i,j}), Y)) = f(g_1 \cdot (X_{i,j}), \det(g_1)^{-1} Y) \in A.$$

We get $\tilde{f} \in \mathcal{I}_G$ because, for $g \in G$, we know from the previous step that $g_1 g \in \mathcal{V}_G$, and then $\tilde{f}(g, \det(g)^{-1}) = f(g_1 g, \det(g_1 g)^{-1}) = 0$, and therefore $\tilde{f}$ vanishes on $\mathcal{V}_G$. In particular, fixing some $g \in \mathcal{V}_G$ and using the fact that $f(g_1 g) = 0$ for all $f \in \mathcal{I}_G$, this means that $g_1 g \in \mathcal{V}_G$.

Thus we have proved that $\mathbf{G} = \mathcal{V}_G$ is a group, which of course contains G.

(2) If $H \supset G$, then any polynomial vanishing on H also vanishes on G, so that $\mathcal{I}_H \subset \mathcal{I}_G$, and then the set $\mathbf{H}$ of zeros of the polynomials in $\mathcal{I}_H$ contains that of zeros of $\mathcal{I}_H$, so $\mathbf{H} \supset \mathbf{G}$.

(3) Now we show that G and its Zariski closure have the same fourth moment. This is a special case of a much more general statement concerning the equality of algebraic invariants of G and $\mathbf{G}$, which is explained in Exercise 7.1.6. We take the definition $\mathrm{M}_4(G) = \dim(\mathrm{End}(\mathbf{C}^n) \otimes \mathrm{End}(\mathbf{C}^n))^G$ (the other invariant space would do just as well). Since $G \subset \mathbf{G}$, we have

$$\dim(\mathrm{End}(\mathbf{C}^n) \otimes \mathrm{End}(\mathbf{C}^n))^{\mathbf{G}} \leqslant \dim(\mathrm{End}(\mathbf{C}^n) \otimes \mathrm{End}(\mathbf{C}^n))^G,$$

and we must therefore check that an element of $F = \mathrm{End}(\mathbf{C}^n) \otimes \mathrm{End}(\mathbf{C}^n)$, which, if G-invariant, is also $\mathbf{G}$-invariant.

The condition that a given element $x \in F$ is G-invariant can be expressed by saying that it satisfies the relations

$$g \cdot x = x$$

for $g \in G$. If we fix a basis $(x_k)_{1 \leqslant k \leqslant n^4}$ of F, each equation $g \cdot x = x$, with unknown $g \in \mathrm{GL}_n(\mathbf{C})$, can be expressed as a set of n^4 equations in terms of the coordinates $(g_{i,j})$ of g, where the k-th equation expresses the fact that the coefficient of x_k of x is equal to the coefficient of x_k of $g \cdot x$. And the point is that this last coefficient is *a polynomial* in the coordinates $(g_{i,j})$, with coefficients depending on x. (This may require a moment's thought to be clear, and it may be easier to think first of the analogous problem for invariant bilinear forms, and to check that a bilinear form b on $\mathbf{C}^n$ is G-invariant if and only if it is $\mathbf{G}$-invariant.)

Thus, to say that $x \in F$ is G-invariant means that all elements of G have the property that their coefficients $g_{i,j}$ are solutions of a certain set of polynomial equations (determined by x). The polynomials determining these equations are therefore elements of $\mathcal{I}_G$, and it follows again from the very definition of $\mathcal{V}_G$ that any element of $\mathbf{G}$ is also a solution of these equations, which means that $x \in F^{\mathbf{G}}$. □

Now we can state a general version of the Larsen alternative:

Theorem 7.1.2 (The Larsen alternative). *Let $n \geqslant 1$ be an integer, and let $G \subset \mathrm{SL}_n(\mathbf{C})$ be any group, with Zariski closure $\mathbf{G} \subset \mathrm{SL}_n(\mathbf{C})$. Assume that G acts irreducibly on $\mathbf{C}^n$. If $\mathrm{M}_4(G) = 2$, then either G is finite or $\mathbf{G} = \mathrm{SL}_n(\mathbf{C})$.*

Sketch of the proof. First note that $\mathbf{G} \subset \mathrm{SL}_n(\mathbf{C})$, because the relation $\det(g) = 1$ is assumed to be valid for all $g \in G$, so that $\det(g) - 1$, viewed as a polynomial of the coordinates $(g_{i,j})$, is an element of the ideal $\mathcal{I}_G$, and therefore every $g \in \mathbf{G}$ must satisfy $\det(g) = 1$ also.

Since $\mathbf{G} \supset G$, the action of $\mathbf{G}$ on $\mathbf{C}^n$ is also irreducible. We also know that $\mathrm{M}_4(G) = \mathrm{M}_4(\mathbf{G})$, and it is relatively easy to check (see Proposition 7.1.7 below) that if G is finite, then $G = \mathbf{G}$, so that $\mathbf{G}$ is also finite. Thus $\mathbf{G}$ satisfies the same assumptions as G, and we must show that either $\mathbf{G}$ is finite or $\mathbf{G} = \mathrm{SL}_n(\mathbf{C})$.

We now appeal without proof to an important result. Since $\mathbf{G}$ is given with the inclusion $\mathbf{G} \subset \mathrm{SL}_n(\mathbf{C})$, which is a faithful irreducible representation (in particular, a faithful semisimple representation), it is known that $\mathbf{G}$ contains a *compact* subgroup $K \subset \mathbf{G}$ such that the Zariski closure of K (computed by the same method as above, with $G = K$ instead) is still $\mathbf{G}$. We get

$$\mathrm{M}_4(K) = \mathrm{M}_4(\mathbf{G}) = \mathrm{M}_4(G) = 2,$$

and by the Larsen alternative (Theorem 6.3.2) for compact groups, it follows that either K is finite, or K contains a conjugate of $\mathrm{SU}_n(\mathbf{C})$. In the first case, $K = \mathbf{G} = G$ is finite, while in the second $\mathbf{G}$ contains (a conjugate of) the Zariski closure of $\mathrm{SU}_n(\mathbf{C})$. But it is known that the Zariski closure of $\mathrm{SU}_n(\mathbf{C})$ is $\mathrm{SL}_n(\mathbf{C})$ (see again Proposition 7.1.7), and hence $\mathbf{G} \supset \mathrm{SL}_n(\mathbf{C})$ (without conjugacy ambiguity, since this group is conjugation-invariant in $\mathrm{GL}_n(\mathbf{C})$). □

At first sight, Theorem 7.1.2 might seem to be a poor substitute for the earlier version of the Larsen alternative, since it involves the group $\mathbf{G}$ which may seem complicated and obscure. In fact, one should think of $\mathbf{G}$ as being typically a much simpler object than G, and of the passage from G to $\mathbf{G}$ as simplifying the problem considerably. This is because $\mathbf{G}$ belongs to the category of *linear algebraic groups*, i.e., it is a subgroup of $\mathrm{GL}_n(\mathbf{C})$ which is the set of common zeros of a set of polynomials (elements of the algebra A).

One may also ask why one does not try to go directly from the given group G to the compact group K which is used to reduce to the earlier version of the Larsen alternative. This seems difficult because it may well happen that G and K have trivial intersection! A basic example is $G = \mathrm{GL}_n(\mathbf{Z})$, $n \geqslant 2$: Then one can show (see Proposition 7.1.7 for $n = 2$ and Exercise 7.1.8 for $n \geqslant 3$) that $\mathbf{G}$ is the set of matrices g in $\mathrm{GL}_n(\mathbf{C})$ with $\det(g)^2 = 1$. The subgroup K is then the group of unitary matrices $g \in \mathrm{U}_n(\mathbf{C})$ with $\det(g)^2 = 1$. The intersection $G \cap K$ is the finite group W_{2n} of signed permutation matrices. This is discussed in Exercise 4.7.13, and indeed, if $g = (g_{i,j})$ is any unitary matrix with integral coefficients, the condition

$$\sum_j g_{i,j}^2 = 1$$

implies that, for a given i, a single $g_{i,j} \in \mathbf{Z}$ is ± 1, and the others are 0. Denoting $j = \sigma(i)$, one sees that σ must be a permutation of $\{1, \ldots, n\}$, so that $g \in W_{2n}$. Since any $W_{2n} \subset \mathrm{GL}_n(\mathbf{Z})$, we get the result. In this case, $G \cap K$ still acts irreducibly on $\mathbf{C}^n$ (as the reader should check), but if we replace G by the subgroup

$$G_3 = \{g \in \mathrm{GL}_n(\mathbf{Z}) \mid g \equiv \mathrm{Id} \,(\mathrm{mod}\, 3)\},$$

which has finite index, it also possible to show that the Zariski closure of G_3, and hence the compact subgroup K, are the same as that for G. However, $G_3 \cap K$ is now trivial since a signed permutation matrix, which is congruent to the identity modulo 3, has to be the identity. We used reduction modulo 3 instead of 2 here to be able to distinguish the two signs.

Remark 7.1.3. One can also bypass the compact case and give a purely algebraic proof of Theorem 7.1.2, as done by Katz in [**32**], but the reader who checks this proof will see that it parallels very closely the argument of Theorem 6.3.2. This is no coincidence, but another illustration of the close links between representations of linear algebraic groups (at least, of the so-called reductive groups, which contain all Zariski closures of subgroups of $\mathrm{GL}_n(\mathbf{C})$ acting irreducibly on $\mathbf{C}^n$) and representations of compact Lie groups.

Exercise 7.1.4 (The Zariski topology). The association of the *big* group $\mathbf{G}$ to the group G has (besides the name itself!) the aspect of a *closure* operation. Indeed, it can be interpreted as taking the closure of G in $\mathrm{GL}_n(\mathbf{C})$ with respect to a certain topology, called the Zariski topology.

Let k be an algebraically closed field, and let $n \geqslant 1$ be an integer. We define $\mathbf{A}^n = k^n$, which is called the affine n-space over k, and we denote $k[X] = k[X_1, \ldots, X_n]$ the algebra of polynomials in n variables. Note that for a polynomial $f \in k[X]$ and $x \in \mathbf{A}^n$, we can evaluate f at x.

(1) For any subset $\mathcal{I} \subset k[X]$, let

$$\mathcal{V}(\mathcal{I}) = \{x \in \mathbf{A}^n \mid f(x) = 0 \text{ for all } f \in \mathcal{I}\}. \tag{7.2}$$

Show that the collection of sets $(\mathcal{V}(\mathcal{I}))_{\mathcal{I} \subset A}$ is the collection of *closed* sets for a topology on $\mathbf{A}^n$. In particular, this allows us to speak of the *Zariski closure* of any subset $V \subset \mathbf{A}^n$: it is the intersection of all Zariski closed subsets of $\mathbf{A}^n$ which contain V.

(2) Show that any polynomial map $f : \mathbf{A}^n \longrightarrow \mathbf{A}^m$ is continuous with respect to the Zariski topologies on the respective affine spaces.

(3) For $n = 1$, show that a subset $V \subset \mathbf{A}^1$ is closed for the Zariski topology if and only if $V = \mathbf{A}^1$ or V is finite. Show that the Zariski topology on $\mathbf{A}^2$ is *not* the product topology on $\mathbf{A}^1 \times \mathbf{A}^1$. Show also that the Zariski topology is *not* Hausdorff, at least for $\mathbf{A}^1$ (the case of $\mathbf{A}^n$ for arbitrary n might be more difficult).

(4) Let $G = \mathrm{GL}_n(k)$, seen as a subset of $\mathbf{A}^{n^2}$ by means of the coefficients of the matrices. Show that G is dense in $\mathbf{A}^{n^2}$ with respect to the Zariski topology. Furthermore, show that the set

$$\tilde{G} = \{(g, x) \in \mathbf{A}^{n^2+1} \mid \det(g)x = 1\} \subset \mathbf{A}^{n^2+1}$$

is Zariski closed in $\mathbf{A}^{n^2+1}$.

(5) For $k = \mathbf{C}$ and $G \subset \mathrm{GL}_n(\mathbf{C})$ with Zariski closure $\mathbf{G}$, show that the Zariski closure of $\tilde{G} \subset \mathbf{A}^{n^2+1}$ is equal to

$$\tilde{\mathbf{G}} = \{(g, x) \in \mathbf{G} \times \mathbf{C} \mid \det(g)x = 1\}.$$

The Zariski topology is a foundational notion in algebraic geometry. Readers interested in learning more may look at [**27**] for the general theory (or should really try to take a course in algebraic geometry, if at all possible!) In the context of linear algebraic groups, the books [**7**] and [**60**] may be more accessible.

We use this language to give a formal definition:

Definition 7.1.5 (Linear algebraic group)**.** Let k be an algebraically closed field. A group G is a *linear algebraic group* over k if there exists some $n \geqslant 1$ such that G is a Zariski closed subgroup of $\mathrm{GL}_n(k)$.

Since we will not talk of any other types of algebraic groups, we will often omit the adjective "linear" below.

Exercise 7.1.6 (Zariski closure and polynomial representations). Let k be an algebraically closed field, and let $G \subset \mathrm{GL}_n(k)$ be any subgroup (not necessarily algebraic). Let

$$\varrho : G \longrightarrow \mathrm{GL}_m(k)$$

be a (matrix) representation of G. We assume that ϱ is a *polynomial representation*, i.e., that the coefficients of $\varrho(g)$ are functions on G which are restrictions of polynomials in the coordinates $g_{i,j}$ and in $\det(g)^{-1}$. (Note that, for instance, the one-dimensional representation $t \mapsto e^{it}$ of $\mathbf{R}$, viewed as the subgroup

$$\left\{ \begin{pmatrix} 1 & t \\ 0 & 1 \end{pmatrix} \mid t \in \mathbf{R} \right\}$$

of $\mathrm{GL}_2(\mathbf{C})$, is *not* polynomial.)

(1) Let $\mathbf{G} \subset \mathrm{GL}_n(k)$ be the Zariski closure of G. Show that there exists a unique representation

$$\boldsymbol{\varrho} : \mathbf{G} \longrightarrow \mathrm{GL}_m(k)$$

such that $\boldsymbol{\varrho}$ coincides with ϱ on G. What is $\boldsymbol{\varrho}$ if ϱ is the injection of G in $\mathrm{GL}_n(k)$?

(2) Show that a subspace $V \subset k^m$ is a subrepresentation of ϱ if and only if it is a subrepresentation of $\boldsymbol{\varrho}$. Deduce that ϱ is G-irreducible (resp. G-semisimple) if and only if $\boldsymbol{\varrho}$ is $\mathbf{G}$-irreducible (resp. $\mathbf{G}$-semisimple).

(3) Show that the subspace $(k^m)^G$ of vectors invariant under G is equal to the subspace of vectors invariant under $\mathbf{G}$.

(4) Show that the representations $\check{\varrho}$ (contragredient), $\mathrm{Sym}^k \varrho$ (k-th symmetric power, for $k \geqslant 0$), and $\bigwedge^k \varrho$ (alternating power, for $k \geqslant 0$) are also polynomial.

(5) Using these ideas, prove that any finite-dimensional complex polynomial representation of $\mathbf{C}^\times$ is semisimple. Show this is false for arbitrary (or even continuous) representations.

Going from G to the Zariski closure $\mathbf{G}$ in the above proof might be a difficult psychological step at first, especially if one thinks of G as being particularly concrete (e.g., $G = \mathrm{GL}_n(\mathbf{Z}) \subset \mathrm{GL}_n(\mathbf{C})$) while $\mathbf{G}$ seems a very abstract object. However, it is a fact that computing the Zariski closure of a group is quite often relatively easy, using known results (which may themselves, of course, be non-trivial). We give here a few simple examples. Parts (1) and (5) fill up some steps we only claimed in the sketch of the proof of Theorem 7.1.2, and item (5) above illustrates also, on a special case, the passage from $\mathbf{G}$ to a compact subgroup $K \subset \mathbf{G}$ with Zariski closure equal to $\mathbf{G}$ which was quoted in that argument.

Proposition 7.1.7 (Examples of Zariski closure). (1) *If $G \subset \mathrm{GL}_n(\mathbf{C})$ is a finite group, then $G = \mathbf{G}$.*

(2) *The Zariski closure of $\mathbf{Z}$ in $\mathbf{C}$ is $\mathbf{C}$.*

(3) *The Zariski closure of $\mathrm{SL}_2(\mathbf{Z})$ in $\mathrm{GL}_2(\mathbf{C})$ is $\mathrm{SL}_2(\mathbf{C})$, and that of $\mathrm{GL}_2(\mathbf{Z})$ is the subgroup of matrices in $\mathrm{GL}_2(\mathbf{C})$ with determinant equal to 1 or -1.*

(4) *The Zariski closure of the unit circle in $\mathbf{C}^\times$ is equal to $\mathbf{C}^\times$.*

(5) *For $n \geqslant 1$, the Zariski closure of $\mathrm{SU}_n(\mathbf{C})$ in $\mathrm{GL}_n(\mathbf{C})$ is $\mathrm{SL}_n(\mathbf{C})$.*

Proof. We will use the language of Zariski topology and Zariski closure applied to more general subsets than subgroups of $\mathrm{GL}_n(\mathbf{C})$, and we will use three very elementary observations:

- If $X \subset \mathbf{C}$ is any subset, then its Zariski closure $\bar{X}$ is either equal to X, if X is finite, or is equal to $\mathbf{C}$ if X is infinite (this is part of Exercise 7.1.4).
- If $X \subset Y$, then the Zariski closure of Y contains the Zariski closure of X (this is a formal property of any topology).
- If $X \subset \mathrm{SL}_n(\mathbf{C})$ is such that its Zariski closure is dense in $\mathrm{SL}_n(\mathbf{C})$ for the usual euclidean topology, then the Zariski closure is equal to $\mathrm{SL}_n(\mathbf{C})$ (this is because a polynomial that vanishes on X must vanish, by continuity, on its euclidean closure, hence on all of $\mathrm{SL}_n(\mathbf{C})$).

For (1), the first observation applies, and similarly for (2). For (3), we first note that since the equation $\det(g) = 1$ is a polynomial equation satisfied by all elements of $\mathrm{SL}_2(\mathbf{Z})$, it is also satisfied by all elements of its Zariski closure, and hence the latter must be contained in $\mathrm{SL}_2(\mathbf{C})$. Now note that $\Gamma = \mathrm{SL}_2(\mathbf{Z})$ contains the subgroups

$$U^+ = \left\{\begin{pmatrix} 1 & n \\ 0 & 1 \end{pmatrix} \mid n \in \mathbf{Z}\right\}, \qquad U^- = \left\{\begin{pmatrix} 1 & 0 \\ n & 1 \end{pmatrix} \mid n \in \mathbf{Z}\right\},$$

each isomorphic in an obvious way to $\mathbf{Z}$ and contained in the corresponding groups with complex coefficients. By (2), the Zariski closure of U^+ and U^- are therefore equal to the groups

$$\bar{U}^+ = \left\{\begin{pmatrix} 1 & t \\ 0 & 1 \end{pmatrix} \mid t \in \mathbf{C}\right\}, \qquad \bar{U}^- = \left\{\begin{pmatrix} 1 & 0 \\ t & 1 \end{pmatrix} \mid t \in \mathbf{C}\right\},$$

and by the second principle, the Zariski closure $\bar{\Gamma}$ contains $\bar{U}^+$ and $\bar{U}^-$, and hence the group they generate. But it is well known that this group is all of $\mathrm{SL}_2(\mathbf{C})$. We can conclude in fact very concretely even without knowing this: For any complex numbers (t, u, v), we compute

$$\begin{pmatrix} 1 & t \\ 0 & 1 \end{pmatrix}\begin{pmatrix} 1 & 0 \\ u & 1 \end{pmatrix}\begin{pmatrix} 1 & v \\ 0 & 1 \end{pmatrix} = \begin{pmatrix} ut+1 & (uv+1)t+v \\ u & uv+1 \end{pmatrix}. \tag{7.3}$$

Since each factor is known to be in $\bar{\Gamma}$, it is then easy to deduce by inspection that $\bar{\Gamma}$ contains the set C of all matrices in $\mathrm{SL}_2(\mathbf{C})$ with non-zero bottom-left coefficient (define u to be this coefficient, and solve for t and v to have the right diagonal coefficients, then check that the top-right must also be correct). One can then also check that C is dense in $\mathrm{SL}_2(\mathbf{C})$ for the usual topology, and apply the third principle to conclude that $\bar{\Gamma} = \mathrm{SL}_2(\mathbf{C})$.

In the case of $\mathrm{GL}_2(\mathbf{Z})$, the Zariski closure must contain $\mathrm{SL}_2(\mathbf{C})$, and it is contained in

$$\tilde{G} = \{g \in \mathrm{GL}_2(\mathbf{C}) \mid \det(g)^2 = 1\}$$

because the polynomial equation $\det(g)^2 = 1$ is satisfied by all elements of $\mathrm{GL}_2(\mathbf{Z})$. Since we have

$$\tilde{G} = \mathrm{SL}_2(\mathbf{C}) \cup \mathrm{SL}_2(\mathbf{C}) \begin{pmatrix} -1 & 0 \\ 0 & 1 \end{pmatrix}$$

and both sets on the right-hand side are in the Zariski closure of $\mathrm{GL}_2(\mathbf{Z})$, we get the converse inclusion.

Case (4) is again a special case of the first principle. For (5), let $K = \mathrm{SU}_n(\mathbf{C})$. Its Zariski closure $\mathbf{G}$ is a subgroup of $\mathrm{SL}_n(\mathbf{C})$ (since the condition $\det(g) = 1$ is a polynomial equation satisfied by K, and hence also by its Zariski closure). Applying the first and second principles, we see that $\mathbf{G}$ contains the diagonal subgroup $A \subset \mathrm{SL}_n(\mathbf{C})$, because for each $i \neq j$, it must contain the subset $A_{i,j} \subset A$ of diagonal matrices g with diagonal coefficients $g_{k,k} = 1$ for $k \notin \{i, j\}$, and hence also the subgroup generated by these $A_{i,j}$, which one sees is equal to A.

We conclude that $\mathbf{G}$ contains KAK, but it is a special case of the *polar decomposition* of matrices that $KAK = \mathrm{SL}_n(\mathbf{C})$, and therefore $\mathbf{G} = \mathrm{SL}_n(\mathbf{C})$. (This is also called the *Iwasawa decomposition* of $\mathrm{SL}_n(\mathbf{C})$.) □

Exercise 7.1.8. Let $n \geqslant 2$ be an integer. Show that the Zariski closure of $\mathrm{SL}_n(\mathbf{Z})$ in $\mathrm{GL}_n(\mathbf{C})$ is equal to $\mathrm{SL}_n(\mathbf{C})$ and that the Zariski closure of $\mathrm{GL}_n(\mathbf{Z})$ is equal to

$$\{g \in \mathrm{GL}_n(\mathbf{C}) \mid \det(g)^2 = 1\} \subset \mathrm{GL}_n(\mathbf{C}).$$

(*Hint*: Find subgroups of $\mathrm{SL}_n(\mathbf{Z})$ isomorphic to $\mathrm{SL}_2(\mathbf{Z})$ which generate it, and use the case $n = 2$.)

There are very general results which make computations such as those in the previous exercise quite easy. For instance, this result is a special case of a much more general theorem of Borel, which computes the Zariski closure of many important discrete groups. We state a special case, though it might not be obvious that $\mathrm{SL}_n(\mathbf{Z})$ satisfies the assumption of this result; see for instance the book of Witte–Morris [**65**, §4.7] for a proof.

Theorem 7.1.9 (Borel density theorem). *Let $n \geqslant 2$, and let $\Gamma \subset \mathrm{SL}_n(\mathbf{R})$ be a discrete subgroup such that there exists a probability measure on the quotient $\mathrm{SL}_n(\mathbf{R})/\Gamma$ which is invariant under the left-action of $\mathrm{SL}_n(\mathbf{R})$. Then the Zariski closure of Γ in $\mathrm{SL}_n(\mathbf{C})$ is equal to $\mathrm{SL}_n(\mathbf{C})$.*

We conclude this section with sketches of two other applications of algebraic groups to rather basic questions of general representation theory. First, we need some further facts about algebraic groups:

Lemma 7.1.10 (Connected component of the identity). *Let $\mathbf{G} \subset \mathrm{GL}_n(\mathbf{C})$ be an algebraic group. Then the connected component $\mathbf{G}^0$ of $\mathbf{G}$ containing 1, for the Zariski topology, is a normal subgroup of* finite *index in $\mathbf{G}$. Furthermore, any Zariski closed subgroup of finite index in $\mathbf{G}$ contains $\mathbf{G}^0$.*

The fact that $[\mathbf{G} : \mathbf{G}^0] < +\infty$ is quite strong and is an important feature of algebraic groups.

Sketch of proof. Rather elementary formal manipulations show that $\mathbf{G}^0$ is a normal subgroup of $\mathbf{G}$. For example, once one knows that $\mathbf{G}^0$ is a subgroup, its image under a conjugation $x \mapsto gxg^{-1}$, for $g \in \mathbf{G}$, is also a connected subgroup of $\mathbf{G}$, hence must be contained in $\mathbf{G}^0$, which implies that $\mathbf{G}^0$ is normal. To show that the index of $\mathbf{G}^0$ in $\mathbf{G}$ is finite, one first notices that $\mathbf{G}$, with the Zariski topology, is a *noetherian topological space*. That is, any descending chain

$$\mathbf{G} \supset V_1 \supset V_2 \supset \cdots \supset V_n \supset \cdots$$

of (Zariski) closed subsets of $\mathbf{G}$ is stationary, i.e., there exists $i \geqslant 1$ such that $V_j = V_i$ if $j \geqslant i$. Indeed, let $A_{\mathbf{G}}$ be the algebra of functions on $\mathbf{G}$ (the restrictions to $\mathbf{G}$ of the polynomial functions on $\mathrm{GL}_n(\mathbf{C})$), and let

$$\mathcal{I}(V_i) = \{f \in A_{\mathbf{G}} \mid f(V_i) = 0\} \subset A_{\mathbf{G}}.$$

Then $\mathcal{I}(V_i)$ is an ideal of $A_{\mathbf{G}}$ and these form an increasing chain

$$\mathcal{I}(V_1) \subset \mathcal{I}(V_2) \subset \cdots \subset \mathcal{I}(V_n) \subset \cdots.$$

Since $A_{\mathbf{G}}$ (which is a quotient of the ring of polynomials on $\mathrm{GL}_n(\mathbf{C})$) is a noetherian ring, this chain of ideals becomes ultimately constant. However, using the notation (7.2), one can see that

$$V_i = \mathcal{V}(\mathcal{I}(V_i))$$

(because V_i is tautologically included in the right-hand side, and one checks that the right-hand side is the Zariski closure of V_i, which was assumed to be closed), and hence all V_i coincide from some point on.

Coming back to $\mathbf{G}^0$, it is an open subset of $\mathbf{G}$ (for the Zariski topology), and any $\mathbf{G}^0$-coset in $\mathbf{G}$ is also open. If $(g_i)_{i \geqslant 1}$ is a sequence of elements such that $g_i\mathbf{G}^0$ are distinct (hence disjoint) $\mathbf{G}^0$-cosets, we obtain an increasing sequence of open sets

$$\mathbf{G}^0 \subset \mathbf{G}^0 \cup g_1\mathbf{G}^0 \subset \cdots \subset \mathbf{G}^0 \cup \bigcup_{1 \leqslant i \leqslant n} g_i\mathbf{G}^0 \subset \cdots$$

whose complements in $\mathbf{G}$ are therefore decreasing closed sets, which means the sequence must in fact be finite by the noetherian property. This implies that there are only finitely many distinct $\mathbf{G}^0$-cosets, as claimed.

Finally, if $\mathbf{H}$ is a closed subgroup of $\mathbf{G}$ of finite index, it is also open in $\mathbf{G}$ (the complement is the union of finitely many cosets, each of which is closed), and hence must contain the connected component of 1 in $\mathbf{G}$. □

We next consider a very simple-looking property of representation theory, which is due to Chevalley: over $\mathbf{C}$, the tensor product of two semisimple representations remains semisimple (with no continuity or any other extra assumption on the group or the representations involved!) No proof of this fact is known without using algebraic groups, and (as mentioned by Serre [**55**, Part II, §1]) it would be interesting to have an elementary proof of this fact.

Theorem 7.1.11 (Chevalley)**.** *Let ϱ_1, ϱ_2 be arbitrary finite-dimensional complex semisimple representations of an arbitrary group G acting on vector spaces E_1 and E_2, respectively. Then $\varrho_1 \otimes \varrho_2$ is a semisimple representation of G.*

Sketch of proof. The first step is a reduction that allow us to use algebraic groups instead of arbitrary groups. We denote by $\tilde{G} \subset \mathrm{GL}(E_1) \times \mathrm{GL}(E_2) \subset \mathrm{GL}(E_1 \oplus E_2)$ the image of G under the homomorphism $g \mapsto (\varrho_1(g), \varrho_2(g))$. We then have two representations of $\tilde{G}$ on E_1 and E_2, respectively, given by

$$\tilde{\varrho}_1(g_1, g_2) = g_1, \qquad \tilde{\varrho}_2(g_1, g_2) = g_2.$$

Note that $\tilde{\varrho}_1$ and $\tilde{\varrho}_2$ are defined on a subgroup of $\mathrm{GL}(E_1 \oplus E_2)$ and are polynomial representations of $\tilde{G}$ (indeed, they are just obtained by "forgetting", by projection, one of the factors).

By construction, the image of G in $\mathrm{GL}(E_1)$ (resp. $\mathrm{GL}(E_2)$) under $\tilde{\varrho}_1$ (resp. under $\tilde{\varrho}_2$) is the same group as the image of $\tilde{G}$ under $\tilde{\varrho}_1$ (resp. $\tilde{\varrho}_2$). This means in particular that $\tilde{\varrho}_1$ and $\tilde{\varrho}_2$ are semisimple, because ϱ_1 and ϱ_2 are semisimple by assumption.

Similarly, the representation $\tilde{\varrho}_1 \otimes \tilde{\varrho}_2$ of $\tilde{G}$ on $E_1 \otimes E_2$ is such that the image of G in $\mathrm{GL}(E_1 \otimes E_2)$ under $\varrho_1 \otimes \varrho_2$ is equal to the image of $\tilde{G}$ under $\tilde{\varrho}_1 \otimes \tilde{\varrho}_2$. Hence $\varrho_1 \otimes \varrho_2$ is semisimple if and only if $\tilde{\varrho}_1 \otimes \tilde{\varrho}_2$ is semisimple.

Since $(\tilde{\varrho}_1 \otimes \tilde{\varrho}_2)(g_1, g_2) = g_1 \otimes g_2$, the representation $\tilde{\varrho}_1 \otimes \tilde{\varrho}_2$ is also polynomial (the formulas for the tensor product of two endomorphisms are polynomial).

Thus, we may replace $(G, \varrho_1, \varrho_2)$ by $(\tilde{G}, \tilde{\varrho}_1, \tilde{\varrho}_2)$, which satisfy the same hypotheses, and if we can prove that $\tilde{\varrho}_1 \otimes \tilde{\varrho}_2$ is semisimple, we will be able to conclude that $\varrho_1 \otimes \varrho_2$ is also semisimple. This reduction means that we may assume from the start that G is a subgroup of $\mathrm{GL}(E_1) \times \mathrm{GL}(E_2)$, with ϱ_1 and ϱ_2 given by the projections on each factor. What is important is that, with this reduction, all three representations ϱ_1, ϱ_2, and $\varrho_1 \otimes \varrho_2$ are polynomial.

After this reduction, let $\mathbf{G}$ be the Zariski closure of G. By Exercise 7.1.6(2), it is enough to prove that the action of $\mathbf{G}$ on $E_1 \otimes E_2$ is semisimple. Furthermore, after passing possibly to the connected component of the identity of $\mathbf{G}$, one can assume that $\mathbf{G}$ is connected for the Zariski topology. This step uses Lemma 7.1.10, Exercise 2.3.4, and the fact that we work with representations over a field of characteristic zero.

We now require one non-trivial ingredient. For any linear algebraic group $\mathbf{H} \subset \mathrm{GL}_n(\mathbf{C})$, one shows that there exists a unique maximal normal connected subgroup $R_u\mathbf{H}$ of $\mathbf{H}$ such that all elements of $R_u\mathbf{H}$ are unipotent as elements of $\mathrm{GL}_n(\mathbf{C})$. This subgroup is unique, and it is called the *unipotent radical* of $\mathbf{H}$. This has the crucial property that, for a connected algebraic group $\mathbf{H}$, a finite-dimensional polynomial complex representation of $\mathbf{H}$ is semisimple *if and only if* its kernel contains $R_u\mathbf{H}$.

We apply this criterion in our case. Since $\mathbf{G}$ is given with a faithful representation on $E_1 \oplus E_2$, which is semisimple by assumption, it follows that the unipotent radical $R_u\mathbf{G}$ must be trivial. But then, $R_u\mathbf{G}$ acts obviously trivially on $E_1 \otimes E_2$, and hence this representation of $\mathbf{G}$ is semisimple. □

Remark 7.1.12. It is known, and it is a more elementary property, that the converse holds: if the tensor product $\varrho_1 \otimes \varrho_2$ is semisimple, then ϱ_1 and ϱ_2 are semisimple. The proof of this does not require the use of algebraic groups. One may guess that this is more elementary because this converse is valid over any algebraically closed field k under the condition that $\dim \varrho_1$ and $\dim \varrho_2$ are both invertible in k; whereas Chevalley's result holds only in this generality if the characteristic p of the field is large enough compared with $\dim \varrho_1$ and $\dim \varrho_2$, or more precisely if

$$p > \dim \varrho_1 + \dim \varrho_2 - 2$$

(see [**55**, Part II, lecture 1, Th. 2] and [**57**]).

Example 7.1.13. Let $\mathbf{G}$ be the group

$$\mathbf{G} = \left\{ \begin{pmatrix} a & b \\ 0 & d \end{pmatrix} \in \mathrm{GL}_2(\mathbf{C}) \mid a,\ b,\ d \in \mathbf{C} \right\}.$$

This is an algebraic subgroup of $\mathrm{GL}_2(\mathbf{C})$ (defined by the polynomial equation that asserts that the bottom-left coordinate of a matrix is 0), and its unipotent radical is the subgroup

$$R_u\mathbf{G} = \left\{ \begin{pmatrix} 1 & b \\ 0 & 1 \end{pmatrix} \mid b \in \mathbf{C} \right\},$$

which we have already seen at work many times. (This subgroup is clearly made of unipotent elements, and it is normal in $\mathbf{G}$; a simple computation shows that it is maximal with this property.) The quotient $\mathbf{G}/R_u\mathbf{G}$ is isomorphic to $\mathbf{C}^\times \times \mathbf{C}^\times$ (mapping the matrix above to (a, d)), and therefore the criterion we used in the proof states that a finite-dimensional polynomial complex representation of $\mathbf{G}$ is semisimple if and only if it factors through the quotient $\mathbf{C}^\times \times \mathbf{C}^\times$. The "if" direction is part of Exercise 7.1.6, and as an example of the "only if" part, note that the tautological faithful two-dimensional representation of $\mathbf{G}$ is indeed not semisimple.

The last application is due to N. Katz [**34**, Lemma 2.18.2 bis], and again it would be interesting to know if there is a more direct proof.

Proposition 7.1.14. *Let k be an algebraically closed field of characteristic zero, let G be a group, and let $\varrho : G \longrightarrow \mathrm{GL}(E)$ be a non-zero finite-dimensional k-representation of G. Let A be the group of one-dimensional k-characters of G such that ϱ and ϱ twisted by χ are isomorphic, i.e.,*

$$A = \{\chi : G \longrightarrow k^\times \mid \varrho \simeq \varrho \otimes \chi\}.$$

Then A is finite.

Sketch of proof. We give the main steps of the proof, which the reader can fill in (or check in the original text), assuming $k = \mathbf{C}$. It is convenient to write $A = A(G, \varrho)$ to indicate the dependency of this group on G and ϱ.

(1) One checks that $A(G, \varrho) = A(G/\mathrm{Ker}(\varrho), \varrho)$, which means that one may assume that ϱ is a faithful representation. We then identify it with its image in $\mathrm{GL}(E)$ and reduce to $G \subset \mathrm{GL}(E)$ acting on E in the standard way, and then write $A(G)$ instead of $A(G, \varrho)$.

(2) Let $\mathbf{G}$ be the Zariski closure of G in $\mathrm{GL}(E)$. The crucial step is to show that $A(G) = A(\mathbf{G})$. To do this, one shows that any $\chi \in A(G)$ extends to a character $\tilde{\chi} \in A(\mathbf{G})$ (see Exercise 7.1.15 below).

(3) Now consider $\chi \in A(\mathbf{G})$. By the definition of isomorphisms of representations, there exists an element $g \in \mathrm{GL}(E)$ such that

$$gxg^{-1} = \chi(x)x$$

for all $x \in \mathbf{G}$. Taking the determinant, it follows that

$$\chi(x)^{\dim(E)} = 1$$

for all $x \in \mathbf{G}$. This means that χ is a character of order dividing $\dim E$, and in particular its kernel is a normal subgroup of finite index. But, by the last part of Lemma 7.1.10 (which is where we use the reduction to algebraic groups), the kernel $\operatorname{Ker}\chi$ contains $\mathbf{G}^0$, which means that χ factors through the finite group $\mathbf{G}/\mathbf{G}^0$. We thus get an injection

$$A(\mathbf{G}) \hookrightarrow \operatorname{Hom}(\mathbf{G}/\mathbf{G}^0, k^\times),$$

and therefore $A(\mathbf{G})$ is finite. □

Exercise 7.1.15. In this exercise, we explain the proof of step (2) in the previous sketch. The notation is as in this proof: $\mathbf{G} \subset \mathrm{GL}(E)$ is the Zariski closure of G. Let Z be the center of $\mathrm{GL}(E)$, i.e., the subgroup of scalar matrices.

(1) Let $\chi \in A(G)$ be given. Show that there exists $g \in \mathrm{GL}(E)$ such that

$$gxg^{-1}x^{-1} = \chi(x)$$

for all $x \in G$, and such that $\mathbf{G}$ is contained in

$$H_g = \{x \in \mathrm{GL}(E) \mid gxg^{-1}x^{-1} \in Z\}.$$

(2) With χ and g as in (1), show that there exists a map

$$\begin{cases} \mathbf{G} & \longrightarrow & k^\times \\ x & \mapsto & \alpha(x), \end{cases}$$

such that

$$gxg^{-1} = \alpha(x)x$$

for all $x \in \mathbf{G}$. Show that $x \mapsto \alpha(x)$ is an element of $A(\mathbf{G})$ and that $\alpha(x) = \chi(x)$ for all $x \in G$.

(3) Conclude that $A(G) = A(\mathbf{G})$.

7.2. Locally compact groups: general remarks

In the remaining sections of this chapter, we will give a short introduction to some aspects of the representation theory of non-compact, but locally compact, groups. We begin with an introductory section to describe, in general terms, some of the new phenomena that occur in this case. We will then illustrate many of them in the next two sections, sometimes with proofs, and otherwise with references to the literature.

We will discuss two type of differences: *global* ones concern the way unitary representations are (or are not) built out of irreducible ones, while *local* ones deal with the existence and other properties of the irreducible unitary representations. The properties of the regular representation sit somehow in the middle, since in the case of compact groups, we saw that it contains (in a relatively well-controlled way) all the irreducible unitary representations.

(1) On the global side, a major difference is that a (non-zero) unitary representation ϱ of a non-compact group G might have no irreducible subrepresentation. An example of this behavior was already given in Example 3.4.15 with the regular representation of $\mathbf{R}$. Certainly, if G is not compact, the *trivial* one-dimensional representation $\mathbf{1}$ is never a subrepresentation of the regular representation of G, since the constant function 1 is not square-integrable when G is not compact.

This shows that if one wishes to uses irreducible unitary representations to describe all representations, different constructions than just direct sums are needed. One can indeed define *direct integrals* of families of representations parametrized by measure spaces and show that any unitary representation is a direct integral of irreducible unitary representations (see, e.g., [**44**, Th. 2.9]). However, even when the subtleties involved in establishing such a theory are dealt with, a major difficulty remains: there is, in general, no uniqueness result for such direct integral decompositions. This applies already to groups like non-abelian free groups, or some solvable Lie groups, or even to a group as innocent looking as $G = \mathbf{Q} \rtimes \mathbf{Q}^{\times}$ seen as a discrete (solvable) group. Indeed, one can show (see [**44**, §3.5]) that the regular representation of G on $L^2(G)$ (note that here the Haar measure is just the counting measure, since G is discrete) decomposes in two *essentially inequivalent* ways as a direct integral of irreducible unitary representations.

Nevertheless, the existence of integral decompositions and the fact that the regular representation is faithful imply the important fact that any locally compact group has sufficiently many irreducible unitary representations to distinguish group elements:

Theorem 7.2.1 (Gelfand and Raikov). *Let G be a locally compact group. For any $g \neq 1$ in G, there exists an irreducible unitary representation ϱ of G such that $\varrho(g) \neq 1$.*

See [**44**, p. 110] for a proof in the case of a separable group (one in which there is a countable dense subset). This result was first proved using different techniques by Gelfand and Raikov.

(2) Concerning the *local* side of the theory, that of properties of the irreducible representations themselves, a major difference is that a non-compact group G may have *no* (non-trivial) finite-dimensional unitary representations. For instance, this is the case of $G = \mathrm{SL}_2(\mathbf{R})$ by Proposition 7.4.1, and therefore also of any G which is generated by subgroups which are

topologically isomorphic to $\mathrm{SL}_2(\mathbf{R})$. A direct consequence is that there is no obvious possibility for defining a character theory for such groups, although one can develop a more refined theory of characters as *distributions* on Lie groups (variants also exist for p-adic groups). One can still define matrix coefficients, since the functions

$$f(g) = \langle \varrho(g)v, w \rangle$$

exist on G and are bounded, but in the non-compact case these functions are not necessarily in $L^2(G)$, which reflects the fact that the irreducible unitary representations are not necessarily to be found as subrepresentations of the regular representation. Both possibilities may arise for the same group, as we will see for $G = \mathrm{SL}_2(\mathbf{R})$, which admits both irreducible unitary representations that are subrepresentations of the regular representations (these form an infinite countable set) and many which are not (an uncountable set). On the other hand, for $\mathrm{SL}_2(\mathbf{C})$, there are no irreducible subrepresentations of the regular representation.

Finally, perhaps the most mysterious aspect of the theory of irreducible representations for Lie groups such as $\mathrm{SL}_2(\mathbf{R})$ is that there may exist irreducible unitary representations which are completely "invisible" from the point of view of the regular representation. To be a bit more precise, even if an irreducible representation ϱ is not contained as a subrepresentation in $L^2(G)$, it may appear in the generalization of the Plancherel formula (or Fourier inversion formula). Indeed, this is particularly clear for abelian groups, say $G = \mathbf{R}$, where for $f \in L^2(\mathbf{R})$, smooth with compact support, we have the expression

$$f(x) = \int_{\mathbf{R}} \hat{f}(t) e^{itx} dt$$

in which all unitary characters $x \mapsto e^{itx}$ occur, although none of them is a subrepresentation of $L^2(\mathbf{R})$. For $G = \mathrm{SL}_2(\mathbf{R})$ (and for other Lie groups) there is a Plancherel formula, due to Harish-Chandra, for suitably smooth functions in $L^2(G)$, but the representations occurring on the right-hand side do not exhaust all irreducible unitary representations (see [**36**, Th. 11.6]).

These few lines should be enough to convince the reader that the representation theory of non-compact groups is a fascinating area of mathematics. We refer to [**36**] for a full-featured account.

7.3. Locally compact abelian groups

In the case of locally compact abelian groups, there is a very satisfactory theory, where the direct integral decompositions of unitary representations which generalize direct sum decompositions are relatively transparent. This is known as the theory of *Pontryagin duality* and generalizes classical Fourier analysis.

Let G be a locally compact abelian group. By Schur's Lemma (Proposition 3.4.17) all irreducible unitary representations of G are of dimension 1. Hence the set $\widehat{G}$ of isomorphism classes of unitary irreducible representations of G is well defined, as the set of continuous homomorphisms $\chi : G \longrightarrow \mathbf{S}^1$. This set is a group, with the operations of pointwise multiplication and inversion.

Definition 7.3.1 (Dual of a locally compact group). The *dual group* of a locally compact abelian group G is the group $\widehat{G}$, with the topology of uniform convergence on compact subsets of G, i.e., with the topology such that a basis of neighborhoods of $\chi_0 \in \widehat{G}$ is given by the sets

$$\{\chi \in \widehat{G} \mid |\chi(x) - \chi_0(x)| \leqslant \varepsilon \text{ for all }, x \in K\},$$

where $\varepsilon > 0$ and $K \subset G$ is compact.

The choice of topology is important, as the next lemma shows.

Lemma 7.3.2. *For any locally compact abelian group G, the dual group $\widehat{G}$ is a locally compact abelian group.*

The most important fact is that $\widehat{G}$ is locally compact (which implies, for instance, that there is a Haar measure on the dual group).

Now we can generalize Theorems 4.5.1 and 4.5.2:

Theorem 7.3.3 (Pontryagin duality). *Let G be a locally compact abelian group, and let μ be a Haar measure on G.*

(1) *The homomorphism*

$$\begin{cases} G & \longrightarrow & \widehat{\widehat{G}} \\ x & \mapsto & e_x, \end{cases}$$

where $e_x(\chi) = \chi(x)$ is an isomorphism of locally compact groups, i.e., a group isomorphism and a homeomorphism.

(2) *There exists a unique Haar measure $\hat{\mu}$ on $\widehat{G}$ such that, for any function $f \in C_c(G)$ continuous and compactly supported on G, the function*

$$\hat{f} : \begin{cases} \widehat{G} & \longrightarrow & \mathbf{C} \\ \chi & \mapsto & \int_G f(x)\overline{\chi(x)} d\mu(x) \end{cases}$$

is in $L^2(G, \mu)$ and satisfies the Plancherel formula

$$\int_G |f(x)|^2 d\mu(x) = \int_{\widehat{G}} |\hat{f}(\chi)|^2 d\hat{\mu}(\chi).$$

In particular, the linear map $f \mapsto \widehat{f}$ *extends to an isometry* $L^2(G, \mu) \longrightarrow L^2(\widehat{G}, \widehat{\mu})$.

(3) *Furthermore, if* $\widehat{f}$ *is in* $L^1(\widehat{G}, \widehat{\mu})$, *we have the inversion formula*

$$f(x) = \int_{\widehat{G}} \widehat{f}(\chi)\chi(x) d\widehat{\mu}(\chi).$$

Exercise 7.3.4. (1) If μ is replaced by $c\mu$ with $c > 0$, how is $\widehat{\mu}$ changed?

(2) If G is a group such that $\widehat{G} \simeq G$ (examples are finite groups or $G = \mathbf{R}^n$ for $n \geqslant 1$, see the example below), show that there exists a unique Haar measure μ on G such that $\widehat{\mu} = \mu$. What is this measure when G is finite? (This measure is sometimes called the *self-dual Haar measure* on G.)

Before giving examples of this result, we explain how to interpret it from the point of view of representation theory as giving a *direct integral decomposition*

$$\varrho_G \simeq \bigoplus\!\!\!\!\!\!\int_{\widehat{G}} \chi \, d\widehat{\mu}(\chi) \tag{7.4}$$

of the regular representation of a locally compact abelian group G. Although we do not give a general definition of such decompositions, since this would involve rather delicate measurability issues, this example and the others below should give a good idea of what is happening. For full details, see, for instance, the book by Mackey [**44**].

Intuitively, an element of the right-hand side of (7.4) should be a family $v = (v_\chi)$, where χ runs over all characters of G and v_χ is a vector in the space on which χ acts. Since χ has dimension 1, we can identify all these spaces with $\mathbf{C}$, so that v can be seen as a function $\widehat{G} \longrightarrow \mathbf{C}$. From a Hilbert space perspective, it is natural to view the underlying space of the direct integral as the collection of these families such that the norm

$$\|v\|^2 = \int_{\widehat{G}} |v_\chi|^2 d\widehat{\mu}(\chi),$$

is finite, which means that the space of the direct integral is $L^2(\widehat{G}, \widehat{\mu})$. From the point of view of representation theory, the group G should act on these families (v_χ) by

$$g \cdot v = (\chi(g)v_\chi)_\chi. \tag{7.5}$$

The map from the regular representation to the direct integral is given by

$$f \mapsto (\widehat{f}(\chi))_\chi,$$

which, by the Plancherel formula, makes sense and gives an isometry between the spaces $L^2(G, \mu)$ and $L^2(\widehat{G}, \widehat{\mu})$. And further, this abstract Fourier transform is an intertwiner: for any $g \in G$, we have

$$\begin{aligned}(\widehat{\varrho_G(g)f})(\chi) &= \int_G (\varrho_G(g)f)(x)\overline{\chi(x)}d\mu(x) \\ &= \int_G f(x+g)\overline{\chi(x)}d\mu(x) = \chi(g)\hat{f}(\chi)\end{aligned}$$

using invariance of Haar measure and the fact that χ is a homomorphism (this is a formal computation and a rigorous one if f is also in $L^1(G,\mu)$). This means that the χ-component of the image of f is multiplied by $\chi(g)$ under the regular action of g on f. This corresponds exactly to the definition (7.5) of the direct integral.

Example 7.3.5. (1) Let $G = \mathbf{R}$. Then, as we have already seen (see Example 3.4.10), the dual group is isomorphic as a group to $\mathbf{R}$ using the map

$$\begin{cases} \mathbf{R} & \longrightarrow & \hat{G} \\ t & \mapsto & e_t, \end{cases}$$

where $e_t(x) = e^{itx}$. It is an elementary exercise that this is also an homeomorphism, so that $\hat{G}$ is isomorphic to $\mathbf{R}$ as a topological group. The corresponding abstract Fourier transform is defined for $f \in L^1(\mathbf{R})$ by

$$\hat{f}(t) = \int_{\mathbf{R}} f(x)e^{-itx}dx$$

and coincides therefore with the "usual" Fourier transform. The Fourier inversion formula, with this normalization, is

$$f(x) = \frac{1}{2\pi}\int_{\mathbf{R}} \hat{f}(t)e^{itx}dt$$

so that the dual Haar measure of the Lebesgue measure is $\frac{1}{2\pi}dt$. (This illustrates that one must be somewhat careful with the normalizations of Haar measure.)

(2) Let $G = \mathbf{Z}$. This is a discrete non-compact locally compact abelian group. A unitary representation ϱ of $\mathbf{Z}$ on a Hilbert space H is simply determined by the data of a unitary operator $\varrho(1)$, and two such representations are isomorphic if and only if the operators are *unitarily equivalent*. This means in particular that $\hat{\mathbf{Z}} \simeq \mathbf{S}^1 \simeq \mathbf{R}/\mathbf{Z}$ as a group (a one-dimensional unitary operator is a complex number of modulus 1), and one can then check furthermore that these isomorphisms are also homeomorphisms.

(3) The dual group of $\mathbf{S}^1 \simeq \mathbf{R}/\mathbf{Z}$ was already determined as isomorphic to $\mathbf{Z}$ (in Example 3.4.10), the isomorphism being

$$\begin{cases} \mathbf{Z} & \longrightarrow & \widehat{\mathbf{R}/\mathbf{Z}} \\ n & \mapsto & e_{2i\pi n}, \end{cases}$$

so that, comparing (2) and (3), we see a special case of Pontryagin duality. Here also, one must of course check that the topology on $\widehat{\mathbf{R}/\mathbf{Z}}$ is indeed discrete.

(4) Let $G = \mathbf{R}^\times$. Then we have an isomorphism

$$\begin{cases} G & \longrightarrow & (\mathbf{Z}/2\mathbf{Z}) \times \mathbf{R} \\ x & \mapsto & \left(\dfrac{x}{|x|}, \log |x|\right) \end{cases}$$

of locally abelian groups. Hence $\widehat{G}$ is isomorphic to $\mathbf{Z}/2\mathbf{Z} \times \mathbf{R}$ (because the dual of a product is the product of the duals), where the character corresponding to $(\varepsilon, t) \in \mathbf{Z}/2\mathbf{Z} \times \mathbf{R}$ is given by

$$\chi(x) = \operatorname{sgn}(x)^\varepsilon |x|^{it}, \qquad \operatorname{sgn}(x) = \frac{x}{|x|}.$$

Exercise 7.3.6. Let G be a compact abelian group. Show that, with the topology of uniform convergence on G, the dual group is discrete.

One can prove that any unitary representation of a locally compact abelian group G has a decomposition into a direct integral of characters. In the special case of $G = \mathbf{Z}$, this is another interpretation of the spectral theorem for unitary operators, which is mentioned at the end of Theorem 3.4.18: a unitary representation ϱ of $\mathbf{Z}$ on a Hilbert space H is uniquely determined by the operator $\varrho(1) \in \mathrm{U}(\mathrm{H})$, and the spectral theorem gives an isomorphism of this representation with a representation $\tilde{\varrho}$ on a space $L^2(X, \mu)$, for some measure space (X, μ), where $\tilde{\varrho}(1)$ is the multiplication operator by some function of the form e^{if}, for a real-valued measurable function f on X. We interpret this as stating that

$$L^2(X, \mu) \simeq \int_X^{\oplus} e_{f(x)} d\mu(x)$$

(where, for $t \in \mathbf{R}/\mathbf{Z}$, the character e_t of $\mathbf{Z}$ is given by $e_t(n) = e^{int}$), with the representation given formally for $v = (v_x)_{x \in X}$ in the direct integral by

$$1 \cdot v = (e_{f(x)} \cdot v_x) = (e^{if(x)} v_x)_x.$$

This intuitive description fits perfectly with the spectral theorem.

There is another important example where the general decomposition has a concrete and relatively classical form. This is when $G = \mathbf{R}$, and the result is due to Stone:

Theorem 7.3.7 (Stone's Theorem). *Let* H *be a separable Hilbert space, and let* $\varrho : \mathbf{R} \longrightarrow \mathrm{U}(\mathrm{H})$ *be a unitary representation, also known as a strongly continuous one-parameter unitary group in* H. *There exists a finite measure* *space* (X, μ) *and a real-valued function* f *on* X *such that* ϱ *is isomorphic to the representation on* $L^2(X, \mu)$ *given by*

$$\tilde{\varrho}(t)\varphi(x) = e^{itf(x)}\varphi(x).$$

For a proof, combine [**49**, Th. VIII.4] with [**49**, Th. VIII.8] (or [**39**, §4.6, §6.1]).

This result plays a very important role in quantum mechanics, as giving the time evolution of a state under the Schrödinger equation; indeed (at least formally), given an initial state $\varphi_0 \in L^2(X, \mu)$ of norm 1, the functions $\varphi_t = \tilde{\varrho}(-t)\varphi_0$ satisfy the differential equation

$$i\frac{d}{dt}\varphi_t = f\varphi_t,$$

which is (a close variant of) the equation (6.18) for the Hamiltonian operator of multiplication by f.

Exercise 7.3.8. Interpret Stone's Theorem as an isomorphism

$$\varrho \simeq \bigoplus\!\!\!\!\!\!\int_X e_{f(x)} d\mu(x),$$

where (as before) $e_t(x) = e^{itx}$ for $t \in \mathbf{R}$ and $x \in \mathbf{R}$.

7.4. A non-abelian example: $\mathrm{SL}_2(\mathbf{R})$

In this section, we will discuss some basic aspects of the irreducible unitary representations of the group $G = \mathrm{SL}_2(\mathbf{R})$ and of its regular representation. Although some technical facts will not be proven in full detail, we will be able to present concretely many aspects of this theory, which will give a first orientation toward its generalization to other similar groups. The results we present were first obtained by Bargmann in 1947.

Before beginning, we recall from Example 5.2.4(6) that the group G is unimodular: its Haar measure is both left- and right-invariant.

We begin by showing that there are no non-trivial finite-dimensional unitary representations of $\mathrm{SL}_2(\mathbf{R})$:

Proposition 7.4.1. *Let* $G = \mathrm{SL}_2(\mathbf{R})$. *Then any finite-dimensional unitary representation of* G *is trivial.*

Proof. It is enough to show that an irreducible unitary representation ϱ of G with $\dim(\varrho) < +\infty$ is trivial. The proof we give is not the shortest, but it is quite enlightening.

Assume that ϱ is not trivial. We consider the restriction of ϱ to the subgroups

$$U^+ = \left\{\begin{pmatrix} 1 & t \\ 0 & 1 \end{pmatrix} \mid t \in \mathbf{R}\right\} \simeq \mathbf{R}, \qquad U^- = \left\{\begin{pmatrix} 1 & 0 \\ t & 1 \end{pmatrix} \mid t \in \mathbf{R}\right\} \simeq \mathbf{R}.$$

Using the identity (7.3) (with real parameters), we see that one of these restrictions must be non-trivial. We may assume that ϱ restricted to U^+ is not trivial, the other case being exactly similar.

Since $\dim(\varrho) < +\infty$, this restriction $\mathrm{Res}_{U^+}^G(\varrho)$ contains a non-trivial irreducible subrepresentation of $U^+ \simeq \mathbf{R}$. By the classification of characters of $\mathbf{R}$, this means that there exists a real number $x \neq 0$ and a non-zero vector v such that

$$\varrho\Big(\begin{pmatrix} 1 & t \\ 0 & 1 \end{pmatrix}\Big)v = e^{itx}v$$

for all $t \in \mathbf{R}$. Now, using the relation

$$\begin{pmatrix} a & 0 \\ 0 & a^{-1} \end{pmatrix}\begin{pmatrix} 1 & t \\ 0 & 1 \end{pmatrix}\begin{pmatrix} a^{-1} & 0 \\ 0 & a \end{pmatrix} = \begin{pmatrix} 1 & a^2t \\ 0 & 1 \end{pmatrix}, \tag{7.6}$$

it follows that for all $a \in \mathbf{R}^\times$, the vector

$$v_a = \varrho\Big(\begin{pmatrix} a & 0 \\ 0 & a^{-1} \end{pmatrix}\Big)v \neq 0$$

satisfies

$$\varrho\Big(\begin{pmatrix} 1 & t \\ 0 & 1 \end{pmatrix}\Big)v_a = e^{ia^2tx}v_a.$$

This means that the restriction $\mathrm{Res}_{U^+}^G(\varrho)$ also contains the character $t \mapsto e^{ia^2xt}$, for all $a > 0$. Since $x \neq 0$, this gives infinitely many distinct characters of U^+ contained in ϱ, which is impossible since $\dim(\varrho) < +\infty$. This contradiction means that ϱ does not exist. ∎

Exercise 7.4.2. This exercise presents a different proof of this proposition. Let ϱ be a finite-dimensional unitary representation of $\mathrm{SL}_2(\mathbf{R})$, and let χ be its character.

(1) Show that $\chi(x) = \dim(\varrho) = \chi(1)$ for all $x \in U^+$ and all $x \in U^-$. (*Hint*: Use (7.6) and continuity; this does not use unitarity.)

(2) Deduce that the kernel of ϱ contains U^+ and U^-, and conclude that ϱ is trivial.

The proof of Proposition 7.4.1 may remind the reader of the argument used in Exercise 4.7.3 to show that $\mathrm{SL}_2(\mathbf{F}_p)$ has no non-trivial irreducible representation *of small dimension*. There are in fact many concrete parallels between aspects of the representation theory of $\mathrm{SL}_2(\mathbf{F}_p)$ and that of $\mathrm{SL}_2(\mathbf{R})$ and $\mathrm{SL}_2(\mathbf{C})$. These are explained to a certain extent by the fact that these groups are *specializations* of the *algebraic group* SL_2—for instance, identities like (7.3) or (7.6) are valid uniformly for coefficients in any field.

This analogy continues with the first concrete examples of infinite-dimensional irreducible representations of these groups, which we now present.

The construction is similar to that of the principal series of the finite groups $\mathrm{GL}_2(\mathbf{F}_q)$ (Section 4.6.4). It proceeds by induction of one-dimensional characters from the subgroup of upper-triangular matrices. Naturally, these representations are also called the *principal series representations*, and many of their properties are formally similar.

We consider $G = \mathrm{SL}_2(\mathbf{R})$ and the subgroups

$$B = \left\{ \begin{pmatrix} \alpha & \beta \\ 0 & \alpha^{-1} \end{pmatrix} \mid \alpha \in \mathbf{R}^{\times},\ \beta \in \mathbf{R} \right\} \subset G,$$

$$B^{+} = \left\{ \begin{pmatrix} \alpha & \beta \\ 0 & \alpha^{-1} \end{pmatrix} \mid \alpha > 0,\ \beta \in \mathbf{R} \right\} \subset B.$$

The argument in Proposition 7.4.1 shows in fact that B has no other finite-dimensional unitary representation than the one-dimensional characters

$$\begin{pmatrix} \alpha & \beta \\ 0 & \alpha^{-1} \end{pmatrix} \mapsto \chi(\alpha)$$

for some $\chi \in \widehat{(\mathbf{R}^{\times})}$. Such a character is of the form

$$\chi(\alpha) = \mathrm{sgn}(\alpha)^{\varepsilon} |\alpha|^{it},$$

where $\varepsilon \in \{0, 1\}$ and $t \in \mathbf{R}$ (see Example 7.3.5(4)).

We wish to define a representation $\mathrm{Ind}_B^G(\chi)$ in such a way that it is unitary and that it satisfies irreducibility properties similar to those of the principal series in Section 4.6.4 and Exercise 4.8.3. The Hilbert space will be a space of functions

$$f : G \longrightarrow \mathbf{C},$$

which transform suitably under multiplication on the left by $b \in B$ and which carries formally the regular representation of G. However, the most naive definition of the space (asking that $f(bg) = \chi(b) f(g)$) and of the norm (the L^2-norm with respect to a Haar measure on G) do not work.

Instead, we denote by $\delta : B \longrightarrow \mathbf{R}^{+,\times}$ the homomorphism

$$\delta\left(\begin{pmatrix} \alpha & \beta \\ 0 & \alpha^{-1} \end{pmatrix}\right) = |\alpha|,$$

and by K the compact subgroup $K = SO_2(\mathbf{R}) \subset G$ (equipped with its probability Haar measure), and we define

$$\mathrm{H}_{\chi} = \{ f : G \longrightarrow \mathbf{C} \text{ measurable} \mid f(bg) = \chi(b)\delta(b) f(g) \text{ for all } b \in B,\ g \in G, \text{ and } \|f\|^2 < +\infty \},$$

where the norm is given by

$$\|f\|^2 = \int_K |f(k)|^2 dk,$$

and (as usual) we will really view elements of H_χ as equivalence classes where functions which are equal almost everywhere on G are identified. The reason we can restrict to K the integral in the definition of the norm will be found in Lemma 7.4.4.

Proposition 7.4.3 (Principal series of $\mathrm{SL}_2(\mathbf{R})$). (1) *For any unitary character $\chi \in \widehat{(\mathbf{R}^\times)}$, the space H_χ with the norm $\|\cdot\|$ above is a Hilbert space, and the action*

$$\varrho_\chi(g)f(x) = f(xg)$$

defines a unitary representation of $G = \mathrm{SL}_2(\mathbf{R})$ on H_χ. It is also denoted $\mathrm{Ind}_B^G(\chi)$ and called a principal series representation. *The representation space H_χ is infinite dimensional.*

(2) *The representation ϱ_χ is irreducible if and only if χ is* not *the character given by $\chi(x) = \mathrm{sgn}(x) = x/|x|$.*

(3) *For χ_1 and χ_2 in $\widehat{(\mathbf{R}^\times)}$, we have $\varrho_{\chi_1} \simeq \varrho_{\chi_2}$ if and only $\chi_1 = \chi_2$ or $\chi_1 = \chi_2^{-1}$.*

(4) *There exists no subrepresentation of the regular representation of G which is isomorphic to ϱ_χ.*

Note how (2) and (3) are formally similar with the properties of the induced representations forming the principal series of the finite linear groups $\mathrm{GL}_2(\mathbf{F}_p)$ (if one makes the adaptation to deal with $\mathrm{SL}_2(\mathbf{F}_p)$ and if one accounts for the appearance of δ, which changes the irreducibility condition). Indeed, there exists a general notion of induction for locally compact groups, satisfying a formalism of intertwiners which is parallel to that of Proposition 4.8.1 and which explains these similarities (see [**36**, VII.3].)

To prove part (1) of Proposition 7.4.3, we use the following lemma:

Lemma 7.4.4. (1) *We have $G = BK$, where $K = \mathrm{SO}_2(\mathbf{R})$, and more precisely any $g \in \mathrm{SL}_2(\mathbf{R})$ can be expressed as $g = bk$ with $b \in B^+$ and $k \in K$, and this representation is unique. In fact, we can write*

$$g = \begin{pmatrix} y^{1/2} & xy^{-1/2} \\ 0 & y^{-1/2} \end{pmatrix} k(\theta), \text{ where } k(\theta) = \begin{pmatrix} \cos(\theta) & \sin(\theta) \\ -\sin(\theta) & \cos(\theta) \end{pmatrix} \tag{7.7}$$

for unique coordinates (x, y, θ) with $x \in \mathbf{R}$, $y > 0$ and $\theta \in \mathbf{R}/2\pi\mathbf{Z}$, which are called the Iwasawa coordinates *of g.*

(2) *For $\varepsilon \in \{0, 1\}$, let*

$$L^2_\varepsilon(K) = \{f \in L^2(K) \mid f(-k) = (-1)^\varepsilon f(k) \text{ for all } k \in K\}$$

be seen as a closed subspace of $L^2(K)$. *Then if* χ *corresponds to the parameters* (ε, t), *the restriction map*

$$\begin{cases} \mathrm{H}_\chi & \longrightarrow & L^2_\varepsilon(K) \\ f & \mapsto & f \mid K \end{cases}$$

is a well-defined isometric isomorphism of Hilbert spaces, and in particular, H_χ *is infinite dimensional.*

Proof. (1) One can of course check the result by brute force computation, but a better understanding arises using the action of G on the complex upper half-plane

$$\mathbf{H} = \{z = x + iy \in \mathbf{C} \mid x \in \mathbf{R}, \quad y > 0\} \subset \mathbf{C}$$

given by

$$\begin{pmatrix} a & b \\ c & d \end{pmatrix} \cdot z = \frac{az + b}{cz + d}.$$

One can easily check that this is indeed an action (in particular, the imaginary part of $g \cdot z$ is > 0 if $z \in \mathbf{H}$) and that K can be identified with the stabilizer of $i \in \mathbf{H}$. In addition, the formula

$$\begin{pmatrix} y^{1/2} & xy^{-1/2} \\ 0 & y^{-1/2} \end{pmatrix} \cdot i = x + iy$$

shows that the subgroup B acts transitively on $\mathbf{H}$. Hence if $g \in G$, we can express $g \cdot i$ in the form $b \cdot i$ for some $b \in B$. Then $b^{-1}g \in K$ so that $g = bk$ for some $k \in K$. And since $B \cap K = \{\pm \mathrm{Id}\}$, we also obtain the uniqueness statement, while the formula (7.7) follows from this computation.

(2) Using (1), we see that the norm on H_χ is indeed a norm, from which it follows at least that H_χ is a pre-Hilbert space: if $\|f\| = 0$, we get $f(k) = 0$ for almost all $k \in K$, and then by definition

$$f(bk) = \delta(b)\chi(b)f(k) = 0$$

for almost all $b \in B$ and $k \in K$, hence $f = 0$ by (1). The restriction map to K is well defined on H_χ since for any $f \in \mathrm{H}_\chi$ and $k \in K$, we have $-\mathrm{Id} \in B$, hence

$$f(-k) = \delta(-\mathrm{Id})\chi(-\mathrm{Id})f(k) = (-1)^\varepsilon f(k),$$

so that f restricted to K is in the space $L^2_\varepsilon(K)$. Now it is immediate by definition that the restriction to K preserves the norm. Further, if $f \in L^2_\varepsilon(K)$, we extend it to a function $\tilde{f} \in \mathrm{H}_\chi$ in the obvious way: we define

$$\tilde{f}(bk) = \delta(b)\chi(b)f(k)$$

for $b \in B$ and $k \in K$. This is an unambiguous definition, again because $B \cap K = \pm\mathrm{Id}$ and $f \in L^2_\varepsilon(K)$, so the right-hand side takes the same value for $(-b, -k)$ as for (b, k).

Clearly the restriction of this function to K is equal to f, and since, for $b \in B$ and $g = b_1k_1 \in G$, we have

$$\begin{aligned}\tilde{f}(bg) &= \tilde{f}(bb_1k_1) = \delta(bb_1)\chi(bb_1)f(k_1)\\ &= \delta(b)\chi(b) \times (\delta(b_1)\chi(b_1)f(k_1)) = \delta(b)\chi(b)\tilde{f}(bk_1),\end{aligned}$$

we see that $\tilde{f} \in \mathrm{H}_\chi$, which finishes the proof of (2). □

Remark 7.4.5. Another view of the decomposition $G = B^+K$ (also called the Iwasawa decomposition) is that it can be seen as a reformulation of the Gram–Schmidt orthogonalization procedure applied to the column vectors of a matrix g (see also the polar decomposition used in the proof of Proposition 7.1.7).

We will use the coordinates (x, y, θ) of (7.7) below for various computations. In particular, these provide a way to speak of differentiable functions on $G = \mathrm{SL}_2(\mathbf{R})$, which correspond to its manifold (or Lie group) structure.

Exercise 7.4.6. (1) Show that the measure $y^{-2}dxdy$ on $\mathbf{H}$ is invariant for the action of G.

(2) Deduce that, in Iwasawa coordinates, a Haar measure on G is given by

$$\frac{1}{y^2}dxdyd\theta.$$

(3) Check directly that this measure is also left-invariant.

It is formally clear that the action of G on H_χ in Proposition 7.4.3 is a representation. The next step is to check that ϱ_χ is unitary, and this is where the tweak to the invariance properties of functions in the definition of H_χ is important. We explain this, and leave to the reader the exercise of checking that $g \mapsto \varrho_\chi(g)$ is strongly continuous; once this is done, we have proved part (1) of the proposition.

Lemma 7.4.7. *Let $g \in G$ be fixed. For $k \in K$, let $\beta(k) \in B^+$, $\kappa(k) \in K$ be the unique elements such that*

$$kg = \beta(k)\kappa(k).$$

Then the maps β and κ are continuous, and for any $f \in L^1(K)$ we have

$$\int_K f(k)dk = \int_K \delta(\beta(k))^2 f(\kappa(k))dk.$$

Proof. We use the fact that a (left or right) Haar measure on G can be decomposed as $dbdk$ where db denotes a *left-invariant* Haar measure on B^+ (note that B^+ is not unimodular).

We first pick a function $\varphi \in C_c(G)$. Since G itself is unimodular, we have

$$\int_{B^+}\int_K \varphi(bk)dbdk = \int_G \varphi(x)dx = \int_G \varphi(xg)dx = \int_{B^+}\int_K \varphi(bkg)dbdk,$$

and using the definition, we find

$$\int_{B^+}\int_K \varphi(bk)dbdk = \int_{B^+}\int_K \varphi(b\beta(k)\kappa(k))dbdk.$$

The point is that $\delta(b)^{-2}$ on B^+ is the modular character linking the left Haar measure db to a right Haar measure. Thus, using Fubini's Theorem to integrate over b first, we obtain

$$\int_{B^+}\int_K \varphi(bk)dbdk = \int_K\int_B \varphi(b\kappa(k))\delta(\beta(k))^2 dbdk.$$

Now we select φ of the type

$$\varphi(bk) = \psi(b)f(k),$$

where f is continuous on K and ψ is continuous and compactly supported on B^+ and satisfies $\int_{B^+}\psi(b)db = 1$. The formula becomes

$$\int_K f(k)dk = \int_K f(\kappa(k))\delta(\beta(k))^2 dk$$

by Fubini's Theorem, which gives the result for f continuous on K. It is now a standard limiting process to extend the result to any integrable function f on K. □

Using this lemma, we obtain unitarity of H_χ as follows: for $g \in G$ and $f \in \mathrm{H}_\chi$, we have

$$\begin{aligned}\|\varrho_\chi(g)f\|^2 &= \int_K |f(kg)|^2 dk = \int_K |f(\beta(k)\kappa(k)|^2 dk \\ &= \int_K |x(\beta(k))\delta(\beta(k))f(\kappa(k))| = \int_K |f(k)|^2 dk = \|f\|^2\end{aligned}$$

(using the notation of the lemma, applied to $|f|^2$).

We will now prove the irreducibility statement in part (2) of Proposition 7.4.3. The method we use illustrates two important general techniques in the representation theory of Lie groups: exploiting the restriction to compact subgroups, and "differentiating" a representation to study it through its Lie algebra (although we do not develop formally these aspects in any depth).

We denote by $\mathfrak{g}$ the Lie algebra of $G = \mathrm{SL}_2(\mathbf{R})$ (the gothic font is customary for Lie algebras). This is the space of real matrices of size 2 with trace 0, and if $X \in \mathfrak{g}$, we have $\exp(tX) \in G$ for all $t \in \mathbf{R}$, since the determinant of $\exp(tX)$ is $\exp(t \operatorname{Tr}(X)) = 1$. We wish to extend the representation ϱ_χ to an action of the Lie algebra by differentiating, as in (3.4), but since elements in H_χ are just L^2-functions (in fact, equivalence classes of functions), this is not always possible. Thus we define a subspace where this can be done. Let

$$\mathrm{H}_\chi^\infty = \{f \in \mathrm{H}_\chi \mid f \text{ is smooth}\},$$

where smoothness refers to the Iwasawa coordinates.

We have obviously $\varrho_\chi(g)(\mathrm{H}_\chi^\infty) \subset \mathrm{H}_\chi^\infty$, and H_χ^∞ is dense in H_χ. The Lie algebra action can be defined on H_χ^∞:

Lemma 7.4.8. *For $X \in \mathfrak{g}$ and $f \in \mathrm{H}_\chi^\infty$, let*

$$\varrho_\chi(X)f = \frac{d}{dt}\varrho_\chi(\exp(tX))\Big|_{t=0} = \lim_{t\to 0} \frac{\varrho_\chi(\exp(tX))f - f}{t}.$$

(1) *The limit above exists in the L^2-sense, i.e., we have*

$$\lim_{t\to 0}\left\| \frac{\varrho_\chi(\exp(tX))f - f}{t} - \varrho_\chi(X)f \right\| = 0$$

and the function $\varrho_\chi(X)f$ is in H_χ^∞. Furthermore, for $X, Y \in \mathfrak{g}$, we have

$$\varrho_\chi([X,Y])f = \varrho_\chi(X)(\varrho_\chi(Y)f) - \varrho_\chi(Y)(\varrho_\chi(X)f).$$

(2) *For any $g \in G$, we have*

$$\varrho_\chi(X)f(g) = \frac{d}{dt} f(g\exp(tX))\Big|_{t=0}.$$

Sketch of proof. We can prove the existence of the limit at the same time as we prove part (2). Thus we are claiming that

$$\lim_{t\to 0}\int_K \left| \frac{\varrho_\chi(\exp(tX))f(k) - f(k)}{t} - \frac{d}{ds} f(k\exp(sX))\Big|_{s=0}\right|^2 dk = 0.$$

The integral is equal to

$$\int_K \left| \frac{f(k\exp(tX)) - f(k)}{t} - \frac{d}{ds} f(k\exp(sX))\Big|_{s=0}\right|^2 dk.$$

The integrand tends to 0 as $t \to 0$ by standard calculus, and the mean value theorem proves that it is bounded for t close enough to 0. Since K is compact, this bound allows us to use the dominated convergence theorem to conclude that the integral converges to 0 as $t \to 0$. The formula for $\varrho_\chi([X,Y])$ is left as an exercise to the reader (see, e.g., [**41**, VI.1]). ☐

The second part allows us to express easily the Lie algebra action in terms of coordinates on G such that (x, y, θ) as in (7.7), using differential calculus.

Lemma 7.4.9. *Let E, H, and F be the following elements of $\mathfrak{g}$:*

$$E = \begin{pmatrix} 0 & 1 \\ 0 & 0 \end{pmatrix}, \quad H = \begin{pmatrix} 1 & 0 \\ 0 & -1 \end{pmatrix}, \quad F = \begin{pmatrix} 0 & 0 \\ 1 & 0 \end{pmatrix}.$$

For a smooth function $f : \mathbf{R}\times]0,+\infty[\times\mathbf{R}/2\pi\mathbf{Z} \longrightarrow \mathbf{C}$, identified with a function on G, we have

$$\frac{d}{dt} f\Big(g\exp(tE)\Big)\Big|_{t=0} = y\cos 2\theta\frac{\partial f}{\partial x} + y\sin 2\theta\frac{\partial f}{\partial y} + \sin^2\theta\frac{\partial f}{\partial\theta},$$

$$\frac{d}{dt} f\Big(g\exp(tH)\Big)\Big|_{t=0} = -y\sin 2\theta\frac{\partial f}{\partial x} + 2y\cos 2\theta\frac{\partial f}{\partial y} + \sin 2\theta\frac{\partial f}{\partial\theta},$$

$$\frac{d}{dt} f\Big(g\exp(tF)\Big)\Big|_{t=0} = y\cos 2\theta\frac{\partial f}{\partial x} + y\sin 2\theta\frac{\partial f}{\partial y} - (\cos 2\theta - \sin^2\theta)\frac{\partial f}{\partial\theta}.$$

Since the three elements (E, H, F) above form a basis of $\mathfrak{g}$ as a vector space, this lemma and the previous one completely describe the Lie algebra action.

Proof. This is an application of differential calculus; we explain the proof of the first formula, and leave the remainder to the reader. Consider

$$g = \begin{pmatrix} a & b \\ c & d \end{pmatrix} \in \mathrm{GL}_2(\mathbf{R})$$

with Iwasawa coordinates (x, y, θ). Then

$$g\exp(tE) = g\begin{pmatrix} 1 & t \\ 0 & 1 \end{pmatrix} = \begin{pmatrix} a & at+b \\ c & ct+d \end{pmatrix},$$

and we write $(x(t), y(t), \theta(t))$, the corresponding Iwasawa coordinates. By the chain rule formula, we have

$$\frac{d}{dt} f(x(t), y(t), \theta(t)) = \frac{\partial f}{\partial x}\frac{dx}{dt} + \frac{\partial f}{\partial y}\frac{dy}{dt} + \frac{\partial f}{\partial\theta}\frac{d\theta}{dt},$$

and in particular

$$\frac{d}{dt} f(g\exp(tE))\Big|_{t=0} = x'(0)\frac{\partial f}{\partial x} + y'(0)\frac{\partial f}{\partial y} + \theta'(0)\frac{\partial f}{\partial\theta}. \tag{7.8}$$

We now compute $(x'(0), y'(0), \theta'(0))$. First, we denote $z(t) = x(t) + iy(t)$ and observe that (using the action on the upper half-plane, as in the proof of Lemma 7.4.4) we have

$$g\exp(tE)\cdot i = x(t) + iy(t) = z(t) = \frac{ai + at + b}{ci + ct + d},$$

hence $z'(0) = (ci + d)^{-2}$ by direct computation (recall $ad - bc = 1$).

Note then that, by multiplying the matrices in the Iwasawa decomposition, solving for c and d, and computing $c + id$, we have the formulas

$$ci + d = y^{-1/2}(-i\sin\theta + \cos\theta) = y^{-1/2}e^{-i\theta}, \qquad y = \frac{1}{c^2 + d^2}, \tag{7.9}$$

hence

$$z'(0) = (ci + d)^{-2} = ye^{2i\theta},$$

which gives

$$x'(0) = y\cos(2\theta), \qquad y'(0) = y\sin(2\theta).$$

For the computation of $\theta'(0)$, we observe that

$$e^{-2i\theta(t)} = \frac{(ct + ci + d)^2}{(ct + d)^2 + c^2}$$

(by the method used to derive (7.9) applied to $g\exp(tE)$ instead of g), and by differentiating, we deduce that

$$-2i\theta'(0)e^{-2i\theta} = \frac{2c(ci + d)(c^2 + d^2) - 2cd(ci + d)^2}{(c^2 + d^2)^2}.$$

Using (7.9) again, the right-hand side is

$$-2\Big\{\sin\theta e^{-i\theta} - \sin\theta\cos\theta e^{-2i\theta}\Big\} = -2e^{-2i\theta}i\sin^2\theta,$$

which leads to $\theta'(0) = \sin^2\theta$. Combining these computations with (7.8), we obtain the formula for the derivative of $f(g\exp(tX))$. ☐

Now we prove that H_χ is irreducible if χ is not the sign character of B. We first note that the restriction to K of H_χ is isomorphic to the subrepresentation $L^2_\varepsilon(K)$ of the regular representation of K on $L^2(K)$ (this is immediate from the definition and from Lemma 7.4.4) and that this subrepresentation is just the Hilbert direct sum of the characters

$$\begin{pmatrix} \cos(\theta) & \sin(\theta) \\ -\sin(\theta) & \cos(\theta) \end{pmatrix} \mapsto e^{2i\pi n\theta}, \tag{7.10}$$

where n runs over integers $n \in \mathbf{Z}$ with parity ε, i.e., such that $n \equiv \varepsilon \pmod 2$. This last fact is a consequence of the decomposition of the regular representation $L^2(K) \simeq L^2(\mathbf{R}/2\pi\mathbf{Z})$ as a direct sum of these characters over all $n \in \mathbf{Z}$. We denote by φ_n the function (7.10) on K.

Let $V \subset \mathrm{H}_\chi$ be a non-zero closed subspace invariant under ϱ_χ. We must check that (if χ is not the sign), the space V is necessarily all of H_χ. First, viewing V as a subrepresentation of $\mathrm{Res}^G_K(\varrho_\chi)$, the decomposition above shows that V is a (Hilbert) direct sum of spaces $\mathbf{C}\varphi_n$ for n in some subset of integers with the parity ε. Since $V \neq 0$, this implies that, for some n, we have $\varphi_n \in V$.

Now we use the Lie algebra action. Note that the function f_n of the coordinates (x, y, θ) that corresponds to φ_n is given by

$$f_n(x, y, \theta) = y^{(it+1)/2} e^{in\theta}.$$

By Lemmas 7.4.8 and 7.4.9, we have

$$\varrho_\chi\left(\begin{pmatrix} 0 & 1 \\ -1 & 0 \end{pmatrix}\right) f_n = \varrho_\chi(E) f_n - \varrho_\chi(F) f_n = \frac{\partial f_n}{\partial \theta} = in f_n, \tag{7.11}$$

and (extending the action to $\mathfrak{g} \otimes \mathbf{C}$ by linearity) we also get

$$\begin{aligned} \varrho_\chi\left(\begin{pmatrix} 1 & -i \\ -i & -1 \end{pmatrix}\right) f_n &= \varrho_\chi(H) f_n - i\varrho_\chi(E) f_n - i\varrho_\chi(F) f_n \\ &= -2iye^{-2i\theta}\Big(\frac{\partial f_n}{\partial x} + i\frac{\partial f_n}{\partial y}\Big) + ie^{-2i\theta}\frac{\partial f_n}{\partial \theta} \\ &= (it + 1 - n) f_{n-2}, \end{aligned} \tag{7.12}$$

and in the same way

$$\varrho_\chi\left(\begin{pmatrix} 1 & i \\ i & -1 \end{pmatrix}\right) f_n = (it + 1 + n) f_{n+2}. \tag{7.13}$$

If we first assume that $t \neq 0$, we note that since the factors $(it+1\pm n)$ are non-zero for all $n \in \mathbf{Z}$, the last relations show (after an immediate induction) that $\varphi_{n\pm 2} \in V$ whenever $\varphi_n \in V$. It follows that the set of integers n, such that $\varphi_n \in V$, coincides with all integers with the given parity ε, and therefore we conclude in that case that $V = \mathrm{H}_\chi$.

If $t = 0$ and $\varepsilon = 0$ (which means that χ is the trivial character of B), the same argument applies because all integers n where $\varphi_n \in V$ are even, and hence $n \pm 1 \neq 0$ for all such n, and we find $\mathrm{H}_1 \simeq L^2_0(K)$ again by induction.

The last remaining case is when $t = 0$ and $\varepsilon = 1$, so that χ is the sign character. Then the argument breaks down. Indeed, note that

$$L^2_1(K) = L^2_{1,+} \oplus L^2_{1,-},$$

where

$$L^2_{1,\pm} \simeq \widehat{\bigoplus_{\substack{n \geqslant 1 \\ n \text{ odd}}}} \varphi_{\pm n} \mathbf{C}$$

(these are the spaces of odd functions on K with, respectively, all positive or negative Fourier coefficients zero), and the following lemma shows that this gives the decomposition of $\mathrm{H}_{\mathrm{sgn}}$ in irreducible components:

Lemma 7.4.10. *The subspaces $\mathrm{H}^\pm_{\mathrm{sgn}}$ of $\mathrm{H}_{\mathrm{sgn}}$ corresponding to $L^2_{1,\pm}$ are non-zero irreducible subrepresentations of ϱ_{sgn}.*

These two representations are known as *limits of discrete series* or *mock discrete series*. We will denote them as ϱ^+_{sgn} and ϱ^-_{sgn}.

Sketch of the proof. It is clear that the spaces are non-zero and are proper subspaces in H_χ. It is also clearly visible from the relations (7.11), (7.12), and (7.13) that the subspaces

$$\mathrm{H}^\infty_{\mathrm{sgn}} \cap L^2_{1,\pm}$$

(which are dense in $L^2_{1,\pm}$) are stable under the action of the Lie algebra. (The point is that for $n = 1$, we get

$$\varrho_{\mathrm{sgn}}\left(\begin{pmatrix} 1 & -i \\ -i & -1 \end{pmatrix}\right) f_1 = 0$$

instead of a non-zero multiple of f_{-1}.) This strongly suggests that these are indeed subrepresentations, but this is not a completely formal consequence (see the remark below).

Let $n \geqslant 1$ be odd. We will show that $\varrho_{\mathrm{sgn}}(g) f_n \in \mathrm{H}^+_{\mathrm{sgn}}$ for any g, and from this it follows easily that $\mathrm{H}^+_{\mathrm{sgn}}$ is stable under the action of G.

It is enough to prove that $\varrho_{\mathrm{sgn}}(g) f_n \in \mathrm{H}^+_{\mathrm{sgn}}$ for g in the subgroups U^+ and U^- since these generate G (see Proposition 7.4.1). We have

$$\varrho_{\mathrm{sgn}}\left(\begin{pmatrix} 1 & t \\ 0 & 1 \end{pmatrix}\right) f_n(g) = f_n(g \exp(tE)),$$

and for a fixed $t \in \mathbf{R}$, we see (from the formula) that this is an *analytic function* of the Iwasawa coordinates (x, y, θ) of g. From this, one deduces the existence of a power series expansion

$$\varrho_{\mathrm{sgn}}\left(\begin{pmatrix} 1 & t \\ 0 & 1 \end{pmatrix}\right) f_n = \sum_{k \geqslant 0} \frac{t^k}{k!} (\varrho_{\mathrm{sgn}}(E))^k f_n, \tag{7.14}$$

which converges for t small enough. The previous observation concerning the Lie algebra show that $(\varrho_{\mathrm{sgn}}(E))^k f_n$ is in $L^2_{1,\pm}$ for all $k \geqslant 0$, and hence so is the sum of the series when it converges. Hence $\mathrm{H}^+_{\mathrm{sgn}}$ is stable under the action of ϱ_{sgn} restricted to *small enough* elements of U^+. But it is easy to see that such elements generate U^+, and hence $\mathrm{H}^+_{\mathrm{sgn}}$ is stable under all of U^+. Arguing similarly for U_-, and then for $\mathrm{H}^-_{\mathrm{sgn}}$, we obtain the result. □

Remark 7.4.11. The passage from a subspace invariant under the Lie algebra action to an actual subrepresentation is not a formal matter. For instance, consider the regular representation ϱ of $\mathbf{R}$ on $L^2(\mathbf{R})$. The corresponding Lie algebra is one dimensional, and a generator X acts on smooth functions with compact support simply by

$$\varrho(X) f(t) = f'(t).$$

Let $V^\infty \subset L^2(\mathbf{R})$ be the subspace of smooth functions with support in the fixed interval $[0, 1]$ (for example). This is of course stable under the derivation action $\varrho(X)$, but the closure of V^∞ in $L^2(\mathbf{R})$ is simply $L^2([0, 1])$, which is *not* a subrepresentation of the regular representation. The difference between this space and the lemma is that the functions in V^∞ are not analytic.

At this point we have proved parts (1) and (2) of Proposition 7.4.3. Concerning (3), we will just prove that two principal series representations are not isomorphic if the characters are not equal or inverse, and leave as an exercise the fact that $\mathrm{H}_\chi \simeq \mathrm{H}_{\chi^{-1}}$. We will use the Lie algebra action (which is simpler) to isolate a vector in H_χ with specific properties that (almost) pin down the parameters (ε, t) of χ, and which must be preserved under an isomorphism.

Precisely, let χ_1 and χ_2 be characters of B, and denote $\mathrm{H}_{\chi_i} = \mathrm{H}_i$ for $i = 1, 2$. Assuming that $\Phi : \mathrm{H}_1 \longrightarrow \mathrm{H}_2$ is an isomorphism of representations, first observe that the linear map Φ also commutes with the Lie algebra action. We have

$$\Phi(\varrho_{\chi_1}(X)f) = \varrho_{\chi_2}(X)\Phi(f) \tag{7.15}$$

for $X \in \mathfrak{g}$ and $f \in \mathrm{H}_1^\infty$ (this is an elementary consequence of the definition). Now assume first that the parity ε is 0, and let $v_1 \in \mathrm{H}_1$ (resp. $v_2 \in \mathrm{H}_2$) be the function denoted f_0 above. (Since we are using two characters, it is better here to be careful with notation: f_0 restricted to K is independent of the character, but as a function on G, the two vectors are different.) Since v_1 and v_2 are characterized up to a non-zero scalar by the fact that they span the respective spaces H_1^K and H_2^K of K-invariant vectors, the vector $\tilde{v}_2 = \Phi(v_1)$ is a (non-zero) multiple of v_2. Now denote

$$E^- = \begin{pmatrix} 1 & -i \\ -i & -1 \end{pmatrix}, \qquad E^+ = \begin{pmatrix} 1 & i \\ i & -1 \end{pmatrix} \in \mathfrak{g} \otimes \mathbf{C}.$$

By (7.12) and (7.13), the Lie algebra action satisfies

$$\begin{aligned} \varrho_{\chi_1}(E^-)\varrho_{\chi_1}(E^+)v_1 &= (it_1 + 1)\varrho_{\chi_1}(E^-)f_2 \\ &= (it_1 + 1)(it_1 + 1 - 2)v_1 \\ &= (1 - t_1^2)v_1, \end{aligned} \tag{7.16}$$

and similarly

$$\varrho_{\chi_2}(E^-)\varrho_{\chi_2}(E^+)v_2 = (1 - t_2^2)v_2$$

(with obvious notation for f_2, t_1 and t_2). Applying Φ and using (7.15) and the fact that $\tilde{v}_2$ is a non-zero multiple of v_2, we deduce that

$$1 - t_1^2 = 1 - t_2^2,$$

which means that an isomorphism can only exist, in the even case, when $t_1 = t_2$ (i.e., $\chi_2 = \chi_1$) or $t_1 = -t_2$ (i.e., $\chi_1 = \chi_2^{-1}$).

The odd case is similar (using the function f_1, for instance), and since there is no K-invariant vector in H_χ, if χ has odd parity, there can be no isomorphism between principal series associated to characters with different parity. This finishes the proof of the "only if" part of (3) in Proposition 7.4.3. The next exercise sketches the remaining part.

Exercise 7.4.12. (1) Construct a linear map

$$\Phi : \mathrm{H}_\chi \longrightarrow \mathrm{H}_{\chi^{-1}},$$

which is a linear isomorphism commuting with the Lie algebra action (i.e., such that (7.15) holds).

(2) Prove that some non-zero multiple of Φ is an isometric intertwiner of ϱ_χ and $\varrho_{\chi^{-1}}$. (*Hint*: Use analyticity as in the proof of Lemma 7.4.10.)

We finally explain the proof of the last part of Proposition 7.4.3: no principal series is isomorphic to a subrepresentation of the regular representation. Here the motivation comes from the basic results we proved concerning isotypic components of the regular representation, both in the algebraic case (Corollary 2.7.30) and in the case of compact groups (see Section 5.4): the isotypic component is described using matrix coefficients, and the reason the principal series do not embed in the regular representation is simply that their matrix coefficients, although they are well defined, are not in $L^2(G)$.

Lemma 7.4.13 (Matrix coefficients and subrepresentations of the regular representation)**.** *Let G be a unimodular locally compact group, and let $\varrho : G \longrightarrow \mathrm{U}(\mathrm{H})$ be an irreducible unitary representation of G. Then ϱ is a subrepresentation of the regular representation ϱ_G if and only if all matrix coefficients*

$$f_{v,w} : g \mapsto \langle \varrho(g)v, w\rangle$$

are in $L^2(G)$. Furthermore, this property holds as soon as some non-zero matrix coefficient is in $L^2(G)$.

Sketch of the proof. If all matrix coefficients are in $L^2(G)$, then for any fixed vector $w \in \mathrm{H}$, the map

$$\Phi_w : v \mapsto f_{v,w}$$

is formally a linear map $\mathrm{H} \longrightarrow L^2(G)$, which intertwines ϱ and the regular representation (as in Corollary 2.7.30). If $w \neq 0$, this map is injective (because $\mathrm{Ker}(\Phi_w)$ is seen to be the orthogonal of the span of the vectors $\varrho(g)w$, which is dense in H by irreducibility of ϱ). The problem (as in Chapter 5) is that its image might not be closed in $L^2(G)$. But it can be proved that this does not happen, using the Closed Graph Theorem (see, e.g., [**49**, Th. III.12]) of functional analysis (to first show that Φ_w is continuous) and a variant of Schur's Lemma to prove that Φ_w is necessarily proportional to an isometry $\mathrm{H} \longrightarrow L^2(G)$. Since an isometry obviously has a closed image, the map Φ_w gives then an embedding of ϱ inside ϱ_G.

To prove that it is enough to have a single non-zero matrix coefficient in $L^2(G)$ in order to conclude that all of them are square-integrable requires another argument, which is where the assumption that G be unimodular is used. We refer, e.g., to [**14**, Prop. 12.2.3, Th. 12.2.5] for details. □

We will now show concretely that the principal series representations ϱ_χ corresponding to $\varepsilon = 0$ have some matrix coefficient which is not in $L^2(G)$. We use the following integration lemma:

Lemma 7.4.14. *Let A be the diagonal subgroup*

$$A = \left\{a(\alpha) = \begin{pmatrix} \alpha & 0 \\ 0 & \alpha^{-1} \end{pmatrix} \mid \alpha > 0\right\} \subset \mathrm{SL}_2(\mathbf{R}).$$

Let $\varphi \in C(G)$ be a non-negative function which is K-invariant on both sides, i.e., such that

$$\varphi(k_1 g k_2) = \varphi(g)$$

for all k_1, $k_2 \in K$. Then φ is in $L^1(G)$ if and only if

$$\int_0^{+\infty} \varphi(a(e^r)) \sinh(2r) dr < +\infty.$$

Sketch of the proof. The point is that the group G satisfies the Cartan (or polar) decomposition

$$G = KA^+K, \qquad A^+ = \{a(\alpha) \mid \alpha \geqslant 1\},$$

which shows that a function which is K-invariant on both sides "lives" on the double coset space $K\backslash G/K$, which can be identified with (a quotient of) $A^+ \simeq [1, +\infty[$. Thus A^+ contains "all the non-compactness" that might prevent a continuous function from being integrable. To obtain the statement, one computes the Jacobian of the change of variable

$$\begin{cases} K \times A^+ \times K & \longrightarrow & G \\ (k_1, a(\alpha), k_2) & \mapsto & k_1 a(\alpha) k_2 \end{cases}$$

to deduce that the integral of a K-bi-invariant function on G is proportional to

$$\int_0^{+\infty} \varphi(a(e^r)) \sinh(2r) dr$$

(see, e.g., [**41**, VII, §2] for details, or [**36**, Prop. 5.28] for a more general version). □

Now let χ be a character of B of the form

$$\chi\Big(\begin{pmatrix} \alpha & \beta \\ 0 & \alpha^{-1} \end{pmatrix}\Big) = |\alpha|^{it}$$

with $t \in \mathbf{R}$. We consider the function f_0 in H_χ corresponding to the characteristic function of K in $L^2_0(K)$. This is a vector of norm 1. Let

$$\varphi(g) = \langle \varrho_\chi(g) f_0, f_0 \rangle. \tag{7.17}$$

We will show that $\varphi \notin L^2(G)$ using the last lemma. Indeed, first of all, the K-invariance on both sides holds: since $\varrho_\chi(k) f_0 = f_0$ and ϱ_χ is unitary, we find

$$\varphi(k_1 g) = \langle \varrho_\chi(k_1 g) f_0, f_0 \rangle = \langle \varrho_\chi(g) f_0, \varrho_\chi(k_1^{-1}) f_0 \rangle = \varphi(g)$$

as well as

$$\varphi(g k_2) = \langle \varrho_\chi(g k_2) f_0, f_0 \rangle = \langle \varrho_\chi(g) f_0, f_0 \rangle = \varphi(g).$$

Now we compute the function $\varphi(a(\alpha))$ for $\alpha \geqslant 1$, with notation as in the lemma. We have

$$\varphi(g) = \langle \varrho_\chi(g) f_0, f_0 \rangle = \int_K f_0(kg) \overline{f_0(k)} dk = \int_K f_0(kg) dk.$$

On the other hand, we know that

$$f_0(g) = y^{(it+1)/2}$$

in terms of Iwasawa coordinates (x, y, θ). We compute the coordinate $y(\alpha, \theta)$ of

$$g = a(\alpha) k(\theta) = \begin{pmatrix} \star & \star \\ -\alpha^{-1} \sin\theta & \alpha \cos\theta \end{pmatrix}$$

by the recipe (7.9), obtaining

$$y(\alpha, \theta) = (\alpha^2 \cos\theta + \alpha^{-2} \sin^2\theta)^{-1}.$$

We therefore have

$$\varphi(a(\alpha)) = |\alpha|^{-1-it} \times \frac{1}{2\pi} \int_0^{2\pi} \frac{d\theta}{(\cos^2\theta + \alpha^{-4} \sin^2\theta)^{(it+1)/2}}$$

for $\alpha > 1$, and

$$\varphi(a(\alpha)) = |\alpha|^{1+it} \times \frac{1}{2\pi} \int_0^{2\pi} \frac{d\theta}{(\sin^2\theta + \alpha^4 \cos^2\theta)^{(it+1)/2}}$$

for $0 < \alpha < 1$.

For $\alpha > 1$, we have

$$|(1 + \alpha^{-4} \cos^2\theta)^{(it+1)/2}| \leqslant \sqrt{2}$$

for all θ, and hence

$$|\varphi(a(\alpha))| \geqslant \frac{1}{\sqrt{2}\alpha}.$$

The integral in the lemma that determines whether φ is in $L^2(G)$ therefore satisfies

$$\int_0^{+\infty} |\varphi(a(e^r))|^2 \sinh(2r)dr \geqslant \frac{1}{2}\int_1^{+\infty} \alpha^{-1}d\alpha = +\infty,$$

proving that the matrix coefficient φ is not in $L^2(G)$.

This concludes the sketch of the proof of Proposition 7.4.3.

Exercise 7.4.15 (Hypergeometric functions and matrix coefficients)**.** Let $\tilde{A}$ be the subgroup

$$\tilde{A} = \left\{\tilde{a}(\tau) = \begin{pmatrix} \cosh(\tau) & \sinh(\tau) \\ \sinh(\tau) & \cosh(\tau) \end{pmatrix} \mid \tau \in \mathbf{R}\right\} \subset \mathrm{SL}_2(\mathbf{R}).$$

(1) Show that a function $\varphi \in C(G)$ which is K-bi-invariant, i.e., such that $\varphi(k_1 g k_2) = \varphi(g)$ for all k_1, k_2 in K, is in $L^2(G)$ if and only if

$$\int_0^{+\infty} |\varphi(\tilde{a}(\tau))|^2 e^{2\tau} d\tau < +\infty.$$

(2) Let χ be a character of B as above (i.e., with parity $\varepsilon = 0$), and let φ be the matrix coefficient (7.17) considered previously. Show that

$$\varphi(\tilde{a}(\tau)) = (\cosh\tau)^{-1-it}\frac{1}{\pi}\int_0^{\pi} \frac{d\theta}{(1 + \tanh^2\tau - 2\tanh\tau\cos\theta)^{(it+1)/2}}.$$

(3) For $a \in \mathbf{C}$ and an integer $k \geqslant 0$, let

$$(a)_k = a(a+1)\cdots(a+k-1).$$

For a, b, $c \in \mathbf{C}$, c not being a non-positive integer, let

$${}_2F_1(a, b; c; z) = \sum_{k \geqslant 0} \frac{(a)_k (b)_k}{(c)_k k!} z^k,$$

seen at first as a formal power series. Show that this power series converges absolutely for $|z| < 1$.

(4) Show that ${}_2F_1(a, b; c; z)$ is a solution of the differential equation

$$z(1-z)y'' + (c - (a+b+1)z)y' - aby = 0$$

(either as a formal power series or as holomorphic function when convergent). This function is known as a *Gauss hypergeometric function*; see for instance [**66**, XIV] for more information.

(5) Show that for $x \in \mathbf{C}$ with $|x| < 1$ and for $\nu \in \mathbf{C}$, we have

$$\frac{1}{\pi}\int_0^{\pi} \frac{d\theta}{(1+x^2-2x\cos\theta)^{\nu}} = {}_2F_1(\nu,\nu;1;x^2).$$

(*Hint*: Factor the denominator in the form $(1-xe^{i\theta})^{\nu}(1-xe^{-i\theta})^{\nu}$, and apply the binomial theorem before integrating termwise.)

(6) Express the matrix coefficient $\varphi(\tilde{a}(\tau))$ in terms of some hypergeometric function. This gives another example of the frequent relations between special functions and representation theory, as we already discussed briefly in Example 5.6.1. The fact that suitable matrix coefficients satisfy differential equations as the hypergeometric function does, is a feature that extends to other Lie groups and which is crucial to the analysis of their unitary representations (see, e.g., [**36**, VIII.7]).

(7) Possibly by looking at references on the behavior of a hypergeometric function as $z \to 1$ (e.g., [**25, 63, 66**]), prove that the matrix coefficient φ is in $L^{2+\varepsilon}(G)$ for all $\varepsilon > 0$.

Proposition 7.4.3 constructs many irreducible unitary representations of $\mathrm{SL}_2(\mathbf{R})$. Are these all of them (up to isomorphism)? The example of $\mathrm{SL}_2(\mathbf{F}_p)$ may suggest that this is unlikely, and indeed there exist two other types of representations, of a very different nature. We will only give their definitions and sketch their construction, but both are very important when considering generalizations.

We begin with the so-called *discrete series* representations. We recall the definition

$$\mathbf{H} = \{z \in \mathbf{C} \mid \mathrm{Im}(z) > 0\}$$

of the upper half-plane, and the fact that $G = \mathrm{SL}_2(\mathbf{R})$ acts on $\mathbf{H}$ by

$$\begin{pmatrix} a & b \\ c & d \end{pmatrix} \cdot z = \frac{az+b}{cz+d}.$$

For $g \in \mathrm{SL}_2(\mathbf{R})$ as above, we will write

$$j(g,z) = cz + d.$$

This function satisfies the relation

$$j(g_1g_2, z) = j(g_1, g_2 \cdot z)j(g_2, z), \tag{7.18}$$

as the reader should check, and in addition we have $j(g,z) \neq 0$ whenever $g \in G$ and $z \in \mathbf{H}$.

Proposition 7.4.16 (Discrete series). *Let $n \geqslant 2$ be an integer, and let*

$$\mathrm{H}_n = \Big\{ f : \mathbf{H} \longrightarrow \mathbf{C} \mid f \text{ is holomorphic and } \|f\|_n^2 = \int_{\mathbf{H}} |f(z)|^2 y^n \frac{dxdy}{y^2} < +\infty \Big\}.$$

(1) *The vector space H_n is an infinite-dimensional Hilbert space, and the representation π_n defined by*

$$\pi_n(g)f(z) = j(g^{-1}, z)^{-n} f(g^{-1} \cdot z)$$

is a unitary representation of $\mathrm{SL}_2(\mathbf{R})$ on H_n.

(2) *The representation π_n is irreducible for $n \geqslant 2$.*

(3) *The representation π_n is isomorphic to a subrepresentation of the regular representation.*

Sketch of proof. (1) The fact that π_n is formally a representation depends on the property (7.18) of the factor $j(g^{-1}, z)$. It is not obvious that H_n is a Hilbert space, but one proves using the Cauchy integral formula that if a sequence (f_n) converges in $L^2(\mathbf{H}, y^{-2}dxdy)$ to some function f, then the convergence is in fact locally uniform. From this, standard arguments of complex analysis show that f is also holomorphic, and one deduces that H_n is complete for the given norm (see, e.g., [**41**, IX, §2, Lemma 1] for details). The unitarity is a consequence of the invariance of the measure $y^{-2}dxdy$ on $\mathbf{H}$ (for the action of $\mathrm{SL}_2(\mathbf{R})$) and of the formula (7.18).

(2) Although it would be more robust to analyze the Lie algebra representation for suitable vectors in H_n (to compare the structure of π_n with the principal series, see the exercise below), we use an elegant trick from [**36**, Prop. 2.7]. We begin by noting that the function

$$f_n : z \mapsto \frac{1}{(z+i)^n}$$

is in H_n (it is not necessarily obvious that $\|f_n\|_n < +\infty$, but the computation that shows this is done below, while proving (3)). Now let $V \subset \mathrm{H}_n$ be a *non-zero* closed invariant subspace. We will show that $f_n \in V$; since V is arbitrary, this implies that $V^\perp = 0$ (otherwise we would get $f_n \in V^\perp$ also…) and hence $V = \mathrm{H}_n$.

The underlying motivation is that f_n transforms according to a character of K, namely

$$\pi_n(k(\theta))f_n = e^{-in\theta} f_n \tag{7.19}$$

for $\theta \in \mathbf{R}$. We will compute the projection operator onto the corresponding isotypic component, using the holomorphic structure, and see that the image of any closed invariant subspace is one dimensional, spanned by f_n. Thus let $P : \mathrm{H}_n \longrightarrow \mathrm{H}_n$ be the linear map defined by

$$P(f) = \frac{1}{2\pi}\int_0^{2\pi} (\pi_n(k(\theta)f)e^{in\theta}d\theta$$

for $f \in \mathrm{H}_n$, which (by Peter–Weyl theory, see Theorem 5.5.1, (2)) is the projection on this isotypic component. Since V is a subrepresentation of H_n, this projection also maps V to V, so that $P(f) \in V$ for any $f \in V$. We claim that

$$P(f) = -(2i)^n f(i) f_n.$$

If this is true, then taking any $f \in V$ such that $f(i) \neq 0$, we deduce that $f_n \in V$, and conclude that H is irreducible. (Such a function f exists because there exists $f \in V$ and $z \in \mathbf{H}$ with $f(z) \neq 0$, and picking $g \in G$ with $g \cdot z = i$, we get $\pi_n(g)f(i) \neq 0$.)

To prove the formula, we fix any $z \in \mathbf{H}$ and express $P(f)(z)$ as a complex line integral. We have

$$\begin{aligned} P(f)(z) &= -\frac{1}{2\pi}\int_0^{2\pi} (\pi_n(k(\theta))f)(z)e^{in\theta}d\theta \\ &= \frac{1}{2\pi}\int_0^{2\pi} (z\sin\theta + \cos\theta)^{-n} f\Big(\frac{z\cos\theta - \sin\theta}{z\sin\theta + \cos\theta}\Big)e^{in\theta}d\theta \\ &= -\frac{1}{2i\pi}\int_{\mathcal{C}} \Big(\frac{iz(w - w^{-1}) + (w + w^{-1})}{2}\Big)^{-n} \\ &\qquad \times f\Big(\frac{z(w + w^{-1}) - i(w - w^{-1})}{iz(w - w^{-1}) + (w + w^{-1})}\Big)\frac{dw}{w^{n+1}} \end{aligned}$$

after putting $w = e^{-i\theta}$, where $\mathcal{C} \subset \mathbf{C}$ is the unit circle (oriented in the usual counterclockwise direction). One now checks that the integrand is a holomorphic function in a neighborhood of the unit disc $|w| = 1$, except for a simple pole at $w = 0$. Indeed the factor

$$\Big(\frac{iz(w - w^{-1}) + (w + w^{-1})}{2}\Big)^{-n} w^{-n-1} = \frac{2^n}{w}\frac{1}{(w^2(1 + iz) + 1 - iz)^n}$$

has poles at $w = 0$ and at w with $w^2 = (1-iz)/(1+iz)$; since it is well known (and easily checked, indeed we will do this below) that $|(z + i)/(z - i)| > 1$ for all $z \in \mathbf{H}$, this rational function has indeed no pole except 0 in a neighborhood of $|w| \leqslant 1$. Furthermore, the argument of f is

$$\frac{z(w + w^{-1}) - i(w - w^{-1})}{iz(w - w^{-1}) + (w + w^{-1})} = \frac{w^2(z - i) + z + i}{w^2(iz + 1) + 1 - iz},$$

and we leave to the reader to check that the imaginary part of this argument is, for all w with $|w| \leqslant 1$, strictly positive (it is best to use continuity arguments...).

We can therefore compute the integral above using Cauchy's residue formula, obtaining

$$\begin{aligned}P(f)(z) &= -\mathrm{Res}_{w=0}\frac{2^n}{w}\frac{1}{(w^2(1+iz)+1-iz)^n}f\Big(\frac{w^2(z-i)+z+i}{w^2(1+iz)+1-iz}\Big)\\ &= -\frac{2^n}{(1-iz)^n}f(i) = -(2i)^n f(i) f_n(z),\end{aligned}$$

as claimed.

(3) By Lemma 7.4.13, it suffices to prove that some matrix coefficient of π_n is in $L^2(G)$. We use

$$\varphi(x) = \langle \pi_n(x) f_n, f_n\rangle.$$

(Note that $\varphi(1) = \|f_n\|_n^2$, so by computing φ, and showing it is finite, we will in passing confirm that $f_n \in \mathrm{H}_n$.)

We observe that $|\varphi|$ is K-invariant on both sides (because of (7.19), although φ itself is not), so we can use Lemma 7.4.14 to determine whether $\varphi \in L^2(G)$. For $\alpha > 0$, we have

$$\pi_n(a(\alpha)) f_n(z) = j(a(\alpha^{-1}), z)^{-n} f_n(a(\alpha^{-1})\cdot z) = (\alpha^{-1}z + \alpha i)^{-n},$$

and therefore

$$\varphi(a(\alpha)) = \int_{\mathbf{H}} (\alpha^{-1}z+\alpha i)^{-n}(\bar{z}-i)^{-n} y^n \frac{dxdy}{y^2}.$$

This integral is rather more transparent if one performs the change of variable $w = (z-i)/(z+i)$, which maps the upper half-plane $\mathbf{H}$ to the unit disc $D = \{w \in \mathbf{C} \mid |w| < 1\}$. The measure corresponding to $y^{-2}dxdy$ on D is $(1-|w|^2)^{-2}dw$, and since the inverse substitution is $z = -i(w+1)/(w-1)$, simple computations give

$$y = \frac{1}{2i}(z-\bar{z}) = \frac{1-|w|^2}{|1-w|^2}$$

(which certainly makes it clear that $|w| < 1$ if and only if $y > 0$) and

$$\begin{aligned}(\bar{z}-i)(\alpha^{-1}z+\alpha i) &= \frac{2i}{\bar{w}-1}\times\frac{i}{w-1}\times(w(\alpha-\alpha^{-1})-(\alpha+\alpha^{-1}))\\ &= -\frac{2}{|1-w|^2}((\alpha-\alpha^{-1})w-(\alpha+\alpha^{-1})).\end{aligned}$$

It follows that

$$\varphi(a(\alpha)) = 2^{-n}\int_D \frac{(1-|w|^2)^{n-2}}{((\alpha+\alpha^{-1})-(\alpha-\alpha^{-1})w)^n}dw.$$

In particular, we get

$$\varphi(a(e^t)) = \int_D \frac{(1-|w|^2)^{n-2}}{(\cosh(t)-w\sinh(t))^n}dw = \frac{1}{\cosh(t)^n}\int_D\frac{(1-|w|^2)^{n-2}}{(1-\tanh(t)w)^n}dw.$$

The last integral can be computed in polar coordinates by integrating over the argument first, namely

$$\int_D \frac{(1-|w|^2)^{n-2}}{(1-\tanh(t)w)^n}dw = \int_0^1 (1-r^2)^n \Big(\int_0^{2\pi} \frac{1}{(1-r\tanh(t)e^{i\theta})^n} d\theta\Big) r dr$$
$$= 2\pi \int_0^1 (1-r^2)^n r dr = \frac{\pi}{n}.$$

(For the integration over θ, use a geometric series expansion and integrate termwise. The point is that only terms with $e^{ik\theta}$, where $k \geqslant 0$, appear and just once $k = 0$.) Hence we have

$$|\varphi(a(e^t))|^2 = \frac{\pi^2}{n^2}\cosh(t)^{-2n},$$

and multiplying by $\sinh(2t) \sim e^{2t}$, we see that

$$\int_0^{+\infty} |\varphi(a(e^t))|^2 \sinh(2t) dt < +\infty$$

for $n \geqslant 2$. □

Exercise 7.4.17. Let $n \geqslant 2$ be an integer, and let π_n be the discrete series representation constructed in Proposition 7.4.16.

Show that the restriction of π_n to K is given by the Hilbert direct sum

$$\widehat{\bigoplus_{j\geqslant 0}} \mathbf{C} f_{n,j},$$

where

$$f_{n,j}(z) = f_n(z)\Big(\frac{z-i}{z+i}\Big)^j$$

satisfies

$$\pi_n(k(\theta))f_{n,j} = e^{-i(n+2j)\theta} f_{n,j}.$$

In particular, this shows that π_n is not isomorphic to π_m for $m \neq n$, since n can be recovered from the abstract representation $\pi = \pi_n$ as the *smallest* integer $h \in \mathbf{Z}$ such that the K-isotypic component of the character $k(\theta) \mapsto e^{-ih\theta}$ is non-zero in π. One says that π_n is the representation of $\mathrm{SL}_2(\mathbf{R})$ with *lowest* K*-type* given by this character.

Exercise 7.4.18 (The other discrete series)**.** For $n \geqslant 2$, let H_n be the same Hilbert space as in Proposition 7.4.16, but define

$$\tilde{\pi}_n(g)f(z) = j(\eta g^{-1}\eta^{-1}, z)^{-n} f((\eta g^{-1}\eta^{-1})\cdot z),$$

where

$$\eta = \begin{pmatrix} 0 & 1 \\ 1 & 0 \end{pmatrix}.$$

Show that $\tilde{\pi}_n$ is an irreducible unitary representation of G, that it is isomorphic to a subrepresentation of the regular representation, and that the restriction to K is the direct sum of the characters $k(\theta) \mapsto e^{i(n+2j)\theta}$, where $j \geqslant 0$. (In particular, no π_n is isomorphic to a $\tilde{\pi}_m$.)

Remark 7.4.19. In contrast with the principal series of unitary representations, which are constructed by means of a very general construction (induced representation from a one-dimensional representation of a maximal solvable subgroup) that can be applied to all semisimple Lie groups (e.g., to $\mathrm{SL}_n(\mathbf{R})$ for all $n \geqslant 2$, or symplectic groups), the existence of discrete series of $\mathrm{SL}_2(\mathbf{R})$ depends on rather specific properties of the group.

Indeed, if we define in general a discrete series of a locally compact group G to be one which is isomorphic to a subrepresentation of its regular representation, then one can show, for instance, that the group $\mathrm{SL}_2(\mathbf{C})$ has no discrete series, and also that $\mathrm{SL}_n(\mathbf{R})$, for $n \geqslant 3$, has no discrete series. There is a beautiful group-theoretic criterion for the existence of discrete series which explains this behavior and which was discovered by Harish-Chandra: a semisimple Lie group G admits discrete series if and only if it contains a connected *compact* abelian subgroup of *the same dimension* as a maximal connected abelian subgroup (see, for instance, [**36**, Th. 12.20]). Since a maximal connected abelian subgroup is called a *Cartan subgroup*, the criterion is that G has discrete series if and only if it has a compact Cartan subgroup.

For $G = \mathrm{SL}_2(\mathbf{R})$, the diagonal subgroup A is a maximal abelian subgroup (it is not connected, but the subgroup with positive diagonal coefficients is) of dimension 1, and the compact subgroup K also has dimension 1. But in $\mathrm{SL}_2(\mathbf{C})$, the diagonal subgroup is isomorphic to $\mathbf{C}^\times$ and has dimension 2 (in terms of real coordinates), while any compact subgroup is contained in a conjugate of $\mathrm{SU}_2(\mathbf{C})$, and it has maximal compact abelian subgroups isomorphic to $\mathrm{SO}_2(\mathbf{R})$ again, hence of smaller dimension.

There remains one series of irreducible unitary representations of G to discuss. These are the most elusive and are called the *complementary series*. We will obtain them by considering induced representations from some one-dimensional *non-unitary* representations of B. The remarkable fact is that, for some of these characters, one may define a (non-obvious!) G-invariant inner product, at least on a dense subspace of the representation space, for which the representation becomes unitary. The complementary series are then defined as the completions of these subspaces.

For simplicity, we consider only a character χ defined by

$$\chi\Big(\begin{pmatrix} \alpha & \beta \\ 0 & \alpha^{-1} \end{pmatrix}\Big) = |\alpha|^s$$

for some fixed $s \in \mathbf{C}$ (i.e., with parity $\varepsilon = 0$), since Exercise 7.4.22 will show that the other characters do not lead to any new unitary representations.

We consider the Hilbert space

$$\mathrm{H}_s = \{f : G \longrightarrow \mathbf{C} \mid f(bg) = \chi(b)\delta(b)f(g) \text{ for all } b \in B,\ g \in G, \text{ and } \|f\|^2 = \int_K |f|^2 dk < +\infty\},$$

and the representation

$$\varrho_s(g)f(x) = f(xg).$$

One then shows (as in Proposition 7.4.3(1)) that ϱ_s is a strongly continuous representation of G on the Hilbert space H_s, and that H_s is (by restriction of functions to K) isometric to the space $L^2_0(K)$ of even square-integrable functions on K. However, one can also see that if $\mathrm{Re}(s) \neq 0$, this representation is not unitary (thus ϱ_s is viewed as a continuous representation of G on the Banach space underlying H_s, as in Definition 3.3.1).

On the other hand, arguing by restriction to K and by differentiation exactly as in (3) of Proposition 7.4.3 (note that the restriction of ϱ_s to K is isomorphic to the regular representation on $L^2_0(K)$, independently of s, and in particular is unitary), we see using the analogues

$$\begin{aligned} \varrho_s\left(\begin{pmatrix} 0 & 1 \\ -1 & 0 \end{pmatrix}\right) f_n &= in f_n, \\ \varrho_s\left(\begin{pmatrix} 1 & -i \\ -i & -1 \end{pmatrix}\right) f_n &= (s+1-n) f_{n-2}, \\ \varrho_s\left(\begin{pmatrix} 1 & i \\ i & -1 \end{pmatrix}\right) f_n &= (s+1+n) f_{n+2} \end{aligned} \tag{7.20}$$

of (7.11), (7.12), and (7.13) that if s is not an odd integer, then ϱ_s is irreducible. (Here n is an even integer and f_n, as before, denotes the function in H_s corresponding to $k(\theta) \mapsto e^{in\theta}$.)

The question we consider is, For which values of s (not integers) is the representation ϱ_s unitarizable? We already know this is the case for $\mathrm{Re}(s) = 0$, but there might conceivably exist other values of s for which this is true. Our strategy will be to find a necessary condition for this property using Lie algebra, and then to check when this condition is sufficient.

Lemma 7.4.20. *Let $s \in \mathbf{C}$ be a complex number, and assume it is not an odd integer. If there exists a positive-definite inner product $(\cdot, \cdot)$ on* H *which is invariant under the action of ϱ_s, then we have $s^2 - 1 < 0$. In other words,* a necessary *condition for the unitarizability, according to Definition* 3.4.1(3) *of the representation ϱ_s is that either $s = it$ with $t \in \mathbf{R}$ or s is a real number with $-1 < s < 1$. In the first case, the representation is a principal series.*

Proof. We assume that the inner product $(\cdot,\cdot)$ exists on H. Because the restriction of ϱ_s to K is the regular representation on $L^2_0(K)$, the functions f_n with n even, which span distinct subrepresentations of $L^2_0(K)$, must be orthogonal with respect to the inner product. Since they span a dense subset of H_s, it follows that the inner product is entirely determined by the norms

$$a_n = \sqrt{(f_n, f_n)} > 0,$$

and we can normalize this inner product (by multiplying by a fixed constant) to ensure that $a_0 = 1$.

We will find relations between the a_n by exploiting the G-invariance of the inner product (this should be compared with the second argument used in Example 5.2.12 to determine the invariant inner product on the irreducible representations of $\mathrm{SU}_2(\mathbf{C})$). As in the proof of Proposition 7.4.3, it is much easier to work with the Lie algebra $\mathfrak{g}$. For the action of the latter, a formal computation shows that the G-invariance of the inner product implies that the operators $\varrho_s(X)$ for $X \in \mathfrak{g} \otimes \mathbf{C}$ are skew-hermitian when acting on H_s^∞, so that

$$(\varrho_s(X)v, w) = (v, -\varrho_s(\bar{X})w)$$

for v, $w \in \mathrm{H}_s^\infty$. (If $X \in \mathfrak{g}$, so that X has real coefficients, $\varrho_s(X)$ is skew-symmetric, but the sesquilinearity of the inner product leads to the extension to $\mathfrak{g} \otimes \mathbf{C}$ being skew-hermitian.)

We apply this to $v = f_n$, $w = f_{n+2}$ and

$$X = \begin{pmatrix} 1 & i \\ i & -1 \end{pmatrix}, \qquad \bar{X} = \begin{pmatrix} 1 & -i \\ -i & -1 \end{pmatrix}$$

so that, using the relations (7.20), we get

$$\begin{aligned} (s+1+n)a_{n+2} &= (\varrho_s(X)f_n, f_{n+2}) \\ &= (f_n, -\varrho_s(\bar{X})f_{n+2}) = -(\bar{s}+1-(n+2))a_n. \end{aligned} \tag{7.21}$$

For $n = 0$, this leads to

$$a_2 = \frac{\bar{s}-1}{s+1} = -\frac{|s-1|^2}{1-s^2},$$

and since $a_2 > 0$, we find the necessary condition $1 - s^2 < 0$. □

As we have already hinted, it turns out that this necessary condition for unitarity is sufficient, or more precisely, it is so for $0 < s < 1$.

Proposition 7.4.21 (Existence of complementary series)**.** *Let s be a non-zero real number such that $0 < s < 1$. There exists a G-invariant inner product $(\cdot,\cdot)$ on H_s so that ϱ_s extends by continuity to an irreducible unitary representation of the completion of H_s with respect to this inner product, which is called the complementary series with parameter s.*

Sketch of proof. We use the notation of the proof of the lemma. The point is that, since s is real, we have the induction relation

$$a_{n+2} = -\frac{s-1-n}{s+1+n}a_n = \frac{n+1-s}{n+1+s}a_n \tag{7.22}$$

(see (7.21)) which allows us to determine uniquely the constants a_n for all $n \in \mathbf{Z}$ even, once $a_0 = 1$ is fixed. For $-1 < s < 1$, the product of the numerator and denominator is

$$(n+1)^2 - s^2 \geqslant 0,$$

so that they are non-zero real numbers of the same sign. It follows that $(a_n)_{n \text{ even}}$ is a sequence of positive real numbers.

We can thus define a positive-definite inner product $(\cdot,\cdot)$ on $L^2_0(K)$ by

$$\Big(\sum_j \lambda_j f_{2j}, \sum_j \mu_j f_{2j}\Big) = \sum_j \lambda_j \overline{\mu_j} a_{2j}, \tag{7.23}$$

which is well defined on all of $L^2_0(K)$ because the a_{2j} are bounded. (One can see by induction that $0 \leqslant a_{2j} \leqslant 1$ for all j; this is where it is important that $0 < s < 1$.) We denote by $\bar{\mathrm{H}}_s$ the completion of H_s with respect to this inner product. (One can check from the recurrence relation that $a_{2j} \to 0$ as $j \to +\infty$, which implies that $\bar{\mathrm{H}}_s$ is strictly larger than H_s, although H_s is of course dense.)

The next step is to check (using the formulas (7.20)) that $\varrho_s(X)$ is skew-hermitian with respect to this inner product, for all $X \in \mathfrak{g}$, when applied to finite linear combinations of the functions f_{2j}, $j \in \mathbf{Z}$. Note that this is not automatic, because we only used these relations partially in order to obtain a necessary condition for the existence of the inner product. We leave this however as an exercise.

We then need to prove that the representation ϱ_s is, in fact, unitary with respect to this new inner product. For this purpose, we use analyticity as in the proof of Lemma 7.4.10. Indeed, for small enough X in the Lie algebra and for v a finite linear combination of basis vectors f_n, we have

$$\varrho_s(\exp(X))v = \sum_{k \geqslant 0} \frac{1}{k!}(\varrho_s(X))^k v;$$

hence for another such vector w, we get

$$\begin{aligned}(\varrho_s(\exp(X))v, w) &= \sum_{k\geqslant 0} \frac{1}{k!}((\varrho_s(X))^k v, w)\\ &= \sum_{k\geqslant 0} \frac{(-1)^k}{k!}(v, \varrho_s(X)^k w)\\ &= (v, \varrho_s(\exp(-X))w).\end{aligned}$$

Since $\exp(-X) = \exp(X)$, this shows (by continuity and density) that $\varrho_s(\exp(X))$ is unitary for X small enough in $\mathfrak{g}$. But we know that such elements of g generate $\mathrm{SL}_2(\mathbf{R})$, and therefore ϱ_s is unitary with respect to $(\cdot,\cdot)$. For instance, we have, as used earlier,

$$\begin{pmatrix} 1 & t \\ 0 & 1 \end{pmatrix} = \exp\Bigl(\begin{pmatrix} 0 & t \\ 0 & 0 \end{pmatrix}\Bigr), \quad \begin{pmatrix} 1 & 0 \\ t & 1 \end{pmatrix} = \exp\Bigl(\begin{pmatrix} 0 & 0 \\ t & 0 \end{pmatrix}\Bigr),$$

and for t in small interval around 0 these generate $\mathrm{SL}_2(\mathbf{R})$.

Finally, since $\bar{\mathrm{H}}_s$ contains $L^2_0(K)$, it is not difficult to see that the extension of ϱ_s to $\bar{\mathrm{H}}_s$ remains (topologically) irreducible (by arguments similar to the proof of irreducibility of H_s). Hence this gives the desired unitarization. □

We now change notation, and denote by ϱ_s the complementary series representation of G.

It is natural to ask what happens when $-1 < s < 0$. The situation is different because, when computing the inner products a_n using (7.22), one sees that $a_n \to +\infty$ as $|n| \to +\infty$, which means that the corresponding inner product (7.23) is not defined on all of $\mathrm{H}_s \simeq L^2_0(K)$, but merely on a dense subspace $\tilde{\mathrm{H}}_s \subset \mathrm{H}$. However, $\tilde{\mathrm{H}}_s$, with this inner product, is a Hilbert space, and one can then show that it is stable under ϱ_s and that this defines a unitary representation of G on $\tilde{\mathrm{H}}_s$. We will also denote it by ϱ_s, and call it a complementary series.

But it turns out that this case does not lead to new representations, up to isomorphism: in fact, we have $\varrho_{-s} \simeq \varrho_s$ for $0 < s < 1$. We can see that this is plausible (and that there are no further isomorphisms of complementary series) by noting that the same computation leading to (7.16) shows that the function f_0 (which is also in the Hilbert space of the complementary representation when $s < 0$) satisfies

$$\varrho_s(E^-)\varrho_s(E^+)f_0 = (s^2 - 1)f_0.$$

Since f_0 is determined, up to a scalar, by the fact that it spans the K-invariant subspace of ϱ_s, it follows as in Proposition 7.4.3(3), that the complementary series ϱ_s can only be isomorphic to ϱ_s or to ϱ_{-s} and that ϱ_s is not isomorphic to any principal series representation. (The complementary series are also different from discrete series since the latter have no K-invariant vectors.)

Exercise 7.4.22 (No odd complementary series). Let χ be a character of B given by

$$\chi\Bigl(\begin{pmatrix} \alpha & \beta \\ 0 & \alpha^{-1} \end{pmatrix}\Bigr) = \Bigl(\frac{\alpha}{|\alpha|}\Bigr)|\alpha|^s,$$

and define H_χ and ϱ_χ by the obvious adaptation of the previous definition. Show that the representation ϱ_χ cannot be unitarized if $\mathrm{Re}(s) \neq 0$.

Exercise 7.4.23 (A matrix coefficient of complementary series). Let $0 < s < 1$ be given, and let ϱ_s be the corresponding complementary series as constructed in the proof of Proposition 7.4.21, in particular with the inner product $(\cdot, \cdot)$.

(1) Show that for any $v \in \mathrm{H}_s$ and $g \in G$, we have

$$(\varrho(g)v, f_0) = \langle \varrho(g)v, f_0 \rangle,$$

where f_0 is the function in H_s with restriction to K equal to 1, and the inner product on the right is the standard one on $\mathrm{H}_s \simeq L^2_0(K)$.

(2) Deduce that there exists $\delta > 0$ such that the matrix coefficient $\varphi(g) = \langle \varrho(g)f_0, f_0 \rangle$ is *not* in $L^{2+\delta}(G)$. (*Hint*: Argue as in the proof of Proposition 7.4.3(4).)

This exercise, together with Exercise 7.4.15(7) and the properties of discrete series, reveals what is, in many applications, the most crucial difference between the complementary series and the other series of representations: for principal series, matrix coefficients are in $L^{2+\varepsilon}(G)$ for any $\varepsilon > 0$, while matrix coefficients of discrete series are in $L^2(G)$, by definition. We will mention briefly below one of the problems where this fact is decisive for applications.

With the complementary series, it turns out that we have now found all irreducible unitary representations of $\mathrm{SL}_2(\mathbf{R})$ (up to isomorphism), as was first proved by Bargmann. We summarize:

Theorem 7.4.24 (Bargmann's classification). *Let $G = \mathrm{SL}_2(\mathbf{R})$. Any irreducible unitary representation ϱ of G is isomorphic to one, and only one, of the following:*

(1) *The principal series ϱ_χ associated to*

$$\chi\Bigl(\begin{pmatrix} \alpha & \beta \\ 0 & \alpha^{-1} \end{pmatrix}\Bigr) = \Bigl(\frac{\alpha}{|\alpha|}\Bigr)^{\varepsilon} |\alpha|^{it},$$

where $t \geqslant 0$ if $\varepsilon = 0$ and $t > 0$ if $\varepsilon = 1$.

(2) *The two mock discrete series ϱ^+_{sgn} and ϱ^-_{sgn}, which are the irreducible components of the principal series ϱ_{sgn}.*

(3) *The discrete series π_n or $\tilde{\pi}_n$ for $n \geqslant 2$ an integer.*[2]

(4) *The complementary series ϱ_s with $0 < s < 1$.*

(5) *The trivial one-dimensional representation.*

[2] The representations $\tilde{\pi}_n$ are described in Exercise 7.4.18.

Sketch of proof. The restrictions on t and s in (1) and (4) are imposed to ensure that the representations are pairwise non-isomorphic, and such that all representations we have found appear in the list. Hence the point of this theorem is that the given representations exhaust the possibilities for irreducible unitary representations of G up to isomorphism.

To prove this, one can begin to classify all possibilities for the combination of the Lie algebra action and the restriction to K (introducing the so-called $(\mathfrak{g}, K)$-modules), which gives rather straightforwardly a rough classification in terms of the set of characters of K occurring in a unitary irreducible representation of G. One can then match each possibility with one of the representations above and deduce that any unitary representation is such that it corresponds to one in the list, as far as this combined $(\mathfrak{g}, K)$-action is concerned. The last step is then to prove that non-isomorphic unitary representations cannot have isomorphic $(\mathfrak{g}, K)$-actions. We refer to [**11**, §2.6] or [**40**, VI.6] for details (in the very similar case of $\mathrm{GL}_2(\mathbf{R})$). □

An important definition in further applications is that of a *tempered* irreducible unitary representations: ϱ is said to be tempered if and only if its matrix coefficients are in $L^{2+\varepsilon}(G)$ for all $\varepsilon > 0$. We see then from Bargmann's classification that the tempered representations are the trivial ones, the discrete and mock discrete series, and the principal series.

Exercise 7.4.25. (1) For each irreducible unitary representation of $\mathrm{SL}_2(\mathbf{R})$, determine its (topological) contragredient.

(2) An irreducible unitary representation ϱ of $\mathrm{SL}_2(\mathbf{R})$ on a Hilbert space H is called *unramified* or *spherical* if H^K is non-zero, i.e., if there exists a non-zero $v \in \mathrm{H}$ invariant under the action of $\mathrm{SO}_2(\mathbf{R})$. Show that ϱ is spherical if and only if it is either a principal series with even parity $\varepsilon = 0$, or a complementary series. Show that in all spherical cases, the dimension of H^K is equal to 1.

We conclude this chapter by describing one problem where the representation theory of $\mathrm{SL}_2(\mathbf{R})$ is important and where the distinction between principal and complementary series appears clearly.

The group $G = \mathrm{SL}_2(\mathbf{R})$ contains many discrete subgroups Γ such that the coset space

$$Y_\Gamma = \Gamma\backslash G$$

is a compact space, or at least has finite measure for the natural measure μ_Γ that is obtained from the Haar measure on G. For instance, the second property holds for $\Gamma = \mathrm{SL}_2(\mathbf{Z})$ and any finite-index subgroup of $\mathrm{SL}_2(\mathbf{Z})$, and there are many examples of the latter, for instance

$$\Gamma(N) = \{g \in \mathrm{SL}_2(\mathbf{Z}) \mid g \equiv \mathrm{Id} \pmod{N}\}$$

is a subgroup of $\mathrm{SL}_2(\mathbf{Z})$ with index equal to $|\mathrm{SL}_2(\mathbf{Z}/N\mathbf{Z})|$. More generally, subgroups of $\mathrm{SL}_2(\mathbf{Z})$ containing a subgroup $\Gamma(N)$ for some $N \geqslant 1$ are called *congruence subgroups*.

Using the measure μ_Γ, one can construct the space $L^2(Y_\Gamma, \mu_\Gamma)$, and since we used a left action of G to define Y_Γ, the (right) regular representation still makes sense on this Hilbert space:

$$(g \cdot f)(x) = f(xg)$$

for $g \in G$ and $f \in L^2(Y_\Gamma, \mu_\Gamma)$. This is a unitary representation (because the Haar-measure is invariant). Then one may ask, What is the structure of this space as a unitary representation? How can it be expressed in terms of irreducible representations?

Remark 7.4.26. Here, there is a certain useful analogy with the case of the discrete subgroup $\mathbf{Z} \subset \mathbf{R}$, in which case $L^2(\mathbf{R}/\mathbf{Z})$ is the same as the regular representation of the compact group $\mathbf{R}/\mathbf{Z}$. In particular, it decomposes as a direct sum of countably many irreducible subspaces with multiplicity one.

The functions on $\mathbf{R}/\mathbf{Z}$ can be seen as $\mathbf{Z}$-periodic functions on $\mathbf{R}$, and their importance is certainly clear. This suggests that spaces like $L^2(Y_\Gamma, \mu_\Gamma)$ should also be interesting objects to study.

One can perform a first decomposition. This goes back to Gelfand, Graev, and Piatetski-Shapiro and shows that

$$L^2(Y_\Gamma, \mu_\Gamma) = \mathrm{H}_1 \oplus \mathrm{H}_2,$$

where H_1 is a Hilbert direct sum of at most countably many irreducible subspaces, each irreducible representation occurring with finite multiplicity and H_2 is a direct integral of the principal series representations ϱ_{it} for $t \geqslant 0$, each occurring with the same constant finite multiplicity $n_\Gamma \geqslant 0$ (where $n_\Gamma = 0$ means that $\mathrm{H}_2 = 0$). One can easily compute n_Γ, given Γ, as the number of "cusps" of Γ. One finds, for instance, that $n_\Gamma = 1$ if $\Gamma = \mathrm{SL}_2(\mathbf{Z})$, and that $n_\Gamma = 0$ whenever Y_Γ is compact (so that $\mathrm{H}_2 = 0$ if Y_Γ is compact). Note in particular that H_2 never contains a complementary series representation.

The space H_1 is the most interesting. If Y_Γ is compact, one can show relatively easily that H_1 is the direct sum of *infinitely many* irreducible components, but the corresponding parameters (in the Bargmann classification) are not explicit, with very few exceptions. Moreover, the subspace H_1 is much more mysterious if Y_Γ is not compact. In some important special cases, such as $\Gamma = \mathrm{SL}_2(\mathbf{Z})$ or a finite-index congruence subgroup, one can show (as first done by Selberg) that H_1 contains infinitely many distinct irreducible subrepresentations, but this is not known (nor expected) in general.

Selberg realized that a crucial issue in applications is whether H_1 contains or not a complementary series representation. It is possible to construct examples of quotients Y_Γ which contain an arbitrarily large (finite) number of complementary series representations. Selberg conjectured, on the other hand, that this does not happen if $\Gamma = \Gamma(N)$ for some $N \geqslant 1$ (for instance, for $\Gamma(1) = \mathrm{SL}_2(\mathbf{Z})$). This conjecture (for which there is a lot of evidence, for instance it is known for $\mathrm{SL}_2(\mathbf{Z})$ itself, or for N small enough) reveals a remarkable feature of these special quotients $Y(N) = Y_{\Gamma(N)}$, which is understood to be related to the arithmetic nature of their definition.

Selberg's conjecture is usually described in more concrete terms. Since a complementary representation ϱ_s (where $0 < s < 1$) is spherical, any subrepresentation H of H_1 isomorphic to ϱ_s contains a vector $v_0 \in \mathrm{H}$ which is K-invariant, and which is unique up to multiplication by a non-zero scalar. Such a vector corresponds in $L^2(Y_\Gamma, \mu_\Gamma)$ to a complex-valued function on

$$X_\Gamma = \Gamma\backslash G/K \simeq \Gamma\backslash\mathbf{H},$$

where $\mathbf{H}$ is the upper half-plane (see the proof of Lemma 7.4.4), i.e., to a function f on $\mathbf{H}$ which satisfies the Γ-invariance relations

$$f(\gamma \cdot z) = f(z) \tag{7.24}$$

for all $\gamma \in \Gamma$ (these are called *automorphic functions*). It is not difficult to check using the Lie algebra action that, because f is related to v_0, it is smooth on $\mathbf{H}$ and satisfies the differential equation

$$\Delta f = \lambda f,$$

where $\lambda = 1/4 - s^2 \in \,]0, 1/4[$ and

$$\Delta f = -y^2\Big(\frac{\partial^2 f}{\partial x^2} + \frac{\partial^2 f}{\partial y^2}\Big).$$

(In other words, f is an eigenfunction of the so-called *hyperbolic Laplace operator* Δ.) This applies also to principal series ϱ_{it} in H_1, except that one finds a function f on $\mathbf{H}$ which is Γ-invariant and satisfies

$$\Delta f = \Big(\frac{1}{4} + t^2\Big) f,$$

with eigenvalue $\lambda = 1/4 + t^2 \geqslant 1/4$.

In fact, this correspondence also goes backward: given a smooth function f on $\mathbf{H}$ satisfying (7.24) such that it "lives" on X_Γ, and satisfying

$$\Delta f = \lambda_f f$$

for some $\lambda_f > 0$, and $f \in L^2(Y_\Gamma, \mu_\Gamma)$ (after extending it to Y_Γ using the quotient map $Y_\Gamma \to X_\Gamma$), one can associate to it an irreducible subrepresentation which is a subrepresentation π_f of H_1 (it is essentially the subrepresentation generated by f). One can then check that π_f is a complementary series if and only if $\lambda_f < 1/4$.

Selberg himself succeeded in proving the first result toward his conjecture. Consider again $N \geqslant 1$ and $\Gamma = \Gamma(N)$. Informally, even if the conjecture could fail (as far as we know), it cannot fail too badly. More precisely, if ϱ_s is a complementary series in H_2 for $\Gamma = \Gamma(N)$, then we have

$$s \leqslant \frac{1}{4}$$

(or equivalently, for an eigenfunction f of Δ as above, we have $\lambda_f \geqslant 3/16$). The main point is that the lower-bound $1/4$ is independent of the integer N, i.e., of the subgroup $\Gamma(N)$ which is considered. This fundamental result turns out to be crucial in the original proof of the seemingly elementary Theorem 1.2.5! (We already mentioned that more direct and elementary proofs are now known; see [**13**].)

Although Selberg's conjecture has a concrete formulation in terms of eigenfunctions of Δ, it turns out that the most recent progress in its direction has relied very extensively on results which are based in a fundamental way on the representation-theoretic version of the statement. We refer to [**52**] for a first hint of these arguments and to [**11, 3**] for introductions to the general theory of automorphic representations which underlies this type of problem.

Example 7.4.27. Here is a very brief example, taken from the work of Bernstein and Reznikov [**4**], of how representation theory may help in solving problems concerning Γ-invariant eigenfunctions of Δ using the correspondence with subrepresentations of the space $L^2(Y_\Gamma, \mu_\Gamma)$. Let

$$j\,:\,\mathrm{H}_{it} \longrightarrow L^2(Y_\Gamma, \mu_\Gamma)$$

be an isometric embedding of a principal series representation ϱ_{it}, and let $\mathrm{H} = \mathrm{Im}(\mathrm{H}_{it})$ be the corresponding closed subspace of $L^2(Y_\Gamma, \mu_\Gamma)$. In [**4**], one issue is estimating the supremum norm of certain functions $f = j(v)$ in H, i.e., estimating

$$N_\infty(v) = \sup_{x \in Y_\Gamma} |j(v)(x)| = \|j(v)\|_\infty$$

(in terms of certain parameters describing the function $f = j(v)$, which we will not describe).

Viewed from the sole point of view of the function f on $\mathbf{H}$, this problem seems delicate. But one can relatively easily succeed in proving a first bound

$$N_\infty(v) \leqslant N(v)$$

if v is in H_{it}^∞, where N is a type of Sobolev norm on H_{it}^∞.

Then Bernstein and Reznikov observe that one can exploit the existence of the underlying representation to improve this first estimate. The point is that

$$N_\infty(g \cdot v) = N_\infty(v)$$

for all $g \in \mathrm{SL}_2(\mathbf{R})$, i.e., the sup norm is invariant, whereas the Sobolev norm is not. One can therefore also claim that

$$N_\infty(v) \leqslant N(g \cdot v)$$

for any $g \in \mathrm{SL}_2(\mathbf{R})$, or in fact that

$$N_\infty(v) \leqslant N(g_1 \cdot v_1) + \cdots + N(g_k \cdot v_k) \tag{7.25}$$

for any decomposition

$$v = v_1 + \cdots + v_k$$

of the vector v with $k \geqslant 1$, $v_i \in \mathrm{H}_{it}^\infty$, and for any $g_i \in \mathrm{SL}_2(\mathbf{R})$. Denoting by $N^G(v)$ the minimum of the right-hand side of (7.25) as one runs over all such decompositions, we deduce therefore that

$$N_\infty(v) \leqslant N^G(v).$$

The new norm N^G is also invariant and depends only on the representation ϱ_{it} (not on the embedding j). Bernstein and Reznikov succeed in estimating it by exploiting suitable decompositions of the vectors they are considering in a convenient model of ϱ_{it}. This is a beautiful illustration of the type of arguments made possible by good use of representation theory, and we refer to the paper [**4**] for more details.

Appendix A

Some useful facts

A.1. Algebraic integers

Readers familiar with algebraic integers will probably not need to read this section. We recall the definition and those properties of algebraic integers which are relevant to the applications in the text, especially in Section 4.7.2. A very good summary for similar purposes (with proofs) is found in [**18**, Section 4.3].

An algebraic integer $z \in \mathbf{C}$ is any complex number such that there exists a non-zero *monic* polynomial $p \in \mathbf{Z}[X]$ such that $p(z) = 0$; we denote by $\bar{\mathbf{Z}}$ the set of algebraic integers.[1] Since, for any $z \in \bar{\mathbf{Z}}$, the set $I_z = \{p \in \mathbf{Z}[X] : p(z) = 0\}$ is an ideal of $\mathbf{Z}[X]$, there exists a unique monic generator p_z of I_z, which is called the *minimal polynomial* of z. This polynomial is irreducible over $\mathbf{Q}$.

For instance, any $n \in \mathbf{Z}$ is a zero of $p = X - n$, and hence $\mathbf{Z} \subset \bar{\mathbf{Z}}$. In fact, we have the following stronger fact, which illustrates one way in which algebraic integers generalize integers.

Proposition A.1.1. *An algebraic integer z, which is also a rational number, is an element of $\mathbf{Z}$, i.e., we have $\bar{\mathbf{Z}} \cap \mathbf{Q} = \mathbf{Z}$.*

Proof. To see this, let $z = a/b$, with a and $b \neq 0$ coprime integers, be a rational number which is also an algebraic integer, so that we have

$$z^n + a_{n-1}z^{n-1} + \cdots + a_1 z + a_0 = 0,$$

for some $n \geqslant 1$ and integral coefficients $a_i \in \mathbf{Z}$.

[1] It is the restriction to monic equations which is crucial to obtaining a generalization of integers; if instead p is allowed to be any non-zero $p \in \mathbf{Q}[X]$, one obtains the notion of *algebraic number*.

Substituting $z = a/b$ and multiplying with the common denominator b^n, one finds

$$a^n + a_{n-1}a^{n-1}b + \cdots + a_1ab^{n-1} + a_0b^n = 0,$$

and therefore $b \mid a^n$, which means $b = 1$ or -1 since a and b are coprime. □

Other important examples of algebraic integers include arbitrary roots of unity (solutions of $X^n - 1 = 0$). Moreover, although this is not entirely obvious, $\bar{\mathbf{Z}}$ is a ring: the sum, difference, and product of algebraic integers remains an integer. We prove this in an ad hoc, yet fairly elegant, manner (this is the approach used in [**18**]):

Proposition A.1.2. (1) *A complex number $z \in \mathbf{C}$ is an algebraic integer if and only there exists a square matrix $A \in \mathrm{M}_n(\mathbf{Z})$, for some $n \geqslant 1$, with integral coefficients, such that z is an eigenvalue of A, i.e., such that we have* $\det(z - A) = 0$.

(2) *If z_1, z_2 are algebraic integers, then so are $z_1 + z_2$ and z_1z_2.*

Proof. (1) The characteristic polynomial $\det(X-A)$ of A is a monic integral polynomial of degree n, and therefore any of its zeros is an algebraic integer. We must check the converse. Thus let z be an algebraic integer, and let

$$p = X^n + a_{n-1}X^{n-1} + \cdots + a_1X + a_0 \in \mathbf{Z}[X]$$

be the minimal polynomial of z. We can write down immediately a suitable matrix, namely

$$A_z = \begin{pmatrix} 0 & \cdots & \cdots & \cdots & -a_0 \\ 1 & 0 & \cdots & \cdots & -a_1 \\ 0 & 1 & 0 & \cdots & -a_2 \\ \vdots & \ddots & \ddots & \ddots & \vdots \\ 0 & \cdots & \cdots & 1 & -a_{n-1} \end{pmatrix}.$$

It is a standard exercise that $\det(A_z - z) = 0$, but the following explains how to *produce* it. Consider the abelian group $M = \mathbf{Z}[z]$ generated by powers of z as a subgroup of $\mathbf{C}$. Because of the relation $p(z) = 0$, this is a free finitely generated abelian group, with basis

$$(1, z, z^2, \ldots, z^{n-1}),$$

hence of rank n (the rank is no smaller because p is the minimal polynomial of z). The map

$$m_z \begin{cases} M \longrightarrow M \\ a \mapsto za \end{cases}$$

is a homomorphism of abelian groups and, with respect to the basis of M above, it is represented by an integral matrix which is precisely A_z. But since $m_z(z) = z^2 = z \times z$, we see that z is an eigenvalue of the homomorphism m_z, hence of the matrix A_z.

(2) We can now prove that $\bar{\mathbf{Z}}$ is stable under product and multiplication using the characterization we just obtained. Indeed, let A_1 and A_2 be integral matrices such that z_i is an eigenvalue of A_i. It is standard that $z_1 z_2$ is an eigenvalue of $A_1 \otimes A_2$, which is also an integral matrix (if e_i are eigenvectors of A_i for the eigenvalue z_i, then $(A_1 \otimes A_2)(e_1 \otimes e_2) = z_1 z_2 (e_1 o \times e_2)$, by definition of tensor products).

The case of sums is a bit less obvious, and the formula is worth remembering: $z_1 + z_2$ is an eigenvalue of

$$A = A_1 \otimes \mathrm{Id} + \mathrm{Id} \otimes A_2. \tag{A.1}$$

Indeed, with e_i as before, we have

$$A(e_1 \otimes e_2) = (A_1 \otimes \mathrm{Id} + \mathrm{Id} \otimes A_2)(e_1 \otimes e_2) = z_1(e_1 \otimes e_2) + z_2(e_1 \otimes e_2),$$

so $e_1 \otimes e_2$ is an eigenvector for the eigenvalue $z_1 + z_2$. □

Remark A.1.3. The last formula $A = A_1 \otimes \mathrm{Id} + \mathrm{Id} \otimes A_2$ may seem mysterious at first. It has a natural explanation in terms of representations of Lie algebras, however: if L is a Lie algebra (over $\mathbf{C}$, say) and

$$\Phi_1 : L \longrightarrow \mathfrak{gl}(E_1), \qquad \Phi_2 : L \longrightarrow \mathfrak{gl}(E_2)$$

are representations of L (as in Definition 3.2.2), then the *correct* definition of the tensor product $\Phi_1 \otimes \Phi_2$ of these representations is the map

$$\Phi : L \longrightarrow \mathfrak{gl}(E_1 \otimes E_2)$$

given by

$$\Phi(x) = \Phi_1(x) \otimes \mathrm{Id} + \mathrm{Id} \otimes \Phi_2(x),$$

which has the same form as (A.1).

We can now see that any linear combination of roots of unity with *integral* coefficients is an algebraic integer. In particular:

Corollary A.1.4. *Let G be a finite group, and let ϱ be a finite-dimensional complex representation of G. Then for any $g \in G$, the character value $\chi_\varrho(g)$ is an algebraic integer.*

Proof. Indeed, $\chi_\varrho(g) = \mathrm{Tr}\,\varrho(g)$ is the sum of the $\dim(\varrho)$ eigenvalues of $\varrho(g)$, and each of them is a root of unity, since $\varrho(g)^{|G|} = 1$ (using the finiteness of G). □

We now discuss quickly the divisibility relation in $\bar{\mathbf{Z}}$. As might be expected, if z_1, z_2 are algebraic integers, one says that z_1 *divides* z_2, denoted

$$z_1 \mid z_2,$$

if and only if $z_2 = z_1 z$ with z also an algebraic integer (in other words, the ratio $z_2/z_1 \in \mathbf{C}$ is in fact in $\bar{\mathbf{Z}}$).

We see clearly that if the same z_1 divides z_2 and z_3 (all in $\bar{\mathbf{Z}}$), it divides their sum or difference, or their product. In particular, if we have a relation

$$\sum_i z_i w_i = 1$$

with z_i, w_i all algebraic integers, and if we fix some positive integer $q \geqslant 2$, we can conclude that some w_i is *not* divisible by q, in view of the fact that $1/q \notin \bar{\mathbf{Z}}$.

One can also define a coprimality relation between algebraic integers: algebraic integers z_1 and z_2 are said to be *coprime*, which is denoted $(z_1, z_2) = 1$, if and only if there exist algebraic integers w_1, w_2 such that

$$z_1 w_1 + z_2 w_2 = 1.$$

This shows in particular that if two ordinary integers in $\mathbf{Z}$ are coprime (in the usual sense), they are also coprime as algebraic integers. On the other hand, suppose that z_1 and z_2 in $\bar{\mathbf{Z}}$ have a *common divisor* $w \in \bar{\mathbf{Z}}$, i.e., we have $w \mid z_1$ and $w \mid z_2$. Then the algebraic integers z_1 and z_2 can be coprime only if w is a *unit*, i.e., if $1/w$ is also in $\bar{\mathbf{Z}}$. Indeed, writing $z_i = wy_i$, we get

$$wy_1 w_1 + wy_2 w_2 = 1,$$

and thus $1/w = y_1 w_1 + y_2 w_2 \in \bar{\mathbf{Z}}$. If $w \in \mathbf{Z}$, of course, the condition $1/w \in \bar{\mathbf{Z}}$ means that $w = \pm 1$.

Remark A.1.5 (Units)**.** There are many examples of units, some of which are complex numbers with modulus not equal to 1, in contrast with the units ± 1 in $\mathbf{Z}$. For instance, the element $\varepsilon = 1 + \sqrt{2}$ satisfies

$$\frac{1}{1+\sqrt{2}} = \frac{1-\sqrt{2}}{1-2} = -1 + \sqrt{2} \in \bar{\mathbf{Z}},$$

so it is a unit, although $|\varepsilon| > 1$.

Other examples of units are roots of unity, since the inverse of a root of unity is also one.

The fundamental link between divisibility and coprimality among integers remains valid:

Proposition A.1.6. *Let z_1, z_2, z_3 be algebraic integers such that $z_1 \mid z_2 z_3$. If z_1 and z_2 are coprime, then z_1 divides z_3.*

Proof. Indeed, from a relation

$$z_1 w_1 + z_2 w_2 = 1,$$

we get $z_3 = z_1 w_1 z_3 + z_2 z_3 w_2$, and since each term of the sum is divisible by z_1, so is z_3. ∎

The last notion we need is the definition of the conjugates of an algebraic integer, and of the norm map. The former is very natural:

Definition A.1.7 (Conjugates and norm of algebraic integers)**.** Let $z \in \bar{\mathbf{Z}}$ be an algebraic integer. Let $p \in \mathbf{Z}[X]$ be the minimal polynomial of z. A *conjugate* w of z is any root of p in $\mathbf{C}$, and the *norm* of z, denoted $N(z)$, is the product of all conjugates of z.

In other words, a conjugate of z is an algebraic integer which satisfies *the same minimal equation* as z. Since any polynomial $p_1 \in \mathbf{Z}[X]$ for which $p_1(z) = 0$ is a multiple of p, it follows that whenever z satisfies a polynomial relation with integral coefficients, so do all its conjugates.

For example, if $n \in \mathbf{Z}$, we have $p = X - n$, and n has no other conjugate than itself. If z is a root of unity of order n, so that $z^n - 1 = 0$, then all its conjugates are also roots of unity of order n (but not all roots of $z^n - 1$ are conjugates of z, e.g., -1 is not a conjugate of the fourth root of unity $z = i$, because the minimal polynomial for i is $X^2 + 1$, and not $X^4 - 1$).

Factoring the minimal polynomial p, we have

$$p(X) = \prod_{w} (X - w),$$

where the product, by definition, is over all conjugates of z. In particular, we find

$$N(z) = \prod_{w} w = (-1)^n p(0)$$

with $n = \deg(p)$. This shows that the norm of z is an integer in $\mathbf{Z}$.

Proposition A.1.8 (Conjugates of sums and products)**.** *Let z_1 and z_2 be algebraic integers. Then any conjugate w of $z_1 z_2$ can be written*

$$w = w_1 w_2$$

with w_1 a conjugate of z_1 and w_2 a conjugate of z_2. Similarly, any conjugate of $z_1 + z_2$ is of the form $w_1 + w_2$ for some conjugate w_i of z_i.

Note, however, that not *all* sums $w_1 + w_2$ are necessarily conjugates of $z_1 + z_2$. (For instance, take $z_1 = \sqrt{2}$, $z_2 = -\sqrt{2}$; then $w_1 = z_1$ and $w_2 = \sqrt{2}$ are conjugates of z_1 and z_2, with $w_1 + w_2 = 2\sqrt{2}$, although $z_1 + z_2 = 0$ has no non-zero conjugate.)

Proof. Although this property is much better understood in terms of Galois theory, there is a cute argument using the criterion in Proposition A.1.2. We present this for $z_1 z_2$, leaving the case of the sum to the reader.

Consider integral matrices A_1 and A_2 with characteristic polynomials p_1 and p_2, the minimal polynomials for z_1 and z_2, respectively. Form the matrix $A = A_1 \otimes A_2$. Since $z_1 z_2$ is an eigenvalue of A, we know that any conjugate of z_1z_2 is among the other eigenvalues of A. Similarly, for any conjugate w_1 of z_1 and w_2 of z_2, there are eigenvectors e_1, e_2 with $A_ie_i = w_ie_i$. Each of these gives an eigenvector $e_1 \otimes e_2$ of A with eigenvalue w_1w_2. If we count, we see that we construct this way n_1n_2 eigenvectors of A, with eigenvalues given by products of conjugates of z_1 and z_2. Since A has size n_1n_2, there can be no other eigenvector, and therefore no other eigenvalue either! □

Corollary A.1.9. *Let z be an algebraic integer, and let $n \in \mathbf{Z}$ be such that $n \mid z$. Then n^r divides $N(z)$ in the ring $\mathbf{Z}$, where r is the degree of the minimal polynomial of z, or equivalently, the number of conjugates of z.*

Proof. The point is that n divides any conjugate w of z. Indeed, if we write $z = nz_1$ for some algebraic integer z_1, we see that w, as a conjugate of nz_1, must be of the form nw_1, where w_1 is a conjugate of z_1 (since n is the only conjugate of itself...). □

Exercise A.1.10. Let z be an algebraic integer. Show that z is a unit in $\bar{\mathbf{Z}}$ if and only if $N(z) = \pm 1$.

A.2. The spectral theorem

In the proof of the general case of the Peter–Weyl theorem, a crucial ingredient is the fact that certain operators constructed using the regular (or left-regular) representation have non-trivial, but finite-dimensional, eigenspaces. The standard statement along these lines is the spectral theorem for compact normal operators. We will state this result, but we will only prove a weaker statement that is sufficient for our purposes.

We start with the definition:

Definition A.2.1 (Compact operator)**.** Let H be a Hilbert space, and let $T : \mathrm{H} \longrightarrow \mathrm{H}$ be a continuous linear operator. Then T is *compact* if and only if there is a sequence (T_n) of continuous operators $T_n : \mathrm{H} \to \mathrm{H}$ such that $\dim \mathrm{Im}(T_n) < +\infty$ for all n, and $T_n \to T$ in the operator norm topology, i.e., uniformly on the unit ball of H.

Example A.2.2. A compact operator should be considered as *small* in some sense. An illustration of this intuitive idea is the fact that if $\lambda \neq 0$, the operator $\lambda\mathrm{Id}$ on H is compact if and only if H is finite dimensional. Indeed, suppose $T_n \to \lambda\mathrm{Id}$ in L(H). Then, for n sufficiently large, we $\|\lambda^{-1}T_n - \mathrm{Id}\| < 1$, and this implies that $\lambda^{-1}T_n$ is invertible. (This is well known: check that the geometric series

$$\sum_{k \geqslant 0} (\mathrm{Id} - \lambda^{-1}T_n)^k$$

converges in L(H), as it should, to the inverse of $\mathrm{Id} - (\mathrm{Id} - \lambda^{-1} T_n) = \lambda^{-1} T_n$.) Then

$$\dim \mathrm{H} = \dim \mathrm{Im}(T_n) < +\infty.$$

The spectral theorem for compact operators is the following:

Theorem A.2.3 (Spectral theorem). *Let* H *be a Hilbert space, and let* $T : \mathrm{H} \longrightarrow \mathrm{H}$ *be a normal compact operator, i.e., a compact operator such that* $TT^* = T^*T$, *for instance a self-adjoint operator with* $T = T^*$.

There exists a subset $S \subset \mathbf{C}$ *which is finite or countable, such that the eigenspace* $\mathrm{Ker}(T - \lambda)$ *is non-zero if and only if* $\lambda \in S$, *and is finite dimensional if* $\lambda \neq 0$. *Furthermore, we have a Hilbert space orthogonal direct sum decomposition*

$$\mathrm{H} = \widehat{\bigoplus_{\lambda \in S}} \mathrm{Ker}(T - \lambda).$$

In particular, if $T \neq 0$, *the set* S *is not reduced to* $\{0\}$.

In other words, a normal compact operator can be diagonalized with at most countably many eigenvalues and, for any non-zero eigenvalue, the corresponding eigenspace is finite dimensional. Note that, if T has finite rank (i.e., $\dim \mathrm{Im}(T) < +\infty$), this is just a form of the spectral theorem for matrices of finite size commuting with their adjoint. In view of the definition of compact operators, the result can therefore be seen as "passing to the limit" with this classical theorem.

Exercise A.2.4. Prove Theorem A.2.3 for self-adjoint compact operators defined on a separable Hilbert space using the spectral theorem for arbitrary self-adjoint operators, which is stated in Theorem 3.4.18. (*Hint*: Recall the following result of functional analysis: a Hilbert space H such that the closed unit ball of H is compact is finite dimensional. Use this to prove that a compact multiplication operator M_f has the property that, for any $\varepsilon > 0$, the function f only takes finitely many values y with $|y| > \varepsilon$ (up to sets of measure zero).)

We will apply this to Hilbert–Schmidt integral operators:

Proposition A.2.5. *Let* (X, μ) *be a measure space, and let*

$$k : X \times X \longrightarrow \mathbf{C}$$

be a function which is in $L^2(X \times X, \mu \times \mu)$. *The linear operator* T_k *acting on* $L^2(X, \mu)$ *by*

$$(T_k \varphi)(x) = \int_X k(x, y) f(y) d\mu(y)$$

is compact. It is self-adjoint if k *satisfies* $k(x, y) = \overline{k(y, x)}$.

The operator T_k is customarily called the *Hilbert–Schmidt (integral) operator with kernel* k. We give the proof in the case when $L^2(X,\mu)$ is a separable Hilbert space, which is the case in most applications (for instance for X a compact metric space and μ a Radon measure on X). The general case is considered, e.g., in [**17**, XI.8.44].

Proof when $L^2(X,\mu)$ is separable. First of all, the Cauchy–Schwarz inequality and Fubini's Theorem give

$$\int_X |T_k(\varphi)(x)|^2 d\mu(x) \leqslant \int_X \Big(\int_X |k(x,y)|^2 d\mu(y)\Big)\Big(\int_X |f(y)|^2 d\mu(y)\Big) d\mu(x)$$
$$= \|k\|^2_{L^2(X\times X)} \|f\|^2_{L^2(X)},$$

which shows that T_k is well defined and continuous with norm

$$\|T_k\| \leqslant \|k\|_{L^2(X\times X)}. \tag{A.2}$$

We can now check quite easily that T_k is compact when $L^2(X,\mu)$ is separable. Fix any orthonormal basis $(\varphi_m)_{k\geqslant 1}$ of $L^2(X)$. Then the functions

$$\psi_{m,n} \,:\, (x,y) \mapsto \varphi_m(x)\varphi_n(y)$$

are known to form an orthonormal basis of $L^2(X \times X)$ (see, e.g., [**49**, II.4, p. 51]), and we can therefore expand k in an L^2-convergent series

$$k = \sum_m \sum_n \alpha(m,n)\psi_{m,n}, \qquad \alpha(m,n) \in \mathbf{C}. \tag{A.3}$$

Given any $N \geqslant 1$, we define the approximation $T^{(N)} = T_{k_N}$, where k_N is the corresponding partial sum

$$k_N = \sum_{m\leqslant N} \sum_{n\leqslant N} \alpha(m,n)\psi_{m,n}.$$

Note that for any $\varphi \in L^2(X)$, we have

$$T^{(N)}\varphi = \sum_{m\leqslant N} \sum_{n\leqslant N} \alpha(m,n)\Big(\int_X \varphi_n(y) f(y) d\mu(y)\Big)\varphi_m,$$

which shows that the image of $T^{(N)}$ is finite dimensional, spanned by the φ_m, $m \leqslant N$. In addition, in the space of operators on $L^2(X)$, we have

$$\|T_k - T^{(N)}\| = \|T_{k-k_N}\| \leqslant \|k - k_N\|_{L^2(X\times X)}$$

by (A.2), which tends to 0 as $N \to +\infty$ (since the series (A.3) converges in $L^2(X \times X)$). So we have found finite rank approximations of T_k, and thus T_k is compact.

Finally, it is a formal computation, left to the reader, to check that

$$T_k^* = T_{\tilde{k}}, \quad \text{where} \quad \tilde{k}(x,y) = \overline{k(y,x)},$$

so that k is self-adjoint if $k = \tilde{k}$. ∎

We are now going to prove part of the spectral theorem in a special case involving Hilbert–Schmidt operators. This statement is enough for the application to the proof of the Peter–Weyl theorem.

Proposition A.2.6. *Let X be a compact space with a Radon measure μ, and let $T = T_k$ be a non-zero self-adjoint Hilbert–Schmidt operator with continuous kernel k such that T_k is non-negative, i.e.,*

$$\langle T_k f, f\rangle \geqslant 0$$

for all $f \in L^2(X,\mu)$. Then there exists a positive eigenvalue $\lambda > 0$ of T, and the corresponding eigenspace is finite dimensional.

Proof when $L^2(X,\mu)$ is separable. We denote $\mathrm{H} = L^2(X)$ in this argument, since parts of it will apply to any positive compact operator. We first show the general fact that if $\lambda \neq 0$ is an eigenvalue of a compact operator T, the eigenspace $V = \mathrm{Ker}(T - \lambda)$ is finite-dimensional (this is false for $\lambda = 0$, e.g., take $T = 0\ldots$). The basic idea is that T, restricted to V, remains compact—but this operator on V is $\lambda\mathrm{Id}$, and we can apply Example A.2.2. To implement this, let P denote the orthogonal projection onto the closed subspace V. Consider a sequence (T_n) of operators with finite rank such that $T_n \to T$ in $\mathrm{L}(\mathrm{H})$, and define $U_n = PT_n$ as operator on V. These are finite rank operators in $\mathrm{L}(V)$, and we have

$$\|U_n - T\|_{\mathrm{L}(V)} \leqslant \|(P - \mathrm{Id})T_n\|_{\mathrm{L}(V)} + \|T_n - T\|_{\mathrm{L}(\mathrm{H})} = \|T_n - T\|_{\mathrm{L}(\mathrm{H})}$$

since $P = \mathrm{Id}$ on V. This shows that, indeed, T is compact when restricted to the stable subspace V. Since $T = \lambda\mathrm{Id}$ on V, we conclude that V is finite dimensional.

Now for the existence of the eigenvalue, which is the most crucial part. The idea is to show the norm $\lambda = \|T\|_{\mathrm{L}(\mathrm{H})}$ itself is an eigenvalue and, in order to prove this it is enough, to show that the supremum

$$\|T\|_{\mathrm{L}(\mathrm{H})} = \sup_{v \neq 0} \frac{\|Tv\|}{\|v\|}$$

is reached. Indeed, assume that $v \in \mathrm{H}$ has norm 1 and satisfies $\|Tv\| = \lambda$. Considering any vector $w \in \mathrm{H}$ and $t \in \mathbf{R}$, we have

$$\|T(v + tw)\|^2 \leqslant \lambda^2 \|v + tw\|^2,$$

and after expanding both sides as polynomials in t of degree 2, the coefficient of t must vanish for this inequality to hold for all t close to 0; one checks that this gives the formula

$$\mathrm{Re}\langle Tv, Tw\rangle = \lambda^2\, \mathrm{Re}\langle v, w\rangle.$$

Since T is self-adjoint and λ is real, we deduce that

$$\mathrm{Re}\langle T^2 v, w\rangle = \mathrm{Re}\langle Tv, Tw\rangle = \mathrm{Re}\langle \lambda^2 v, w\rangle$$

for all $w \in \mathrm{H}$. Taking $w = T^2v - \lambda^2 v$, we obtain $T^2 v = \lambda^2 v$. Now we *take the square root* as follows. We write $Tv = \lambda v + v_1$, and apply T, getting

$$\lambda^2 v = T^2 v = \lambda T v + T v_1 = \lambda^2 v + \lambda v_1 + T v_1,$$

so $Tv_1 = -\lambda v_1$. But $\lambda > 0$ and T is positive, so this is impossible unless $v_1 = 0$, and hence $Tv = \lambda v$, as desired.

Thus we are reduced to proving the existence of a vector achieving the norm of T; indeed, this reduction did not use anything about T itself! Here the idea is that this property holds for a finite-rank operator—and we hope to get it to carry through the limiting process. Readers familiar with weak convergence and the alternative definition of compact operators based on the relative compactness of the image of the unit ball will find the following arguments rather naive, but one should remember that the Peter–Weyl theorem predates such notions.

We argue from scratch, starting with a sequence (v_n) of unit eigenvectors of T_n for the norm $\lambda_n = \|T_n\|$. Since $\lambda_n \to \lambda$, it would of course be enough to know that (v_n), or a subsequence of (v_n), converges. But since the unit ball of an infinite-dimensional Hilbert space is not compact, we cannot claim that such a subsequence exists. However, we can fix a countable dense subset $\mathcal{H}$ of H, consisting of the $\mathbf{Q}$-linear span of an orthonormal basis (ψ_m) of H. Then, by a diagonal argument, we can find a subsequence $w_j = v_{n_j}$ of (v_n) such that

$$\langle \psi, w_j \rangle \longrightarrow \alpha(\psi)$$

for all $\psi \in \mathcal{H}$, where $\alpha(\psi) \in \mathbf{C}$. We apply this to the elements ψ_m of the chosen orthonormal basis, and we derive

$$\sum_{m \leqslant M} |\langle \psi_m, w_j \rangle|^2 \longrightarrow \sum_{m \leqslant M} |\beta_m|^2$$

for any $M \geqslant 1$, where $\beta_m = \alpha(\psi_m)$. The left-hand side is the norm of the projection of w_j on the span of $(\psi_1, \ldots, \psi_M)$, and as such is $\leqslant 1$. Hence we get

$$\sum_{m \leqslant M} |\beta_m|^2 \leqslant 1$$

for all $M \geqslant 1$, which shows that the vector

$$v = \sum_{m \geqslant 1} \beta_m \psi_m$$

exists in H and has norm $\leqslant 1$. But then, since the $\psi \in \mathcal{H}$ are finite linear combinations of the ψ_m, we get

$$\langle w_j, \psi \rangle \longrightarrow \langle v, \psi \rangle$$

as $j \to +\infty$, for all $\psi \in \mathcal{H}$. Since $\mathcal{H}$ is dense in H, it follows (formally) that

$$\langle w_j, w\rangle \longrightarrow \langle v, w\rangle \tag{A.4}$$

as $j \to +\infty$, for all $w \in \mathrm{H}$.[2]

We now argue that, due to the compactness of T, this weaker convergence property suffices to ensure that $Tw_j \to Tv$ in H. If this is true, we are done: since $Tw_j = T_{n_j}v_{n_j} + (T - T_{n_j})v_{n_j}$, we have

$$\|Tw_j\| = \lambda_{n_j} + o(1) \to \lambda,$$

and therefore $\|Tv\| = \lambda$, as desired (note that this also shows that $\|v\| = 1$).

At last, we can use the explicit form of T to prove that $Tw_j \to Tv$. We write $k(x, y) = k_x(y)$, so that (using our simplifying assumption that k is continuous) each k_x is a well-defined continuous (hence bounded and square-integrable) function on X. Then we can write

$$(Tw_j - Tv)(x) = \langle w_j - v, \overline{k_x}\rangle,$$

and hence

$$\|Tw_j - Tv\|^2 = \int_X |\langle w_j - v, \overline{k_x}\rangle|^2 d\mu(x).$$

We can now apply the dominated convergence theorem to conclude, by (A.4), that the integrand converges pointwise to 0, and moreover

$$|\langle w_j - v, \overline{k_x}\rangle|^2 \leqslant (\|w_j\| + \|v\|)^2\|k\|_\infty^2 \leqslant 2\|k\|_\infty^2,$$

which is an integrable function on X, so that the dominated convergence theorem does apply. □

Remark A.2.7. In the language of weak convergence, we can summarize the last steps of the argument as follows: (1) any sequence of unit vectors in H contains a weakly convergent subsequence; (2) a compact operator maps a weakly convergent sequence to a norm-convergent sequence. In general, (1) is a consequence of the Banach–Alaoglu Theorem (see, e.g., [**49**, Th. IV.21]) and the self-duality of Hilbert spaces, while (2) is most easily seen as coming from the alternate characterization of compact operators (on a Hilbert space H) as those mapping the unit ball to a relatively compact subset H (see, e.g., [**49**, VI.5, Th. VI.11] for this fact).

[2] One can think of this property as saying that "every coordinate" of w_j, captured by any linear functional on H, converges to the corresponding coordinate of v. To clarify the meaning of this, note that the formula $\|v\| = \sup_{\|w\|\leqslant 1} |\langle v, w\rangle|$ shows that the convergence in norm is equivalent to a uniform convergence (over w of norm 1) of the corresponding "coordinates".

A.3. The Stone–Weierstrass Theorem

We conclude with the statement of the general Stone–Weierstrass approximation theorem, which is used in Exercises 5.4.4 and 5.4.6 for an alternative proof of the Peter–Weyl Theorem.

Theorem A.3.1 (Stone–Weierstrass approximation theorem)**.** *Let X be a compact topological space, and let $A \subset C(X)$ be a subspace of continuous functions on X such that*

(1) *A is an algebra—if f, $g \in A$, the product fg is also in A;*

(2) *A is stable under complex conjugation—if $f \in A$, then $\bar{f}$ is in A, where $\bar{f}$ maps x to $\overline{f(x)}$;*

(3) *A separates points, i.e., for any elements x, y in X with $x \neq y$, there exists some $f \in A$ such that $f(x) \neq f(y)$.*

Then A is dense in $C(X)$ for the uniform topology, i.e., for any $f \in C(X)$ and $\varepsilon > 0$, there exists $g \in A$ with

$$\sup_{x \in X} |f(x) - g(x)| < \varepsilon.$$

Bibliography

[1] J. Baez and J. Huerta: *The algebra of Grand Unified Theories*, Bulletin of the American Mathematical Society 47 (2010), 483–552.

[2] B. Bekka, P. de la Harpe, and A. Valette: *Kazhdan's Property* (T), New Math. Monographs 11, Cambridge University Press, 2008.

[3] J. Bernstein and S. Gelbart (eds): *An introduction to the Langlands program*, Birkhaüser, 2003.

[4] J. Bernstein and A. Reznikov: *Analytic continuation of representations and estimates of automorphic forms*, Annals of Math. 150 (1999), 329–352.

[5] N. Berry, A. Dubickas, N. Elkies, B. Poonen, and C. J. Smyth: *The conjugate dimension of algebraic numbers*, Quart. J. Math. 55 (2004), 237–252.

[6] E. D. Bolker: *The spinor spanner*, American Math. Monthly 80 (1973), 977–984.

[7] A. Borel: *Linear algebraic groups*, 2nd edition, Graduate Texts in Mathematics 126, Springer, 1991.

[8] W. Bosma, J. Cannon, and C. Playoust: *The Magma algebra system, I. The user language* J. Symbolic Comput. 24 (1997), 235–265; also `http://magma.maths.usyd.edu.au/magma/`

[9] N. Bourbaki: *Algebra I*, Chapter 2, Springer, 1989.

[10] A. E. Brouwer and M. Popoviciu: *The invariants of the binary nonic*, J. Symb. Computation 45 (2010) 709–720.

[11] D. Bump: *Automorphic forms and representations*, Cambridge Studies in Adv. Math. 55, Cambridge University Press, 1998.

[12] C. W. Curtis and I. Reiner: *Representation theory of finite groups and associative algebras*, AMS Chelsea Publishing, 1962.

[13] G. Davidoff, P. Sarnak, and A. Valette: *Elementary number theory, group theory and Ramanujan graphs*, London Mathematical Society Student Text 55, Cambridge University Press, 2003.

[14] A. Deitmar and S. Echterhoff: *Principles in harmonic analysis*, Universitext, Springer, 2009.

[15] P. Deligne: *La conjecture de Weil, II*, Publ. Math. I.H.É.S 52 (1980), 137–252.

[16] P. Diaconis: *Group representations in probability and statistics*, Inst. Math. Stat. Lecture Notes 11, Institute of Math. Statistics, 1988.

[17] N. Dunford and J. T. Schwarz: *Linear operators, part II: spectral theory*, Wiley, 1963.

[18] P. Etingof, O. Golberg, S. Hensel, T. Liu, A. Schwendner, D. Vaintrob, and E. Yudovina: *Introduction to representation theory*, Student Math. Library 59, American Mathematical Society, 2011; also `arXiv:0901.0827`.

[19] G. Folland: *Real Analysis*, Wiley, 1984.

[20] W. Fulton and J. Harris: *Representation theory, a first course*, Universitext, Springer, 1991.

[21] The GAP Group: *GAP – Groups, Algorithms, and Programming, Version 4.4.9*, 2007; also `www.gap-system.org`

[22] K. Girstmair: *Linear relations between roots of polynomials*, Acta Arithmetica 89 (1999), 53–96.

[23] W. T. Gowers: *Quasirandom groups*, Comb. Probab. Comp. 17 (2008), 363–387.

[24] W. T. Gowers (editor): *The Princeton companion to mathematics*, Princeton University Press, 2008.

[25] I. S. Gradshteyn and I. M. Ryzhik: *Table of integrals, series, and products*, 6th Ed., Academic Press, 2000.

[26] P. de la Harpe: *Topics in geometric group theory*, Chicago Lectures in Mathematics, Chicago University Press, 2000.

[27] R. Hartshorne: *Algebraic Geometry*, Grad. Texts in Math. 52, Springer Verlag, 1977.

[28] I. M. Isaacs: *Character theory of finite groups*, Academic Press, 1976.

[29] I. M. Isaacs: *Finite group theory*, Graduate Studies in Math. 92, American Mathematical Society, 2008.

[30] H. Iwaniec and E. Kowalski: *Analytic number theory*, Coll. Publ. 53, American Mathematical Society, 2004.

[31] F. Jouve, E. Kowalski and D. Zywina: *An explicit integral polynomial whose splitting field has Galois group $W(E_8)$*, Journal de ThÉorie des Nombres de Bordeaux 20 (2008), 761–782.

[32] N. M. Katz: *Larsen's alternative, moments and the monodromy of Lefschetz pencils*, in "Contributions to automorphic forms, geometry, and number theory (collection in honor of J. Shalika's 60th birthday)", Johns Hopkins University Press (2004), 521–560.

[33] N. M. Katz: *Gauss sums, Kloosterman sums and monodromy groups*, Annals of Math. Studies, 116, Princeton University Press, 1988.

[34] N. M. Katz: *Rigid local systems*, Annals of Math. Studies, 139, Princeton University Press, 1996.

[35] A. Kirillov: *Introduction to Lie groups and Lie algebras*, Cambridge Studies in Advanced Math. 113, 2008.

[36] A. W. Knapp: *Representation theory of semisimple groups*, Princeton Landmarks in Math., Princeton University Press, 2001.

[37] A. W. Knapp: *Lie groups beyond an introduction*, Progress in Math. 140, Birkhaüser, 2002.

[38] E. Kowalski: *The large sieve, monodromy, and zeta functions of algebraic curves, II: independence of the zeros*, International Math. Res. Notices 2008, doi:10.1093/imrn/rnn091.

[39] E. Kowalski: *Spectral theory in Hilbert spaces*, Lectures notes for ETH course (2009), www.math.ethz.ch/~kowalski/spectral-theory.pdf

[40] S. Lang: *Algebra*, 3rd edition, Graduate Texts in Mathematics 211, Springer, 2002.

[41] S. Lang: $\mathrm{SL}_2(\mathbf{R})$, Addison Wesley, 1974.

[42] M. Larsen: *The normal distribution as a limit of generalized Sato-Tate measures*, preprint (http://mlarsen.math.indiana.edu/~larsen/papers/gauss.pdf).

[43] M. W. Liebeck, E. A. O'Brien, A. Shalev, and P. H. Tiep: *The Ore conjecture*, J. Eur. Math. Soc. (JEMS) 12 (2010), 939–1008.

[44] G. W. Mackey: *The theory of unitary group representations*, Chicago Lectures in Math., Chicago University Press, 1976.

[45] W. Magnus: *Residually finite groups*, Bull. Amer. Math. Soc. 75 (1969), 305–316.

[46] B. Nica: *Linear groups: Malcev's theorem and Selberg's lemma*, arXiv:1306.2385.

[47] N. Nikolov and L. Pyber: *Product decompositions of quasirandom groups and a Jordan-type theorem*, J. European Math. Soc., to appear.

[48] I. Piatetstki-Shapiro: *Complex representations of* $\mathrm{GL}(2, K)$ *for finite fields* K, Contemporary Math., Vol. 16, American Mathematical Society, 1983.

[49] M. Reed and B. Simon: *Methods of modern mathematical physics, I: Functional analysis*, Academic Press, 1980.

[50] M. Reed and B. Simon: *Methods of modern mathematical physics, II: Self-adjointness and Fourier theoretic techniques*, Academic Press, 1980.

[51] J. Rotman: *An introduction to the theory of groups*, 4th ed., Graduate Texts in Mathematics 148, Springer, 1995.

[52] P. Sarnak: *Selberg's eigenvalue conjecture*, Notices Amer. Math. Soc. 42 (1995), 1272–1277.

[53] J.-P. Serre: *Linear representations of finite groups*, Graduate Texts in Mathematics 42, Springer, 1977.

[54] J.-P. Serre: *A course in arithmetic*, Graduate Texts in Mathematics 7, Springer, 1973.

[55] J.-P. Serre: *Moursund lectures*, arXiv:math.0305257

[56] J.-P. Serre: *Topics in Galois theory*, Res. Notes in Math., Jones and Bartlett, 1992.

[57] J.-P. Serre: *Semisimplicity and tensor products of group representations: converse theorems*, with an Appendix by W. Feit, J. Algebra 194 (1997), 496–520.

[58] S. F. Singer: *Linearity, symmetry, and prediction in the hydrogen atom*, Undergraduate Texts in Mathematics, Springer, 2005.

[59] T. A. Springer: *Invariant theory*, Lecture Notes in Mathematics 585, Springer, 1977.

[60] T. A. Springer: *Linear algebraic groups*, 2nd edition, Progr. Math. 9, Birkhaüser, 1998.

[61] S. Sternberg: *Group theory and physics*, Cambridge University Press, 1994.

[62] L. Takhtajan: *Quantum mechanics for mathematicians*, American Mathematical Society Graduate Studies in Mathematics, 95, Providence, RI, 2008.

[63] N. J. Vilenkin: *Special functions and the theory of group representations*, Translations of Mathematical Monographs 22, American Mathematical Society, 1968.

[64] H. Weyl: *The theory of groups and quantum mechanics*, Dover, 1950.

[65] D. Witte Morris: *Ratner's Theorems on unipotent flows*, University of Chicago Press, 2005.

[66] E. T. Whittaker and G. N. Watson: *A course of modern analysis*, Fourth Ed., Cambridge University Press, 1927; reprinted 1988.

[67] A. Zygmund: *Trigonometric series*, 3rd Edition, Cambridge Mathematics Library, Cambridge University Press, 2002.

Index

G-finite vector, 309
L^1-action, 289, 297, 304
L^1-approximation, 293
L^p-space, 281
SL_2, 4, 6, 71, 88, 118, 135, 137, 151, 155, 169, 233, 235, 236, 245, 278, 314, 329, 363, 376, 403, 404
SO_3, 315, 343, 348
SU_2, 71, 74, 90, 277, 285, 310, 329, 340, 343
k-th moment, 342
p-adic integers, 321
p-group, 164, 213

abelian group, 7, 93, 164, 182, 201, 203, 276
abelianization, 10, 43, 129, 164, 214
absolutely irreducible representation, 66, 100
action, 14
action map, 140
adjoint function, 290, 293
adjoint operator, 11, 146, 290, 337
adjoint representation, 91, 134, 336, 337, 339, 343
adjointness, 48
affine n-space, 361
algebra, 127
algebraic group, 124, 365, 377
algebraic integer, 249, 250, 409–411
algebraic number, 409
algebraically closed field, 98, 110
algebraically closed field, 123
algebraically closed field, 65, 80, 91, 93, 95, 101, 106, 107, 112, 121, 163–165, 168, 169, 173, 177, 178, 185, 188, 189, 258, 361, 367
alternating bilinear form, 320, 325, 326, 329, 338
alternating group, 248, 250
alternating power, 35, 362
angular momentum quantum number, 5, 350
approximation by convolution, 293
arithmetic-geometric mean inequality, 207
Artin's Theorem, 267
associative algebra, 4, 132
automorphic function, 406
automorphic representations, 34
automorphism, 67
averaging, 160, 249, 276, 284

Banach isomorphism theorem, 140
Banach space, 139, 142, 144, 145
Banach–Alaoglu Theorem, 419
Banach–Steinhaus theorem, 285
Bargmann's classification, 403, 405
basis, 30, 41, 111, 174, 410
beta function, 287
bilinear form, 133, 325, 332
bilinear map, 9, 97
binary
  cubic forms, 75
  forms, 70
  quadratic forms, 74

block-diagonal matrix, 69, 85, 98
block-permutation matrix, 117
block-triangular matrix, 69, 85
Borel density theorem, 365
Borel measure, 11
Borel set, 11, 273
bounded linear operator, 11, 139
Brauer character, 168, 173
Brauer's Theorem, 268
Britton's Lemma, 79, 80
Bruhat decomposition, 266
Burnside irreducibility criterion, 2, 41, 97, 102, 107, 110, 189, 255
Burnside's $p^a q^b$ theorem, 7, 246
Burnside's inequality, 206

Cartan decomposition, 390
Catalan numbers, 342
category, 17, 21, 360
Cauchy integral formula, 394
Cauchy's residue formula, 396
center, 93, 128, 131, 186, 188–190, 210, 214, 222, 247–250, 369
central character, 93, 234, 248
central class, 232
centralizer, 191, 211, 222, 223, 229
character, 29, 67, 71, 107, 111, 113, 114, 121, 141, 150, 163, 170, 171, 177–181, 192, 194, 196, 199, 200, 203, 205, 208, 215, 219, 221, 228, 243, 261, 266, 267, 304, 310, 324, 326, 411
character table, 232
character ring, 120, 267
character table, 208, 209, 211–214, 220, 221, 233, 235, 245–247, 308, 332, 338
character theory, 74, 159, 304, 311, 315, 327, 330, 334, 342, 371
characteristic p, 71, 367
characteristic function, 191, 192, 205
characteristic polynomial, 251, 410, 413
Chebychev inequality, 242
Chebychev polynomial, 313
circle group, 269, 297, 363
class function, 115, 167, 173, 175, 186, 189, 190, 197, 208, 229, 230, 251, 292, 304, 312, 340
classical mechanics, 344
Clebsch–Gordan formula, 72, 74, 120, 310, 314
Clifford theory, 121
closed graph theorem, 389
cofactor matrix, 357
coinvariants, 23, 24, 36
cokernel, 23
commutation relations, 137
commutator, 10, 192–196, 213, 214, 218, 235
commutator group, 10, 195, 225
compact group, 4, 125, 155, 262, 269, 273, 282, 284, 294, 295, 297, 299, 300, 302, 304, 307, 309, 310, 316, 324, 325, 332, 341, 360, 382
compact Lie group, 322, 341
compact operator, 299, 307, 414, 415
complementary series, 398, 400, 403, 404, 406
completely reducible representation, 31, 34, 153
complex conjugate of a representation, 67, 71, 114, 119
complex type, 325, 327
composition factor, 81, 84, 85, 92, 95, 114, 115, 125
composition series, 81, 83
conjugacy class, 115, 118, 119, 165, 167, 168, 174, 189–192, 196, 197, 208–211, 213, 214, 216, 217, 222–224, 237, 238, 247, 249, 255, 308, 311, 340
conjugate, 194, 245, 253, 413
conjugate of a representation, 67
conjugate space, 150
conjugation, 41, 90, 335, 336
connected component, 322
connected component of the identity, 365, 367
continuous representation, 140, 145
contragredient, 21, 37, 38, 70, 98, 102, 115, 144, 150, 234, 326, 328, 362, 404
contravariant functor, 37
convolution, 272, 293
convolution operator, 272, 298
copies, 41, 95
coprimality, 248, 412
counting measure, 276
Coxeter group, 261
cusp, 405
cuspidal representation, 237
cuspidal representation, 229, 233, 235, 245

cycle decomposition, 237
cyclic group, 202, 206, 220, 254, 267
cyclic representation, 29, 189
cyclic vector, 28, 29, 131

degree, 13, 15
Deligne's equidistribution theorem, 340
derivation, 133, 134
derived subgroup, 10, 43
determinant, 73, 225, 229, 369
dihedral group, 220, 221, 343
dimension, 13, 51, 120, 168, 173, 181, 183, 215, 252, 264, 322, 324
direct integral, 125, 370, 373
direct product, 65, 107, 179, 316
direct sum, 10, 24, 32, 41, 87, 125, 268
Dirichlet character, 205, 206
discrete series, 229, 393, 397, 398, 403
discrete subgroup, 404
discrete topology, 139
discriminant, 75
divisibility, 173, 250, 252, 411, 412
division algebra, 93
dominated convergence theorem, 148, 419
double coset, 130, 265, 390
doubly transitive action, 180, 181, 228
dual Banach space, 144
dual basis, 100, 111, 174
dual group, 201, 372
dual space, 21, 24, 37, 98
duality bracket, 8
dynamics, 346
Dynkin diagram, 261

eigenfunction, 406, 407
eigenvalue, 92
elementary subgroup, 268
equidistribution, 340, 343
Euler function, 205
Euler product, 203, 205
exact sequence, 9
exponential, 16, 202, 276, 313
exponential of a matrix, 137, 338
extension, 27
external tensor product, 64, 65, 107, 109, 111, 149, 166, 305

faithful representation, 15, 79, 183, 184, 195, 212, 234, 260, 281, 297, 302, 320, 321, 323, 338, 359, 367, 368
finite abelian group, 202, 206
finite field, 6, 10
finite-dimensional representation, 38, 80, 83, 95, 100, 111, 144, 162, 178, 211, 284, 295, 296, 299, 320, 323, 333, 376, 411
finite-index subgroup, 54, 115
finite-rank operator, 418
finitely generated abelian group, 410
finitely generated group, 78, 79
fixed points, 114, 118, 164, 219, 260
Fourier analysis, 372
Fourier coefficient, 270, 271, 309
Fourier decomposition, 201
Fourier integrals, 7, 125
Fourier series, 7, 269–271, 312
Fourier transform, 155, 236
fourth moment, 333, 338, 343, 355, 358
free abelian group, 256
free group, 78
free product, 78
Frobenius reciprocity, 2, 48, 51, 55, 60–62, 77, 94, 110, 181, 182, 197, 203, 227, 234, 262, 265–267, 283, 303, 348
Frobenius–Schur indicator, 262, 324, 325, 328, 332, 333
functions on cosets, 199
functor, 21, 42
functorial, 34, 37, 42, 64, 131, 185
functorial representation, 21
functoriality, 17, 22, 47, 60

Galois group, 256–261, 339
Galois theory, 207, 259, 413
Gamma function, 277
Gauss hypergeometric function, 392
Gaussian elimination, 266
Gelfand–Graev representation, 233
generalized character, 120
generating series, 183
generators, 235, 261
generators and relations, 77, 78
graph theory, 240
group algebra, 128, 131, 169, 175, 184, 186, 188–190, 210, 250, 288
group ring, 131

Haar measure, 272, 273, 275–279, 281, 282, 284, 372, 381, 404
Hamiltonian, 346
harmonic analysis, 190–192
harmonics, 205

Heisenberg group, 213
hermitian form, 10, 284, 286, 324
hermitian matrix, 337
highest weight vector, 90, 135, 138
Hilbert space, 10, 139, 145, 146, 149, 152, 155, 175, 190, 191, 283, 285, 289, 344, 347, 394, 415
Hilbert–Schmidt operator, 299, 416, 417
HNN extension, 80
homogeneous polynomials, 137, 285, 310, 329
homomorphism of Lie algebras, 133
Hydrogen, 5, 125, 315, 344, 352
hydrogen, 4
hyper-Kloosterman sum, 343
hyperbolic Laplace operator, 406
hypergeometric function, 392

ideal, 129, 188
idempotent, 188
Identitätssatz, 79
identity, 17, 22, 62, 63, 155
image, 23, 28
image measure, 11, 156, 276
induced representation, 121
induced character, 117
induced representation, 44, 46, 47, 51, 53, 61, 62, 76, 110, 115, 118, 181, 182, 212, 215, 227–229, 231, 232, 262, 264, 265, 267, 282, 283, 303, 398
induction, 44, 48, 55, 58, 61, 168, 219, 225, 268, 379
induction in stages, 47, 58
infinite cardinal, 8, 51
infinite tensor products, 34
infinite-dimensional representation, 144, 154, 377, 379, 394
inner automorphism, 46, 336
inner product, 10, 139, 149, 157, 171, 283, 284
integral matrix, 251, 410
integrality, 246, 250, 251, 308, 409
intersection, 22, 33
intertwiner, 17, 19, 25, 40, 54, 73, 76, 84, 105, 108, 123, 146, 155, 157, 186, 262, 264, 272, 292, 298, 303, 307, 313, 331, 374, 389
intertwining operator, 17
invariant bilinear form, 324, 331
invariant theory, 74, 75
invariant vector, 76, 160
invariants, 18, 24, 41, 45, 74, 88, 91, 115, 130, 134, 169–171, 184, 196, 304, 306, 324, 326, 334, 362
inversion formula, 373
involution, 260
irreducibility criterion, 178, 181, 208, 264
irreducible representation, 189
irreducible character, 190
irreducible representation, 261
irreducible character, 111, 168, 172, 181, 190, 198, 206–208, 229–231, 267, 304, 313
irreducible components, 227, 233, 235
irreducible polynomial, 257, 259, 260
irreducible representation, 27–29, 35, 38, 44, 53, 65–67, 71, 73, 76, 77, 80, 83, 86, 88, 91, 94, 98, 100, 101, 106–108, 121, 122, 130, 134, 135, 137, 144, 155, 157, 163, 165, 169, 172, 173, 177, 180–183, 185, 189, 200, 211–213, 216, 218, 220, 225, 227, 228, 234, 235, 237, 238, 241, 247, 249, 251, 252, 254, 262, 264, 266, 267, 270, 294, 300, 305, 307, 311, 315, 316, 320, 322, 327, 331, 336, 359, 362, 389, 394, 402
irreducible subrepresentation, 33, 258, 406
isomorphism of representations, 23
isotypic component, 258
isotypic component, 87, 88, 94–97, 104, 110, 122, 157, 163, 184, 185, 187, 189, 209, 271, 294, 302, 304–306, 309, 352, 389, 394, 397
isotypic projection, 185, 193, 210, 251
isotypic representation, 27, 121, 122, 124
Iwasawa coordinates, 379, 381, 384
Iwasawa decomposition, 364

Jacobi identity, 132–134
joint eigenspace, 210
Jordan–Hölder–Noether Theorem, 28, 30, 80, 99, 109

kernel, 23, 28, 211, 367
Kloosterman sum, 340
Kummer polynomial, 257, 259

Langlands Program, 125

Larsen alternative, 332, 333, 337, 341, 355, 359
Lebesgue measure, 271, 272, 276, 277, 282
left-regular representation, 15, 243, 290, 298, 300
Lie algebra, 4, 132, 138, 287, 336, 382, 384, 387, 399, 411
Lie bracket, 132
Lie group, 132, 138, 319, 322, 336, 371, 381
limits of discrete series, 386
linear algebraic group, 360, 361, 367
linear form, 23
linear group, 221, 319, 356
linear independence of matrix coefficients, 101, 105
linear representation, 3, 13
locally compact group, 281
locally compact abelian group, 371, 372
locally compact group, 152, 272, 273, 275, 276, 278, 279, 369, 370, 389, 398
lowest K-type, 397

magnetic quantum number, 5, 350
Malcev's Theorem, 79, 80
manifold, 319, 322
Maschke's Theorem, 159, 161, 163, 168, 169, 258
matrix algebra, 188
matrix coefficient, 29, 100, 101, 110–112, 115, 150, 165, 166, 170, 171, 174, 190, 192, 194, 200, 296, 297, 300, 302, 305, 308, 313, 317, 371, 389, 393, 396, 403
matrix representation, 67, 69, 72, 77, 85, 102, 211, 260, 343
measure space, 155
measure-preserving action, 282
minimal dimension, 241, 245, 377
minimal index, 168
minimal polynomial, 409, 413
mock discrete series, 386, 403
model, 47, 53, 62, 114, 263
modular character, 275, 278
module, 127
modules over the group algebra, 128
modulus of a group, 275
momentum, 344
morphism of representations, 17, 22, 133
multilinear operations, 35
multiplication operator, 155, 156, 375
multiplicity, 81, 84, 89, 94, 113, 115, 163, 166, 169, 177–179, 181–183, 243, 305, 307, 342, 348, 405

natural homomorphisms, 60, 62
nilpotent group, 213, 217
non-abelian simple group, 195, 221, 338
non-linear group, 78
non-semisimple class, 222
non-split semisimple class, 223, 226, 229, 232
non-unitary representation, 398
norm, 10, 139, 413
norm map, 230, 236
norm topology, 140, 142, 280, 414
normal operator, 415
normal subgroup, 19, 43, 118, 121, 123, 183, 197, 199, 212, 213, 246, 248, 365, 369
normed vector space, 10

observable, 344, 347, 350
Odd-order Theorem, 8
one-dimensional character, 218, 225, 264
one-dimensional representation, 16, 19, 29, 35, 43, 88, 93, 101, 104, 123, 200, 214, 266, 286
one-parameter unitary group, 375
one-relator group, 79
orbits, 77, 179, 261
Ore conjecture, 195
orthogonal complement, 153
orthogonal direct sum, 152, 154, 294, 348, 397
orthogonal group, 278, 320, 332
orthogonal polynomials, 314
orthogonal projection, 157, 185, 271, 304, 345
orthogonal type, 325, 327–329, 331
orthogonality, 157, 170
orthogonality of characters, 171, 173, 178, 179, 187, 198
orthogonality of matrix coefficients, 173, 174, 309
orthogonality relation, 200, 201, 203, 205, 249, 252
orthonormal basis, 11, 176, 177, 190, 192, 196, 197, 199, 267, 271, 304, 306, 308, 312, 314

orthonormality of characters, 177, 254, 307, 310, 325
orthonormality of matrix coefficients, 176, 197

palindromic polynomial, 260, 261
Parseval formula, 270
particle, 344, 350
partition, 237, 238, 240
perfect group, 43, 195
periodic functions, 405
permutation, 237, 239
permutation group, 75
permutation matrix, 114, 118, 266
permutation representation, 76, 77, 118, 179, 180, 212, 219, 228, 239, 240, 257, 260, 282
Peter–Weyl theory, 269, 283, 288, 294, 297, 300, 302, 309, 311, 317, 320, 323, 395
Plancherel formula, 201, 302, 371–373
Planck's constant, 346
polar decomposition, 364, 390
polynomial representation, 362
Pontryagin duality, 372, 375
position, 344, 347, 348
power series, 204
pre-unitary representation, 153, 283
presentation of a group, 77, 237
prime number, 203, 321
primes in arithmetic progressions, 5, 203
principal quantum number, 5, 352
principal series, 227, 232, 266, 378, 379, 388, 390, 402, 403, 405, 406
probability, 12, 192, 241, 345
probability Haar measure, 273, 276, 294, 302, 318
probability measure, 12
product topology, 316
projection, 8, 160, 169–171, 185–187, 394
projection formula, 55, 58, 118, 121, 326
pure tensors, 9, 34, 40, 56, 70, 108, 149

quadratic form, 207
quantum mechanics, 5, 344, 376
quantum number, 5
quantum system, 347
quasirandom group, 240, 245, 246
quaternion, 221
quaternionic type, 325
quotient, 82
quotient representation, 22, 23, 30, 81, 95, 144, 148, 177

Radon measure, 11, 273, 275, 282, 416, 417
rank, 13, 256
rank 1 linear map, 39, 41, 170, 171, 175
real type, 325
reduction modulo a prime, 68, 80, 249
reductive group, 360
regular polygon, 220
regular representation, 19, 20, 44, 45, 47, 59, 68, 76, 104, 106, 107, 110, 114, 117, 142, 148, 152, 155, 161–163, 176, 209, 212, 243, 271, 272, 279, 281, 283, 289, 294, 296, 309, 313, 373, 379, 385, 387, 394, 398, 405
relations, 235, 255, 261
representation, 13
representation generated by a vector, 29
representation of a Lie algebra, 133
representation of an algebra, 127
representations of a quotient, 43
residually finite group, 79
restriction, 42, 43, 48, 55, 58, 148, 182, 268, 285, 376
Riemann hypothesis for curves over finite fields, 340
Riesz representation theorem, 150, 291
right-ideal, 188
right-regular representation, 14
roots of unity, 202, 253–255, 259, 410, 412

Sarnak's philosophy, 1
scalar class, 222, 226
Schrödinger equation, 346, 352, 376
Schur's Lemma, 28, 41, 74, 87, 91, 93, 94, 97, 99, 103, 107–109, 123, 134, 154, 155, 158, 166, 170, 175, 186, 227, 264, 295, 298, 306, 331, 372, 390
second orthogonality formula, 191, 193, 308, 332
Selberg's conjecture, 406, 407
self-adjoint operator, 11, 155, 243, 298, 344, 415, 417
self-dual Haar measure, 373
self-dual representation, 328, 332
self-reciprocal polynomial, 260

semisimple representation, 106
semisimple conjugacy class, 254
semisimple representation, 27, 30, 35, 43, 67, 68, 71, 72, 83, 85, 86, 89, 92–94, 102–104, 112, 114, 121, 130, 153, 154, 159, 161, 177, 284, 359, 362, 366–368
semisimplicity criterion, 31
separable, 11
short exact sequence, 9, 27, 69
signature, 208, 233, 239, 240
signed permutation, 260
signed permutation matrix, 195, 360
skew-hermitian matrix, 337, 343
skew-hermitian operator, 400, 401
small subgroup, 322
Sobolev norm, 407
solvable group, 7, 213, 217, 246, 247, 255, 398
Specht module, 237, 239, 240
spectral measure, 345
spectral theorem, 4, 155, 156, 299, 375, 414
spectrum, 345
spherical representation, 404
spin, 5, 352
split exact sequence, 27
split semisimple class, 222
split semisimple class, 226, 228, 229
stabilizer, 179, 239
stable complement, 26, 73, 77, 153, 161, 335
stable lattice, 68, 77
stable subspace, 18, 394
state, 5, 344, 346, 350
Steinberg representation, 228, 233, 247, 254
Stone's Theorem, 375
Stone–Weierstrass Theorem, 270, 297, 300, 302, 420
strong continuity, 146, 153, 279, 280, 283, 291, 294, 399
strong topology, 147, 281, 293
submodule, 130
subquotient, 81, 95
subrepresentation, 18, 19, 22, 24–26, 38, 47, 66, 71, 72, 81, 85, 87, 91, 98, 100, 122, 136, 144, 148, 156, 157, 160, 177, 239, 256–258, 286, 290, 326, 327, 334, 337, 362, 379, 385, 389, 394
subrepresentation generated by a vector, 309
sum, 22
support of a measure, 11, 156, 273
surface Lebesgue measure, 348
symmetric bilinear form, 326
symmetric bilinear form, 171, 267, 325, 329, 338
symmetric group, 36, 168, 237, 238, 257, 331
symmetric power, 35, 71, 362
symmetry, 3, 347
symplectic type, 325, 327–329

tableau, 239
tabloid, 239
tangent space, 336
tangent vector, 338
tautological, 14, 72, 333, 355, 368
tempered representation, 404
tensor power, 183, 307, 313
tensor product, 8, 34, 70, 72, 97, 149, 207, 334, 366, 411
topological contragredient, 145, 404
topological group, 47, 139, 144–146, 153, 155, 157
topologically irreducible representation, 144
torus, 320
trace, 10, 115, 117, 169, 172, 196, 211
trace map, 236
transitive action, 180, 239, 348
transitivity, 58, 62
translates, 13, 73, 162
transpose, 37, 235, 331, 357
triangle inequality, 212
trigonometric polynomial, 270
trivial representation, 15, 26, 30, 51, 88, 134, 165, 169, 180, 183, 191, 195, 201, 208, 219, 240, 317, 342, 403
trivial subrepresentation, 18
twisting, 35, 123, 149, 214, 220, 368
two-sided ideal, 188
Tychonov Theorem, 316

unbounded self-adjoint operator, 346
unimodular group, 276–278, 389
unipotent element, 8, 89, 113, 120, 367
unipotent radical, 367, 368
unit, 412
unit disc, 396
unit vector, 344

unitarizability criterion, 151
unitarizable representation, 146, 162, 284, 324, 399
unitary group, 4, 281, 322
unitary matrix, 150, 184, 191, 311
unitary matrix coefficient, 176, 295
unitary operator, 11, 146, 155
unitary representation, 145, 146, 148–150, 152, 155, 157, 162, 175, 185, 187, 211, 271, 272, 278, 279, 282, 289, 307, 309, 316, 348, 401, 405
unitary symplectic group, 320, 332
universal endomorphisms, 131
universal endomorphisms, 131, 186
unramified representation, 404
upper half-plane, 380, 393, 396
upper-triangular matrix, 120, 217, 266

variance, 241
velocity, 344
virtual character, 120, 181, 268

weak convergence, 418
weak integral, 289
Weil representation, 78, 235, 236
Weyl group, 261
Weyl integration formula, 312
Whittaker functional, 234

Young diagram, 238, 240

Zariski closure, 356–359, 361, 362, 364, 368
Zariski topology, 360
Zorn's Lemma, 32, 34, 301

Selected Published Titles in This Series

155 **Emmanuel Kowalski,** An Introduction to the Representation Theory of Groups, 2014
152 **Gábor Székelyhidi,** An Introduction to Extremal Kähler Metrics, 2014
151 **Jennifer Schultens,** Introduction to 3-Manifolds, 2014
150 **Joe Diestel and Angela Spalsbury,** The Joys of Haar Measure, 2013
149 **Daniel W. Stroock,** Mathematics of Probability, 2013
148 **Luis Barreira and Yakov Pesin,** Introduction to Smooth Ergodic Theory, 2013
147 **Xingzhi Zhan,** Matrix Theory, 2013
146 **Aaron N. Siegel,** Combinatorial Game Theory, 2013
145 **Charles A. Weibel,** The K-book, 2013
144 **Shun-Jen Cheng and Weiqiang Wang,** Dualities and Representations of Lie Superalgebras, 2012
143 **Alberto Bressan,** Lecture Notes on Functional Analysis, 2013
142 **Terence Tao,** Higher Order Fourier Analysis, 2012
141 **John B. Conway,** A Course in Abstract Analysis, 2012
140 **Gerald Teschl,** Ordinary Differential Equations and Dynamical Systems, 2012
139 **John B. Walsh,** Knowing the Odds, 2012
138 **Maciej Zworski,** Semiclassical Analysis, 2012
137 **Luis Barreira, Claudia Valls, Luis Barreira, and Claudia Valls,** Ordinary Differential Equations, 2012
136 **Arshak Petrosyan, Henrik Shahgholian, and Nina Uraltseva,** Regularity of Free Boundaries in Obstacle-Type Problems, 2012
135 **Pascal Cherrier and Albert Milani,** Linear and Quasi-linear Evolution Equations in Hilbert Spaces, 2012
134 **Jean-Marie De Koninck and Florian Luca,** Analytic Number Theory, 2012
133 **Jeffrey Rauch,** Hyperbolic Partial Differential Equations and Geometric Optics, 2012
132 **Terence Tao,** -, 2012
131 **Ian M. Musson,** Lie Superalgebras and Enveloping Algebras, 2012
130 **Viviana Ene and Jürgen Herzog,** Gröbner Bases in Commutative Algebra, 2011
129 **Stuart P. Hastings and J. Bryce McLeod,** Classical Methods in Ordinary Differential Equations, 2012
128 **J. M. Landsberg,** Tensors: Geometry and Applications, 2012
127 **Jeffrey Strom,** Modern Classical Homotopy Theory, 2011
126 **Terence Tao,** An Introduction to Measure Theory, 2011
125 **Dror Varolin,** Riemann Surfaces by Way of Complex Analytic Geometry, 2011
124 **David A. Cox, John B. Little, and Henry K. Schenck,** Toric Varieties, 2011
123 **Gregory Eskin,** Lectures on Linear Partial Differential Equations, 2011
122 **Teresa Crespo and Zbigniew Hajto,** Algebraic Groups and Differential Galois Theory, 2011
121 **Tobias Holck Colding and William P. Minicozzi II,** A Course in Minimal Surfaces, 2011
120 **Qing Han,** A Basic Course in Partial Differential Equations, 2011
119 **Alexander Korostelev and Olga Korosteleva,** Mathematical Statistics, 2011
118 **Hal L. Smith and Horst R. Thieme,** Dynamical Systems and Population Persistence, 2011
117 **Terence Tao,** An Epsilon of Room, I: Real Analysis, 2010
116 **Joan Cerdà,** Linear Functional Analysis, 2010

For a complete list of titles in this series, visit the
AMS Bookstore at **www.ams.org/bookstore/gsmseries/**.